W9-AAZ-194

Games and Information

GAMES AND INFORMATION

An Introduction to Game Theory

Second Edition

Eric Rasmusen
Indiana University

First published 1989
Second Edition 1994

Blackwell Publishers
238 Main Street
Cambridge, Massachusetts 02142
USA

108 Cowley Road
Oxford OX4 1JF
UK

Library of Congress Cataloging-in-Publication Data

ISBN 1-55786-502-7

British Library Cataloguing in Publication Data
A CIP catalogue record for this book is available from the British Library.

Typeset in 10 on 12 pt Century Expanded
by Compset, Inc.
Printed in the United States of America

This book is printed on acid-free paper

Contents

Preface

Contents and Purpose

This book is about noncooperative game theory and asymmetric information. In the Introduction, I will say why I think these subjects are important. Here in the Preface I will try to help you decide whether this is the appropriate book to read if they do interest you.

I write as an applied theoretical economist, not as a game theorist, and readers in anthropology, law, physics, accounting, and management science have helped me to be aware of the provincialisms of economics and game theory. My aim is to present the game theory and information economics that currently exists in journal articles and oral tradition in a way that shows how to build simple models using a standard format. Journal articles are more complicated and less clear than seems necessary in retrospect; precisely because it is original, even the discoverer rarely understands a truly novel idea. After a few dozen successor articles have appeared, we all understand it and marvel at its simplicity. But journal editors are unreceptive to new articles that admit to containing exactly the same idea as the old articles, just presented more clearly. At best, the clarification is hidden in some new article's introduction or condensed to a paragraph in a survey. Students, who find every idea as complex as the idea's originators did when it was new, must learn either from the confused original articles or from the oral tradition of a top economics department. This book provides an alternative.

Using the Book

The book is divided into three parts: Part I on game theory; Part II on information economics; and Part III on applications to particular subjects. Parts I and II, but not III, are ordered sets of chapters.

Part I by itself would be appropriate for a course on game theory, and sections from Part III could be added for illustration. If students are already familiar with basic game theory, Part II can be used for a course on information economics. The entire book would be useful as a secondary text for a course on industrial organization. I teach material from every chapter in a semester-long course for first- and second-year doctoral students at the Indiana University School of Business, including more or fewer chapter sections depending on the progress of the class.

Exercises and notes follow the chapters. It is useful to supplement a book like this with original articles, but I leave it to my readers or their instructors to follow up on the topics that interest them. I also recommend that readers try attending a seminar presentation of current research on some topic from the book; while most of the seminar may be incomprehensible, there is a real thrill in hearing someone attack the speaker with "Are you sure that the equilibrium is perfect?" after just learning the week before what "perfect" means.

Some of the exercises at the end of the chapters put slight twists on concepts in the text while others introduce new concepts. Answers to odd-numbered questions are given at the end of the book. I particularly recommend working through the problems for those trying to learn this material without an instructor.

The endnotes to each chapter include substantive material as well as references. Unlike the notes in many books, they are not meant to be skipped, since many of them are important but tangential and some qualify statements in the main text. Less important notes supply additional examples or list technical results for reference. A mathematical appendix at the end of the book supplies technical references, defines certain mathematical terms, and lists some items for reference which are not used in the main text.

The Level of Mathematics

In surveying the prefaces of previous books on game theory, I see that advising readers how much mathematical background they need exposes an author to charges of being out of touch with reality. The mathematical level here is about the same as in Luce & Raiffa (1957), and I can do no better than to quote the advice on page 8 of their book:

> Probably the most important prerequisite is that ill-defined quality: mathematical sophistication. We hope that this is an ingredient not required in large measure, but that it is needed to some degree there can be no doubt. The reader must be able to accept conditional statements, even though he feels the suppositions to be false; he must be willing to make concessions to mathematical simplicity; he must be patient enough to follow along with the peculiar kind of construction that mathematics is; and, above all, he must have sympathy with the method—a sympathy based upon his knowledge of its past successes in various of the empirical sciences and upon his realization of the necessity for rigorous deduction in science as we know it.

If you do not know the terms "risk-averse," "first-order condition," "utility function," "probability density," and "discount rate," you will not fully understand this book. Flipping through it, however, you will see that the equation density is much lower than in first-year graduate microeconomics texts. In a sense, game theory is less abstract than price theory, because it deals with individual agents rather than aggregate markets and is oriented towards explaining stylized facts rather than supplying econometric specifications.

Changes in the Second Edition

George Stigler used to say that it was a great pity Alfred Marshall spent so much time on the eight editions of *Principles of Economics* that appeared between 1890 and 1920, given the opportunity cost of the other books he might have written. I am no Marshall, so I have been willing to sacrifice a Rasmusen article or two for this new edition, though I doubt I will keep it up till 2019.

Games and Information has been quite successful, selling well throughout the world and appearing in Italian and Japanese translations. Given this success, I have not changed the book's style or organization. I have, however, added a number of new topics, increased the number of exercises (and provided detailed answers), updated the references, changed the terminology here and there, and reworked the entire book for clarity, since a book, like a poem, is never finished—only abandoned. The one section I have dropped is the first edition's discussion of existence theorems. The new topics include auditing games, nuisance suits, recoordination in equilibria, renegotiation in contracts, supermodularity, signal jamming, market microstructure, and government procurement. The discussion of moral hazard has been reorganized, and the total number of chapters has been increased by two.

I will try the experiment of putting errata and problem set answers in a file on my NeXT computer. The machine name is rasmusen.bus.indiana.edu, the IP number is 129.79.122.177, the account is 'guest' and the password is 'guest'. This is a Unix account, so remember to use lowercase letters, and use the command 'ls' to list files. It can be reached by telnet using the command "telnet rasmusen.bus.indiana.edu" or by telephoning 812-855-4211 to reach Indiana University's 2400 baud modem and typing 'connect rasmusen.' (The number for the 9600 baud modem is 812-855-9681.) The text files are in LaTeX, which uses only ASCII characters. I encourage readers to notify me of errors by Internet e-mail at Erasmuse@Indiana.edu.

Other Books

At the time of the first edition of this book, most of the topics covered were absent from existing books on either game theory or information economics. Older books on game theory include Davis (1970), Harsanyi (1977), Luce & Raiffa (1957), Moulin (1986a, b), Ordeshook (1986), Owen (1982), Rapoport (1970), Shubik (1982), Szep & Forgo (1985), Thomas (1984), and Williams (1966). Books on information in economics were mainly concerned with decision-making under uncertainty rather than asymmetric information. Exceptions include Bamberg & Spremann (1987) on agency theory, Hess (1983) on organization, and J. Green & Laffont (1979) on mechanism design. Binmore & Dasgupta (1986), Diamond & Rothschild (1978), and the immense Rubinstein (1990) are collections of key articles in the game theory and information economics literature. For articles from the prehistory of mathematical economics, see the collection by Baumol & Goldfeld (1968).

Fudenberg & Tirole (1986a) on oligopoly and McMillan (1986) on international economics are excellent short books of applications.

Since the first edition, a spate of books on game theory have appeared, and I have come across a few others that I had earlier missed.

Recent Books on Game Theory and its Applications

Aumann, Robert (1988) *Lectures on Game Theory.* (Underground Classics in Economics) Boulder, Colorado: Westview Press, 1989.

Aumann, Robert & Sergiu Hart (1992) *Handbook of Game Theory with Economic Applications.* New York: North-Holland, 1992.

Baird, Douglas, Robert Gertner, & Randal Picker (forthcoming) *Strategic Behavior and the Law: The Role of Game Theory and Information Economics in Legal Analysis.*

Banks, Jeffrey (1990) *Signalling Games in Political Science.* Chur, Switzerland: Harwood Publishers, 1990.

Basu, Kaushik (1993) *Lectures in Industrial Organization Theory.* Oxford: Blackwell Publishers, 1993.

Bierman, H. Scott & Luis Fernandez (1993) *Game Theory with Economic Applications.* Reading, Massachusetts: Addison Wesley, 1993.

Binmore, Ken (1992) *Fun and Games: A Text on Game Theory.* Lexington, Mass.: D.C. Heath, 1992.

Dixit, Avinash K. & Barry J. Nalebuff (1991) *Thinking Strategically: The Competitive Edge in Business, Politics, and Everday Life.* New York: Norton, 1991.

Eatwell, John, Murray Milgate & Peter Newman, eds. (1989) *The New Palgrave: Game Theory.* New York: W.W. Norton & Co., 1989.

Friedman, James (1990) *Game Theory with Applications to Economics.* New York: Oxford University Press, 1990. (First edition, 1986)

Fudenberg, Drew & Jean Tirole (1986) *Dynamic Models of Oligopoly.* Chur, Switzerland: Harwood Academic Publishers, 1986.

Fudenberg, Drew & Jean Tirole (1991) *Game Theory.* Cambridge, Mass.: MIT Press, 1991.

Gibbons, Robert (1992) *Game Theory for Applied Economists.* Princeton: Princeton University Press, 1992.

Harris, Milton (1987) *Dynamic Economic Analysis.* Oxford, Oxford University Press, 1987.

Jacquemin, Alex (1985) *The New Industrial Organization.* Cambridge, Mass.: MIT Press, 1987.

Kreps, David (1990) *A Course in Microeconomic Theory.* Princeton: Princeton University Press, 1990.

Kreps, David (1990) *Game Theory and Economic Modeling.* Oxford: Clarendon Press; New York: Oxford University Press, 1990.

Krouse, Clement (1990) *Theory of Industrial Economics.* Oxford: Blackwell Publishers, 1990.

Laffont, Jean-Jacques and Michel Moreaux, editors, translated by Francois Laisney (1991) *Dynamics, Incomplete Information, and Industrial Economics.* Oxford: Blackwell Publishing, 1991.

Laffont, Jean-Jacques & Jean Tirole (1993) *A Theory of Incentives in Procurement and Regulation.* Cambridge, Mass.: MIT Press, 1993.

McMillan, John (1986) *Game Theory in International Economics.* Chur, Switzerland: Harwood Academic Publishers, 1986.

McMillan, John (1992) *Games, Strategies, and Managers: How Managers can use Game Theory to Make Better Business Decisions.* Oxford: Oxford University Press, 1992.

Milgrom, Paul and John Roberts (1991) *Economics of Organization and Management.* Englewood Cliffs, New Jersey: Prentice-Hall, 1991.

Myerson, Roger (1991) *Game Theory: Analysis of Conflict.* Cambridge, Mass.: Harvard University Press, 1991.

Palfrey, Thomas & Sanjay Srivastava (1993) *Bayesian Implementation.* New York: Harwood Academic Publishers, 1993.

Phlips, Louis (1988) *The Economics of Imperfect Information.* Cambridge: Cambridge University Press, 1988.

Rasmusen, Eric (1993) *Games and Information.* Oxford: Blackwell Publishers, 1993 (1st edition, 1989).

Schmalensee, Richard & Robert Willig, eds. (1989) *The Handbook of Industrial Organization.* New York: North-Holland, 1989.

Tirole, Jean (1988) *The Theory of Industrial Organization.* Cambridge, Mass.: MIT Press, 1988.

Van Damme, Eric (1983) *Refinements of the Nash Equilibrium Concept.* Berlin: Springer-Verlag, 1983.

Van Damme, Eric (1987) *Stability and Perfection of Nash Equilibrium.* Berlin: Springer-Verlag, 1987.

Varian, Hal (1992) *Microeconomic Analysis.* Third edition. New York: W. W. Norton, 1992.

Acknowledgments

I would like to thank Dean Amel, Dan Asquith, Sushil Bikhchandani, Patricia Hughes Brennan, Paul Cheng, Luis Fernandez, David Hirshleifer, Jack Hirshleifer, Steven Lippman, Ivan Png, Benjamin Rasmusen, Marilyn Rasmusen, Ray Renken, Richard Silver, Yoon Suh, Brett Trueman, and Barry Weingast. D. Koh, Jeanne Lamotte, In-Ho Lee, Loi Lu, Patricia Martin, Timothy Opler, Sang Tran, Jeff Vincent, Tao Yang, Roy Zerner, and especially Emmanuel Petrakis helped me with research assistance at one stage or another.

I would also like to thank Robert Boyd, Mark Ramseyer, Ken Taymor, and John Wiley for their regular comments as each chapter appeared.

Jonathan Berk, Mark Burkey, Craig Holden, Peter Huang, Michael Katz, Thomas Lyon, Steve Postrel, Herman Quirmbach, H. Shifrin, George Tsebelis, Thomas Voss, and Jong-Shin Wei made useful comments on sections of the second edition, and Alexander Butler and John Spence provided research assistance. My students in Management 200 at UCLA and Business Economics G601 at Indiana University provided invaluable help, especially in suffering through the first drafts of the homework problems.

Eric Rasmusen
Erasmuse@indiana.edu

Introduction

History

Not long ago, the scoffer could say that econometrics and game theory were like Japan and Argentina. In the late 1940s both disciplines and both economies were full of promise, poised for rapid growth and ready to make a profound impact on the world. We all know what happened to the economies of Japan and Argentina. Of the disciplines, econometrics became an inseparable part of economics, while game theory languished as a subdiscipline, interesting to its specialists but ignored by the profession as a whole. The specialists in game theory were generally mathematicians, who cared about definitions and proofs rather than applying the methods to economic problems. Game theorists took pride in the diversity of the disciplines to which their theory could be applied, but in none of these disciplines had it become indispensable.

In the 1970s, the analogy with Argentina broke down. At the same time that Argentina was inviting back its former dictator, Juan Peron, economists were beginning to discover what they could achieve by combining game theory with the structure of complex economic situations. Innovation in theory and application was especially useful for situations with asymmetric information and a temporal sequence of actions, the two major themes of this book. During the 1980s, game theory became dramatically more important to mainstream economics. Indeed, it seems to be swallowing up microeconomics altogether, just as econometrics swallowed up empirical economics.

Game theory is generally considered to have begun with the publication of von Neumann and Morgenstern's *The Theory of Games and Economic Behavior* in 1944. Although very little of the game theory in that thick volume is relevant to the present book, it did introduce the idea that conflict could be mathematically analyzed and provided the terminology with which to do it. The development of the Prisoner's Dilemma (Tucker [unpub]) and Nash's papers on the definition and existence of equilibrium (Nash [1950b, 1951]) laid the foundations for modern noncooperative game theory. At the same time, cooperative game theory reached important results in papers by Nash (1950a) and Shapley (1953b) on bargaining games and Gillies (1953) and Shapley (1953a) on the core.

By 1953 virtually all the game theory that was to be used by economists for the next 20 years had been developed. Until the mid-1970s, game theory remained an autonomous field with little relevance to mainstream economics, important exceptions being Schelling's 1960 book, *The Strategy of Conflict*, which introduced the focal point, and a series of papers (of which Debreu & Scarf [1963] is typical) that showed the relationship of the core of a game to the general equilibrium of an economy.

In the 1970s, information became the focus of many models as economists started to put the emphasis on individuals who act rationally but with limited information. When attention was given to individual agents, the time ordering in which they carried out actions began to be explicitly incorporated. With this addition, games had enough structure to reach interesting and non-obvious results. Important "toolbox" references include the earlier but long unapplied articles of Selten (1965) (on perfectness) and Harsanyi (1967) (on incomplete information), the papers by Selten (1975) and Kreps & Wilson (1982b) extending perfectness, and the article by Kreps, Milgrom, Roberts & Wilson (1982) on incomplete information in repeated games. Most of the applications in the present book were developed after 1975, and the flow of research shows no signs of diminishing.

The Method of Game Theory

Game theory has been successful in recent years because it fits so well into the new methodology of economics. In the past, macroeconomists began with broad behavioral relationships such as the consumption function, and microeconomists often began with precise but irrational behavioral assumptions such as sales maximization. Now all economists start with primitive assumptions about utility functions, production functions, and endowments of the actors in the models (to which the available information must often be added). The reason is that it is usually easier to judge whether primitive assumptions are sensible than to evaluate high-level assumptions about behavior. Having accepted the primitive assumptions, the modeller figures out what happens when the actors maximize their utility subject to the constraints imposed by their information, endowments, and production functions. This is exactly the paradigm of game theory: the modeller assigns payoff functions and strategy sets to his players, and sees what happens when they pick strategies to maximize their payoffs. The approach is a combination of the Maximization Subject to Constraints of MIT, and the No Free Lunch of Chicago. We shall see, however, that game theory relies only on the spirit of these two approaches: it has moved away from maximization by calculus, and inefficient allocations are common. The players act rationally, but the consequences are often bizarre, which makes application to a world of intelligent men and ludicrous outcomes appropriate.

Exemplifying Theory

Along with the trend towards primitive assumptions and maximizing behavior has been a trend toward simplicity. I called this "no-fat modelling" in the first edition, but the term "exemplifying theory" (see Fisher [1989]) is more apt. This has also been called "modelling by example," "MIT-style theory," or, less modestly in its double meaning, "exemplary theory." The heart of the approach is to discover the simplest assumptions needed to generate an interesting conclusion—the starkest, barest, model that has the desired result. This desired result is the answer to some relatively narrow question. Could education be just a signal of ability? Why might bid-ask spreads exist? Is predatory pricing ever rational?

The modeller starts with a vague idea such as "People go to college to show they're smart." He then models the idea formally in a simple way. The idea might survive intact, it might be found formally meaningless, it might survive with qualifications, or its opposite might turn out to be true. The modeller then uses the model to come up with precise propositions, whose proofs may tell him still more about the idea. After the proofs, he goes back to thinking in words, trying to understand more than just whether the proofs are mathematically correct.

Good theory of any kind uses the notion of Occam's razor, which cuts out superfluous explanations, and the *ceteris paribus* assumption, which restricts attention to one issue at a time. Exemplifying theory goes a step further by providing, in the theory, only a narrow answer to the question. As Fisher says, "Exemplifying theory does not tell us what *must* happen. Rather it tells us what *can* happen." In the same vein, at Chicago I have heard the style called "Stories that Might be True." This is not destructive criticism if the modeller is modest, since there are also a great many "Stories that Can't be True". The aim should be to come up with one or more stories that might apply to a particular situation, and then try to sort out which story gives the best explanation. In this, economics combines the deductive reasoning of mathematics with the analogical reasoning of law.

A critic of the mathematical approach in biology has compared it to an hourglass (Slatkin [1980]). First, a broad and important problem is introduced. Second, it is reduced to a very special but tractable model that is intended to capture its essence. Finally, in the most perilous part of the process, the results are expanded to apply to the original problem. Exemplifying theory does the same thing.

The process is one of setting up "if-then" statements, whether in words or symbols. To apply such statements, their premises and conclusions need to be verified, either by casual or careful empiricism. If the required assumptions seem contrived or the assumptions and implications contradict reality, the idea should be discarded. If "reality" is not immediately obvious and data is available, econometric tests may help show whether the model is valid. Predictions can be made about future events, but that is not usually the primary motivation: most of us are more interested in explaining and understanding than in predicting.

The method just described is close to how, according to Lakatos (1976), mathematical theorems are developed. It contrasts sharply with the common view that the researcher starts with a hypothesis and proves or disproves it. Instead, the process of proof helps show how the hypothesis should be formulated.

An important part of exemplifying theory is what Kreps & Spence (1984) have called "blackboxing": treating unimportant subcomponents of a model in a cursory way. The game Entry for Buyout of section 14.4, for example, asks whether a new entrant would be bought out by the industry's incumbent producer, something that depends on duopoly pricing and bargaining. Both pricing and bargaining are complicated games in themselves, but if the modeller does not wish to deflect attention to those topics, he can use the simple Cournot and Nash solutions to those games and go on to analyze buyout. If the entire focus of the model were duopoly pricing, using the Cournot solution would be open to attack, but as a simplifying assumption, rather than one that "drives" the model, it is acceptable.

Despite the style's drive towards simplicity, a certain amount of formalism and mathematics is required to pin down the modeller's thoughts. Exemplifying theory treads a middle path between mathematical generality and nonmathematical vagueness. Proponents of either alternative may complain that exemplifying theory is too narrow. But beware of calls for more "rich," "complex," or "textured" descriptions; these often lead to theory which is either too incoherent or too incomprehensible to be applied to real situations.

Some readers will think that exemplifying theory uses too little mathematical technique, but others, especially non-economists, will think it uses too much. Intelligent laymen have objected to the amount of mathematics in economics since at least the 1880s, when George Bernard Shaw said that as a boy he (1) let someone assume that $a = b$, (2) permitted several steps of algebra, and (3) found he had accepted a proof that $1 = 2$. Forever after, Shaw distrusted assumptions and algebra. Despite the effort to achieve simplicity (or perhaps because of it), mathematics is essential to exemplifying theory. The conclusions can be retranslated into words, but rarely can they be found by verbal reasoning. The economist Philip Wicksteed put this nicely in his reply to Shaw's criticism:

> Mr Shaw arrived at the sapient conclusion that there "was a screw loose somewhere"—not in his own reasoning powers, but—"in the algebraic art"; and thenceforth renounced mathematical reasoning in favour of the literary method which enables a clever man to follow equally fallacious arguments to equally absurd conclusions *without seeing that they are absurd.* This is the exact difference between the mathematical and literary treatment of the pure theory of political economy. (Wicksteed [1885] p. 732)

In exemplifying theory, one can still rig a model to achieve a wide range of results, but it must be rigged by making strange primitive assumptions. Everyone familiar with the style knows that the place to look for the source of suspicious results is the description at the start of the model. If that description is not clear, the reader deduces that the model's counterintuitive results arise from bad assumptions concealed in poor writing. Clarity is therefore important, and the somewhat inelegant Players-Actions-Payoffs presentation used in this book is useful not only for helping the writer, but for persuading the reader.

This Book's Style

Substance and style are closely related. The difference between a good model and a bad one is not just whether the essence of the situation is captured, but also how much froth covers the essence. In this book, I have tried to make the games as simple as possible. They often, for example, allow each player a choice of only two actions. Our intuition works best with such models, and continuous actions are technically more troublesome. Other assumptions, such as zero production costs, rely on trained intuition. To the layman, the assumption that output is costless seems very strong, but a little experience with these models teaches that it is the constancy of the marginal cost that usually matters, not its level.

What matters more than what a model says is what we understand it to say. Just as an article written in Sanskrit is useless to me, so is one that is excessively mathematical or poorly written, no matter how rigorous it seems to the author. Such an article leaves me with some new belief about its subject, but that belief is not sharp, or precisely correct. Overprecision in sending a message creates imprecision when it is received, because precision is not clarity. The result of an attempt to be mathematically precise is sometimes to overwhelm the reader, in the same way that someone who requests the answer to a simple question in the discovery process of a lawsuit is overwhelmed when the other side responds with 70 boxes of documents. The quality of the author's input should be judged not by some abstract standard but by the output in terms of reader understanding.

In this spirit, I have tried to simplify the structure and notation of models while giving credit to their original authors, but I must ask pardon of anyone whose model has been oversimplified or distorted, or whose model I have inadvertently replicated without crediting them. In trying to be understandable, I have taken risks with respect to accuracy. My hope is that the impression left in the reader's mind will be more accurate than if a style more cautious and obscure had left him to devise his own errors.

Readers may be surprised to find occasional references to newspaper and magazine articles in this book. I hope these references will be reminders that models ought eventually to be applied to specific facts, and that a great many interesting situations are waiting for our analysis. The principal-agent problem is not found only in back issues of *Econometrica*: it can be found on the front page of today's *Wall Street Journal* if one knows what to look for.

I make the occasional joke here and there, and game theory is a subject intrinsically full of paradox and surprise. I want to emphasize, though, that I take game theory seriously, in the same way that Chicago economists take price theory seriously. It is not just an academic artform: people do choose actions deliberately and trade off one good against another, and game theory will help you understand how they do that. If it did not, I would not advise you to study such a difficult subject. As it is, I think it important that every educated person have some contact with the ideas in this book, just as they should have some idea of the basic principles of price theory.

I have been forced to exercise more discretion over definitions than I had hoped. Many concepts have been defined on an article-by-article basis in the literature, with no consistency and little attention to euphony or usefulness. Other concepts, such as "asymmetric information" and "incomplete information," have been considered so basic as to not need definition, and hence have been used in contradictory ways. I use existing terms whenever possible, and synonyms are listed.

I have often named the players Smith and Jones rather than 1 or 2 so that the reader's memory will be less taxed in remembering which is a player and which is a time period. I hope also to reinforce the idea that a model is a story made precise; we begin with Smith and Jones, even if we quickly descend to s and j. Keeping this in mind, the modeller is less likely to build mathematically correct models with absurd action sets, and his descriptions are more pleasant to read. In the same vein, labelling a curve "$U = 83$" sacrifices no generality: the phrase "$U = 83$ and $U = 66$" has virtually the same content as "$U = \alpha$ and $U = \beta$, where $\alpha > \beta$," but uses less short-term memory.

A danger of this approach is that readers may not appreciate the complexity of some of the material. While journal articles make the material seem harder than it really is, this approach makes it seem easier (a statement that can be true even if you find this book difficult). The better the author does his job, the worse this problem becomes. Keynes says of Alfred Marshall's *Principles*,

> The lack of emphasis and of strong light and shade, the sedulous rubbing away of rough edges and salients and projections, until what is most novel can appear as trite, allows the reader to pass too easily through. Like a duck leaving water, he can escape from this douche of ideas with scarce a wetting. The difficulties are concealed; the most ticklish problems are solved in footnotes; a pregnant and original judgement is dressed up as a platitude. (Keynes [1933] p. 212)

This book may well be subject to the same criticism, but I have tried to face up to difficult points, and the problems at the end of each chapter will keep the reader's progress from becoming too easy. Only a certain amount of understanding can be expected from a book, however. The efficient way to learn how to do research is to start doing it, not to read about it, and after reading this book many readers will want to build their own models. My purpose here is to show them the big picture, to help them understand the models intuitively and give them a feel for the modelling process.

Notes

- Perhaps the most important contribution of von Neumann & Morgenstern (1944) is the theory of expected utility. Although they developed the theory because they needed it to find the equilibria of games, it is today heavily used in all branches of economics. In game theory proper, they contributed the framework to describe games, and the concept of mixed strategies.
- On method, see the dialogue by Lakatos (1976), or Davis & Hersh (1981), chapter 6 of which is a shorter dialogue in the same style. M. Friedman (1953) is the classic essay on a different methodology: evaluating a model by testing its predictions. Kreps & Spence (1984) is a discussion of exemplifying theory.
- Because style and substance are so closely linked, how one writes is important. For advice on writing, see McCloskey (1985, 1987) (on economics), Basil Blackwell (1985) (on books), Bowersock (1985) (on footnotes), Fowler (1965), Fowler & Fowler (1949), Halmos (1970) (on mathematical writing), Strunk & White (1959), Weiner (1984), and Wydick (1978).
- **A Fallacious Proof that 1 = 2.** Suppose that $a = b$. Then $ab = b^2$ and $ab - b^2 = a^2 - b^2$. Factoring the last equation gives us $b(a - b) = (a + b)(a - b)$, which can be simplified to $b = a + b$. But then, using our initial assumption, $b = 2b$ and $1 = 2$. (The fallacy is division by zero.)

Part I

Game Theory

1 The Rules of the Game

1.1 Basic Definitions

Game theory is concerned with the actions of decision makers who are conscious that their actions affect each other. When the only two publishers in a city choose prices for their newspapers, aware that their sales are determined jointly, they are players in a game with each other. They are not in a game with the readers who buy their newspapers, because each reader ignores his effect on the publisher. Game theory is not useful when decisions are made that ignore the reactions of others or treat them as impersonal market forces.

The best way to understand which situations can be modelled as games and which cannot is to think about examples like the following:

(1) OPEC members choosing their annual output.
(2) General Motors purchasing steel from USX.
(3) Two manufacturers, one of nuts and one of bolts, deciding whether to use metric or American standards.
(4) A board of directors setting up a stock option plan for the Chief Executive Officer.
(5) United Fruit Company hiring workers in Honduras in the 1930s.
(6) An electric company deciding whether to order a new power plant given its estimate of demand for electricity in ten years.

The first four examples are games. In (1) OPEC members are playing a game because Saudi Arabia knows that Kuwait's oil output is based on Kuwait's forecast of Saudi output, and the output from both countries matters to the world price. In (2) a significant portion of American trade in steel is between General Motors and USX, companies which realize that the quantities traded by each of them affect the price. One wants the price low, the other high, so this is a game with conflict between the two players. In (3) the nut and bolt manufacturers are not in conflict, but the actions of one do affect the desired actions of the other, so the situation is a game nonetheless. In (4) the board of directors chooses a stock option plan anticipating the effect on the actions of the CEO.

Game theory is inappropriate for modelling the final two examples. In (5) each individual worker affects United Fruit insignificantly, and each worker makes his employment decision without regard for the impact on United Fruit's behavior. In (6) the electric company faces a complicated decision, but it does not face another rational agent. Changes in the important economic variables could turn examples (5) and (6) into games. The appropriate model changes if United Fruit faces a plantation workers' union or if the public utility commission pressures the utility to change its generating capacity.

Game theory as it will be presented in this book is a modelling tool, not an axiomatic system. The presentation in this chapter is unconventional. Rather than starting with mathematical definitions or simple little games of the kind used later in the chapter, we will start with a situation to be modelled, and build a game from it step by step.

Describing a Game

The essential elements of a game are **players**, **actions**, **information**, **strategies**, **payoffs**, **outcomes**, and **equilibria**. At a minimum, the game's description must include the players, strategies, and payoffs, for which the actions and information are building blocks. The players, actions, and outcomes are collectively referred to as the **rules of the game**, and the modeller's objective is to use the rules of the game to determine the equilibrium.

We will define these terms using a game called OPEC Model I as an example.

> **Players** *are the individuals who make decisions. Each player's goal is to maximize his utility by choice of actions.*

For OPEC Model I, let us specify the players to be Saudi Arabia (S) and Other Producers (O) (i.e., the other oil producers in the world). Passive individuals like the American consumer, who react predictably to oil price changes without any thought of trying to change anyone's behavior, are not players, but environmental parameters.

Sometimes it is useful to explicitly include individuals in the model called **pseudo-players** whose actions are taken in a purely mechanical way.

> **Nature** *is a pseudo-player who takes random actions at specified points in the game with specified probabilities.*

In OPEC Model I, we will assume that the strength of world demand for oil, denoted D, can take one of two permanent values. At the beginning of the game, Nature randomly decides whether oil demand will be *Weak* or *Strong*, assigning probabilities of 70 and 30 percent. Even if the players always took the same actions, this random move means that the model would yield more than just one prediction. We say that there are different **realizations** of a game depending on the results of random moves.

> *An* **action** *or* **move** *by player i, denoted a_i, is a choice he can make.*

Player i's **action set**, $A_i = \{a_i\}$, *is the entire set of actions available to him.*

An **action profile** *is an ordered set* $a = \{a_i\}$, $(i = 1, \ldots, n)$ *of one action for each of the n players in the game.*

For our model we specify the same action sets for both Saudi Arabia and Other Producers: an oil output that is either *High* or *Low* in each year. We will use the notation $Q_{country,year} = level$. Thus, if Saudi output in 1988 is *High*, we say $Q_{S,8} = H$.

Besides specifying the actions available to a player, one must specify when they are available. This is called the **order of play**. We will say that a country chooses its outputs afresh each year rather than choosing both years' outputs at the start of the game. The order of play is therefore

(0) Nature picks demand, D, to be *Weak* or *Strong*.
(1) Saudi Arabia and Other Producers simultaneously choose their individual 1988 outputs from the action sets

$$[\{Q_{S,8} = L, Q_{S,8} = H\} \text{ and } \{Q_{O,8} = L, Q_{O,8} = H\}].$$

(2) Saudi Arabia and Other Producers simultaneously choose their individual 1989 outputs from the action sets

$$[\{Q_{S,9} = L, Q_{S,9} = H\} \text{ and } \{Q_{O,9} = L, Q_{O,9} = H\}].$$

An alternative specification, appropriate if the technology of oil production requires advance planning, is that a country chooses its output for both years at the start of the game. Then the order of play would have just two elements:

(0) Nature picks demand, D, to be *Weak* or *Strong*.
(1) Saudi Arabia chooses its individual 1988 and 1989 outputs from the action set

$$\left\{ \begin{matrix} (Q_{S,8} = L, Q_{S,9} = L), (Q_{S,8} = L, Q_{S,9} = H) \\ (Q_{S,8} = H, Q_{S,9} = L), (Q_{S,8} = H, Q_{S,9} = H) \end{matrix} \right\}$$

Other Producers simultaneously chooses actions from its equivalent action set.

Information is modelled using the concept of the **information set**, a concept which will be defined precisely in section 2.3. For now, think of a player's information set as his knowledge at a particular time of the values of different variables. The elements of the information set are the different values the player thinks are possible. If the information set has many elements, there are many values the player cannot rule out; if it has one element, he knows the value precisely. Let us specify that after Nature moves, Saudi Arabia knows whether world oil demand is *Strong* or *Weak*, but Other Producers cannot rule out either possibility. The information sets are

Other Producers: $\{D = Strong, D = Weak\}$;
Saudi Arabia: $\{D = Strong\}$ or $\{D = Weak\}$, depending on demand.

A player's information set includes not only distinctions between the values of variables such as the strength of oil demand, but also knowledge of what actions have previously been taken, so his information set changes over the course of the game.

Player i's **strategy** s_i *is a rule that tells him which action to choose at each instant of the game, given his information set.*

Player i's **strategy set** *or* **strategy space** $S_i = \{s_i\}$ *is the set of strategies available to him.*

A **strategy profile** $s = (s_1, \ldots, s_n)$ *is an ordered set consisting of one strategy for each of the n players in the game.*

Since the information set includes whatever the player knows about the previous actions of other players, the strategy tells him how to react to their actions. In OPEC Model I the actions are to produce *High* or *Low* in 1988 and 1989. One strategy in Saudi Arabia's strategy set is

$$\left\{ \begin{array}{c} Q_{S,8}(D) = \begin{cases} L & \text{if } D = Weak \\ H & \text{if } D = Strong \end{cases} \\ \\ Q_{S,9}(D, Q_{S,8}, Q_{O,8}) = \begin{cases} L & \text{if } D = Weak,\ Q_{S,8} = L, \text{ and } Q_{O,8} = \text{L} \\ H & \text{otherwise} \end{cases} \end{array} \right\}$$

This strategy gives Saudi Arabia's 1988 action as a function of the strength of demand alone (since the action of Other Producers is not yet known), and its 1989 action as a function of demand, its own 1988 action, and the 1988 action of Other Producers. A strategy is a function only of observed history, not of current actions or of another player's strategy. Saudi Arabia's strategy cannot be specified to give its 1989 action as a function of the 1989 action of Other Producers or of its strategy. Such misspecification is a common source of confused thinking. In the simple games of the next few sections the distinction between actions and strategies is not important, but in later chapters it will be quite helpful. The concept of the strategy is useful because the action a player wishes to pick depends on the past actions of Nature and the other players. Only rarely can we predict a player's actions unconditionally, but often we can predict how he will respond to the outside world.

Keep in mind that a player's strategy is a complete set of instructions for him, which tells him what actions to pick in every conceivable situation, even if he does not expect to reach that situation. Strictly speaking, even if a player's strategy instructs him to commit suicide in 1989, it ought also to specify what actions he takes if he is still alive in 1990. This kind of care will be crucial in the discussion in Chapter 4 of subgame perfect equilibrium. The completeness of the description also means that strategies, unlike actions, are unobservable. An action is physical, but a strategy is only mental.

By player i's **payoff** $\pi_i(s_1, \ldots, s_n)$, *we mean either:*

(1) *The utility player i receives after all players and Nature have picked their strategies and the game has been played out; or*

(2) *The expected utility player i receives as a function of the strategies chosen by himself and the other players.*

Definitions (1) and (2) are distinct, but in the literature and in this book the term "payoff" is used for both the actual payoff and the expected payoff. The context will make clear which is meant. Here, let us assume that the payoffs for Saudi Arabia and Other Producers are the expected sums of their oil revenues over the two years of production.

The **outcome** *of a game is the set of interesting elements that the modeller picks from the values of the actions, payoffs, and other variables after the game is played out.*

The definition of the outcome for any particular model depends on what variables the modeller finds interesting. One outcome of OPEC Model I is

$$Q_{S,8} = L,\ Q_{S,9} = H,\ Q_{O,8} = H,\ Q_{O,9} = L,\ D = L,\ \pi_S = 100,\ \pi_O = 80, \quad (1)$$

where 100 and 80 are the values specified by the payoff functions. The outcome could be more narrowly defined as just the set of payoffs or the levels of output. Which definition is chosen depends on what the modeller thinks is interesting about OPEC.

This entire model, in fact, is only one of the many possible models of OPEC. For contrast, another game representing OPEC is OPEC Model II.

OPEC Model II

Players

Saudi Arabia, Libya, Venezuela, Kuwait, Nigeria.

Information

All players know the value of demand, but they choose their actions simultaneously.

Order of Play

(1) The players simultaneously choose supply schedules consisting of the quantity each will supply at each possible market price in 1988 (this model does not concern itself with 1989).

(2) Nature picks world demand for oil to be either *Weak* or *Strong*, with equal probability.

Strategies

The strategies are the same as actions, since no information is revealed that might affect the action a player chooses.

Payoffs

For each country the payoff is +100 if its oil revenue remains above a country-specific amount necessary to avoid a military coup, −100 if revenue falls below this amount.

Outcomes

The outcome includes the quantities supplied, the state of demand, the resulting revenues and market price, and whether or not each country has a coup.

Constructing the model, not solving it, is where most of the talent of the modeller is displayed, since he must trade off realism, ease of solution, and clarity of presentation. How would you decide between the two OPEC models?

Equilibrium

To predict the outcome of a game, the modeller focusses on the possible strategy profiles, since it is the interaction of the different players' strategies that determines what happens. The distinction between strategy profiles, which are sets of strategies, and outcomes, which are sets of values of whichever variables are considered interesting, is a common source of confusion. Often different strategy profiles lead to the same outcome. In OPEC Model I, the single outcome

$$(Q_{S,8} = L,\ Q_{O,8} = L,\ Q_{S,9} = L,\ Q_{O,9} = L,\ D = Strong,\ \pi_S = 100,\ \pi_O = 80) \tag{2}$$

is produced by either of the following two strategy profiles, the Golden Rule and the Silver Rule.

The Golden Rule (*Low output no matter what happens*):

$$\left\{\begin{matrix} \text{Saudi Arabia:} & (Q_{S,8} = L;\ Q_{S,9} = L) \\ \text{Other Producers:} & (Q_{O,8} = L;\ Q_{O,9} = L) \end{matrix}\right\}.$$

The Silver Rule (*Retaliate*):

$$\left\{\begin{matrix} \text{Saudi Arabia:} & (Q_{S,8} = L;\ Q_{S,9} = L \text{ if } Q_{O,8} = L,\ Q_{S,9} = H \text{ otherwise}) \\ \text{Other Producers:} & (Q_{O,8} = L;\ Q_{O,9} = L \text{ if } Q_{S,8} = L,\ Q_{O,9} = H \text{ otherwise}) \end{matrix}\right\}$$

Under the Golden Rule, both players always choose low output, so output is low in both years. Under the Silver Rule (a variant of the Tit-for-Tat strategy of section 5.2), both players choose low output in 1988, and in 1989 each chooses the output the other had chosen the previous year, so output is also low.[1]

> *An* **equilibrium** $s^* = (s_1^*, \ldots, s_n^*)$ *is a strategy profile consisting of a best strategy for each of the n players in the game.*

The **equilibrium strategies** are the strategies players pick in trying to maximize their individual payoffs, as distinct from the many possible strategy profiles obtainable by arbitrarily choosing one strategy per player. Equilibrium is understood differently in game theory than in other areas of economics. In a general equilibrium model, for example, an equilibrium is a set of prices resulting from optimal behavior by the individuals in the economy. In game theory, that set of prices would be the **equilibrium outcome**, but the equilibrium itself would be the strategy profile—the individuals' rules for buying and selling—that generated the outcome.

To find the equilibrium it is not enough to specify the players, strategies, and payoffs, because the modeller must also decide what "best strategy" means. He does this by defining an equilibrium concept.

> *An* **equilibrium concept** *or* **solution concept** $F: \{S_1, \ldots, S_n, \pi_1, \ldots, \pi_n\} \rightarrow s^*$ *is a rule that defines an equilibrium based on the possible strategy profiles and the payoff functions.*

Only a few equilibrium concepts are generally accepted. The remaining sections of this chapter are devoted to finding an equilibrium using the two best-known concepts: dominant strategy and Nash equilibrium.

People often carelessly say "equilibrium" when they mean "equilibrium outcome," and "strategy" when they mean "action." The difference is not very important in most of the games that will appear in this chapter, but it is absolutely fundamental to thinking like a game theorist. Consider Germany's decision on whether to remilitarize the Rhineland in 1936. France adopted the strategy: *Do not fight,* and Germany responded by remilitarizing, leading to World War II a few years later. If France had adopted the strategy: *Do not fight unless Germany remilitarizes; otherwise, fight,* the action would still have been that France would not have fought, but Germany would not have remilitarized the Rhineland. Perhaps it was because he thought along these lines that John von Neumann was such a hawk in the Cold War (as MacRae describes in his excellent 1992 biography). This difference between actions and strategies, outcomes and equilibria, is one of the hardest ideas to teach in game theory, even though it is trivial to state.

[1] Note that I was careful to cover all possible contingencies in describing these strategies. Each, for example, specifies what Saudi Arabia does in 1989 if both countries choose *High* in 1988, even though Saudi Arabia does not intend to choose *High* in 1988.

Uniqueness

Accepted solution concepts do not guarantee uniqueness, and lack of a unique equilibrium is a major problem in game theory. Often the solution concept employed leads us to believe that the players will pick one of the two strategy profiles A or B, not C or D, but we cannot say whether A or B is more likely. Sometimes we have the opposite problem and the game has no equilibrium at all. By this is meant either that the modeller sees no good reason why one strategy profile is more likely than another, or that some player wants to pick an infinite value for one of his actions.

A model with no equilibrium or multiple equilibria is underspecified. The modeller has failed to provide a full and precise prediction for what will happen. One option is to admit that the theory is incomplete. This is not a shameful thing to do; an admission of incompleteness like the Folk Theorem of section 5.2 is a valuable negative result. Or perhaps the situation being modelled really is unpredictable. Another option is to renew the attack by changing the game's description or the solution concept. Preferably it is the description that is changed, since economists look to the rules of the game for the differences between models, and not to the solution concept. If an important part of the game is concealed under the definition of equilibrium, in fact, the reader is likely to feel tricked and to charge the modeller with intellectual dishonesty.

1.2 Dominant Strategies: The Prisoner's Dilemma

In discussing equilibrium concepts, it is useful to have shorthand notation for "all the other players' strategies."

For any vector $y = (y_1, \ldots, y_n)$, denote by y_{-i} the vector $(y_1, \ldots, y_{i-1}, y_{i+1}, \ldots, y_n)$, which is the portion of y not associated with player i.

Using this notation, s_{-Smith}, for instance, is the profile of the strategies of every player except player *Smith*. That profile is of great interest to Smith, because he uses it to help choose his own strategy, and the new notation helps define his best response.

Player i's **best response** *or* **best reply** *to the strategies s_{-i} chosen by the other players is the strategy s_i^* that yields him the greatest payoff; that is,*

$$\pi_i(s_i^*, s_{-i}) \geq \pi_i(s_i', s_{-i}) \ \forall s_i' \neq s_i^*. \qquad (3)$$

The best response is strongly best if no other strategies are equally good, and weakly best otherwise.

The first important equilibrium concept is the dominant-strategy equilibrium.

The strategy s_i^* is a **dominant strategy** *if it is a player's strictly best response to* any *strategies the other players might pick, in the sense that whatever strategies they pick, his payoff is highest with s_i^*. Mathematically,*

$$\pi_i(s_i^*, s_{-i}) > \pi_i(s_i', s_{-i}) \ \forall s_{-i}, \ \forall s_i' \neq s_i^*. \quad (4)$$

His inferior strategies are **dominated strategies.**

A **dominant-strategy equilibrium** *is a strategy profile consisting of each player's dominant strategy.*

A player's dominant strategy is his strictly best response even to wildly irrational actions by the other players. Most games do not have dominant strategies, and the players must try to figure out each others' actions in order to arrive at their own strategies.

OPEC Model I incorporated considerable complexity in the rules of the game to illustrate such things as information sets and the time sequence of actions. To illustrate equilibrium concepts, we will use simpler games, such as The Prisoner's Dilemma. In The Prisoner's Dilemma, two prisoners, Messrs Row and Column, are being interrogated separately. If both confess, each is sentenced to eight years in prison; if both deny their involvement, each is sentenced to one year.[2] If just one confesses, he is released but the other prisoner is sentenced to ten years. The Prisoner's Dilemma is an example of a **2-by-2 game**, because each of the two players—Row and Column—has two possible actions in his action set—*Confess* and *Deny*. Table 1.1 gives the payoffs.[3]

Table 1.1 The Prisoner's Dilemma

		Column		
		Deny		*Confess*
Row:	*Deny*	$-1, -1$	$\rightarrow$	$-10, 0$
		$\downarrow$		$\downarrow$
	Confess	$0, -10$	$\rightarrow$	$-8, -8$

Payoffs to: (Row, Column)

Each player has a dominant strategy. Consider Row. Row does not know which action Column is choosing, but if Column chooses *Deny*, Row faces a *Deny* payoff of -1 and a *Confess* payoff of 0, whereas if Column chooses *Confess*, Row faces a *Deny* payoff of -10 and a *Confess* payoff of -8. In either case Row does better with *Confess*. Since the game is symmetric, Column's incentives are the same. The dominant-strategy equilibrium is (*Confess*, *Confess*), and the equilibrium payoffs are $(-8, -8)$, which is worse for both players than $(-1, -1)$. Sixteen, in fact, is the greatest possible combined total of years in prison.

The result is even stronger than it seems, because it is robust to substantial changes in the model. Because the equilibrium is a dominant-strategy equilibrium, the information structure of the game does not matter. If Column is allowed

[2] Another way to tell the story is to say that if both deny, then, with probability 0.1, they are convicted anyway and serve ten years, for an expected payoff of $(-1, -1)$.

[3] The arrows represent a player's preference between actions, as will be explained in section 1.4.

to know Row's move before taking his own, the equilibrium is unchanged. Row still chooses *Confess*, knowing that Column will surely choose *Confess* afterwards.

The Prisoner's Dilemma seems perverse and unrealistic to many people who have never encountered it before (although friends who are prosecutors assure me that it is a standard crime-fighting tool). If the outcome does not seem right to you, you should realize that very often the chief usefulness of a model is to induce discomfort. Discomfort is a sign that your model is not what you think it is—that you left out something essential to the result you expected and didn't get. Either your original thought or your model is mistaken; finding such mistakes is a real if painful benefit of model-building. To refuse to accept surprising conclusions is to reject logic.

The Prisoner's Dilemma crops up in many different situations, including oligopoly pricing, auction bidding, salesman effort, political bargaining, and arms races. Whenever you observe individuals in a conflict that hurts them all, your first thought should be of The Prisoner's Dilemma.

Cooperative and Noncooperative Games

What difference would it make if the two prisoners could talk to each other before making their decisions? It depends on the strength of promises. If promises are not binding, then although the two prisoners might agree to *Deny*, they would *Confess* anyway when the time came to choose actions.

> *A* **cooperative game** *is a game in which the players can make binding commitments, as opposed to a* **noncooperative game**, *in which they cannot.*

This definition draws the usual distinction between the two theories of games, but the real difference lies in the modelling approach. Both theories start off with the rules of the game, but they differ in the kinds of solution concepts employed. Cooperative game theory is axiomatic, frequently appealing to Pareto optimality, fairness, and equity. Noncooperative game theory is economic in flavor, with solution concepts based on players maximizing their own utility functions subject to stated constraints. Except for section 11.2 of the chapter on bargaining, this book is concerned exclusively with noncooperative games.

In applied economics, the most commonly encountered use of cooperative games is to model bargaining. The Prisoner's Dilemma is a noncooperative game, but it could be modelled as cooperative by allowing the two players not only to communicate but to make binding commitments. Cooperative games often allow players to split the gains from cooperation by making **side-payments**—transfers between themselves that change the prescribed payoffs. Cooperative game theory generally incorporates commitments and side-payments via the solution concept, which can become very elaborate, while noncooperative game theory does this by adding extra actions. The distinction between cooperative and noncooperative games does *not* lie in conflict versus the absence of conflict, as is shown by the following examples:

A cooperative game without conflict: Members of a work force choose which of several equally arduous tasks to undertake so as to best coordinate with each other.

A cooperative game with conflict: Bargaining over price between a monopolist and a monopsonist.

A noncooperative game with conflict: The Prisoner's Dilemma.

A noncooperative game without conflict: Two companies set a product standard without communication.

1.3 Iterated Dominance: Battle of the Bismarck Sea

Very few games have a dominant-strategy equilibrium, but sometimes dominance can still be useful even when it does not resolve things quite so neatly as in the Prisoner's Dilemma. The game Battle of the Bismarck Sea is set in the South Pacific in 1943. Rear Admiral Kimura has been ordered to transport Japanese troops across the Bismarck Sea to New Guinea, and Admiral Kenney wants to bomb the troop transports. Kimura must choose between a shorter northern route or a longer southern route to New Guinea, and Kenney must decide where to send his planes to look for the Japanese. If Kenney sends his planes to the wrong route he can recall them, but the number of days of bombing is reduced.

The players are Kenney and Kimura, and they each have the same action set, {*North*, *South*}, but their payoffs, given by table 1.2, are never the same. Kimura loses exactly what Kenney gains. Because of this special feature, the payoffs could be represented using just four numbers instead of eight, but listing all eight payoffs in table 1.2 saves the reader a little thinking.

Table 1.2 Battle of Bismarck Sea

		Kimura		
		North		*South*
Kenney	*North*	2, −2	↔	2, −2
		↑		↓
	South	1, −1	←	3, −3

Payoffs to: (Kenney, Kimura)

Strictly speaking, neither player has a dominant strategy. Kenney would choose *North* if he thought Kimura would choose *North*, but *South* if he thought Kimura would choose *South*. Kimura would choose *North* if he thought Kenney would choose *South*, and he would be indifferent between actions if he thought Kenney would choose *North*. This is what the arrows are showing, and what they showed in table 1.1 for The Prisoner's Dilemma.

But we can still find a plausible equilibrium, using the concept of weak dominance.

Strategy s_i' is **weakly dominated** *if there exists some other strategy s_i'' for player i which is possibly better and never worse, yielding a higher payoff in some strategy profile and never yielding a lower payoff. Mathematically, s_i' is weakly dominated if there exists s_i'' such that*

$$\begin{aligned} \pi_i(s_i'', s_{-i}) &\geq \pi_i(s_i', s_{-i}) \quad \forall s_{-i}, \text{ and} \\ \pi_i(s_i'', s_{-i}) &> \pi_i(s_i', s_{-i}) \quad \text{for some } s_{-i}. \end{aligned} \tag{5}$$

One might define a **weak dominant strategy equilibrium** as the strategy profile found by deleting all the weakly dominated strategies of each player. Eliminating weakly dominated strategies does not help much in the Battle of the Bismarck Sea, however. Kimura's strategy of *South* is weakly dominated by the strategy *North* because his payoff from *North* is never smaller than his payoff from *South*, and it is greater if Kenney picks *South*. For Kenney, however, neither strategy is even weakly dominated. The modeller must therefore go a step further in defining equilibrium, to the iterated dominance equilibrium.

An **iterated dominance equilibrium** *is a strategy profile found by deleting a weakly dominated strategy from the strategy set of one of the players, recalculating to find which remaining strategies are weakly dominated, deleting one of them, and continuing the process until only one strategy remains for each player.*

Applied to Battle of the Bismarck Sea, this equilibrium concept implies that Kenney decides that Kimura will pick *North* because it is weakly dominant, so Kenney eliminates {Kimura chooses *South*} from consideration. Having deleted one column of table 1.2, Kenney has a strongly dominant strategy: he chooses *North*, which achieves payoffs strictly greater than *South*. The strategy profile (*North*, *North*) is an iterated dominance equilibrium, and despite all the qualifying adjectives it seems a good prediction. And indeed (*North*, *North*) was the outcome in 1943.

It is interesting to consider modifying the order of play or the information structure in Battle of the Bismarck Sea. If Kenney moved first, rather than simultaneously with Kimura, (*North*, *North*) would remain an equilibrium, but (*North*, *South*) would become an equilibrium as well. The payoffs would be the same for both equilibria, but the outcomes would be different.

If Kimura moved first, (*North*, *North*) would be the only equilibrium. What is important about a player moving first is not the literal timing of the moves but that it gives the other player more information before he acts. If Kenney has cracked the Japanese code and knows Kimura's plan, it no longer matters that the two players move literally simultaneously; the game is better modelled as a sequential game. Whether Kimura literally moves first or whether his code is cracked, Kenney's information set becomes either {Kimura moved *North*} or {Kimura moved *South*} after Kimura's decision, so Kenney's equilibrium strategy is specified as (*North* if Kimura moved *North*, *South* if Kimura moved *South*).

Game theorists often differ in their terminology, and the terminology applied to the idea of eliminating dominated strategies is particularly diverse. The equilibrium concept used in Battle of the Bismarck Sea might be called **iterated-dominance equilibrium** or **iterated-dominant strategy equilibrium**, or one might say that the game is **dominance-solvable**, that it can be **solved by iterated dominance**, or that the equilibrium strategy profile is **serially undominated**, and each of these might mean either deletion of strictly dominated strategies or deletion of weakly dominated strategies.

The significant difference is between strong and weak domination. Everyone agrees that no rational player would use a strictly dominated strategy, but it is harder to argue against weakly dominated strategies. In economic models, firms and individuals are often indifferent about their behavior in equilibrium. In standard models of perfect competition, firms earn zero profits, but it is crucial that some firms be active in the market and some stay out and produce nothing. If a monopolist knows that customer Smith is willing to pay up to ten dollars for a widget, the monopolist will charge exactly ten dollars to Smith in equilibrium, which makes Smith indifferent about buying and not buying, yet there is no equilibrium unless Smith buys. It is impractical, therefore, to rule out equilibria in which a player is indifferent about his actions. This should be kept in mind later when we discuss the similar "open-set problem" in section 4.3.

Another difficulty is multiple equilibria. The dominant strategy equilibrium is unique, if it exists. Each player has at most one strategy whose payoff in any strategy profile is strictly higher than the payoff from any other strategy, so only one strategy profile can be formed out of dominant strategies. An iterated dominance equilibrium may not be unique, because the order in which strategies are deleted can matter to the final solution. The Iteration Path Game of table 1.3 shows this. The strategy profiles (r_1, c_1) and (r_1, c_3) are both iterated dominance equilibria, because each of them can be found by iterated deletion. The deletion can proceed in the order (r_3, c_3, c_2, r_2) or in the order (c_2, r_2, c_1, r_3).

Table 1.3 The Iteration Path Game

		Column		
		c_1	c_2	c_3
	r_1	**2,12**	1,10	**1,12**
Row:	r_2	0,12	0,10	0,11
	r_3	**0,12**	0,10	0,13

Payoffs to: (Row, Column)

Despite these problems, deletion of weakly dominated strategies is a useful tool, and it is part of many more complicated equilibrium concepts, such as the subgame perfectness concept of section 4.1.

Battle of the Bismarck Sea is special because the payoffs of the players always sum to zero. This feature is important enough to deserve a name.

> *A* **zero-sum game** *is a game in which the sum of the payoffs of all the players is zero whatever strategies each chooses. A game which is not zero-sum is* **variable sum.**

In a zero-sum game, what one player gains, another player must lose. Battle of the Bismarck Sea is a zero-sum game, but The Prisoner's Dilemma and OPEC Model I and OPEC Model II are not, and there is no way that the payoffs in those games can be rescaled to make them zero-sum without changing the essential character of the games. Although zero-sum games have fascinated game theorists for many years, they are uncommon in economics. One of the few examples is the bargaining game between two players who divide a surplus, but even this is often modelled nowadays as a variable-sum game in which the surplus shrinks as the players spend more time deciding how to divide it. In reality, even simple division of property can result in loss—just think of how much some lawyers take out when a divorcing couple bargain over dividing their possessions.

Although the 2-by-2 games in this chapter may seem facetious, they are simple enough for use in modelling economic situations. Battle of the Bismarck Sea, for example, can be turned into a game of corporate strategy. Two firms, Kenney Company and Kimura Incorporated, are trying to maximize their shares of a market of constant size by choosing between the two product designs *North* and *South*. Kenney has a marketing advantage, and would like to compete head-to-head, while Kimura would rather carve out its own niche. The equilibrium is (*North*, *North*).

1.4 Nash Equilibrium: Boxed Pigs, Battle of the Sexes, and Ranked Coordination

For the vast majority of games, which lack even iterated dominance equilibria, modellers use the Nash equilibrium, which is the most important and most frequently encountered equilibrium concept. To introduce the Nash equilibrium we will use the game of Boxed Pigs. Two pigs are put in a Skinner box with a special panel at one end and a food dispenser at the other. When a pig presses the panel, at a utility cost of two units, ten units of food are dispensed. One pig is dominant (let us assume he is bigger), and if he gets to the dispenser first, the other pig will only get his leavings, worth one unit. If, instead, the small pig arrives first, he eats four units, and even if they arrive at the same time the small pig gets three units. Table 1.4 summarizes the payoffs for the strategies *Press* the panel and *Wait* by the dispenser.

Table 1.4 Boxed Pigs

		Small Pig		
		Press		*Wait*
	Press	5 , 1	→	[4] , [4]
Big Pig		↓		↑
	Wait	[9] , −1	→	0 , 0

Payoffs to: (Big Pig, Small Pig)

Boxed Pigs has no dominant-strategy equilibrium, because what the big pig chooses depends on what he thinks the small pig will choose. If he believed that the small pig would press the panel, the big pig would wait by the dispenser, but if he believed that the small pig would wait, the big pig would press. There does exist an iterated dominance equilibrium, (*Press, Wait*), but we will use a different line of reasoning to justify that outcome: Nash equilibrium.

Nash equilibrium is the standard equilibrium concept in economics. It is less obviously correct than the dominant-strategy equilibrium but more often applicable. Nash equilibrium is so widely accepted that the reader can assume that if a model does not specify which equilibrium concept is being used it is Nash or some refinement of Nash.

A strategy profile s^* is a **Nash equilibrium** *if no player has incentive to deviate from his strategy given that the other players do not deviate. Formally,*

$$\forall i \; \pi_i(s_i^*, s_{-i}^*) \geq \pi_i(s_i', s_{-i}^*) \; \forall s_i' . \tag{6}$$

The strategy profile (*Press, Wait*) is a Nash equilibrium. The way to approach Nash equilibrium is to propose a strategy profile and test whether each player's strategy is a best response to the others' strategies. If the big pig picks *Press*, the small pig, who faces a choice between a payoff of 1 from pressing and 4 from waiting, is willing to wait. If the small pig picks *Wait*, the big pig, who has a choice between a payoff of 4 from pressing and 0 from waiting, is willing to press. This confirms that (*Press, Wait*) is a Nash equilibrium, and in fact it is the unique Nash equilibrium.[4]

It is useful to draw arrows in the tables when trying to solve for the equilibrium, since the number of calculations is great enough to soak up quite a bit of mental RAM. Another solution tip, illustrated in Boxed Pigs, is to circle payoffs

[4] This game, too, has its economic analog. If Bigpig, Inc. introduces granola bars, at considerable marketing expense in educating the public, then Smallpig Ltd. can imitate profitably without ruining Bigpig's sales completely. If Smallpig introduces them at the same expense, however, an imitating Bigpig would hog the market.

that dominate other payoffs (or box them, as is especially suitable here). Double arrows or dotted circles indicate weakly dominant payoffs. Any payoff profile in which every payoff is circled, or which has arrows pointing towards it from every direction, is a Nash equilibrium. I like using arrows better in 2-by-2 games, but circles are better for bigger games, since arrows become confusing when payoffs are not lined up in order of magnitude in the table (see table 2.2 in the next chapter).

The pigs in this game have to be smarter than the players in The Prisoner's Dilemma. They have to realize that the only set of strategies supported by self-consistent beliefs is (*Press*, *Wait*). The definition of Nash equilibrium lacks the "$\forall s_{-i}$" of a dominant-strategy equilibrium, so a Nash strategy need only be a best response to the other Nash strategies, not to all possible strategies. And although we talk of "best responses," the moves are actually simultaneous, so the players are predicting each others' moves. If the game were repeated, or the players communicated, Nash equilibrium would be especially attractive, because it is even more compelling that beliefs should be consistent.

Like a dominant-strategy equilibrium, a Nash equilibrium can be either weak or strong. The definition above is for a weak Nash equilibrium. To define the strong Nash equilibrium, make the inequality strict; that is, require that no player is indifferent in choosing between his equilibrium strategy and some other strategy.

Every dominant-strategy equilibrium is a Nash equilibrium, but not every Nash equilibrium is a dominant-strategy equilibrium. If a strategy is dominant, it is a best response to *any* strategies the other players pick, including their equilibrium strategies. If a strategy is part of a Nash equilibrium, it need only be a best response to the other players' *equilibrium* strategies.

The Modeller's Dilemma of table 1.5 illustrates this feature of Nash equilibrium. The situation it models is the same as The Prisoner's Dilemma, with one major exception: although the police have enough evidence to arrest the suspects as the "probable cause" of the crime, they will not have enough evidence to convict them of even a minor offense if neither prisoner confesses. The northwest payoff profile becomes (0, 0) instead of $(-1, -1)$.

Table 1.5 The Modeller's Dilemma

		Column		
		Deny		*Confess*
Row:	*Deny*	[0], [0]	↔	−10, [0]
		↕		↓
	Confess	[0], −10	→	[−8], [−8]

Payoffs to: (Row, Column)

The Modeller's Dilemma does not have a dominant-strategy equilibrium. It does have a weak dominant-strategy equilibrium, because *Confess* is still a weakly dominant strategy for each player. Moreover, using this fact, it can be seen that (*Confess*, *Confess*) is an iterated dominance equilibrium, and it is a strong Nash equilibrium as well. So the case for (*Confess*, *Confess*) still being the equilibrium outcome seems very strong.

There is, however, another Nash equilibrium in The Modeller's Dilemma: (*Deny*, *Deny*), which is a weak Nash equilibrium. This equilibrium is weak and the other Nash equilibrium is strong, but (*Deny*, *Deny*) has the advantage that its outcome is Pareto-superior: (0, 0) is uniformly greater than $(-8, -8)$. Moreover, if the modeller is wrong about the order of play, and the game is actually sequential, (*Deny*, *Deny*) is much more compelling. This makes it difficult to know which behavior to predict.

The Modeller's Dilemma illustrates a common difficulty for modellers: what to predict when two Nash equilibria exist. The modeller could add more details to the rules of the game, or he could use an **equilibrium refinement**, adding conditions to the basic equilibrium concept until only one strategy profile satisfies the refined equilibrium concept. There is no single way to refine Nash equilibrium. The modeller might insist on a strong equilibrium, or rule out weakly dominated strategies, or use iterated dominance. All of these lead to (*Confess*, *Confess*) in The Modeller's Dilemma. Or he might rule out Nash equilibria that are Pareto dominated by other Nash equilibria, and end up with (*Deny*, *Deny*). Neither approach is completely satisfactory.

Battle of the Sexes

The third game we will use to illustrate Nash equilibrium is Battle of the Sexes, a conflict between a man who wants to go to a prize fight and a woman who wants to go to a ballet. While selfish, they are deeply in love, and would, if necessary, sacrifice their preferences to be with each other. Less romantically, their payoffs are given by table 1.6.

Table 1.6 Battle of the Sexes[5]

		Woman		
		Prize Fight		*Ballet*
	Prize Fight	**2,1**	←	0,0
Man		↑		↓
	Ballet	0,0	→	**1,2**

Payoffs to: (Man, Woman)

Battle of the Sexes does not have an iterated dominance equilibrium. It has two Nash equilibria, one of which is the strategy profile (*Prize Fight*, *Prize Fight*). Given that the man chooses *Prize Fight*, so does the woman; given that the woman chooses *Prize Fight*, so does the man. The strategy profile (*Ballet*, *Ballet*) is another Nash equilibrium, by the same line of reasoning.

[5] Many game theory books have presented bowdlerized versions of this game, presumably out of political correctness. This is the original, unexpurgated game.

How do the players know which Nash equilibrium to choose? Going to the fight and going to the ballet are both Nash strategies, but for different equilibria. Nash equilibrium assumes correct and consistent beliefs. If they do not talk beforehand, the man might go to the ballet and the woman to the fight, each mistaken about the other's beliefs. But even if the players do not communicate, Nash equilibrium is sometimes justified by repetition of the game. If the couple do not talk, but repeat the game night after night, one may suppose that eventually they'll settle on one or another of the Nash equilibria.

Each of the Nash equilibria in Battle of the Sexes is Pareto efficient; no other strategy profile increases the payoff of one player without decreasing that of the other. In many games the Nash equilibrium is not Pareto efficient: (*Confess*, *Confess*), for example, is the unique Nash equilibrium of the Prisoner's Dilemma, although its payoffs of $(-8, -8)$ are Pareto inferior to the $(-1, -1)$ generated by (*Deny*, *Deny*).

Who moves first is important in Battle of the Sexes, unlike the Prisoner's Dilemma. If the man could buy the fight ticket in advance, his commitment would induce the woman to go to the fight. In many games, but not all, the player who moves first (which is equivalent to commitment) has a **first-mover advantage**.

Battle of the Sexes has many economic applications. One is the choice of an industry-wide standard when two firms have different preferences but both want a common standard so as to encourage consumers to buy the product. A second is to the choice of language in a contract when two firms want to formalize a sales agreement even though they prefer different terms.

Coordination Games

Sometimes one can use the size of the payoffs to choose between Nash equilibria. In the following game, players Smith and Jones are trying to decide whether to design the computers they sell to use large or small floppy disks. Both players will sell more computers if their disk drives are compatible. The payoffs are given by table 1.7.

Table 1.7 Ranked Coordination

		Jones		
		Large		*Small*
Smith	*Large*	**2, 2**	←	**−1, −1**
		↑		↓
	Small	**−1, −1**	→	**1, 1**

Payoffs to: (Smith, Jones)

The strategy profiles (*Large*, *Large*) and (*Small*, *Small*) are both Nash equilibria, but (*Large*, *Large*) Pareto dominates (*Small*, *Small*). Both players prefer (*Large*, *Large*), and most modellers would use the Pareto-efficient equilibrium to predict the actual outcome. We could imagine that it arises from pre-game communication between Smith and Jones taking place outside of the specification of the model, but the interesting question is what happens if communication is impossible. Is the Pareto-efficient equilibrium still more plausible? The question is really one of psychology rather than economics.

Ranked Coordination is one of a large class of games called **coordination games** which share the common feature that the players need to coordinate on one of multiple Nash equilibria. Ranked Coordination has the additional feature that the equilibria can be Pareto ranked. Table 1.8 shows another coordination game, Dangerous Coordination, which has the same equilibria as Ranked Coordination, but differs in the off-equilibrium payoffs. If an experiment were conducted in which students played Dangerous Coordination against each other, I would not be surprised if (*Large*, *Large*), the Pareto-dominated equilibrium, were the one that was played out. This is true even though (*Small*, *Small*) is still a Nash equilibrium; if Smith thinks that Jones will pick *Small*, Smith is quite willing to pick *Small* himself. The problem is that if the assumptions of the game are violated, and Smith cannot trust Jones to be rational, unconfused, and well-informed about the payoffs of the game, then Smith will be reluctant to pick *Small* because his payoff if Jones picks *Large* is then −1,000. People make mistakes, and with such an extreme difference in payoffs, even a small probability of a mistake is important, so (*Small*, *Small*) would be a bad prediction.

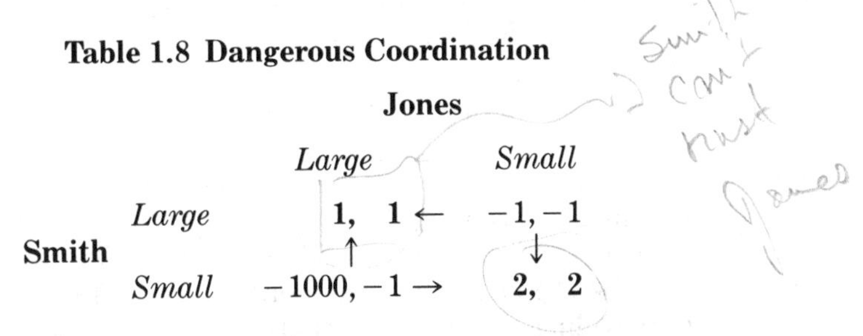

Table 1.8 Dangerous Coordination

		Jones	
		Large	*Small*
Smith	*Large*	1, 1 ←	−1, −1
		↑	↓
	Small	−1000, −1 →	2, 2

Payoffs to: (Smith, Jones)

Games like Dangerous Coordination are a major concern in a 1988 book by John Harsanyi and Reinhard Selten, two of the giants in the field of game theory, but my view is different from theirs. I do not consider the fact that one of the Nash equilibria of Dangerous Coordination is a bad prediction to be a heavy blow against Nash equilibrium. The bad prediction is based on two items: the use of the Nash equilibrium concept, and the use of the game of Dangerous Coordination. If Jones might be confused about the payoffs of the game, the game actually being played out is not Dangerous Coordination, so it is not surprising that it gives poor predictions. The rules of the game ought to describe the probabilities that the players are confused, as well as the payoffs if they take particular actions. If confusion is an important feature of the situation, the two-by-two game of table 1.8 is the wrong model to use, and a more complicated game of incomplete information of the kind described in chapter 2 is more appropriate. Again, as with The Prisoner's Dilemma, the modeller's first thought on finding that the model predicts an odd result should not be "Game theory is bunk," but the more modest

"Maybe I'm not describing the situation correctly" (or, even, "Maybe I should not trust common sense about what will happen").

1.5 Focal Points

Thomas Schelling's book, *The Strategy of Conflict*, is a classic in game theory, even though it contains no equations or Greek letters. Despite being published some 30 years ago, it is surprisingly modern in spirit. Schelling is not a mathematician but a strategist, and he examines things such as threats, commitments, hostages, and delegation which we will examine in a more formal way in the remainder of this book. He is perhaps best known for his coordination games. Take a moment to decide on a strategy in each of the following games, adapted from Schelling, which you win by matching your response to those of as many of the other players as possible.

(1) Name Heads or Tails.
(2) Name Tails or Heads.
(3) Circle one of the following numbers 7, 100, 13, 261, 99, 666.
(4) You are to meet somebody in New York City. When? Where?
(5) You are to split a pie, and get nothing if your proportions add to more than 100 percent.
(6) Circle one of the following numbers: 14, 15, 16, 17, 18, 100.

Each of the games above has many Nash equilibria: if I think you will choose 666, and you think I will choose 666, we both choose it. But to a greater or lesser extent they also have Nash equilibria that seem more likely. Certain of the strategy profiles are **focal points**: Nash equilibria which for psychological reasons are particularly compelling. Formalizing what makes a strategy profile a focal point is not an easy task and depends on the context. In example (3), Schelling found 7 to be the most common strategy, but in a group of Satanists, 666 might be focal. In repeated games, focal points are often provided by past history. If we split a pie once, we are likely to agree on 50:50. But if last year we split a pie in the ratio 60:40, that provides a focal point for this year.

The **boundary** is a particular kind of focal point. If player Russia chooses the action of putting troops anywhere from one inch to 100 miles away from the Chinese border, player China does not react. If he chooses to put troops from one inch to 100 miles *beyond* the border, China declares war. There is an arbitrary discontinuity in behavior at the boundary. Another example, quite vivid in its arbitrariness, is the rallying cry, "Fifty-Four Forty or Fight!," a reference to the geographic parallel claimed as the boundary between the United States and Canada by jingoist Americans in the Oregon dispute between Britain and the United States in the 1840s.[6]

Once the boundary is established, it takes on additional significance, because behavior with respect to the boundary conveys information. When Russia crosses

[6] The threat was not credible: that parallel is now deep in British Columbia.

the established boundary, that tells China that Russia intends to make a serious incursion further into China. Boundaries must be sharp and well-known if they are not to be violated, and a large part of both law and diplomacy is devoted to clarifying them. Boundaries can also arise in business: two companies producing an unhealthful product might agree not to mention relative healthfulness in their advertising, but a boundary rule like "Mention unhealthfulness if you like, but do not stress it," would not work.

Mediation and **communication** are both important in the absence of a clear focal point. If players can communicate, they can tell each other what actions they will take. Sometimes, as in Ranked Coordination, this works, as the players have no motive to lie. If the players cannot communicate, a mediator may be able to help by suggesting an equilibrium to all the players. The players have no reason not to take this suggestion, and might use the mediator even if his services were costly. Mediation in cases like this is as effective as arbitration, in which an outside party imposes a solution.

One disadvantage of focal points is that they lead to inflexibility. Suppose the Pareto-superior equilibrium (*Small*, *Small*) is chosen as a focal point in Ranked Coordination, and the game is repeated over a long interval of time. The numbers in the payoff matrix might slowly change until (*Small*, *Small*) and (*Large*, *Large*) both had payoffs of 1.5, and (*Large*, *Large*) started to dominate. When, if ever, would the equilibrium switch?

In Ranked Coordination, we would expect that after some time one firm would switch and the other would follow. If there were communication, the switch point would be at the payoff of 1.5. But what if the first firm to switch is penalized more? Such is the problem in oligopoly pricing. If costs rise, so should the monopoly price, but whichever firm raises its price first suffers a loss of market share.

Notes

N1.1 Basic Definitions

- The standard description helps both the modeller and his readers. For the modeller, the names are useful because they help ensure that the important details of a game have been fully specified. For his readers, they make the game easier to understand, especially if, as with most technical papers, the paper is first skimmed quickly to see if it is worth reading. The less clear a writer's style, the more closely he should adhere to the standard names, which means that most of us ought to adhere very closely indeed.

  Think of writing a paper as a game between author and reader, rather than as a single-player production process. The author, knowing that he has valuable information but imperfect means of communication, is trying to convey the information to the reader. The reader does not know whether the information is valuable, and he must choose whether to read the paper closely enough to find out. What are the possible equilibria?
- In OPEC Model I, the notation "$Q_{S,8} = H$" was used to denote "Saudi Arabian oil output in 1988 is High." A logically equivalent model uses the notation "$X_{1,1} = 2$" to denote "Country 1's oil output in Period 1 is 2." Does it make any difference which notation is used?
- The term "strategy profile" is not completely established. The first edition of this book used the more sensible "strategy combination," which has less of the flavor of business-speak, but since then the trend has been to use "strategy profile."
- The term "Silver Rule" for the class of strategies also called "Tit-for-Tat" seems to originate with J. Hirshleifer (1982).

N1.2 Dominant Strategies: The Prisoner's Dilemma

- The Prisoner's Dilemma was named by Albert Tucker in an unpublished paper, although the particular 2-by-2 matrix, discovered by Dresher and Flood, was already well-known. Tucker was asked to give a talk on game theory to the psychology department at Stanford, and invented a story to go with the matrix. Straffin (1980) tells the story.
- In The Prisoner's Dilemma the notation *cooperate* and *defect* is often used for the moves. This is bad notation, because it easy to confuse with *cooperative* games and with *deviations*. It is also often called The Prisoners' Dilemma. Whether one looks at it from the point of the individual or of the group, the prisoners have a problem.
- Herodotus (Herodotus [1947], p. 234) describes an early example of the reasoning in The Prisoner's Dilemma in the conspiracy of Darius against the Persian emperor. A group of nobles met and decided to overthrow the emperor, and it was proposed to adjourn till another meeting. Darius then spoke up and said that if they adjourned, he knew that one of them would go straight to the emperor and reveal the conspiracy, because if nobody else did, he would himself. Darius also suggested a solution—that they immediately go to the palace and kill the emperor.

  The conspiracy also illustrates a way out of coordination games. After killing the emperor, the nobles wished to select one of themselves as the new emperor. Rather than fight, they agreed to go to a certain hill at dawn, and whoever's horse neighed first would become emperor. Herodotus also tells how Darius' groom manipulated this randomization scheme to make him the new emperor.
- Philosophers are intrigued by The Prisoner's Dilemma: see Campbell & Sowden (1985), a collection of articles on The Prisoner's Dilemma and the related Newcombe's Paradox. Game theory has even been applied to theology: if one player is omniscient or omnipotent, what kind of equilibrium behavior can we expect? (see Brams (1983)).
- If we consider only the ordinal ranking of the payoffs in 2-by-2 games, there are 78 distinct games in which each player has strict preference ordering over the four outcomes (listed and described in Rapoport & Guyer [1966]), and 726 distinct games (Guyer & Hamburger [1968]) if we allow ties in the payoffs.
- The Prisoner's Dilemma is not always defined the same way. If we consider just ordinal payoffs, the game in Table 1.9 is a Prisoner's Dilemma if T*(temptation)* $>$ R*(revolt)* $>$ P*(punishment)* $>$ S*(Sucker)*, where the terms in parentheses are mnemonics.[7] If the game is repeated, the cardinal values of the payoffs can be important. The requirement $2R > T + S > 2P$ should be added if the game is to be a standard Prisoner's Dilemma, in which (*Deny, Deny*) and (*Confess, Confess*) are the best and worst possible outcomes in terms of the sum of payoffs. Section 5.3 will show that an asymmetric game called One-Sided Prisoner's Dilemma has properties similar to the standard Prisoner's Dilemma, but does not fit this definition.

  Sometimes the game in which $2R < T + S$ is also called a Prisoner's Dilemma, but here the sum of the players' payoffs is maximized when one confesses and the other denies. If the game were repeated, or if the prisoners could use the correlated equilibria as defined in section 3.3, they would prefer taking turns being confessed against, which would make the game a coordination game similar to Battle of the Sexes. David Shimko has suggested the name Battle of the Prisoners for this (or, perhaps, the Sex Prisoners' Dilemma).

Table 1.9 General Prisoner's Dilemma

		Column		
		Deny		*Confess*
	Deny	R,R	$\rightarrow$	S,T
Row:		$\downarrow$		$\downarrow$
	Confess	T,S	$\rightarrow$	$\mathbf{P,P}$

Payoffs to: (Row, Column)

[7] This is standard notation; see Rapoport, Guyer & Gordon (1976), p. 400.

- Many economists are reluctant to use the concept of cardinal utility, and even more reluctant to compare utility across individuals (see Cooter & Rappoport [1984]). Noncooperative game theory never requires interpersonal utility comparisons, and only ordinal utility is needed to find the equilibrium in The Prisoner's Dilemma. So long as each player's rank ordering of payoffs in different outcomes is preserved, the payoffs can be altered without changing the equilibrium. In general, the dominant-strategy and the pure-strategy Nash equilibria of games depend only on the ordinal ranking of the payoffs, but the mixed-strategy equilibria depend on the cardinal values. See section 3.2, and compare Chicken (section 3.3) and the Hawk-Dove Game (section 5.6).

N1.3 Iterated Dominance: Battle of the Bismarck Sea

- Battle of the Bismarck Sea can be found in Haywood (1954).
- The 2-by-2 form with just four entries that could be used for Battle of the Bismarck Sea and other zero-sum games is a **matrix game**, while the equivalent table with eight entries is a **bimatrix game**. Games can be represented as bimatrix games even if they have more than two moves, so long as the number of moves is finite.
- If a game is zero-sum, the utilities of the players can be represented so as to sum to zero under any outcome. Since utility functions are, to some extent, arbitrary, the sum can also be represented to be variable even if the game is zero-sum. Often modellers will refer to a game as zero-sum even when the payoffs do not add up to zero so long as the payoffs add up to some constant amount. The difference is a trivial normalization.
- If outcome X **strongly Pareto-dominates** outcome Y, then all players have higher utility under outcome X. If outcome X **weakly Pareto dominates** outcome Y, some player has higher utility under X, and no player has lower utility. A zero-sum game does not have outcomes that even weakly Pareto dominate other outcomes. All its equilibria are Pareto efficient, because no player gains without another player losing.

  It is often said that strategy profile x "Pareto dominates" or "dominates" strategy profile y. Taken literally, this is meaningless, since strategies do not necessarily have any ordering at all—one could define *Deny* as being bigger than *Confess*, but that would be arbitrary. Rather, the statement is shorthand for "the payoff profile resulting from strategy profile x Pareto dominates the payoff profile resulting from strategy profile y."

N1.4 Nash Equilibrium: Boxed Pigs, Battle of the Sexes, and Ranked Coordination

- I invented the payoffs for Boxed Pigs from the description of one of the experiments in Baldwin & Meese (1979). They do *not* think of this as an experiment in game theory, and they describe the result in terms of "reinforcement." Battle of the Sexes is taken from p. 90 of Luce & Raiffa (1957).
- Some people prefer "equilibrium point" to "Nash equilibrium," but the latter is convenient, since the name is "Nash" and not "Mazurkiewicz."
- Bernheim (1984a) and Pearce (1984) use the idea of mutually consistent beliefs to arrive at a different equilibrium concept than Nash. They define a **rationalizable strategy** to be a strategy which is a best response for some set of rational beliefs in which a player believes that the other players choose their best responses. The difference from Nash is that not all players need have the same beliefs concerning which strategies will be chosen, nor need their beliefs be consistent.

  This idea is attractive in the context of Bertrand games (see section 13.2). The Nash equilibrium in the Bertrand game is weakly-dominated—by picking any other price above marginal cost, which yields the same profit of zero as does the equilibrium. Rationalizability rules that out.
- J. Hirshleifer (1982) uses the name The Tender Trap for a game essentially the same as Ranked Coordination; the name The Assurance Game has also been used.
- One oft-cited coordination problem is that of the QWERTY typewriter keyboard, developed in the 1870s when typists had to proceed slowly to avoid jamming. QWERTY became the standard, although a US Navy study found in the 1940s that the faster speed possible with the Dvorak keyboard would amortize the cost of retraining full-time typists within ten days (David [1985]). Why large companies have not retrained their typists is difficult to explain under this

story, and Liebowitz & Margolis (1990) argue that economists have been too quick to accept claims that this is a coordination inefficiency. The spelling of the English language is perhaps a better example.
- O. Henry's story, "The Gift of the Magi," is about a coordination game noteworthy for the reason communication is ruled out. A husband sells his watch to buy his wife combs for Christmas, while she sells her hair to buy him a watch fob. Communication would spoil the surprise, a worse outcome than discoordination.
- Macroeconomics has more game theory in it than is readily apparent. The macroeconomic concept of *rational expectations* faces the same problems of multiple equilibria and consistency of expectations as does Nash equilibrium. Game theory is now explicitly used in macroeconomics: see the 1991 book by Canzoneri & Henderson.
- The literature on standard setting includes M. Katz & Shapiro (1985) and Farrell & Saloner (1985).
- Section 3.3 returns to problems of coordination to discuss the concepts of "correlated strategies" and "cheap talk."
- The game of Dangerous Coordination is essentially the same as the Assurance Game of Sen (1967).

N1.5 Focal Points

- Besides his 1960 book, Schelling has written books on diplomacy (1966) and the oddities of aggregation (1978). Political scientists are now looking at the same issues more technically; see Brams & Kilgour (1988) and Ordeshook (1986). Riker (1986) and Muzzio's 1982 book, *Watergate Games*, are absorbing examples of how game theory can be used to analyze specific historical episodes.
- In chapter 12 of *The General Theory*, Keynes suggests that the stock market is a game with multiple equilibria, like a contest in which a newspaper publishes the faces of 20 girls, and contestants submit the name of the one they think most people would submit as the prettiest. When the focal point changes, big swings in predictions about beauty and value result.
- Focal points such as standard weights and measures can contribute to the wealth of an economy. See Kindleberger (1983) for an intriguing discussion of historical standards.
- Not all of what we call boundaries have an arbitrary basis. If the Chinese cannot defend themselves so easily once the Russians cross the boundary at the Amur River, they have a clear reason to fight there.
- Crawford & Haller (1990) take a careful look at focalness in repeated coordination games by asking which equilibria are objectively different from other equilibria, and by trying to see how a player can learn through repetition which equilibrium the other players intend to play. If on the first repetition the players choose strategies that are Nash with respect to each other, it seems focal for them to continue playing those strategies, but what happens if they begin in disagreement? See their article.

Problems

1.1: 2-by-2 Games.

Find examples of 2-by-2 games with the following properties:

(1.1a) No Nash equilibrium (you can ignore mixed strategies).

(1.1b) No weakly Pareto-dominant strategy combination.

(1.1c) At least two Nash equilibria, including one equilibrium that Pareto dominates all other strategy profiles.

(1.1d) At least three Nash equilibria.

1.2: Nash and Iterated Dominance.

(1.2a) Show that every iterated dominance equilibrium s^* is Nash.
(1.2b) Show by counterexample that not every Nash equilibrium can be generated by iterated dominance.
(1.2c) Does every iterated dominance equilibrium exclude weakly dominated strategies?

1.3: Scarface and Timmy.

Scarface and Timmy are caught in a game like the Prisoner's Dilemma, except that Scarface already has a criminal record, so he will always get a prison term at least 5 years greater than Timmy will get, regardless of who finks and who denies. Construct an outcome matrix (with Scarface as Row) and find the Nash equilibrium for this game. (Note: There are at least two games that reasonably fit this story.)

1.4: Pareto Dominance.[8]

(1.4a) If a strategy combination s^* is a dominant-strategy equilibrium, does that mean it weakly Pareto dominates all other strategy combinations?
(1.4b) If a strategy combination s strongly Pareto dominates all other strategy combinations, does that mean it is a dominant-strategy equilibrium?
(1.4c) If s weakly Pareto dominates all other strategy combinations, must it be a Nash equilibrium?

1.5: Discoordination.

Suppose that a man and a woman each choose whether to go to a prize fight or to a ballet. The man would rather go to the prize fight, and the woman to the ballet. What is more important to them, however, is that the man wants to show up at the same event as the woman and that the woman wants to avoid him.
(1.5a) Construct a game matrix to illustrate this game, choosing numbers to fit the preferences described verbally.
(1.5b) If the woman moves first, what will happen?
(1.5c) Does the game have a first-mover advantage?
(1.5d) Show that there is no Nash equilibrium if the players move simultaneously.

1.6: Drawing Outcome Matrices.

It can be surprisingly difficult to look at a game using new notation. In this exercise, redraw the outcome matrix in a different form than in the main text. In

[8] This question is based on notes by Jong-shin Wei.

each case, read the description of the game and draw the outcome matrix as instructed. You will learn more if you do this from the description, without looking at the conventional outcome matrix.

(1.6a) Battle of the Sexes (table 1.6). Put (*Prize Fight*, *Prize Fight*) in the northwest corner, but make the woman the row player.

(1.6b) The Prisoner's Dilemma (table 1.1). Put (*Confess*, *Confess*) in the northwest corner.

(1.6c) Battle of the Sexes (table 1.6). Make the man the row player, but put (*Ballet*, *Prize Fight*) in the northwest corner.

2 Information

2.1 Strategic and Extensive Forms of a Game

If half of strategic thinking is predicting what the other player will do, the other half is figuring out what he knows. Most of the games in chapter 1 assumed that the moves were simultaneous, so the players did not have a chance to learn each other's private information by observing each other. Information becomes central as soon as players move in sequence. The important difference, in fact, betweeen simultaneous-move games and sequential-move games is that in sequential-move games the second player acquires the information on how the first player moved before he has to make his own decision.

Section 2.1 shows how to use the strategic form and the extensive form to describe games with sequential moves. Section 2.2 shows how the extensive form, or game tree, can be used to describe the information available to a player at each point in the game. Section 2.3 classifies games based on the information structure. Section 2.4 shows how to redraw games with incomplete information so that they can be analyzed using the Harsanyi transformation, and derives Bayes' Rule for combining a player's prior beliefs with the information which he acquires in the course of the game.

The Strategic Form and the Outcome Matrix

Games with moves in sequence require more care in presentation than single-move games. In section 1.4 we used the 2-by-2 form, which for the game Ranked Coordination is shown in table 2.1.

Because strategies are the same as actions in Ranked Coordination and the outcomes are simple, the 2-by-2 form in table 2.1 accomplishes two things: it relates strategy profiles to payoffs, and action profiles to outcomes. These two mappings are called the strategic form and the outcome matrix, and in more complicated games they are distinct from each other. The strategic form shows what payoffs result from each possible strategy profile, while the outcome matrix shows what outcome results from each possible action profile. The definitions below use n to denote the number of players, k the number of variables in the

Table 2.1 Ranked Coordination

		Jones		
		Large		*Small*
	Large	**2,2**	←	−1, −1
Smith		↑		↓
	Small	−1, −1	→	**1,1**

Payoffs to: (Smith, Jones)

outcome vector, p the number of strategy profiles, and q the number of action profiles.

The **strategic form** *(or* **normal form***) consists of*

(1) All possible strategy profiles $s^1, s^2, \ldots, s^p$.

(2) Payoff functions mapping s^i *onto the payoff n-vector* π^i, $(i = 1, 2, \ldots, p)$.

The **outcome matrix** *consists of*

(1) All possible action profiles $a^1, a^2, \ldots, a^q$.

(2) Outcome functions mapping a^i *onto the outcome k-vector* z^i, $(i = 1, 2, \ldots, q)$.

Consider the following game based on Ranked Coordination, which we will call Follow-the-Leader I, since we will create several variants of the game. The difference from Ranked Coordination is that Smith moves first, committing himself to a certain disk size no matter what size Jones chooses. The new game has an outcome matrix identical to Ranked Coordination, but its strategic form is different, because Jones' strategies are no longer single actions. Jones' strategy set has four elements,

$$\left\{\begin{array}{l}\text{(If Smith chose } \textit{Large}\text{, choose } \textit{Large}\text{; if Smith chose } \textit{Small}\text{, choose } \textit{Large}\text{)}\\ \text{(If Smith chose } \textit{Large}\text{, choose } \textit{Large}\text{; if Smith chose } \textit{Small}\text{, choose } \textit{Small}\text{)}\\ \text{(If Smith chose } \textit{Large}\text{, choose } \textit{Small}\text{; if Smith chose } \textit{Small}\text{, choose } \textit{Large}\text{)}\\ \text{(If Smith chose } \textit{Large}\text{, choose } \textit{Small}\text{; if Smith chose } \textit{Small}\text{, choose } \textit{Small}\text{)}\end{array}\right\}$$

which we will abbreviate as

$$\left\{\begin{array}{l}(L|L,\ L|S)\\ (L|L,\ S|S)\\ (S|L,\ L|S)\\ (S|L,\ S|S)\end{array}\right\}$$

Follow-the-Leader I illustrates how adding a little complexity can make the strategic form too obscure to be very useful. The strategic form is shown in table 2.2, with equilibria boldfaced and labelled E_1, E_2, and E_3.

Table 2.2 Follow-the-Leader

		Jones			
		$L\|L, L\|S$	$L\|L, S\|S$	$S\|L, L\|S$	$S\|L, S\|S$
Smith	*Large*	2, 2, (E_1)	2, 2, (E_2)	−1, −1	−1, −1
	Small	−1, −1	1, 1	−1, −1	1, 1 (E_3)

Payoffs to: (Smith, Jones)

Equilibrium	Strategies	Outcome
E_1	{*Large*, ($L\|L$)($L\|S$)}	Both pick *Large*.
E_2	{*Large*, ($L\|L$, $S\|S$)}	Both pick *Large*.
E_3	{*Small*, ($S\|L$, $S\|S$)}	Both pick *Small*.

Consider why E_1, E_2, and E_3 are Nash equilibria. In Equilibrium E_1, Jones will respond with *Large* regardless of what Smith does, so Smith quite happily chooses *Large*. Jones would be irrational to choose *Large* if Smith chose *Small* first, but that event never happens in equilibrium. In Equilibrium E_2, Jones will choose whatever Smith chose, so Smith chooses *Large* to make the payoff 2 instead of 1. In Equilibrium E_3, Smith chooses *Small* because he knows that Jones will respond with *Small* whatever he does, and Jones is willing to respond with *Small* because Smith chooses *Small* in equilibrium. Equilibria E_1 and E_3 are not completely sensible, but, except for a little discussion in the context of the game tree, we will defer to chapter 4 the discussion of how to redefine the equilibrium concept to rule them out.

The Extensive Form and the Game Tree

Two other ways to describe a game are the extensive form and the game tree. First we need to define their building blocks. As you read the definitions, you may wish to refer to figure 2.1 as an example.

A **node** *is a point in the game at which some player or Nature takes an action, or the game ends.*

A **successor** *to node X is a node that may occur later in the game if X has been reached.*

A **predecessor** *to node X is a node that must be reached before X can be reached.*

A **starting node** *is a node with no predecessors.*

An **end node** *or* **end point** *is a node with no successors.*

A **branch** *is one action in a player's action set at a particular node.*

Figure 2.1 Follow-the-Leader I in Extensive Form

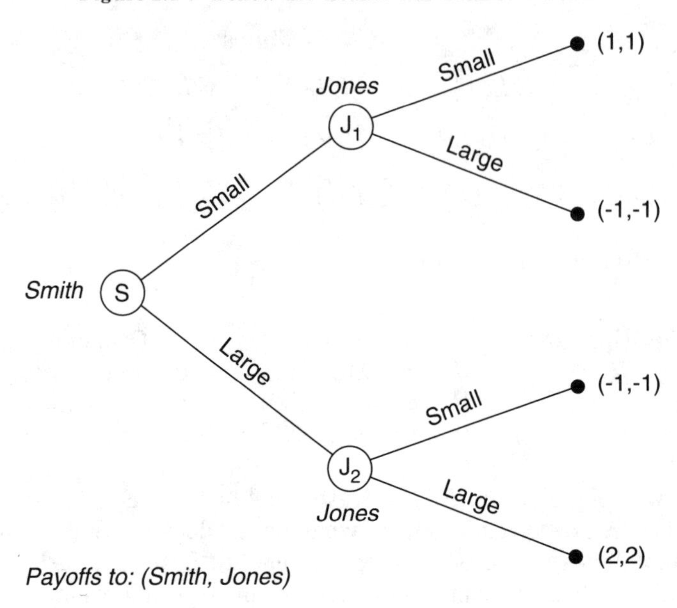

A **path** *is a sequence of nodes and branches leading from the starting node to an end node.*

These concepts can be used to define the extensive form and the game tree.

The **extensive form** *is a description of a game consisting of:*

(1) A configuration of nodes and branches running without any closed loops from a single starting node to its end nodes.

(2) An indication of which node belongs to which player.

(3) The probabilities that Nature uses to choose different branches at its nodes.

(4) The information sets into which each player's nodes are divided.

(5) The payoffs for each player at each end node.

The **game tree** *is the same as the extensive form except that (5) is replaced with:*

(5′) The outcomes at each end node.

"Game tree" is a looser term than "extensive form." If the outcome is defined as the payoff profile, the extensive form is the same as the game tree.

The extensive form for Follow-the-Leader I is shown in figure 2.1. We can see why Equilibria E_1 and E_3 of table 2.2 are unsatisfactory even though they are Nash equilibria. If the game actually reached nodes J_1 or J_2, Jones would have dominant actions, *Small* at J_1 and *Large* at J_2, but E_1 and E_3 specify other actions at those nodes. In section 4.2 we will return to this game and show how the Nash concept can be refined to make E_2 the only equilibrium.

The extensive form for Ranked Coordination shown in figure 2.2 adds dotted lines to the extensive form for Follow-the-Leader I. Each player makes a single decision between two actions. The moves are simultaneous, which we show by letting Smith move first, but not letting Jones know how he moved. The dotted line shows that Jones' knowledge stays the same after Smith moves. All Jones knows is that the game has reached some node within the information set defined by the dotted line; he does not know the exact node reached.

The Time Line

The **time line**, a line showing the order of events, has become popular in recent years as a way to help describe games. Time lines are particularly useful for games with continuous strategies, exogenous arrival of information, and multiple periods, games that are frequently used in the accounting and finance literature. A typical time line is shown in figure 2.3a, which represents a game that will be described in section 10.5.

The time line illustrates the order of actions and events, not necessarily the passage of time. Certain events occur in an instant, others over an interval. In figure 2.3a, Events 2 and 3 occur immediately after Event 1, but Events 4 and 5 might occur ten years later. We sometimes refer to the sequence in which decisions are made as **decision time** and the interval over which physical actions are taken as **real time**. A major difference is that players put higher value on payments received earlier in real time because of time preference, something we will explore further in section 4.5.

Figure 2.2 The Game of Ranked Coordination in Extensive Form

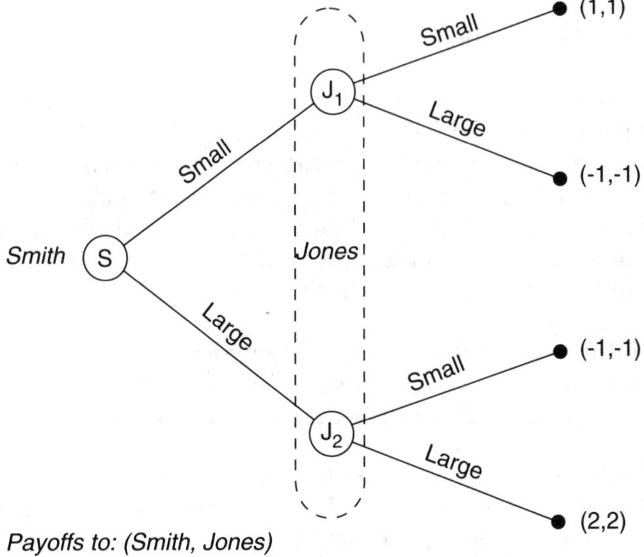

Figure 2.3 The Time Line for the Stock Underpricing Game: (a) a good time line; (b) a bad time line

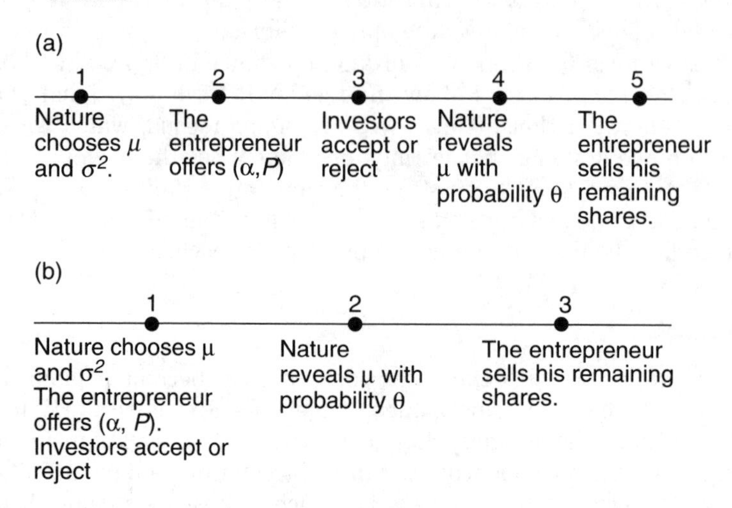

Source: Grinblatt & Hwang (1989)

A common and bad modelling habit is to restrict the use of the dates on the time line to separating events in real time. Events 1 and 2 in figure 2.3a are not separated by real time: as soon as the entrepreneur learns the project's value, he offers to sell stock. The modeller might foolishly decide to depict his model by a picture like Figure 2.3b in which both events happen at Date 1. Figure 2.3b is badly drawn, because readers might wonder which event occurs first or whether they occur simultaneously. In more than one seminar, 20 minutes of heated and confusing debate could have been avoided by 10 seconds' care to delineate the order of events.

2.2 Information Sets

A game's information structure, like the order of its moves, is often obscured in the strategic form. During the Watergate affair, Senator Baker became famous for the question: "How much did the president know, and when did he know it?" In games, as in scandals, these are the big questions. To make this precise, however, requires technical definitions so that one can describe who knows what, and when. This is done using the "information set," the set of nodes a player thinks the game might have reached, as the basic unit of knowledge.

> *Player i's* **information set** ω_i *at any particular point of the game is the set of different nodes in the game tree that he knows might be the actual node, but between which he cannot distinguish by direct observation.*

As defined here, the information set for player i is a set of nodes belonging to one player but on different paths. This captures the idea that player i knows whose turn it is to move, but not the exact location the game has reached in the game tree. Historically, player *i*'s information set has been defined to include only nodes at which player *i* moves, which is appropriate for single-person decision theory, but leaves a player's knowledge undefined for most of any game with two or more players. The broader definition allows comparison of information across players (which, under the older definition, is a comparison of apples and oranges).

In the game in figure 2.4, Smith moves at node S_1 in 1984 and Jones moves at nodes J_1, J_2, J_3, and J_4 in 1985 or 1986. Smith knows his own move, but Jones can tell only whether Smith has chosen the moves which lead to J_1, J_2, or "other"; he cannot distinguish between J_3 and J_4. If Smith has chosen the move leading to J_3, his own information set is simply $\{J_3\}$, but Jones' information set is $\{J_3, J_4\}$.

One way to show information sets on a diagram is to put dashed lines around or between nodes in the same information set. The resulting diagrams can be very cluttered, so it is often more convenient to just draw dashed lines around the information sets of the player making the move at a node. The dashed lines in figure 2.4 show that J_3 and J_4 are in the same information set for Jones, even though they are in different information sets for Smith. An expressive synonym for information set which is based on the appearance of these diagrams is **cloud**; one would say that nodes J_3 and J_4 are in the same cloud, so that while Jones can tell that the game has reached that cloud, he cannot pierce the fog to tell exactly which node has been reached.

One node cannot belong to two different information sets of a single player. If node J_3 belonged to information sets $\{J_2, J_3\}$ and $\{J_3, J_4\}$ (unlike figure 2.4), then if the game reached J_3, Jones would not know whether he was at a node in $\{J_2, J_3\}$ or a node in $\{J_3, J_4\}$—which would imply that they were really the same information set.

Figure 2.4 Information Sets and Information Partitions

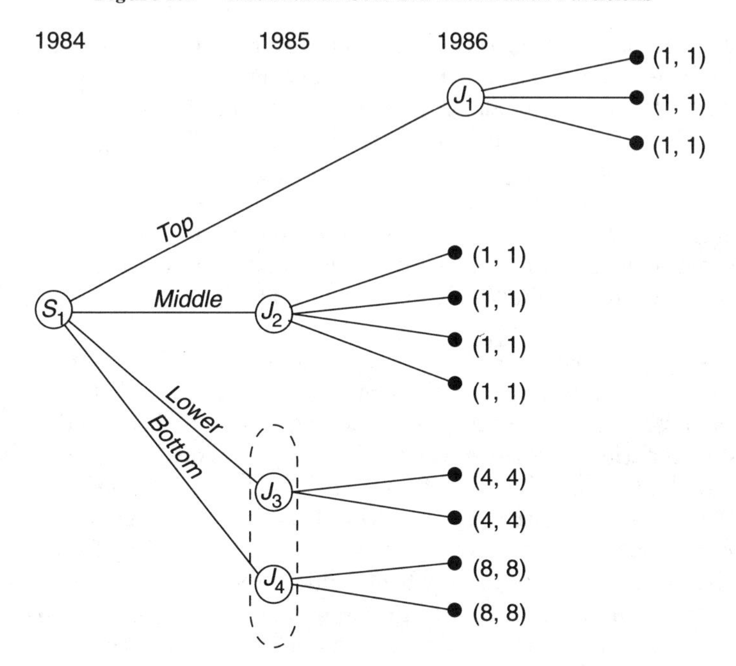

If the nodes in one of Jones' information sets are nodes at which he moves, his action set must be the same at each node, because he knows his own action set (though his actions might differ later on in the game depending on whether he advances from J_3 or J_4). Jones has the same action sets at nodes J_3 and J_4, because if he had some different action available at J_3 he would know he was there and his information set would reduce to just $\{J_3\}$. For the same reason, nodes J_1 and J_2 could not be put in the same information set; Jones must know whether he has three or four moves in his action set. We also require end nodes to be in different information sets for a player if they yield him different payoffs.

With these exceptions, we do not include in the information structure of the game any information acquired by a player's rational deductions. In figure 2.4, for example, it seems clear that Smith would choose *Bottom*, because that is a dominant strategy— his payoff is 8 instead of the 4 from *Lower*, regardless of what Jones does. Jones should be able to deduce this, but even though this is an uncontroversial deduction, it is nonetheless a deduction, not an observation, so the game tree does not split J_3 and J_4 into separate information sets.

Information sets also show the effects of unobserved moves by Nature. In figure 2.4, if the initial move had been made by Nature instead of by Smith, Jones' information sets would be depicted the same way.

Player i's **information partition** *is a collection of his information sets such that*

(1) *Each path is represented by one node in a single information set in the partition, and*

(2) *The predecessors of all nodes in a single information set are in one information set.*

The information partition represents the different positions that the player knows he will be able to distinguish from each other at a given stage of the game, thereby carving up the set of all possible nodes into the subsets called information sets. One of Smith's information partitions is $(\{J_1\}, \{J_2\},\{J_3\}, \{J_4\})$. The definition rules out information set $\{S_1\}$ being in that partition, because the path through S_1 and J_1 would be represented by two nodes. Instead, $\{S_1\}$ is a separate information partition, all by itself. The information partition refers to a stage of the game, not chronological time. The information partition $(\{J_1\}, \{J_2\}, \{J_3, J_4\})$ includes nodes in both 1985 and 1986, but they are all immediate successors of node S_1.

Jones has the information partition $(\{J_1\}, \{J_2\}, \{J_3, J_4\})$. There are two ways to see that his information is worse than Smith's. First is the fact that one of his information *sets*, $\{J_3, J_4\}$, contains *more* elements than Smith's, and second, that one of his information *partitions*, $(\{J_1\}, \{J_2\}, \{J_3, J_4\})$, contains *fewer* elements.

Table 2.3 shows a number of different information partitions for this game. Partition I is Smith's partition and partition II Jones' partition. We say that partition II is **coarser**, and partition I is **finer**. A combinination of two or more of the information sets in a partition, which reduces the number of information sets and increases the numbers of nodes in one or more of them is a **coarsening**. A splitting of one or more of the information sets in a partition, which increases the number of information sets and reduces the number of nodes in one or more of them, is a **refinement**. Partition II is thus a coarsening of partition I, and partition I a refinement of partition II. The ultimate refinement is for each information set to be a **singleton**, containing one node, as in the case of partition I. As in bridge, having a singleton can either help or hurt a player. The ultimate coarsening is for a player not to be able to distinguish between any of the nodes, which is partition III in table 2.3[1].

A finer information partition is the formal definition for "better information." Not all information partitions are refinements or coarsenings of each other, however, so not all information partitions can be ranked by the quality of their information. In particular, just because one information partition contains more information sets does not mean it is a refinement of another information partition. Consider partitions II and IV in table 2.3. Partition II separates the nodes into three information sets, while partition IV separates them into just two information sets. Partition IV is not a coarsening of partition II, however, because it cannot be reached by combining information sets from partition II, and one cannot say that a player with partition IV has worse information. If the node reached is J_1, partition II gives more precise information, but if the node reached is J_4, partition IV gives more precise information.

Table 2.3 Information Partitions

		Partitions			
		I	**II**	**III**	**IV**
Nodes	J_1	$\{J_1\}$	$\{J_1\}$	$\{J_1$	$\{J_1$
	J_2	$\{J_2\}$	$\{J_2\}$	J_2	J_2
	J_3	$\{J_3\}$	$\{J_3$	J_3	$J_3\}$
	J_4	$\{J_4\}$	$J_4\}$	$J_4\}$	$\{J_4\}$

Information quality is defined independently of its utility to the player: it is possible for a player's information to improve and for his equilibrium payoff to fall as a result. Game theory has many paradoxical models in which a player prefers having worse information, not a result of wishful thinking, escapism, or blissful ignorance, but of cold rationality. Coarse information can have a number of advantages. It may permit a player to engage in trade because other players do not fear his superior information. It may give a player a stronger strategic position because he usually has a strong position and is better off not knowing that in a particular realization of the game his position is weak. Or, as in the more traditional economics of uncertainty, poor information may permit players to insure each other.

Discussion of the first two points will be deferred to later chapters (models of entry deterrence in section 6.3 and of used car markets in chapter 9). The point

[1]Note, however, that partitions III and IV are not really allowed in this game, because Jones could tell the node from the actions available to him, as explained earlier.

about insurance, however, is best discussed here, as an example showing that even when information is symmetric and behavior nonstrategic, it may still be the case that better information, in the sense of a finer information partition, actually reduces the utility of all the players in a game—a Pareto worsening.

To see this, suppose that Smith and Jones, both risk averse, work for the same employer, and that both know that one of them will be randomly fired at the end of the year, while the other will be promoted. The one who is fired will end with a wealth of 0 and the one who is not fired will end with a wealth of 100. Smith and Jones will agree to insure each other by pooling their wealth. That is, they will agree that whoever is promoted will pay 50 to whoever is fired. If someone offers to tell them who will be fired before they make this agreement, they should cover their ears and refuse to listen. This refinement of their information would make both worse off, in expectation, because it would ruin their insurance arrangement. Once they knew who would be promoted, they would no longer agree to pool their wealth. Better information would reduce the expected utility of both players.

Common Knowledge

We have been implicitly assuming that the players know what the game tree looks like. In fact, we have assumed that the players also know that the other players know what the game tree looks like. The term, "common knowledge," is used to avoid spelling out the infinite recursion to which this leads.

> *Information is* **common knowledge** *if it is known to all the players, if each player knows that all the players know it, if each player knows that all the players know that all the players know it, and so forth ad infinitum.*

Because of this recursion (the importance of which will be seen in section 6.3), the assumption of common knowledge is stronger than the assumption that players have the same beliefs about where they are in the game tree. J. Hirshleifer & Riley (1992, p. 169) use the term **concordant beliefs** to describe a situation where players share the same belief about the probabilities that Nature has chosen different states of the world, but where they do not necessarily know they share the same beliefs. (Brandenburger (1992) uses the term **mutual knowledge** for the same idea.)

For clarity, models are set up so that information partitions are common knowledge. Every player knows how precise the other players' information is, however ignorant he himself may be as to which node the game has reached. Modelled this way, the information partitions are independent of the equilibrium concept. Making the information partitions common knowledge is important for clear modelling, and restricts the kinds of games that can be modelled less than one might think. This will be illustrated in section 2.4 when the assumption will be imposed on a situation in which one player does not even know which of three games he is playing.

2.3 Perfect, Certain, Symmetric, and Complete Information

We categorize the information structure of a game in four different ways, so a particular game might have perfect, complete, certain, and symmetric information. The categories are summarized in table 2.4.

Table 2.4 Information Categories

Information category	Meaning
perfect	Each information set is a singleton
certain	Nature does not move after any player moves
symmetric	No player has information different from other players when he moves, or at the end nodes
complete	Nature does not move first, or her initial move is observed by every player

The first category divides games into those with perfect and those with imperfect information.

In a game of **perfect information** *each information set is a singleton. Otherwise the game is one of* **imperfect information**.

The strongest informational requirements are met by a game of perfect information, because in such a game each player always knows exactly where he is in the game tree. No moves are simultaneous, and all players observe Nature's moves. Ranked Coordination is a game of imperfect information because of its simultaneous moves, but Follow-the-Leader I is a game of perfect information. Any game of incomplete or asymmetric information is also a game of imperfect information.

A game of **certainty** *has no moves by Nature after any player has moved. Otherwise the game is one of* **uncertainty**.

The moves by Nature in a game of uncertainty may or may not be revealed to the players immediately. A game of certainty can be a game of perfect information if it has no simultaneous moves. The notion "game of uncertainty" is new with this book, but I doubt it would surprise anyone. The only quirk in the definition is that it allows an initial move by Nature in a game of certainty, because in a game of incomplete information Nature moves first to select a player's "type." Most modellers do not think of this situation as uncertainty.

We have already talked about information in Ranked Coordination, a game of imperfect, complete, and symmetric information with certainty. The Prisoner's Dilemma falls into the same categories. Follow-the-Leader I, which does not have simultaneous moves, is a game of perfect, complete, and symmetric information with certainty.

We can easily modify Follow-the-Leader I to add uncertainty, creating the game Follow-the-Leader II (figure 2.5). Imagine that if both players pick *Large* for their disks, the market yields either zero profits or very high profits, depending on the state of demand, but demand would not affect the payoffs in any other strategy profile. We can quantify this by saying that if (*Large*, *Large*) is picked, the payoffs are (10, 10) with probability 0.2, and (0, 0) with probability 0.8, as shown in figure 2.5.

When players face uncertainty, we need to specify how they evaluate their uncertain future payoffs. The obvious way to model their behavior is to say that the players maximize the expected values of their utilities. Players who behave in this way are said to have **von Neumann-Morgenstern utility functions**, a name chosen to underscore von Neumann & Morgenstern's (1944) development of a rigorous justification of such behavior.

Maximizing their expected utilities, the players would behave exactly the same as in Follow-the-Leader I. Often, a game of uncertainty can be transformed into a game of certainty without changing the equilibrium, by eliminating Nature's moves and changing the payoffs to their expected values based on the probabilities of Nature's moves. Here we could eliminate Nature's move and replace the payoffs 10 and 0 with the single payoff 2 ($= 0.2[10] + 0.8[0]$). This cannot be done, however, if the actions available to a player depend on Nature's moves, or if information about Nature's move is asymmetric.

The players in figure 2.5 might be either risk averse or risk neutral. Risk aversion is implicitly incorporated in the payoffs as they are in units of utility, not dollars. When players maximize their expected utility, they are not necessarily maximizing their expected dollars. Moreover, the players can differ in how they map money to utility. It could be that (0, 0) represents ($0, $5,000), (10, 10) rep-

Figure 2.5 Follow-the-Leader II

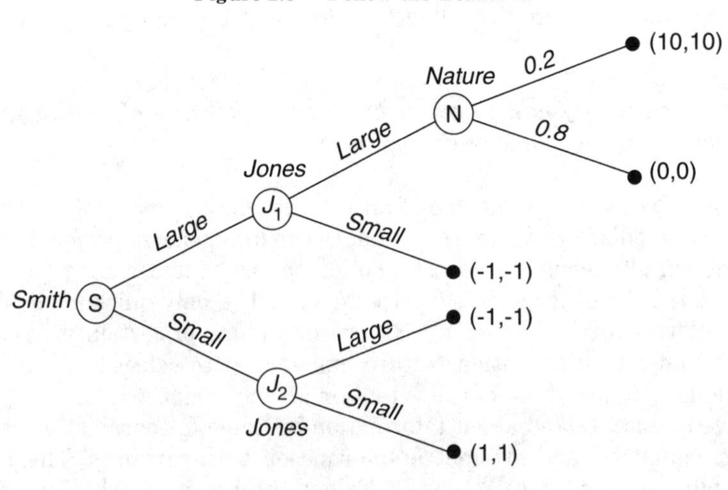

resents ($100,000, $100,000), and (2, 2), the expected utility, is equivalent to a non-risky ($3,000, $7,000).

In a game of **symmetric information**, *a player's information set at*

(1) any node where he chooses an action,
or

(2) an end node

contains at least the same elements as the information sets of every other player. Otherwise the game is one of **asymmetric information**.

In a game of asymmetric information, the information sets of the players differ in ways relevant to their behavior, or differ at the end of the game. Such games have imperfect information, since information sets which differ across players cannot be singletons. The notion of "asymmetric information," which is used in the present book for the first time, is intended for capturing a vague meaning commonly used today. The essence of asymmetric information is that some player has useful **private information**: an information partition that is different and not worse than any other player's.

A game of symmetric information can have moves by Nature or simultaneous moves, but no player ever has an informational advantage. The one point at which information may differ is when the player *not* moving has superior information because he knows what his own move *was*; for example, if the two players move simultaneously. Such information does not help the informed player, since by definition it cannot affect his move.

A game has asymmetric information if information sets differ at the end of the game because we conventionally think of such games as ones in which information differs, even though no player takes an action after the end nodes. The principal-agent model of chapter 7 is an example. The principal moves first, then the agent, and finally Nature. The agent observes the agent's move, but the principal does not, although he may be able to deduce it. This would be a game of symmetric information except for the fact that information continues to differ at the end nodes.

In a game of **incomplete information**, *Nature moves first and is unobserved by at least one of the players. Otherwise the game is one of* **complete information**.

A game with incomplete information also has imperfect information, because some player's information set includes more than one node. Two kinds of games have complete but imperfect information: games with simultaneous moves, and games where, late in the game, Nature makes moves not immediately revealed to all players.

Many games of incomplete information are games of asymmetric information, but the two concepts are not equivalent. If there is no initial move by Nature, but Smith takes a move unobserved by Jones and Smith moves again later in the game, the game has asymmetric but complete information. The principal-agent games of chapter 7 are also examples: the agent knows how hard he worked, but his principal never learns, not even at the end nodes. A game can also have incomplete but symmetric information: let Nature, unobserved by either player, move first and choose the payoffs for (*Confess*, *Confess*) in The Prisoner's Dilemma to be either $(-6, -6)$ or $(-100, -100)$.

A more interesting example of incomplete but symmetric information may be found in Harris & Holmstrom (1982). Here Nature assigns different abilities to workers, but when workers are young their ability is known neither to employers nor to themselves. As time passes, the abilities become common knowledge, and if workers are risk averse and employers risk neutral, the model shows that equilibrium wages are constant or rising over time.

Poker Examples of Information Classification

In the game of poker, the players make bets on who will have the best hand of cards at the end, where a ranking of hands has been pre-established. How would the following rules for behavior before betting be classified? (Answers may be found in note N2.3.)

(1) All cards are dealt face up.
(2) All cards are dealt face down and a player cannot look even at his own cards before he bets.
(3) All cards are dealt face down, and a player can only look at his own cards.
(4) All cards are dealt face up, but each player then scoops up his hand and secretly discards one card.
(5) All cards are dealt face up, the players bet, and then each player receives one more card face up.
(6) All cards are dealt face down, but then each player scoops up his cards without looking at them and holds them against his forehead so all the *other* players can see them (Indian poker).

2.4 The Harsanyi Transformation and Bayesian Games

The Harsanyi Transformation: Follow-the-Leader III

The term, "incomplete information," is used in two quite different senses in the literature, usually without explicit definition. The definition in section 2.3 is what economists commonly *use*, but if asked to *define* the term, they might come up with the following, older, definition.

Old definition

In a game of **complete information**, *all players know the rules of the game. Otherwise the game is one of* **incomplete information**.

The old definition is not meaningful, since the game itself is ill-defined if it does not specify exactly what the players' information sets are. Until 1967, game theorists spoke of games of incomplete information only to point out that they could not be analyzed. Then John Harsanyi pointed out that any game that had incomplete information under the old definition could be remodelled as a game of complete but imperfect information without changing its essentials, simply by adding an initial move in which Nature chooses between different sets of rules. In the transformed game, all players know the new meta-rules, including the fact that Nature has made an initial move unobserved by them. Harsanyi's suggestion trivialized the definition of incomplete information, and people began using the term to refer to the transformed game instead. Under the old definition, a game of incomplete information was transformed into a game of complete information. Under the new definition, the original game is ill-defined, and the transformed version is a game of incomplete information.

Follow-the-Leader III serves to illustrate the Harsanyi transformation. Suppose that Jones does not know the game's payoffs precisely. He does have some idea of the payoffs, and we represent his beliefs by a subjective probability distribution. He places a 70 percent chance on the game being game (A) in figure 2.6 on the next page (which is the same as Follow-the-Leader I), a 10 percent chance on game (B), and a 20 percent on game (C). In reality the game has a particular set of payoffs, and Smith knows what they are. This is a game of incomplete information (Jones does not know the payoffs), asymmetric information (when Smith moves, Smith knows something Jones does not), and certainty. (Nature does not move after the players do.)

The game cannot be analyzed in the form shown in figure 2.6. The natural way to approach such a game is to use the Harsanyi transformation. We can remodel the game to look like figure 2.7, in which Nature makes the first move and chooses the payoffs of game (A), (B), or (C), in accordance with Jones' subjective probabilities. Smith observes Nature's move, but Jones does not. Figure 2.7 depicts the same game as figure 2.6, but now we can analyze it. Both Smith and Jones know the rules of the game, and the difference between them is that Smith has observed Nature's move. Whether Nature actually makes the moves with the indicated probabilities or Jones just imagines them is irrelevant, so long as Jones' initial beliefs or fantasies are common knowledge.

Often what Nature chooses at the start of a game is the strategy set, information partition, and payoff function of one of the players. We say that the player can be any one of several "types," a term to which we will return in later chapters. When Nature moves, especially if she affects the strategy sets and payoffs of both players, it is often said that Nature has chosen a particular "state of the world." In figure 2.7 Nature chooses the state of the world to be (A), (B), or (C).

A player's **type** *is the strategy set, information partition, and payoff function which Nature chooses for him at the start of a game of incomplete information.*

A **state of the world** *is a move by Nature.*

As I have already said, it is good modelling practice to assume that the structure of the game is common knowledge, so that though Nature's choice of Smith's type may really just represent Jones' opinions about Smith's possible type, Smith

Figure 2.6 Follow-the-Leader III: Original Game

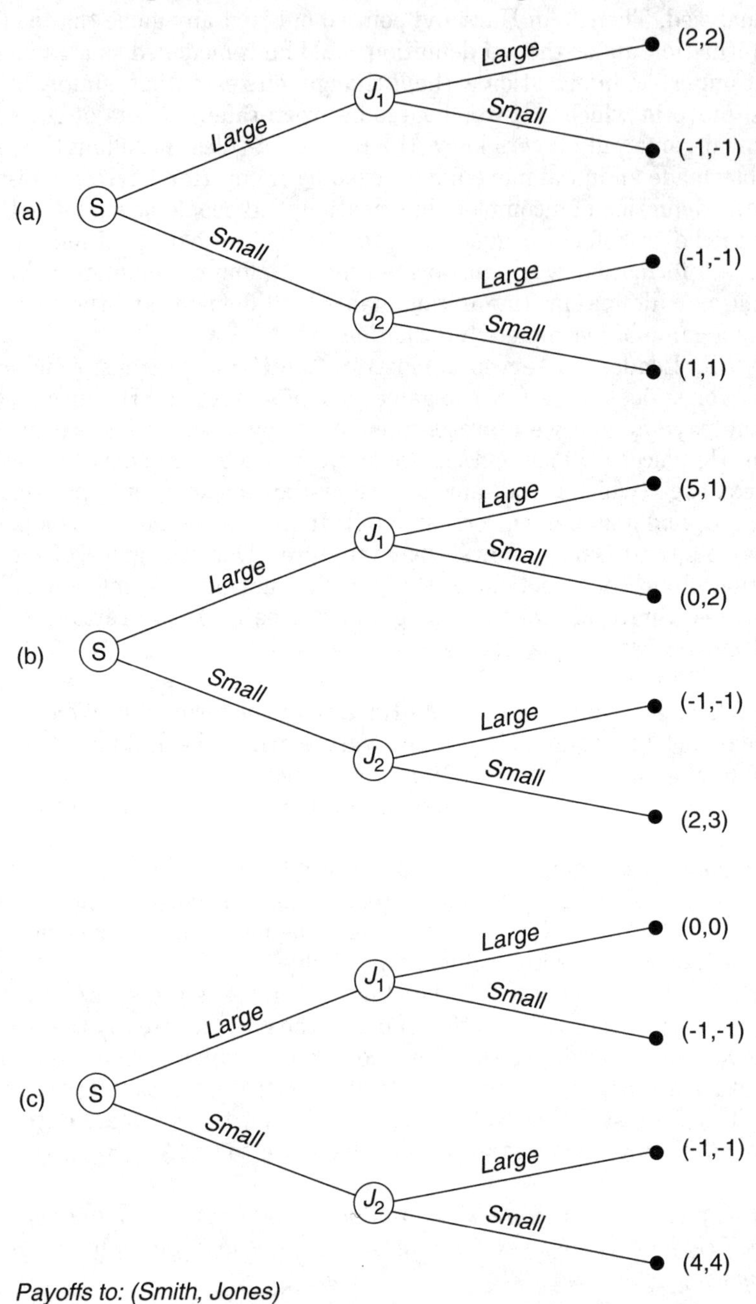

knows what Jones' possible opinions are and Jones knows that they are just opinions. The players may have different beliefs, but that is modelled as the effect of their observing different moves by Nature. All players begin the game with the same beliefs about the probabilities of the moves Nature will make—the same priors, to use a term that will shortly be introduced. This modelling assumption

Figure 2.7 Follow-the-Leader III: After the Harsanyi Transformation

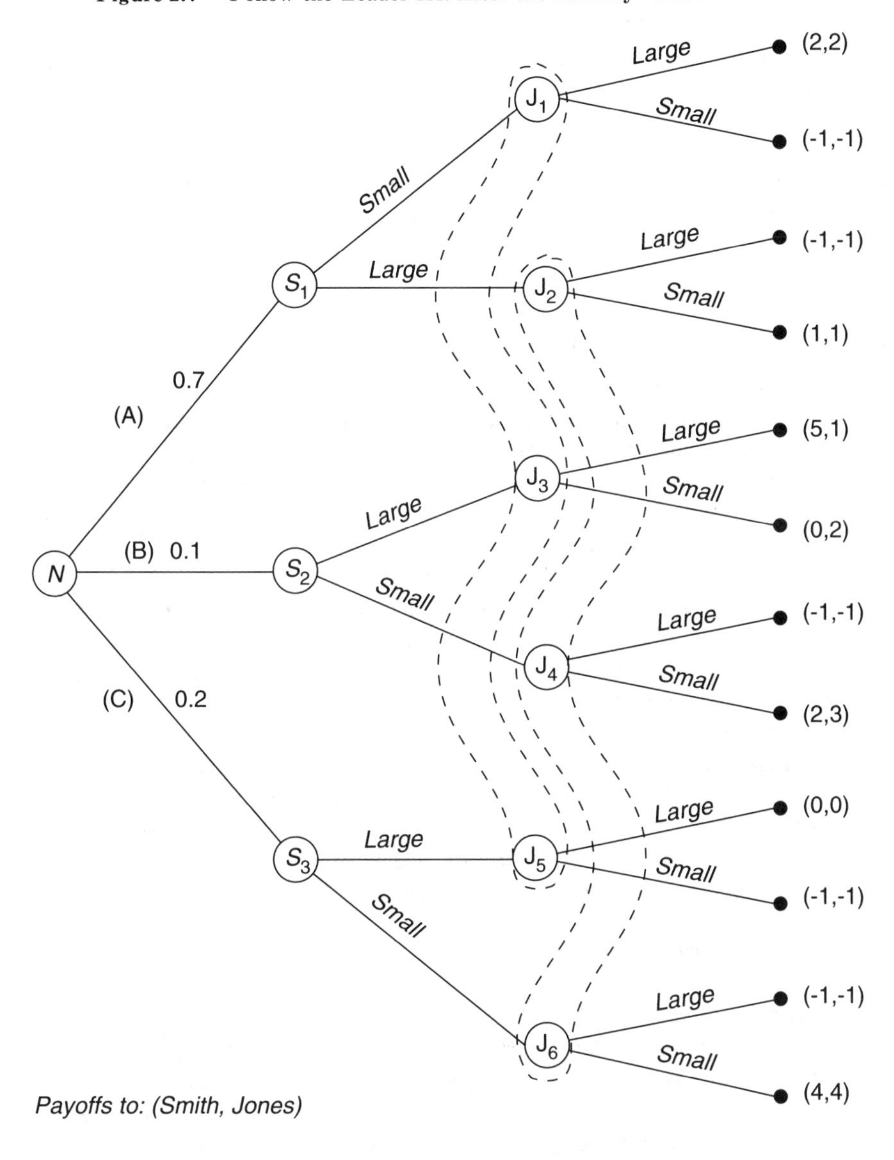

is known as the **Harsanyi doctrine**. If the modeller is following it, his model can never reach a situation where two players possess exactly the same information but disagree as to the probability of some past or future move of Nature. A model cannot, for example, begin by saying that Germany believes its probability of winning a war against France is 0.8 and France believes it is 0.4, so they are both willing to go to war. Rather, he must assume that beliefs begin the same but diverge because of private information. Both players initially think that the probability of a German victory is 0.4 but that if General Schmidt is a genius the probability rises to 0.8, and then Germany discovers that Schmidt is indeed a genius. If it is France that has the initiative to declare war, France's mistaken beliefs may lead to a conflict that is avoidable if Germany can credibly reveal its private information about Schmidt's genius.

An implication of the Harsanyi doctrine is that players are at least slightly open-minded about their opinions. If Germany indicates that it is willing to go to war, France must consider the possibility that Germany has discovered Schmidt's genius and update the probability that Germany will win (keeping in mind that Germany might be bluffing). Our next topic is how a player updates his beliefs upon receiving new information, whether it be by direct observation of Nature or by observing the moves of another player who might be better informed.

Updating Beliefs with Bayes' Rule

When we classify a game's information structure we do not try to decide what a player can deduce from the other players' moves. Player Jones might deduce, upon seeing Smith choose *Large*, that Nature has chosen state (A), but we do not draw Jones' information set in figure 2.7 to take this into account. In drawing the game tree we want to illustrate only the exogenous elements of the game, uncontaminated by the equilibrium concept. But to find the equilibrium we do need to think about how beliefs change over the course of the game.

One part of the rules of the game is the collection of **prior beliefs** (or **priors**) held by the different players, beliefs that they update in the course of the game. A player holds prior beliefs concerning the types of the other players, and as he sees them take actions he updates his beliefs under the assumption that they are following equilibrium behavior.

The term **Bayesian equilibrium** is used to refer to a Nash equilibrium in which players update their beliefs according to Bayes' Rule. Since Bayes' Rule is the natural and standard way to handle imperfect information, the adjective, "Bayesian," is really optional. But the two-step procedure of checking a Nash equilibrium has now become a three-step procedure:

(1) Propose a strategy profile.
(2) See what beliefs the strategy profile generates when players update their beliefs in response to each others' moves.
(3) Check that, given those beliefs together with the strategies of the other players, each player is choosing a best response for himself.

The rules of the game specify each player's initial beliefs, and Bayes' Rule is the rational way to update beliefs. Suppose, for example, that Jones starts with a particular prior belief, *Prob(Nature chose* (*A*)). In Follow-the-Leader III, this equals 0.7. He then observes Smith's move—*Large*, perhaps. Seeing *Large* should make Jones update to the **posterior** belief, *Prob(Nature chose* (*A*)) | *Smith chose Large*), where the symbol "|" denotes "conditional upon," or "given that."

Bayes' Rule shows how to revise a prior belief in the light of new information such as Smith's move. It uses two pieces of information, the likelihood of seeing Smith choose *Large* given that Nature chose state of the world (*A*), *Prob*(*Large* | (*A*)), and the likelihood of seeing Smith choose *Large* given that Nature did not choose state (*A*), *Prob(Large* | (*B*) or (*C*)). From these numbers,

Smith can calculate *Prob(Smith chooses Large)*, the **marginal likelihood** of seeing *Large* as the result of one or another of the possible states of the world that Nature might choose.

$$\begin{aligned} Prob(Smith\ chooses\ Large) &= Prob(Large|A)Prob(A) \\ &\quad + Prob(Large|B)Prob(B) \\ &\quad + Prob(Large|C)Prob(C). \end{aligned} \tag{1}$$

To find his posterior, *Prob(Nature chose (A))|Smith chose Large)*, Jones uses the likelihoods and his priors. The joint probability of both seeing Smith choose *Large* and Nature having chosen (A) is

$$\begin{aligned} Prob(Large, A) &= Prob(A|Large)Prob(Large) \\ &= Prob(Large|A)Prob(A). \end{aligned} \tag{2}$$

Since what Jones is trying to calculate is *Prob(A|Large)*, rewrite the last part of (2) as follows:

$$Prob(A|Large) = \frac{Prob(Large|A)Prob(A)}{Prob(Large)}. \tag{3}$$

Jones needs to calculate his new belief—his posterior—using *Prob(Large)*, which he calculates from his original knowledge using equation (1). Substituting the expression for *Prob(Large)* from (1) into equation (3) gives the final result, a version of Bayes' Rule:

$$Prob(A|Large) = \frac{Prob(Large|A)Prob(A)}{Prob(Large|A)Prob(A) + Prob(Large|B)Prob(B) + Prob(Large|C)Prob(C)}. \tag{4}$$

More generally, for Nature's move x and the observed data,

$$Prob(x|data) = \frac{Prob(data|x)Prob(x)}{Prob(data)}. \tag{5}$$

Equation (6) is a verbal form of Bayes' Rule, which is useful for remembering the terminology, summarized in table 2.5.

$$(Posterior\ for\ Nature's\ Move) = \frac{(Likelihood\ of\ Player's\ Move) \cdot (Prior\ for\ Nature's\ Move)}{(Marginal\ likelihood\ of\ Player's\ Move)}. \tag{6}$$

Bayes' Rule is not purely mechanical, but rather the only way to rationally update beliefs. The derivation is worth understanding because Bayes' Rule is hard to memorize, but easy to rederive.

Table 2.5 Bayesian Terminology

Name	Meaning
likelihood	*Prob(data|event)*
marginal likelihood	*Prob(data)*
conditional likelihood	*Prob*(data X|*data Y, event)*
prior	*Prob(event)*
posterior	*Prob(event|data)*

Updating Beliefs in Follow-the-Leader III

Let us now return to the numbers in Follow-the-Leader III to use the belief-updating rule that was just derived. Jones has a prior belief that the probability of event {Nature picks state (*A*)} is 0.7 and he needs to update that belief on seeing the data {Smith picks *Large*}. His prior is *Prob(A)* = 0.7 and we wish to calculate *Prob(A|Large)*.

To use Bayes' Rule from equation (4), we need the values of *Prob(Large|A)*, *Prob(Large|B)*, and *Prob(Large|C)*. These values depend on what Smith does in equilibrium, so Jones' beliefs cannot be calculated independently of the equilibrium. This is the reason for the three-step procedure suggested above, for what the modeller must do is propose an equilibrium and then use it to calculate the beliefs. Afterwards, he must check that the equilibrium strategies are indeed the best responses given the beliefs they generate.

A candidate for equilibrium in Follow-the-Leader III is for Smith to choose *Large* if the state is (*A*) or (*B*) and *Small* if it is (*C*), and for Jones to respond to *Large* with *Large* and to *Small* with *Small*. This can be abbreviated as (*L|A*, *L|B*, *S|C*; *L|L*, *S|S*). Let us test that this is an equilibrium, starting with the calculation of *Prob(A|Large)*.

If Jones observes *Large*, he can rule out state (*C*), but he does not know whether the state is (*A*) or (*B*). Bayes' Rule tells him that the posterior probability of state (*A*) is

$$Prob(A|Large) = \frac{(1)(0.7)}{(1)(0.7) + (1)(0.1) + (0)(0.2)} = 0.875. \quad (7)$$

The posterior probability of state (*B*) must then be 1 − 0.875 = 0.125, which could also be calculated from Bayes' Rule, as follows:

$$Prob(B|Large) = \frac{(1)(0.1)}{(1)(0.7) + (1)(0.1) + (0)(0.2)} = 0.125. \quad (8)$$

Figure 2.8 shows a graphic intuition for Bayes' Rule. The first line shows the total probability, 1, which is the sum of the prior probabilities of states (*A*), (*B*), and (*C*). The second line shows the probabilities, summing to 0.8, which remain after *Large* is observed and state (*C*) is ruled out. The third line shows that state

Figure 2.8 Bayes' Rule

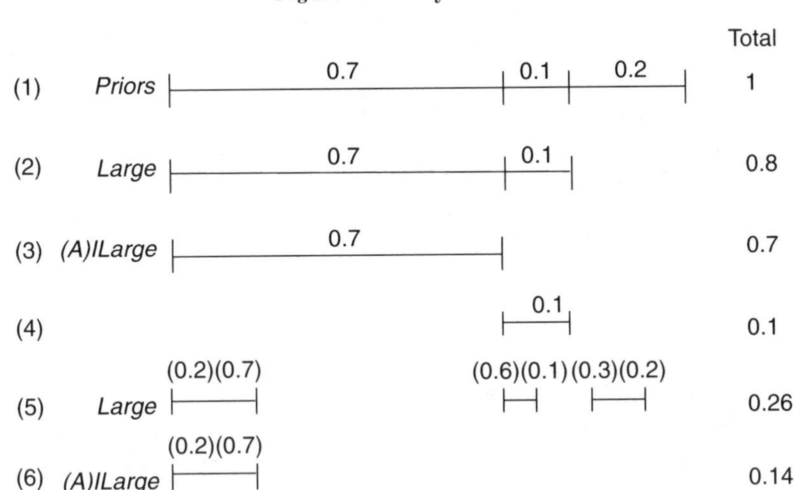

(A) represents an amount 0.7 of that probability, a fraction of 0.875. The fourth line shows that state (B) represents an amount 0.1 of that probability, a fraction of 0.125.

Jones must use Smith's strategy in the proposed equilibrium to find numbers for *Prob(Large*|*A)*, *Prob(Large*|*B)*, and *Prob(Large*|*C)*. As always in Nash equilibrium, the modeller assumes that the players know which equilibrium strategies are being played out, even though they do not know which particular actions are being chosen.

Given that Jones believes that the state is (A) with probability 0.875 and state (B) with probability 0.125, his best response is *Large*, even though he knows that if the state were actually (B) the better response would be *Small*. Given that he observes *Large*, Jones' expected payoff from *Small* is -0.625 ($= 0.875[-1] + 0.125[2]$), but from *Large* it is 1.875 ($= 0.875[2] + 0.125[1]$). Thus, the strategy profile ($L|A$, $L|B$, $S|C$; $L|L$, $S|S$) is a perfect Bayesian equilibrium.

A similar calculation can be done for *Prob(A*|*Small)*. Using Bayes' Rule, equation (4) becomes

$$Prob(A|Small) = \frac{(0)(0.7)}{(0)(0.7) + (0)(0.1) + (1)(0.2)} = 0. \qquad (9)$$

Given that he believes the state is (C), Jones' best response to *Small* is *Small*, which agrees with our proposed equilibrium.

The calculations are relatively simple because Smith uses a nonrandom strategy in equilibrium, so, for instance, $Prob(Small|A) = 0$ in equation (9). Consider what happens if Smith uses a random strategy of picking *Large* with probability 0.2 in state (A), 0.6 in state (B), and 0.3 in state (C) (we will analyze such "mixed" strategies in section 3.1). The equivalent of equation (7) is

$$Prob(A|Large) = \frac{(0.2)(0.7)}{(0.2)(0.7) + (0.6)(0.1) + (0.3)(0.2)} = 0.54 \; (rounded). \qquad (10)$$

If he sees *Large*, Jones' best guess is still that Nature chose state (*A*), even though in state (*A*) Smith has the smallest probability of choosing *Large*, but Jones' subjective posterior probability, $Pr(A|Large)$, has fallen to 0.54 from his prior of $Pr(A) = 0.7$.

The last two lines of figure 2.8 illustrate this case. The second-to-last line shows the total probability of *Large*, which is formed from the probabilities in all three states and sums to 0.26 (=0.14 + 0.06 + 0.06). The last line shows the component of that probability arising from state (*A*), which is the amount 0.14 and the fraction 0.54 (rounded).

Regression to the Mean

Regression to the mean is an old statistical idea that has a Bayesian interpretation. Suppose that each student's performance on a test results partly from his ability and partly from random error because of his mood the day of the test. The teacher does not know the individual student's ability, but does know that the average student will score 70 out of 100. If a student scores 40, what should be the teacher's estimate of his ability?

It should not be 40. A score of 30 points below the average score could be the result of two things: (1) the student's ability is below average, or (2) the student was in a bad mood the day of the test. Only if mood is completely unimportant should the teacher use 40 as his estimate. More likely, both ability and luck matter to some extent, so the teacher's best guess is that the student has an ability below average but was also unlucky. The best estimate lies somewhere between 40 and 70, reflecting the influence of both ability and luck. Of the students who score 40 on the test, more than half can be expected to score above 40 on the next test. Since the scores of these poorly performing students tend to float up towards the mean of 70, this phenomenon is called "regression to the mean." Similarly, students who score 90 on the first test will tend to score less well on the second test.

This is "regression to the mean" ("towards" would be still more accurate) not "regression beyond the mean." A low score does indicate low ability, on average, so the predicted score on the second test is still below average. Regression to the mean merely recognizes that both luck and ability are at work.

In Bayesian terms, the teacher in this example has a prior mean of 70, and is trying to form a posterior estimate using the prior and one piece of data, the score on the first test. For typical distributions, the posterior mean lies between the prior mean and the data point, so the posterior mean is between 40 and 70.

In a business context, regression to the mean can be used to explain business conservatism. It is sometimes claimed that businesses pass up profitable investments because they have an excessive fear of risk. Let us suppose that the business is risk neutral, because the risk associated with the project and the uncertainty over its value are nonsystematic—that is, they are risks that a widely held corporation can distribute in such a way that each shareholder's risk becomes trivial. Suppose that the firm will not spend \$100,000 on an investment with a present value of \$105,000. This is easily explained if the \$105,000 is an estimate and the \$100,000 is cash. If the average value of a new project of this kind is less than \$100,000—as is likely to be the case, since profitable projects are not easy to find—the best estimate of the value will lie between the measured value of

\$105,000 and that average value, unless the staffer who came up with the \$105,000 figure has already adjusted his estimate. Regressing the \$105,000 to the mean may regress it past \$100,000. Put a bit differently, if the prior mean is, let us say, \$80,000 and the data point \$105,000, the posterior may well be less than \$100,000.

It is important to keep regression to the mean in mind as an alternative to strategic behavior in explaining odd phenomena. In analyzing test scores, one might try to explain the rise in the scores of poor students by changes in their effort level in an attempt to achieve a target grade in the course with minimum work. In analyzing business decisions, one might try to explain why apparently profitable projects are rejected because of managers' dislike for innovations that would require them to work harder. These explanations might be valid, but models based on Bayesian updating or regression to the mean might explain the situation just as well, and with fewer hard-to-verify assumptions about the utility functions of the individuals involved.

2.5 Example: The Png Settlement Game

The Png (1983) model of out-of-court settlement is an example of a game with a fairly complicated extensive form.[2] The plaintiff alleges that the defendant was negligent in providing safety equipment at a chemical plant, a charge which is true with probability q. The plaintiff files suit, but the case is not decided immediately. In the meantime, the defendant and the plaintiff can settle out of court.

What are the moves in this game? It is really made up of two games: the one in which the defendant is liable for damages, and the one in which he is blameless. We therefore start the game tree with a move by Nature, who makes the defendant either liable or blameless. At the next node, the plaintiff takes an action: *Sue* or *Grumble*. If he decides on *Grumble* the game ends with zero payoffs for both players. If he decides to *Sue*, we go to the next node. The defendant then decides whether to *Resist* or *Offer* to settle. If the defendant chooses *Offer*, the plaintiff can *Settle* or *Refuse*; if the defendant chooses to *Resist*, the plaintiff can *Try* the case or *Drop* it.

The dashed curves in figure 2.9 enclose the information sets of the plaintiff. His information partition is rather coarse—he must make decisions without knowing exactly which node the game has reached. Maybe the defendant is liable, but maybe not. The plaintiff has three actions open to him, and there are three dashed ovals showing that his information remains coarse.

We assume that the settlement amount, S, and the amounts spent on legal fees are exogenous. Except in the infinitely long games without end nodes that will appear in chapter 5, an extensive form should incorporate all costs and benefits into the payoffs at the end nodes, even if costs are incurred along the way. If the court required a \$100 filing fee (which it does not in this game, although a fee will be required in the similar game of Nuisance Suits in section 4.3), it would be subtracted from the plaintiff's payoffs at every end node except those resulting from his choice of *Grumble*. Such consolidation makes it easier to analyze the

[2] "Png," by the way, is pronounced the same way it is spelled.

Figure 2.9 The Game Tree for The Png Settlement Game

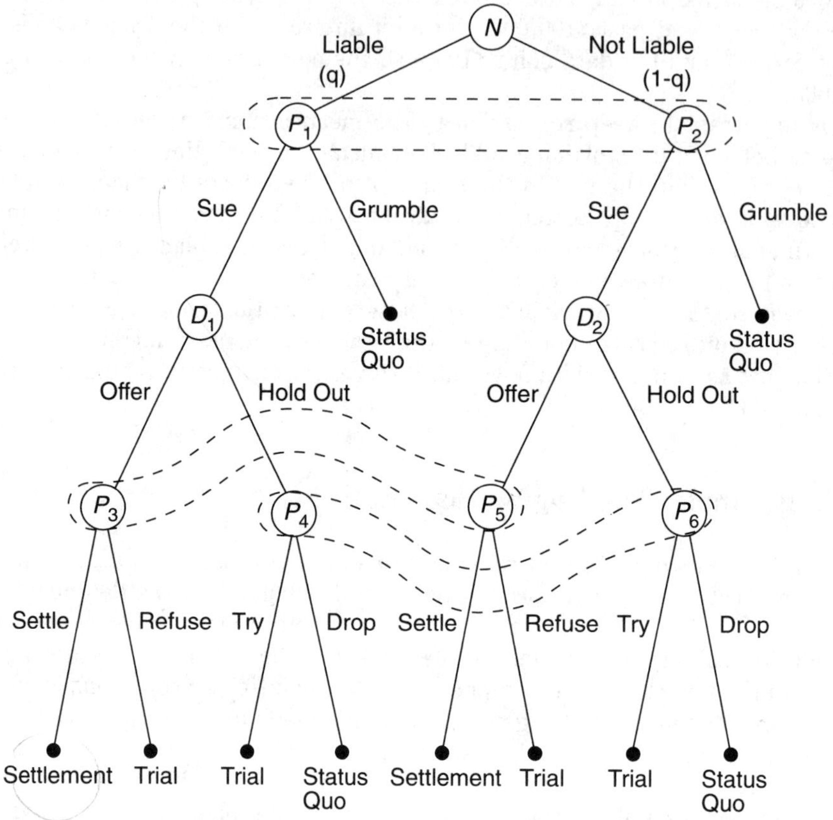

Players: Plaintiff, Defendant

game and would not affect the equilibrium strategies unless payments along the way revealed information, in which case what matters is the information, not the fact that payoffs change.

We assume that if the case reaches the court, justice is done. In addition to his legal fees D, the defendant pays damages W only if he is liable. We also assume that the players are risk neutral, so they only care about the expected dollars they will receive, not the variance. Without this assumption we would have to translate the dollar payoffs into utility, but the game tree would be unaffected.

This is a game of certain, asymmetric, imperfect, and incomplete information. We have assumed that the defendant knows whether he is liable, but we could modify the game by assuming that he has no better idea than the plaintiff of whether the evidence is sufficient to prove him so. The game would become one of symmetric information and we could reasonably simplify the extensive form by eliminating the initial move by Nature and setting the payoffs equal to the expected values. We cannot perform this simplification in the original game, because the fact that the defendant, and only the defendant, knows whether he is liable strongly affects the behavior of both players.

While the extensive form is the natural way to depict the game, we can also bypass it and associate strategies with payoffs using the strategic form. This is a little different from the 2-by-2 strategic forms, because the defendant gets a different payoff depending on whether he is really liable or not. We need three entries for each pair of strategies, one for the Plaintiff, one for the Liable Defendant, and one for the Blameless Defendant. In complicated games like this one, the distinction between action and strategy is important. A typical action is *Settle*. A typical strategy is (*Sue*; *Settle*[if offered], *Try*[if the defendant holds out])). The strategies are shown in table 2.6.

Using dominance we can rule out one of the plaintiff's strategies immediately—*Grumble*—which is dominated by (*Sue, Settle, Drop*).

Whether a strategy profile is a Nash equilibrium depends on the parameters of the model—here, S, W, P, D, and q, which are the settlement amount, the damages, the court costs for the plaintiff and defendant, and the probability the defendant is liable. Depending on the parameter values, three outcomes are possible: settlement (if the settlement amount is low), trial (if expected damages are high and the plaintiff's court costs low), and the plaintiff dropping the action (if expected damages minus court costs are negative). For some parameter values there are multiple Nash equilibria.

Consider the parameter values $S = 0.15$, $D = 0.2$, $W = 1$, $q = 0.13$, and $P = 0.1$. Table 2.7 shows the strategic form in this case. The two Nash equilibria marked in boldface are both weak, and the outcomes of either trial or settlement are possible in equilibrium. We will only trace through the outcome with settlement.

Consider the strategy profile {(*Sue, Settle, Try*), (*Offer, Offer*)}. The plaintiff sues, the defendant offers to settle (whether liable or not), and the plaintiff agrees to settle. Both players know that if the defendant did not offer to settle, the plaintiff would go to court and try the case. Such **out-of-equilibrium** behavior is specified by the equilibrium, because the threat of trial is what induces the defendant to offer to settle, even though trials never occur in equilibrium. This strategy profile is a Nash equilibrium because, given that the plaintiff chooses (*Sue, Settle, Try*), the defendant can do no better than (*Offer, Offer*), settling for a payoff of -0.15 whether he is liable or not; and, given that the defendant chooses (*Offer, Offer*), the plaintiff can do no better than the payoff of 0.15 from (*Sue, Settle, Try*).

A final observation on The Png Settlement Game: the game illustrates the Harsanyi doctrine in action, because while the plaintiff and defendant differ in their beliefs as to the probability the plaintiff will win, they do so because the defendant has different information, not because the modeller assigns them different beliefs at the start of the game. This seems awkward compared to the everyday way of approaching this problem in which we simply note that potential litigants have different beliefs, and will go to trial if they both think they can win. It is very hard to make the everyday story consistent, however, because if the differing beliefs are common knowledge, both players know that one of them is wrong, and each has to believe that he is correct. This may be fine as a "reduced form," in which the attempt is to simply describe what happens without explaining it in any depth. After all, even in The Png Settlement Game, if a trial occurs it is because the players differ in their beliefs, so one could simply chop off the

Table 2.6 The Strategic Form of The Png Settlement Game: General Parameters

Defendant's Strategy

Plaintiff's Strategy	*(Offer, Offer)*		*(Resist, Resist)*		*(Offer, Resist)*		*(Resist, Offer)*	
Grumble	0	(0,0)	0	(0,0)	0	(0,0)	0	(0,0)
(Sue; Settle, Try)	S	$(-S, -S)$	$qW - P$	$(-W - D, -D)$	$qS - (1 - q)P$	$(-S, -D)$	$q(W - P) + (1 - q)S$	$(-W - D, -S)$
(Sue; Refuse, Try)	$qW - P$	$(-W - D, -D)$	$qW - P$	$(-W - D, -D)$	$qW - P$	$(-W - D, -D)$	$qW - P$	$(-W - D, -D)$
(Sue; Refuse, Drop)	$qW - P$	$(-W - D, -D)$	0	(0,0)	$q(W - P)$	$(-W - D, 0)$	$-(1 - q)P$	$(0, -D)$
(Sue, Settle, Drop)	S	$(-S, -S)$	0	(0,0)	qS	$(-S, 0)$	$(1 - q)S$	$(0, -S)$

Payoffs to: *Plaintiff, (Liable Defendant, Blameless Defendant)*

Table 2.7 The Strategic Form of The Png Settlement Game: Particular Parameters

Plaintiff's Strategy	Defendant's Strategy: *(Offer, Offer)*		*(Resist, Resist)*		*(Offer, Resist)*		*(Resist, Offer)*	
Grumble	0	(0,0)	0	(0,0)	0	(0,0)	0	(0,0)
(Sue; Settle, Try)	0.15	(0.15, −0.15)	0.03	(−1.2, −0.2)	−0.0675	(−0.15, −0.2)	0.2475	(−1.2, −0.15)
(Sue; Refuse, Try)	0.03	(−1.2, −0.2)	0.03	(−1.2, −0.2)	0.03	(−1.2, −0.2)	0.03	(−1.2, −0.2)
(Sue; Refuse, Drop)	0.03	(−1.2, −0.2)	0	(0,0)	0.117	(−1.2,0)	−0.187	(0, −0.2)
(Sue; Settle, Drop)	0.15	(−0.15, −0.15)	0	(0,0)	0.0195	(−0.15,0)	0.1305	(0, −0.15)

Payoffs to: *Plaintiff, (Liable Defendant, Blameless Defendant)* first part of the game tree. But that is also the problem with violating the Harsanyi doctrine: one cannot analyze how the players react to each other's moves if the modeller simply assigns them inflexible beliefs. In The Png Settlement Game, a settlement is rejected and a trial can occur under certain parameters because the plaintiff weighs the probability that the defendant knows he will win versus the probability that he is bluffing, and sometimes decides to risk a trial. Without the Harsanyi doctrine it is very hard to evaluate such an explanation for trials.

Notes

N2.1 The Strategic and Extensive Forms of a Game

- The term, "outcome matrix," is used in Shubik (1982, p. 70), but never formally defined there.
- The term, "node," is sometimes defined to include only points at which a player or Nature makes a decision, which excludes the end points.

N2.2 Information Sets

- If you wish to depict a situation in which a player does not know whether the game has reached node A_1 or A_2 and he has different action sets at the two nodes, restructure the game. If you wish to say that he has action set (X, Y, Z) at A_1 and (X, Y) at A_2, first add action Z to the information set at A_2. Then specify that at A_2, action Z simply leads to a new node A_3 at which the choice is between X and Y.
- The term, "common knowledge," comes from Lewis (1969). Recent discussions include Brandenburger (1992) and Geanakoplos (1992). For rigorous but nonintuitive definitions of common knowledge, see Aumann (1976) and Milgrom (1981a) . Following Milgrom, let (Ω, p) be a probability space, let P and Q be partitions of Ω representing the information of two agents, and let R be the finest common coarsening of P and Q. Let ω be an event (an item of information) and $R(\omega)$ be that element of R which contains ω.

  An event A is **common knowledge** *at ω if $R(\omega) \subset A$.*

N2.3 Perfect, Certain, Symmetric, and Complete Information

- Tirole (1988, p. 431) (and Fudenberg & Tirole [1991a, p. 82], more precisely) have defined games of *almost perfect* information. They use this term to refer to repeated simultaneous-move games (of the kind studied here in chapter 5) in which at each repetition all players know the results of all the moves, including those of Nature, in previous repetitions. It is a pity they use such a general-sounding term to describe so narrow a class of games; it could be usefully extended to cover all games which have perfect information except for simultaneous moves.
- **Poker Classifications.** (1) Perfect, certain. (2) Incomplete, symmetric, certain. (3) Incomplete, asymmetric, certain. (4) Complete, asymmetric, certain. (5) Perfect, uncertain. (6) Incomplete, asymmetric, certain.
- For an explanation of von Neumann-Morgenstern utility, see Varian (1992) or Kreps (1990a). Expected utility and Bayesian updating are the two foundations of standard game theory, partly because they seem realistic and partly because they are so simple to use. Sometimes they do not explain people's behavior well, however, and there exist extensive literatures (a) pointing out anomalies, and (b) suggesting alternatives. So far no alternatives have proven to be big enough improvements to justify replacing the standard techniques, given the tradeoff between descriptive realism and added complexity in modelling. The standard response is to admit and ignore the anomalies in theoretical work, and not to press any theoretical models too hard in situations where the anomalies are likely to make a significant difference. On anomalies, see the volume edited by Kahneman, Slovic & Tversky (1982). On alternatives to expected utility, see Machina (1982). For an overview, see the book by Thaler (1992).

- Mixed strategies (section 3.1) are allowed in a game of perfect information, because that is an aspect of the game's equilibrium, not of its exogenous structure.
- Although the word, "perfect," appears in both "perfect information" (section 2.3) and "perfect equilibrium" (section 4.1), the concepts are unrelated.
- An unobserved move by Nature in a game of symmetric information can be represented in any of three ways: (1) as the last move in the game; (2) as the first move in the game; or (3) by replacing the payoffs with the expected payoffs and not using any explicit moves by Nature.

N2.4 The Harsanyi Transformation and Bayesian Games

- Mertens & Zamir (1985) probes the mathematical foundations of the Harsanyi transformation. The transformation requires the extensive form to be common knowledge, which raises subtle questions of recursion.
- A player always has some idea of what the payoffs are, so we can always assign him a subjective probability for each possible payoff. What would happen if he had no idea? Such a question is meaningless, because people always have some notion, and when they say they do not, they generally mean that their prior probabilities are low but positive for a great many possibilities. You, for instance, probably have as little idea as I do of how many cups of coffee I have consumed in my lifetime, but you would admit it to be a nonnegative number less than 3,000,000, and you could make a much more precise guess than that. On the topic of subjective probability, a classic reference is Savage (1954).
- If two players have common priors and their information partitions are finite, but they each have private information, iterated communication between them will lead to the adoption of a common posterior. This posterior is not always the posterior they would reach if they directly pooled their information, but it is almost always that posterior (Geanakoplos & Polemarchakis [1982]).
- For formal analysis of regression to the mean and business conservatism, see Rasmusen (1992b). This can also explain why even after discounting revenues further in the future, businesses favor projects that offer quicker returns, if more distant revenue forecasts are less accurate.

Problems

2.1: The Monty Hall Problem.

You are a contestant on the TV show, "Let's Make a Deal." You face three curtains, labelled A, B and C. Behind two of them are toasters, and behind the third is a Mazda Miata car. You choose A, and the TV showmaster says, pulling curtain B aside to reveal a toaster, "You're lucky you didn't choose B, but before I show you what is behind the other two curtains, would you like to change from curtain A to curtain C?" Should you switch? What is the exact probability that curtain C hides the Miata?

2.2: Elmer's Appetite.

Mrs Jones has made an apple pie for her son, Elmer, and she is trying to figure out whether the pie tasted divine, or merely good. Her pies turn out divinely a third of the time. Elmer might be ravenous, or merely hungry, and he will eat either 2, 3, or 4 pieces of pie. Mrs Jones knows that he is ravenous half the time (but not which half). If the pie is divine, then, if Elmer is hungry, the probabilities of the three levels of consumptions are (0, 0.6, 0.4), but if he is ravenous the probabilities are (0, 0, 1). If the pie is just good, then the probabilities are (0.2, 0.4, 0.4) if he is hungry and (0.1, 0.3, 0.6) if he is ravenous.

Elmer is a sensitive, but useless, boy. He will always say that the pie is divine and his appetite weak, regardless of his true inner feelings.

(2.2a) What is the probability that he will eat four pieces of pie?

(2.2b) If Mrs Jones sees Elmer eat four pieces of pie, what is the probability that he is ravenous and the pie is merely good?

(2.2c) If Mrs Jones sees Elmer eat four pieces of pie, what is the probability that the pie is divine?

2.3: Cancer Tests.

Imagine that you are being tested for cancer, using a test that is 98% accurate. If you indeed have cancer, the test shows positive (indicating cancer) 98% of the time. If you do not have cancer, it shows negative 98% of the time. You have heard that 1 in 20 people in the population actually have cancer. Now your doctor tells you that you tested positive, but you shouldn't worry because his last 19 patients all died. How worried should you be? What is the probability you have cancer?

2.4: The Battleship Problem.

The Pentagon has the choice of building one battleship or two cruisers. One battleship costs the same as two cruisers, but a cruiser is sufficient to carry out the navy's mission—if the cruiser survives to get close enough to the target. The battleship has a probability p of carrying out its mission, whereas a cruiser only has probability $p/2$. Whatever the outcome, the war ends and any surviving ships are scrapped. Which option is superior?[3]

2.5: Joint Ventures.

Software Inc. and Hardware Inc. have formed a joint venture. Each can exert either high or low effort, which is equivalent to costs of 20 and 0. Hardware moves first, but Software cannot observe his effort. Revenues are split equally at the end, and the two firms are risk neutral. If both firms exert low effort, total revenues are 100. If the parts are defective, the total revenue is 100; otherwise, if both exert high effort, revenue is 200, but if only one player does, revenue is 100 with probability 0.9 and 200 with probability 0.1. Before they start, both players believe that the probability of defective parts is 0.7. Hardware discovers the truth about the parts by observation before he chooses effort, but Software does not.

(2.5a) Draw the extensive form and put dotted lines around the information sets of Software at any of the nodes at which he moves.

(2.5b) What is the Nash equilibrium?

(2.5c) What is Software's belief, in equilibrium, as to the probability that Hardware chooses low effort?

[3] Adapted from McMillan (1992).

(2.5d) If Software sees that revenue is 100, what probability does he assign to defective parts if he himself exerted high effort and he believes that Hardware chose low effort?

2.6: California Drought.

California is in a drought and the reservoirs are running low. The probability of rainfall in 1991 is ½, but with probability 1 there will be heavy rainfall in 1992. The state uses rationing rather than the price system, and it must decide how much water to consume in 1990 and how much to save till 1991. Each Californian has a utility function $U = \log(w_{90}) + \log(w_{91})$. Show that the state should allocate twice as much water to 1990 as to 1991.[4]

[4] Adapted from McMillan (1992).

3 Mixed and Continuous Strategies

3.1 Mixed Strategies: The Welfare Game

The games we have looked at have so far been simple in at least one respect: the number of moves in the action set has been finite. In this chapter we allow a continuum of moves, such as when a player chooses a price between 10 and 20 or a purchase probability between 0 and 1. Chapter 3 begins by showing how to find mixed-strategy equilibria for a game with no pure-strategy equilibrium. In section 3.2 the mixed-strategy equilibria are found by the payoff-equating method and mixed strategies are applied to a dynamic game, the war of attrition. Section 3.3 takes a more general look at mixed-strategy equilibria and distinguishes between mixed strategies and random actions in auditing games. Section 3.3 begins the analysis of continuous action spaces, and this is continued in section 3.4 in the Cournot duopoly model, where the discussion focuses on the example of two firms choosing output from the continuum between zero and infinity. These sections introduce other ideas that will be built upon in later chapters—dynamic games in chapter 4, auditing and agency in chapters 7 and 8, and Cournot oligopoly in chapter 13.

We invoked the concept of Nash equilibrium to provide predictions of outcomes without dominant strategies, but some games lack even a Nash equilibrium. It is often useful and realistic to expand the strategy space to include random strategies, in which case a Nash equilibrium almost always exists. These random strategies are called "mixed strategies."

> *A* **pure strategy** *maps each of a player's possible information sets to one action.* $s_i : \omega_i \rightarrow a_i$.

> *A* **mixed strategy** *maps each of a player's possible information sets to a probability distribution over actions.*

$$s_i : \omega_i \rightarrow m(a_i), \text{ where } m \geq 0, \int_{A_i} m(a_i) da_i = 1.$$

A **completely mixed** *strategy puts positive probability on every action, so $m > 0$.*

The version of a game expanded to allow mixed strategies is called the **mixed extension** *of the game.*

A pure strategy constitutes a rule that tells the player what action to choose, while a mixed strategy constitutes a rule that tells him what dice to throw in order to choose an action. If a player pursues a mixed strategy, he might choose any of several different actions in a given situation, an unpredictability which can be helpful to him. Mixed strategies occur frequently in the real world. In American football games, for example, the offensive team has to decide whether to pass or to run. Passing generally gains more yards, but what is most important is to choose an action that is not expected by the other team. Teams decide to run part of the time and pass part of the time in a way that seems random to observers but rational to game theorists.

The Welfare Game

The Welfare Game models a government that wishes to aid a pauper if he searches for work but not otherwise, and a pauper who searches for work only if he cannot depend on government aid. This is a well-known problem in public policy, called the Samaritan's Dilemma by Tullock (1983, p. 59), who attributes it to James Buchanan. The same problem arises on a private level when parents decide how much to help a lazy child.

Table 3.1 shows payoffs which represent the situation. Neither player has a dominant strategy, and with a little thought we can see that no Nash equilibrium exists in pure strategies either.

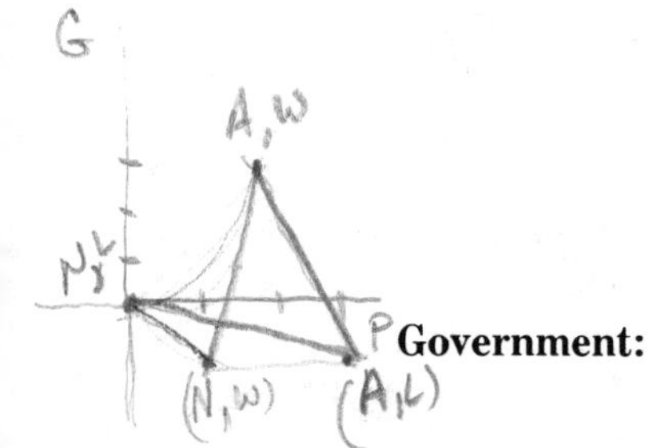

Table 3.1 The Welfare Game

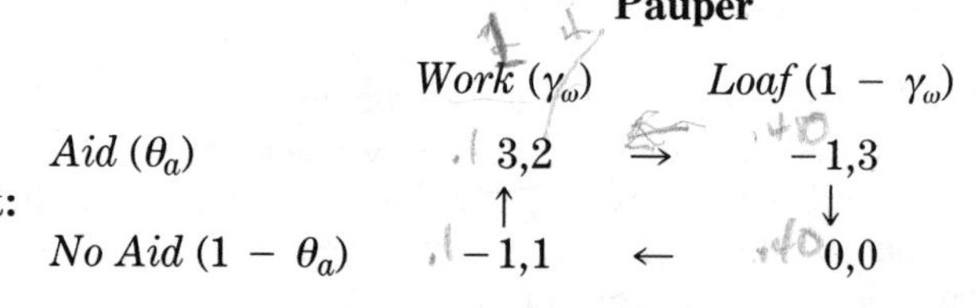

		Pauper		
		Work (γ_ω)		*Loaf* ($1 - \gamma_\omega$)
	Aid (θ_a)	3,2	$\rightarrow$	-1,3
Government:		$\uparrow$		$\downarrow$
	No Aid ($1 - \theta_a$)	-1,1	$\leftarrow$	0,0

Payoffs to: (Government, Pauper)

Each strategy profile must be examined in turn to check for Nash equilibria.

(1) The strategy profile (*Aid, Work*) is not a Nash equilibrium, because the Pauper would respond with *Loaf* if the Government picked *Aid*.
(2) (*Aid, Loaf*) is not Nash, because the Government would switch to *No Aid*.
(3) (*No Aid, Loaf*) is not Nash, because the Pauper would switch to *Work*.
(4) (*No Aid, Work*) is not Nash, because the Government would switch to *Aid*, which brings us back to (1).

The Welfare Game does have a mixed-strategy Nash equilibrium, which we can calculate. The players' payoffs are the expected values of the payments from table 3.1. If the government plays *Aid* with probability θ_a, and the Pauper plays *Work* with probability γ_w, the Government's expected payoff is

$$\begin{aligned} \pi_{Government} &= \theta_a[3\gamma_w + (-1)(1 - \gamma_w)] + (1 - \theta_a)[-1\gamma_w + 0(1 - \gamma_w)] \\ &= \theta_a[3\gamma_w - 1 + \gamma_w] - \gamma_w + \theta_a\gamma_w \\ &= \theta_a[5\gamma_w - 1] - \gamma_w. \end{aligned} \tag{1}$$

If only pure strategies are allowed, θ_a equals zero or one, but in the mixed extension of the game the Government's action of θ_a lies on the continuum from zero to one, the pure strategies being the extreme values. Following the usual procedure for solving a maximization problem, we differentiate the payoff function with respect to the choice variable to obtain the first-order condition:

$$\begin{aligned} 0 &= \frac{d\pi_{Government}}{d\theta_a} = 5\gamma_w - 1 \\ \Rightarrow \gamma_w &= 0.2. \end{aligned} \tag{2}$$

In the mixed-strategy equilibrium, the Pauper selects *Work* 20 percent of the time. The way we obtained the number might seem strange: to obtain the Pauper's strategy, we differentiated the Government's payoff. Understanding why requires several steps.

(1) I assert that an optimal mixed strategy exists for the Government.
(2) If the Pauper selects *Work* more than 20 percent of the time, the Government always selects *Aid*. If the Pauper selects *Work* less than 20 percent of the time, the Government never selects *Aid*.
(3) If a mixed strategy is to be optimal for the Government, the Pauper must therefore select *Work* with probability exactly 20 percent.

To obtain the probability of the Government choosing *Aid*, we must turn to the Pauper's payoff function, which is

$$\begin{aligned} \pi_{Pauper} &= \theta_a(2\gamma_w + 3[1 - \gamma_w]) + (1 - \theta_a)(1\gamma_w + 0[1 - \gamma_w]), \\ &= 2\theta_a\gamma_w + 3\theta_a - 3\theta_a\gamma_w + \gamma_w - \theta_a\gamma_w, \\ &= -\gamma_w(2\theta_a - 1) + 3\theta_a. \end{aligned} \tag{3}$$

The first-order condition is

$$\begin{aligned} \frac{d\pi_{Pauper}}{d\gamma_w} &= -(2\theta_a - 1) = 0, \\ \Rightarrow \theta_a &= 1/2. \end{aligned} \tag{4}$$

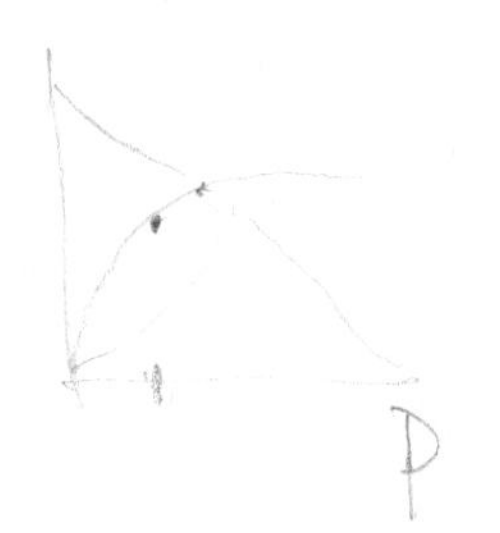

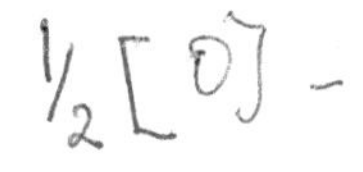

If the Pauper selects *Work* with probability 0.2, the Government is indifferent among selecting *Aid* with probability 100 percent, 0 percent, or anything in between. If the strategies are to form a Nash equilibrium, however, the Government must choose $\theta_a = 0.5$. In the mixed-strategy Nash equilibrium, the Government selects *Aid* with probability 0.5 and the Pauper selects *Work* with probability 0.2. The equilibrium outcome could be any of the four entries in the outcome matrix. The entries having the highest probability of occurrence are (*No Aid*, *Loaf*) and (*Aid*, *Loaf*), each with probability 0.4 ($=0.5[1 - 0.2]$).

Interpreting Mixed Strategies

Mixed strategies are not as intuitive as pure strategies, and many modellers prefer to restrict themselves to pure-strategy equilibria in games which have them. One objection to mixed strategies is that people in the real world do not take random actions. That is not a compelling objection, because all that a model with mixed strategies requires to be a good description of the world is that the actions appear random to observers, even if the player himself has always been sure what action he would take. Even explicitly random actions are not uncommon, however— the Internal Revenue Service randomly selects which tax returns to audit, and telephone companies randomly monitor their operators' conversations to discover whether they are being polite.

A more troubling objection is that a player who selects a mixed strategy is always indifferent between two pure strategies. In The Welfare Game, the Pauper is indifferent between his two pure strategies and a whole continuum of mixed strategies, given the Government's mixed strategy. If the Pauper were to decide not to follow the particular mixed strategy $\gamma_w = 0.2$, the equilibrium would collapse because the Government would change its strategy in response. Even a small deviation in the probability selected by the Pauper, a deviation that does not change his payoff if the Government does not respond, destroys the equilibrium completely because the Government does respond. A mixed-strategy Nash equilibrium is weak in the same sense as the (*North*, *North*) equilibrium in Battle of the Bismarck Sea: to maintain the equilibrium, a player who is indifferent between strategies must pick a particular strategy from out of the set of strategies.

One way to reinterpret The Welfare Game is to imagine that instead of a single pauper there are many, with identical tastes and payoff functions, all of whom must be treated alike by the Government. In the mixed-strategy equilibrium, each of the paupers chooses *Work* with probability 0.2, just as in the one-pauper game. But the many-pauper game has a pure-strategy equilibrium: 20 percent of the paupers choose the pure strategy *Work* and 80 percent choose the pure strategy *Loaf*. The problem persists of how an individual pauper, indifferent between the pure strategies, chooses one or the other, but it is easy to imagine that individual characteristics outside the model could determine which actions are chosen by which paupers.

The number of players needed so that mixed strategies can be interpreted as pure strategies in this way depends on the equilibrium probability γ_w, since we cannot speak of a fraction of a player. The number of paupers must be a multiple of 5 in The Welfare Game in order to use this interpretation, since the equilibrium mixing probability is a multiple of ⅕. For the interpretation to apply no matter how we vary the parameters of a model we would need a *continuum* of players.

Another interpretation of mixed strategies, which works even in the single-pauper game, assumes that the pauper is drawn from a population of paupers, and that the Government does not know his characteristics. The Government only knows that there are two types of paupers, in the proportions (0.2, 0.8): those who pick *Work* if the Government picks $\theta_a = 0.5$, and those who pick *Loaf*. A pauper drawn randomly from the population might be of either type. Harsanyi (1973) gives a careful interpretation of this situation.

Existence of Equilibrium

Sometimes the modeller cannot find a particular equilibrium easily, but is able to show what characteristics an equilibrium must have if it exists, and would like to prove that it does indeed exist. One of the strong points of Nash equilibria is that they exist, in mixed strategies if not in pure, in practically every game one is likely to encounter.

One feature of a game that favors existence of Nash equilibrium is continuity of the payoffs in the strategies. A payoff is continuous in a player's strategy if a small change in his strategy causes a small or zero change in the payoffs. With continuity, the players' strategies can adjust more finely and come closer to being best responses to each other. In The Welfare Game, the payoffs are discontinuous in pure strategies, so no compromise between the players is possible and no pure-strategy equilibrium exists. Once mixed strategies are incorporated, the payoffs are continuous in the mixing probability, so an equilibrium does exist.

A second feature promoting existence is a closed and bounded strategy set. Suppose in a stock market game that Smith can borrow money and buy as many shares of stock as he likes, so his strategy set, the amount of stock he can buy, is $[0, \infty)$, a set which is unbounded above (we assume he can buy fractional shares but cannot sell short). If Smith knows that the price today is lower than it will be tomorrow, he wants to buy an infinite number of shares, which is not an equilibrium purchase. If the amount he buys is restricted to be strictly less than 1,000, then the strategy set is bounded (by 1,000) but not closed (because 1,000 is not included), and no equilibrium purchase exists, since he wants to buy 999.999 . . . shares. An open set such as $\{x : 0 \leq x < 1000\}$ has no maximum, because there is no closest real number to 1000. If the strategy set is closed and bounded, on the other hand, its minimum and maximum are well-defined. If Smith can buy up to but no more than 1,000 shares, then 1000 shares is his equilibrium purchase. Sometimes, as in The Cournot Game, discussed later in this chapter, the unboundedness of the strategy sets does not matter because the optimum is an interior solution, but in other games it is good modelling practice to use closed sets instead of open sets.

3.2 Chicken, The War of Attrition, and Correlated Strategies

Chicken and the Payoff-Equating Method

The next game illustrates why we might decide that a mixed-strategy equilibrium is best even if pure-strategy equilibria also exist. In the game of Chicken, the players are two Malibu teenagers, Smith and Jones. Smith drives a hot rod south down the middle of Route 1, and Jones drives north. As collision threatens, each decides whether to *Continue* in the middle or *Swerve* to the side. If a player is the only one to *Swerve*, he loses face, but if neither player picks *Swerve* they are both killed, which has an even lower payoff. If a player is the only one to *Continue*, he is covered with glory, and if both *Swerve* they are both embarrassed. (We will assume that to *Swerve* means by convention to *Swerve* right; if one swerved to the left and the other to the right, the result would be both death and humiliation.) Table 3.2 assigns numbers to these four outcomes.

Table 3.2 Chicken

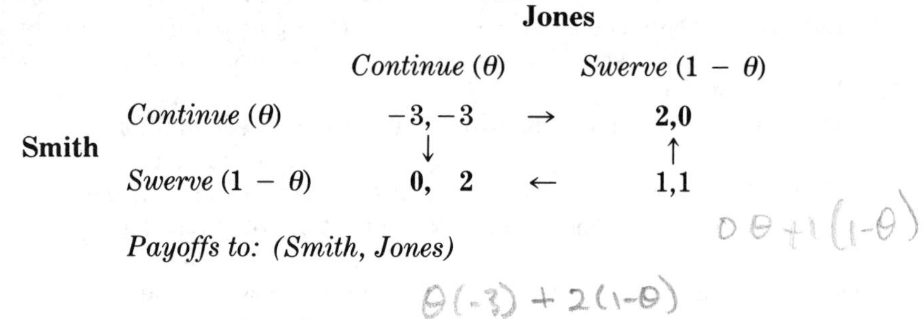

		Jones		
		Continue (θ)		*Swerve* ($1-\theta$)
Smith	*Continue* (θ)	$-3, -3$	$\rightarrow$	**2,0**
		$\downarrow$		$\uparrow$
	Swerve ($1-\theta$)	**0, 2**	$\leftarrow$	1,1

Payoffs to: (Smith, Jones)

Chicken has two pure-strategy Nash equilibria, (*Swerve*, *Continue*) and (*Continue*, *Swerve*), but they have the defect of asymmetry. How do the players know which equilibrium is the one that will be played out? Even if they talk before the game started, it is not clear how they could arrive at an asymmetric result. We encountered the same dilemma in choosing an equilibrium for Battle of the Sexes. As in that game, the best prediction in Chicken is perhaps the mixed-strategy equilibrium, because its symmetry makes it a focal point of sorts, and does not require any differences between the players.

The **payoff-equating** method used here to calculate the mixing probabilities for Chicken will be based on the logic followed in section 3.1, but it does not use the calculus of maximization. In the mixed-strategy equilibrium, Smith must be indifferent between *Swerve* and *Continue*. Moreover, Chicken, unlike The Welfare Game, is a symmetric game, so we can guess that in equilibrium each player will choose the same mixing probability. If that is the case, then, since the payoffs from each of Jones' pure strategies must be equal in a mixed-strategy equilibrium, it is true that

$$\begin{aligned}\pi_{Jones}(Swerve) &= (\theta_{Smith}) \cdot (0) + (1 - \theta_{Smith}) \cdot (1) \\ &= (\theta_{Smith}) \cdot (-3) + (1 - \theta_{Smith}) \cdot (2) = \pi_{Jones}(Continue).\end{aligned} \tag{5}$$

From equation (5) we can conclude that $1 - \theta_{Smith} = 2 - 5\theta_{Smith}$, so $\theta_{Smith} = 0.25$. In the symmetric equilibrium, both players choose the same probability, so we can replace θ_{Smith} with simply θ. As for the question which represents the greatest interest to their mothers, the two teenagers will survive with probability $1 - (\theta \cdot \theta) = 0.9375$.

The payoff-equating method is easier to use than the calculus method if the modeller is sure which strategies will be mixed, and it can also be used in asymmetric games. In The Welfare Game, it would start with $V_g(Aid) = V_g(No\ Aid)$ and $V_p(Loaf) = V_p(Work)$, yielding two equations for the two unknowns, θ_a and γ_w, which, when solved, give the same mixing probabilities as were found earlier for that game. The reason why the payoff-equating and calculus maximization methods reach the same result is that the expected payoff is linear in the possible payoffs, so differentiating the expected payoff equalizes the possible payoffs. The only difference from the symmetric-game case is that two equations are solved for two different mixing probabilities instead of a single equation for the one mixing probability that both players use.

It is interesting to see what happens if the payoff of -3 in the northwest corner of table 3.2 is generalized to x. Solving the analog of equation (5) then yields

$$\theta = \frac{1}{1 - x}. \tag{6}$$

If $x = -3$, this yields $\theta = 0.25$, as was just calculated, and if $x = -9$, it yields $\theta = 0.10$. This makes sense; increasing the loss from crashes reduces the equilibrium probability of continuing down the middle of the road. But what if $x = 0.5$? Then the equilibrium probability of continuing appears to be $\theta = 2$, which is impossible; probabilities are bounded by zero and one.

When a mixing probability is calculated to be greater than one or less than zero, the implication is either that the modeller has made an arithmetic mistake or, as in this case, that he is wrong in thinking that the game has a mixed-strategy equilibrium. If $x = 0.5$, one can still try to solve for the mixing probabilities, but, in fact, the only equilibrium is in pure strategies—(*Continue*, *Continue*) (the game has become a Prisoner's Dilemma). The absurdity of probabilities greater than one or less than zero is a valuable aid to the fallible modeller, because such results show that he is wrong about the qualitative nature of the equilibrium—it is pure, not mixed. Or, if the modeller is not sure whether the equilibrium is mixed or not, he can use this approach to prove that the equilibrium is not in mixed strategies.

The War of Attrition

The War of Attrition is a game something like Chicken stretched out over time, where both players start with *Continue*, and the game ends when the first one picks *Swerve*. Until the game ends, both earn a negative amount per period, and when one exits, he earns 0 and the other player earns a reward for outlasting him.

We will look at a war of attrition in discrete time. We will continue with Smith and Jones, who have both survived to maturity and now play games with more expensive toys: they control two firms in an industry which is a natural monopoly, with demand strong enough for one firm to operate profitably, but not two. The possible actions are to *Exit* or to *Continue*. In each period during which both *Continue*, each earns -1. If a firm exits, its losses cease and the remaining firm obtains the value of the market's monopoly profit, which we set equal to 3. We will set the discount rate equal to $r > 0$, although that is inessential to the model, even if the possible length of the game is infinite (discount rates will be discussed in detail in section 4.3).

The War of Attrition has a continuum of Nash equilibria. One simple equilibrium is for Smith to choose (*Continue* regardless of what Jones does) and for Jones to choose (*Exit* immediately), which are best responses to each other. But we will solve for a symmetric equilibrium in which each player chooses the same mixed strategy: a constant probability θ that the player picks *Exit* given that the other player has not yet exited.

We can calculate θ as follows, adopting the perspective of Smith. Denote the expected discounted value of Smith's payoffs by V_{stay} if he stays and V_{exit} if he exits immediately. These two pure-strategy payoffs must be equal in a mixed-strategy equilibrium (which was the basis for the payoff-equating method). If Smith exits, he obtains $V_{exit} = 0$. If Smith stays in, his payoff depends on what Jones does. If Jones stays in too, which has probability $(1 - \theta)$, Smith gets -1 currently, and his expected value for the following period, which is discounted using r, is unchanged. If Jones exits immediately, which has probability θ, Smith receives a payment of 3. In symbols,

$$V_{stay} = \theta \cdot (3) + (1 - \theta)\left(-1 + \left[\frac{V_{stay}}{1 + r}\right]\right), \tag{7}$$

which, after a little manipulation, becomes

$$V_{stay} = \left(\frac{1 + r}{r + \theta}\right)(4\theta - 1). \tag{8}$$

Once we equate V_{stay} to V_{exit}, which equals zero, equation (8) tells us that $\theta = 0.25$ in equilibrium, and that this is independent of the discount rate r.

Returning from arithmetic to ideas, why does Smith *Exit* immediately with positive probability, given that Jones will exit first if Smith waits long enough? The reason is that Jones might choose to continue for a long time and both players would earn -1 for each period until Jones exited. The equilibrium mixing probability is calculated so that both of them are likely to stay in long enough so that their losses soak up the gain from being the survivor. This is an example of a "rent-seeking" welfare loss. As Posner (1975) and Tullock (1967) have pointed out, the real costs of acquiring rents can be much bigger than the second-order triangle losses from allocative distortions, and the war of attrition shows that the big loss from a natural monopoly might be not the reduced trade that results from higher prices, but the cost of the battle to gain the monopoly.

In wars of attrition, the reward goes to the player who does not choose the move which ends the game, and a cost is paid for each period that both players refuse to end it. Various other **timing games** also exist. The opposite of a war of attrition is a **preemption game**, in which the reward goes to the player who chooses the move which ends the game, and a cost is paid if both players choose that move, but no cost is incurred in a period when neither player chooses it. The game of **Grab the Dollar** is an example. A dollar is placed on the table between Smith and Jones, who each must decide whether to grab for it or not. If both grab, each is fined one dollar. This could be set up as a one-period game, a T period game, or an infinite-period game, but the game definitely ends when someone grabs the dollar. Table 3.3 shows the payoffs.

Table 3.3 Grab The Dollar

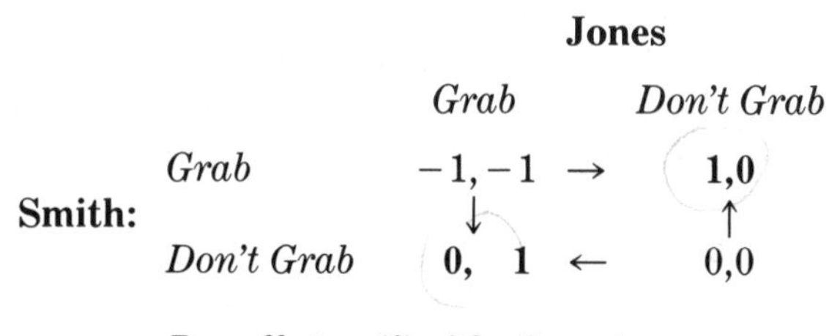

		Jones		
		Grab		*Don't Grab*
Smith:	*Grab*	$-1,-1$	→	**1,0**
		↓		↑
	Don't Grab	**0, 1**	←	0,0

Payoffs to: (Smith, Jones)

Like The War of Attrition, Grab The Dollar has asymmetric equilibria in pure strategies, and a symmetric equilibrium in mixed strategies. In the infinite-period version, the equilibrium probability of grabbing is 0.5 per period in the symmetric equilibrium.

Still another class of timing games are duels, in which the actions are discrete occurrences which the players locate at particular points in continuous time. Two players with guns approach each other and must decide when to shoot. In a **noisy duel**, if a player shoots and misses, the other player observes the miss and can kill the first player at his leisure. An equilibrium exists in pure strategies for the noisy duel. In a **silent duel**, a player does not know when the other player has fired, and the equilibrium is in mixed strategies. Karlin (1959) has details on duelling games, and chapter 4 of Fudenberg & Tirole (1991a) has an excellent discussion of games of timing in general.

Correlated Strategies

One example of a war of attrition is setting up a market for a new security, which may be a natural monopoly for reasons to be explained in section 8.5. Certain stock exchanges have avoided the destructive symmetric equilibrium by using lotteries to determine which of them would trade newly listed stock options, under a system similar to the football draft.[1] Rather than waste resources fighting, these exchanges use the lottery as a coordinating device, even though it might not be a binding agreement.

Aumann (1974) has pointed out that it is often important whether players can use the same randomizing device for their mixed strategies. If they can, we refer to the resulting strategies as **correlated strategies**. Consider the game of Chicken. The only mixed-strategy equilibrium is the symmetric one in which each player chooses *Continue* with probability 0.25 and the expected payoff is 0.75. A correlated equilibrium would be for the two players to flip a coin and for Smith to choose *Continue* if it comes up heads and for Jones to choose *Continue* otherwise. Each player's strategy is a best response to the other's, the probability of each choosing *Continue* is 0.5, and the expected payoff to each is 1.0, which is better than the 0.75 achieved without correlated strategies.

Usually the randomizing device is not modelled explicitly when a model refers to correlated equilibrium. If it is, uncertainty over variables that do not affect preferences, endowments, or production is called **extrinsic uncertainty**. Extrinsic uncertainty is the driving force behind **sunspot models**, so-called because the random appearance of sunspots might cause macroeconomic changes via correlated equilibria (Maskin & Tirole [1987]) or bets made between players (Cass & Shell [1983]).

One way to model correlated strategies is to specify a move in which Nature gives each player the ability to commit first to an action such as *Continue* with equal probability. This is often realistic, because it amounts to a zero probability of both players entering the industry at exactly the same time, without anyone knowing in advance who will be the lucky starter. Neither firm has an *a priori* advantage, but the outcome is efficient.

The population interpretation of mixed strategies cannot be used for correlated strategies. In ordinary mixed strategies, the mixing probabilities are statistically independent, whereas in correlated strategies they are not. In Chicken, the usual mixed strategy can be interpreted as populations of Smiths and Joneses, each population consisting of a certain proportion of pure swervers and pure stayers. The correlated equilibrium has no such interpretation.

Another coordinating device, useful in games that, like Battle of the Sexes, have a coordination problem, is **cheap talk** (Crawford & Sobel [1982], Farrell [1987]). Cheap talk refers to costless communication before the game proper begins. In Ranked Coordination, cheap talk instantly allows the players to make the desirable outcome a focal point. In Chicken, cheap talk is useless, because it is dominant for each player to announce that he will choose *Continue*. But in Battle of the Sexes, coordination and conflict are combined. Without communication, the only symmetric equilibrium is in mixed strategies. If both players know that making inconsistent announcements will lead to the wasteful mixed-strategy outcome, then they will be willing to mix announcing whether they will go to the ballet or the prize fight. With many periods of announcements before the final decision, their chances of coming to an agreement are high. Thus, communication can help reduce inefficiency even if the two players are in conflict.

[1] "Big Board will Begin Trading of Options on 4 Stocks it Lists," *Wall Street Journal*, 4 October 1985, p. 15.

3.3 Mixed Strategies with General Parameters and *N* Players: The Civic Duty Game

Having looked at a number of specific games with mixed-strategy equilibria, let us now apply the method to the general game of table 3.4.

Table 3.4 General 2-by-2 Game

		Column	
		Left (θ)	*Right* ($1-\theta$)
Row:	*Up* (γ)	a,w	b,x
	Down ($1-\gamma$)	c,y	d,z

Payoffs to: (Row, Column)

To find the game's equilibrium, equate the payoffs from the pure strategies. For Row, this yields

$$\pi_{Row}(Up) = \theta a + (1-\theta)b \tag{9}$$

and

$$\pi_{Row}(Down) = \theta c + (1-\theta)d. \tag{10}$$

Equating (9) and (10) gives

$$\theta(a + d - b - c) + b - d = 0, \tag{11}$$

which yields

$$\theta^* = \frac{d-b}{(d-b)+(a-c)}. \tag{12}$$

Similarly, equating the payoffs for Column gives

$$\pi_{Column}(Left) = \gamma w + (1-\gamma)\,y = \pi_{Column}(Right) = \gamma x + (1-\gamma)z, \tag{13}$$

which yields

$$\gamma^* = \frac{z-y}{(z-y)+(w-x)}. \tag{14}$$

The equilibrium represented by (12) and (14) illustrates a number of features of mixed strategies.

First, it is possible, but wrong, to follow the payoff-equating method for finding a mixed strategy even if no mixed strategy equilibrium actually exists. Suppose, for example, that *Up* is a strongly dominant strategy for Row, so $c > a$ and $d > b$. Row is unwilling to mix, so the equilibrium is not in mixed strategies. Equation (14) would be misleading, though some idiocy would be required to stay misled for very long, since it implies that $\theta^* > 1$ in this case.

Second, the exact features of the equilibrium in mixed strategies depend heavily on the cardinal values of the payoffs, not just on their ordinal values as with pure strategy equilibria in other 2-by-2 games. Ordinal rankings are all that are needed to know that an equilibrium exists in mixed strategies, but cardinal values are needed to know the exact mixing probabilities. If the payoff to *Column* from (*Confess, Confess*) is changed slightly in The Prisoner's Dilemma, it makes no difference at all to the equilibrium. If the payoff of z to *Column* from (*Down, Right*) is increased slightly in the General 2-by-2 Game, equation (14) says that the mixing probability γ^* will change also.

Third, the payoffs can be changed by affine transformations without changing the game substantively, even though cardinal payoffs do matter (which is to say that monotonic but non-affine transformations do make a difference). Let each payoff π in table 3.4 become $\alpha + \beta\pi$. Equation (14) then becomes

$$\begin{aligned} \gamma^* &= \frac{\alpha + \beta z - \alpha - \beta y}{(\alpha + \beta z - \alpha - \beta y) + (\alpha + \beta w - \alpha - \beta x)} \\ &= \frac{z - y}{(z - y) + (w - x)}. \end{aligned} \tag{15}$$

The affine transformation has left the equilibrium strategy unchanged.

Fourth, as was mentioned earlier in connection with The Welfare Game, each player's mixing probability depends only on the payoff parameters of the other player. Row's strategy γ^* in equation (14) depends on the parameters w, x, y, and z, which are the payoff parameters for Column, and have no direct relevance for Row.

Categories of Games with Mixed Strategies

Table 3.5 uses the players and actions of table 3.4 to depict three major categories of 2-by-2 games in which mixed-strategy equilibria are important. Some games fall in none of these categories—those with tied payoffs such as the Swiss Cheese Game, in which all 8 payoffs equal zero—but the three games in table 3.5 encompass a wide variety of economic phenomena.

Table 3.5 2-by-2 Games with Mixed-Strategy Equilibria

Discoordination Games				Discoordination Games				Coordination Games				Contribution Games		
a,w	→	b,x		a,w	←	b,x		**a,w**	←	b,x		a,w	→	**b,x**
↑		↓		↓		↑		↑		↓		↓		↑
c,y	←	d,z		c,y	→	d,z		c,y	→	**d,z**		**c,y**	←	d,z

Discoordination games have a single equilibrium, in mixed strategies. The payoffs are such that either (a) $a > c$, $d > b$, $x > w$, and $y > z$, or (b) $c > a$, $b > d$, $w > x$, and $z > y$. The Welfare Game is a discoordination game, as is Auditing Game I in the next section and Matching Pennies in problem 3.3.

Coordination games have three equilibria: two symmetric equilibria in pure strategies and one symmetric equilibrium in mixed strategies. The payoffs are such that $a > c$, $d > b$, $w > x$, and $z > y$. Ranked Coordination and Battle of the Sexes are two varieties of coordination games in which the players have the same and opposite rankings of the pure-strategy equilibria.

Contribution games have three equilibria: two asymmetric equilibria in pure strategies and one symmetric equilibrium in mixed strategies. The payoffs are such that $c > a$, $b > d$, $x > w$, and $y > z$. Also, it must be true that $c < b$ and $y > x$.

I have invented the name, "contribution game," for the occasion, since the type of game described by this term is often used to model a situation in which two players each have a choice of taking some action that contributes to the public good, though each would like the other to bear the cost. The difference from The Prisoner's Dilemma is that each player in a contribution game is willing to bear the cost alone if necessary.

Contribution games appear to be quite different from Battle of the Sexes, but they are essentially the same. Both of them have two pure-strategy equilibria, ranked oppositely by the two players. In mathematical terms, the fact that contribution games have the equilibria in the southwest and northeast corners of the outcome matrix whereas coordination games have them in the northwest and southeast, is unimportant; the location of the equilibria could be changed by just switching the order of Row's strategies. We do view real situations differently, however, depending on whether players choose the same actions or different actions in equilibrium.

Let us take a look at a particular contribution game to show how to extend two-player games to games with several players. A notorious example in social psychology is the murder of Kitty Genovese, who was killed in New York City while 38 neighbors looked on without calling the police. Even so hardened an economist as myself finds it somewhat distasteful to call this a "game," but game theory does explain what happened. Let us use a slightly less appalling situation for our model. In The Civic Duty Game of table 3.6, Smith and Jones observe a burglary taking place. Each would like someone to call the police and stop the burglary, because having the burglary stopped adds 10 to his payoff, but neither wishes to make the call himself, because the effort subtracts 3. If Smith can be assured that Jones will call, Smith himself will ignore the burglary. Table 3.6 shows the payoffs.

The Civic Duty Game has two asymmetric pure-strategy equilibria and a symmetric mixed-strategy equilibrium. In solving for the mixed-strategy equilibrium, let us move from two players to N players. In the N-player version of the game, the payoff to Smith is 0 if nobody calls, 7 if he himself calls, and 10 if one or more of the other $N - 1$ players calls. This game also has an asymmetric pure-strategy and a symmetric mixed-strategy equilibrium. If all players use the same probability γ of *Ignore*, the probability that the other $N - 1$ players besides Smith all choose *Ignore* is γ^{N-1}, so the probability that one or more of them

Table 3.6 The Civic Duty Game

		Jones		
		Ignore (γ)		*Telephone* $(1 - \gamma)$
Smith:	*Ignore* (γ)	0,0	→	**10,7**
		↓		↑
	Telephone $(1 - \gamma)$	**7,10**	←	7,7

Payoffs to: (Smith, Jones)

chooses *Telephone* is $1 - \gamma^{N-1}$. Thus, equating Smith's pure-strategy payoffs using the payoff-equating method of equilibrium calculation yields

$$\pi_{Smith}\,(Telephone) = 7 = \pi_{Smith}\,(Ignore) = \gamma^{N-1}\,(0) + (1 - \gamma^{N-1})(10). \tag{16}$$

Equation (16) tells us that

$$\gamma^{N-1} = 0.3 \tag{17}$$

and

$$\gamma^* = 0.3^{\frac{1}{N-1}}. \tag{18}$$

If $N = 2$, Smith chooses *Ignore* with probability 0.30, and the probability that neither player phones the police is $\gamma^{*2} = 0.09$. As N increases, Smith's expected payoff remains constant at 7, since his expected payoff always equals his payoff from the pure strategy of *Telephone*. The probability γ^* of *Ignore*, however, increases with N. When there are more players, each player relies more on somebody else calling. The probability that nobody calls is γ^{*N}. Equation (17) shows that $\gamma^{*N-1} = 0.3$, so $\gamma^{*N} = 0.3\gamma^*$, which is increasing in N. When there are 38 players, the probability that nobody calls the police is about 0.29, because γ^* is about 0.97. The more people that watch a crime, the less likely it is to be reported.

Like The Prisoner's Dilemma, the disappointing result in The Civic Duty Game suggests a role for active policy. The mixed-strategy outcome is clearly bad. The expected payoff per player remains equal to 7 whether there is 1 player or 38, whereas if the equilibrium played out was the equilibrium in which one and only one player called the police, the average payoff would rise from 7 with one player to about 9.9 with 38 players (= [1(7) + 37(10]/38). A situation like this requires something to make one of the pure-strategy equilibria a focal point. The problem is divided responsibility. One person must be made responsible for calling the police, whether by tradition (e.g., the oldest person on the block always calls the police) or direction (e.g., Smith shouts to Jones: "Call the police!").

3.4 Randomizing versus Mixing: The Auditing Game

The next three games will illustrate the difference between mixed strategies and random actions, a subtle but important distinction. In all three games, the Internal Revenue Service must decide whether to audit a certain class of suspect tax returns to discover whether they are accurate or not. The goal of the IRS is to either prevent or catch cheating at minimum cost. The suspects want to cheat only if they will not be caught. Let us assume that the benefit of preventing or catching cheating is 4, the cost of auditing is C, where $C < 4$, the cost to the suspects of obeying the law is 1, and the cost of being caught is the fine $F > 1$.

Even with all of this information, there are several ways to model the situation. Table 3.7 shows one way: as a 2-by-2 simultaneous-move game.

Table 3.7 Auditing Game I

		Suspects		
		Cheat (θ)		*Obey* ($1-\theta$)
	Audit (γ)	$4-C,-F$	$\rightarrow$	$4-C,-1$
IRS:		$\uparrow$		$\downarrow$
	Trust ($1-\gamma$)	0,0	$\leftarrow$	$4,-1$

Payoffs to: (IRS, Suspects)

Auditing Game I is a discoordination game, with only a mixed strategy equilibrium. Equations (12) and (14) or the payoff-equating method tell us that

$$Probability(Cheat) = \theta^* = \frac{4-(4-C)}{(4-(4-C))+((4-C)-0)} = \frac{C}{4} \tag{19}$$

and

$$Probability(Audit) = \gamma^* = \frac{-1-0}{(-1-0)+(-F--1)} = \frac{1}{F}. \tag{20}$$

Using (19) and (20), the payoffs are

$$\pi_{IRS} = \pi_{IRS}(Trust) = \theta^*(0) + (1-\theta^*)(4) = 4-C. \tag{21}$$

and

$$\pi_{Suspect} = \pi_{Suspect}(Cheat) = \gamma^*(-F) + (1 - \gamma^*)(0) \tag{22}$$
$$= -1.$$

A second way to model the situation is as a sequential game. Let us call this Auditing Game II. The simultaneous game implicitly assumes that both players choose their actions without knowing what the other player has decided. In the sequential game, the IRS chooses government policy first, and the suspects react to it. The equilibrium in Auditing Game II is in pure strategies, a general feature of sequential games of perfect information. In equilibrium, the IRS chooses *Audit*, anticipating that the suspect will then choose *Obey*. The payoffs are $4 - C$ for the IRS and -1 for the suspects, the same for both players as in Auditing Game I, although now there is more auditing and less cheating and fine-paying.

We can go a step further. Suppose the IRS does not have to adopt a policy of auditing or trusting every suspect, but instead can audit a random sample. This is not necessarily a mixed strategy. In Auditing Game I, the equilibrium strategy was to audit all suspects with probability $1/F$ and none of them otherwise. That is different from the IRS announcing in advance that it will audit a random sample of $1/F$ of the suspects. For Auditing Game III, suppose the IRS moves first, but let its move consist of the choice of the proportion α of tax returns which are to be audited.

We know that the IRS is willing to deter the suspects from cheating, since it would be willing to choose $\alpha = 1$ and replicate the result in Auditing Game II if it had to. It chooses α so that

$$\pi_{suspect}(Obey) \geq \pi_{suspect}(Cheat), \tag{23}$$

i.e.,

$$-1 \geq \alpha(-F) + (1 - \alpha)(0). \tag{24}$$

In equilibrium, therefore, the IRS chooses $\alpha = 1/F$ and the suspects respond with *Obey*. The IRS payoff is $4 - \alpha C$, which is better than the $4 - C$ in the other two games, and the suspect's payoff is -1, exactly the same as before.

The equilibrium of Auditing Game III is in pure strategies, even though the IRS's action is random. It is different from Auditing Game I because the IRS must go ahead with the costly audit even if the suspect chooses *Obey*. Auditing Game III is different in another way also: its action set is continuous. In Auditing Game I and Auditing Game II the action set is {*Audit*, *Trust*}, although the strategy set becomes $\gamma \in [0, 1]$ once mixed strategies are allowed. In Auditing Game III, the action set is $\alpha \in [0, 1]$, and the strategy set would allow mixing of any of the elements in the action set, although mixed strategies are pointless for the IRS because the game is sequential.

Games with mixed strategies are like games with continuous strategies, because a probability is drawn from the continuum between zero and one. Auditing Game III makes this very clear, since it is a game with a strategy drawn from the continuous interval between zero and one. It illustrates one difference between mixed and continuous strategies, since the interpretation of the auditing probability in Auditing Game III was as an irreversible choice, whereas a mixed strategy is the result of a player's indifference between pure strategies. The next section will show another difference between mixed strategies and continuous strategies: although the payoffs are linear in the mixed-strategy probability, as is evident from the payoff equations (9) and (10), they can be nonlinear in continuous strategies more generally.

3.5 Continuous Strategies: The Cournot Game

With the exception of Auditing Game III, the actions in the games so far in the book have been discrete: *Aid* or *No Aid*, *Confess* or *Deny*. Quite often when strategies are discrete and moves are simultaneous, no pure-strategy equilibrium exists. The only sort of compromise possible in The Welfare Game, for instance, is to choose *Aid* sometimes and *No Aid* sometimes, a mixed strategy. If *A Little Aid* were a possible action, maybe there would be a pure-strategy equilibrium. The simultaneous-move game we discuss next, The Cournot Game, has a continuous strategy space even without mixing.

In presenting this game, new notation will be useful. If a game has a continuous strategy set, it is not always easy to depict the strategic form and outcome matrix as tables, or the extensive form as a tree. The tables would require a continuum of rows and columns, and the tree a continuum of branches. A new format for game descriptions of the players, actions, and payoffs will be used for the rest of the book. The new format will be similar to the way the rules of OPEC Model II were presented in section 1.1. The Cournot Game models a duopoly in which two firms choose output levels in competition with each other.

The Cournot Game

Players

Firms Apex and Brydox.

Order of Play

Apex and Brydox simultaneously choose quantities q_a and q_b from the set $[0, \infty)$.

Payoffs

Production costs are zero. Demand is a function of the total quantity sold, $Q = q_a + q_b$.

$$P(Q) = 120 - q_a - q_b. \tag{25}$$

Payoffs are profits, which are given by a firm's price times its quantity, i.e.,

$$\begin{aligned} \pi_{Apex} &= 120q_a - q_a^2 - q_a q_b \\ \pi_{Brydox} &= 120q_b - q_a q_b - q_b^2. \end{aligned} \tag{26}$$

Figure 3.1 Reaction Curves in The Cournot Game

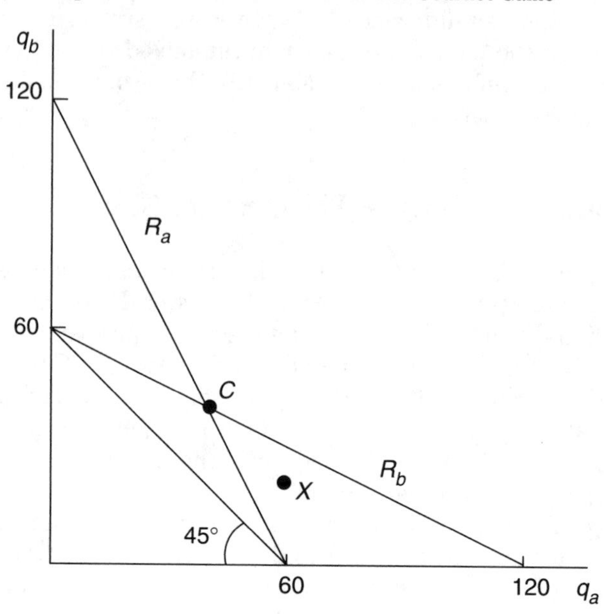

The format first assigns the game a title, after which it lists the players, the information classification, the order of play (together with who observes what), and the payoff functions. Listing the players and the information classification is redundant, strictly speaking, since they can be deduced from the order of play, but it is useful for letting the reader know what kind of model to expect. The format includes very little explanation; that is postponed, lest it obscure the description. This exact format is not standard in the literature, but every good article begins its technical section by specifying the same information, if in a less structured way, and the novice is strongly advised to use all the structure he can.

Cournot (1838) noted that this game has a unique equilibrium when demand curves are linear. If the game were cooperative (see section 1.2), firms would end up producing somewhere on the 45° line in figure 3.1 where total output is the monopoly output and maximizes the sum of the payoffs. More specifically, the monopoly output maximizes $PQ = (120 - Q)Q$ with respect to the total output of Q, resulting in the first-order condition

$$120 - 2Q = 0, \tag{27}$$

which implies a total output of 60 and a price of 60. Deciding how much of that output of 60 should be produced by each firm—where the firm's output should be located on the 45° line—would be a zero-sum cooperative game, an example of bargaining. But since The Cournot Game is noncooperative, the strategy profiles such that $q_a + q_b = 60$ are not necessarily equilibria despite their Pareto optimality.[2]

[2] Pareto optimality is defined from the viewpoint of the players. When total output is 60 neither of them can be made better off without hurting the other. If consumers were added to the game and side payments from consumers to firms were allowed, the monopoly output of 60 would be inefficient.

To find the Nash equilibrium, we need to refer to the **best response functions** for the two players. If Brydox produced 0, Apex would produce the monopoly output of 60. If Brydox produced $q_b = 120$ or greater, the market price would fall to zero and Apex would choose to produce zero. The best response function is found by maximizing Apex' payoff, given in equation (26), with respect to his strategy, q_a. This generates the first order condition $120 - 2q_a - q_b = 0$, or

$$q_a = 60 - q_b/2. \tag{28}$$

Another name for the best response function, the name usually used in the context of The Cournot Game, is the **reaction function**. Both names are somewhat misleading, since the players move simultaneously, with no chance to reply or react, but they are useful in imagining what a player would do if the rules of the game did allow him to move second. The reaction functions of the two firms are labelled R_a and R_b in figure 3.1. Where they cross, point C, is the **Cournot-Nash equilibrium**, which is simply the Nash equilibrium when the strategies consist of quantities. Algebraically, it is found by solving the two reaction functions for q_a and q_b, which generates the unique equilibrium, ($q_a = 40$, $q_b = 40$). The equilibrium price is also 40, coincidentally.

In The Cournot Game, the Nash equilibrium has the particularly nice property of **stability**: we can imagine how starting from some other strategy profile the players might reach the equilibrium. If the initial strategy profile is point X in figure 3.1, for example, Apex' best response is to decrease q_a and Brydox' is to increase q_b , which moves the profile closer to the equilibrium. But this is special to The Cournot Game, and Nash equilibria are not always stable in this way.

We might still be dissatisfied with the Cournot equilibrium. One problem is the assumption implicit in Nash equilibrium that Apex believes that if he changes q_a, Brydox will not respond by changing q_b, an assumption which might be questioned. Another objection is that the strategy sets are specified to be quantities. If the strategies are prices, rather than quantities, the Nash equilibrium is much different. Both objections are discussed in chapter 13. The Cournot model also comes up again in problem 4.1, which asks what happens when one firm's costs are positive and information is incomplete.

Stackelberg Equilibrium

There are many ways to model duopoly, but while we will defer discussion of most of them to chapter 13, we will make an exception for Stackelberg equilibrium. Stackelberg equilibrium differs from Cournot in that one firm gets to choose its quantity first. If Apex moved first, what output would it choose? Apex knows how Brydox will react to its choice, so it picks the point on Brydox' reaction curve that maximizes Apex' profit.

In a two-player game, the player moving first is the **Stackelberg leader** and the other player is the **Stackelberg follower**. The distinguishing characteristic of a Stackelberg equilibrium is that one player gets to commit himself first. In figure 3.2, Apex moves first intertemporally. If moves were simultaneous but Apex could commit himself to a certain strategy, the same equilibrium would be reached as long as Brydox was not able to commit himself. Algebraically, since

Figure 3.2 Stackelberg Equilibrium

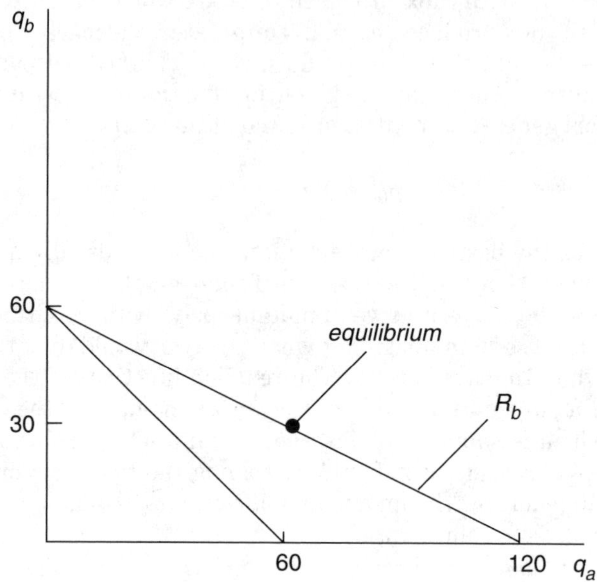

Apex forecasts Brydox' output to be $q_b = 60 - q_a/2$ from the analog of equation (28), Apex can substitute this into his payoff function in (26), obtaining

$$\pi_a = 120q_a - q_a^2 - q_a(60 - q_a/2). \tag{29}$$

Maximizing with respect to q_a yields the first-order condition

$$120 - 2q_a - 60 + q_a = 0, \tag{30}$$

which generates $q_a = 60$. Once Apex chooses this output, Brydox chooses his output to be $q_b = 30$. (That Brydox chooses exactly half the monopoly output is coincidental.) The market price is 30 for both firms, so Apex has benefited from his status as Stackelberg leader.

Notes

N3.1 Mixed Strategies: The Welfare Game

- For a very early reference to mixed strategies, see Waldegrave (1713).
- The January 1992 issue of *Rationality and Society* is devoted to attacks on and defenses of the use of game theory in the social sciences, with considerable discussion of mixed strategies and multiple equilibria. Contributors include Harsanyi, Myerson, Rapaport, Tullock, and Wildavsky. The Spring 1989 issue of the *RAND Journal of Economics* has an exchange on the use of game theory between Franklin Fisher and Carl Shapiro. For an attack on the game theory approach to industrial organization, see Peltzman (1991).
- In this book it will always be assumed that players remember their previous moves. Without this assumption of **perfect recall**, the definition in the text is not that for a mixed strategy, but for a **behavior strategy**. As historically defined, a player pursues a mixed strategy when he

randomly chooses between pure strategies at the starting node, but he plays a pure strategy thereafter. Under that definition, the modeller cannot talk of random choices at any but the starting node. Kuhn (1953) showed that the definition of mixed strategy given in the text is equivalent to the original definition if the game has perfect recall. Since all important games have perfect recall and the new definition of mixed strategy is better in keeping with the modern spirit of sequential rationality, I have abandoned the old definition.

The classic example of a game without perfect recall is **bridge**, where the four players of the actual game can be cutely modelled as two players who forget what half their cards look like at any one time in the bidding. A more useful example is a game that has been simplified by restricting players to Markov strategies (see section 5.4), but usually the modeller sets up such a game with perfect recall and then rules out non-Markov equilibria after showing that the Markov strategies form an equilibrium for the general game.

- For more examples of calculating mixed-strategy equilibria, see sections 4.6, 5.2, 11.5, 13.2, and 14.2. The model in section 14.1 of a patent race shows how to compute a continuous mixed-strategy distribution, giving the probability that each pure strategy on a continuum is played.
- It is *not* true that when two pure-strategy equilibria exist a player would be just as willing to use a strategy mixing the two even when the other player is using a pure strategy. In Battle of the Sexes, for instance, if the man knows the woman is going to the ballet he is not indifferent between the ballet and the prize fight.
- A continuum of players is useful not only because the modeller need not worry about fractions of players, but because he can use more modelling tools from calculus—taking the integral of the quantities demanded by different consumers, for example, rather than the sum. But using a continuum is also mathematically more difficult: see Aumann (1964a, 1964b).

N3.2 Chicken, The War of Attrition, and Correlated Strategies

- Papers on the war of attrition include Fudenberg & Tirole (1986b), Ghemawat & Nalebuff (1985), Maynard Smith (1974), Nalebuff & Riley (1985), and Riley (1980).
- The game of Chicken discussed in the text is simpler than the game acted out in the movie *Rebel Without a Cause*, in which the players race towards a cliff and the winner is the player who jumps out of his car last. The pure-strategy space in the movie game is continuous and the payoffs are discontinuous at the cliff's edge, which makes the game more difficult to analyze technically. (Looking ahead to section 4.1, recall the importance of a "tremble" in the movie.)
- Technical difficulties arise in some models with a continuum of actions and mixed strategies. In The Welfare Game, the government chose a single number, a probability, on the continuum from zero to one. If we allowed the government to mix over a continuum of aid levels, it would choose a function, a probability density, over the continuum. The original game has a finite number of elements in its strategy set, so its mixed extension still has a strategy space in $\mathbf{R}^n$. But with a continuous strategy set extended by a continuum of mixed strategies for each pure strategy, the mathematics becomes difficult. A finite number of mixed strategies can be allowed without much problem, but usually that is not satisfactory.

  Games in continuous time frequently run into this problem. Sometimes it can be avoided by clever modelling, as in Fudenberg & Tirole's (1986b) continuous-time war of attrition with asymmetric information. They specify as strategies the length of time firms would proceed to *Continue* given their beliefs about the type of the other player, in which case there is a pure-strategy equilibrium.

N3.4 Randomizing versus Mixing: The Auditing Game

- Auditing Game I is similar to a game called The Police Game. Care must be taken in such games that one does not use a simultaneous-move game when a sequential game is appropriate. Also, discrete strategy spaces can be misleading. In general, economic analysis assumes that costs rise convexly in the amount of an activity and benefits rise concavely. Modelling a situation with a 2-by-2 game uses just two discrete levels of the activity, so the concavity or convexity is lost in the simplification. If the true functions are linear, as in auditing costs which rise linearly with the probability of auditing, this is no great loss. If the true costs rise convexly, as in the case where the hours a policeman must stay on the street each day are increased, then a 2-by-2 model can be misleading. Be especially careful not to press the idea of a mixed-strategy equilibrium too hard if a pure-strategy equilibrium would exist when intermediate strategies are allowed. See Tsebelis (1989) and the criticism of it in J. Hirshleifer & Rasmusen (1992).

- D. Diamond (1984) shows the implications of monitoring costs for the structure of financial markets. A fixed cost to monitoring investments motivates the creation of a financial intermediary to avoid repetitive monitoring by many investors.
- Baron & Besanko (1984) studies auditing in the context of a government agency which can, at some cost, collect information on the true production costs of a regulated firm.
- Mookherjee & Png (1989) and Border & Sobel (1987) have examined random auditing in the context of taxation. They find that if a taxpayer is audited he ought to be more than compensated for his trouble, if it turns out he was telling the truth. Under the optimal contract, the truth-telling taxpayer should be delighted to hear that he is being audited. The reason is that a reward for truthfulness widens the differential between the agent's payoff when he tells the truth versus occasions when he lies.
- Government action strongly affects what information is available as well as what is contractible. In 1988, for example, the United States passed a law sharply restricting the use of lie detectors for testing or monitoring. Previous to the restriction, about two million workers had been tested each year. ("Law Limiting Use of Lie Detectors is Seen Having Widespread Effect," *Wall Street Journal*, July 1, 1988, p. 13).

N3.5 Continuous Strategies: The Cournot Game

- An interesting class of simple continuous-payoff games are the **Colonel Blotto games** (Tukey [1949]). In these games, two military commanders allocate their forces to m different battlefields, and a battlefield contributes more to the payoff of the commander with the greater forces there. A distinguishing characteristic is that player i's payoff increases with the value of player i's particular action relative to player j's, and i's actions are subject to a budget constraint. Except for the budget constraint, this is similar to the tournament in section 8.5.
- Considerable work has been done characterizing the Cournot model. A recent article is Gaudet & Salant (1991) on conditions which ensure a unique equilibrium.
- **Differential Games** are played in continuous time. The action is a function describing the value of a state variable at each instant, so the strategy maps the game's past history to such a function. Differential games are solved using dynamic optimization. A book-length treatment is Bagchi (1984).
- Fudenberg & Levine (1986) show circumstances under which the equilibria of games with infinite strategy spaces can be found as the limits of equilibria of games with finite strategy spaces.
- "Stability" is a word used in many different ways in game theory and economics. The natural meaning of a stable equilibrium is that it has dynamics which cause the system to return to that point after being perturbed slightly, and the discussion of the stability of Cournot equilibrium was in that spirit. The uses of the term by von Neumann & Morgenstern (1944) and Kohlberg & Mertens (1986) are entirely different.
- The term, "Stackelberg equilibrium," is not clearly defined in the literature. It is sometimes used to denote equilibria in which players take actions in a given order, but since that is just the perfect equilibrium (see section 4.1) of a well-specified extensive form, I prefer to reserve the term for the Nash equilibrium of the duopoly quantity game in which one player moves first, which is the context of Stackelberg (1934).

  An alternative definition is that a Stackelberg equilibrium is a strategy profile in which players select strategies in a given order and in which each player's strategy is a best response to the fixed strategies of the players preceding him and the yet-to-be-chosen strategies of players succeeding him, i.e., a situation in which players precommit to strategies in a given order. Such an equilibrium would not generally be either Nash or perfect.
- Stackelberg (1934) suggested that sometimes the players are confused about which of them is the leader and which the follower, resulting in the disequilibrium outcome called **Stackelberg warfare**.
- With linear costs and demand, total output is greater in the Stackelberg equilibrium than in Cournot. The slope of the reaction curve is less than one, so Apex' output expands more than Brydox' contracts. Total output being greater, the price is less than in the Cournot equilibrium.

- A useful application of Stackelberg equilibrium is to an industry with a dominant firm and a **competitive fringe** of smaller firms that sell at capacity if the price exceeds their marginal cost. These smaller firms act as Stackelberg leaders (not followers), since each is small enough to ignore its effect on the behavior of the dominant firm. The oil market could be modelled this way with OPEC as the dominant firm and producers such as Britain on the fringe.

Problems

3.1: Presidential Primaries.

Smith and Jones are fighting it out for the Democratic nomination for President of the United States. The more months they keep fighting, the more money they spend, because a candidate must spend one million dollars a month in order to stay in the race. If one of them drops out, the other one wins the nomination, which is worth 11 million dollars. The discount rate is r per month. To simplify the problem, you may assume that this battle could go on forever if neither of them drops out. Let θ denote the probability that an individual player will drop out each month in the mixed-strategy equilibrium.

(3.1a) In the mixed-strategy equilibrium, what is the probability θ each month that Smith will drop out? What happens if r changes from 0.1 to 0.15?

(3.1b) What are the two pure-strategy equilibria?

(3.1c) If the game lasts only one period, and the Republican wins the general election (for Democrat payoffs of zero) if both Democrats refuse to exit, what is the probability γ with which each candidate exits in a symmetric equilibrium?

3.2: Running from the Gestapo.

Two risk-neutral men, Schmidt and Braun, are walking south along a street in Nazi Germany when they see a single Gestapo agent coming to check their papers. Only Braun has his papers (unknown to the Gestapo, of course). The Gestapo agent will catch both men if both or neither of them run north, but if just one runs, he must choose which one to stop—the walker or the runner. The penalty for being without papers is 24 months in prison. The penalty for running away from an agent of the state is 24 months in prison, on top of the sentences for any other charges, but the conviction rate for this offense is only 25 percent. The two friends want to maximize their joint welfare, which the Gestapo man wants to minimize. Braun moves first, then Schmidt, then the Gestapo.

(3.2a) What is the outcome matrix for outcomes that might be observed in equilibrium? (Use θ for the probability that the Gestapo chases the runner and γ for the probability that Braun runs.)

(3.2b) What is the probability that the Gestapo agent chases the runner (call it θ^*)?

(3.2c) What is the probability that Braun runs (call it γ^*)?

(3.2d) Since Schmidt and Braun share the same objectives, is this a cooperative game?

3.3: Uniqueness in Matching Pennies.

In the game Matching Pennies, Smith and Jones each show a penny with either heads or tails up. If they choose the same side of the penny, Smith gets both pennies; otherwise, Jones gets them.
(3.3a) Draw the outcome matrix for Matching Pennies.
(3.3b) Show that there is no Nash equilibrium in pure strategies.
(3.3c) Find the mixed-strategy equilibrium, denoting Smith's probability of *Heads* by γ and Jones' by θ.
(3.3d) Prove that there is only one mixed-strategy equilibrium.

3.4: Mixed Strategies in Battle of the Sexes.

Refer back to Battle of the Sexes and Ranked Coordination in section 1.4. Denote the probabilities that the man and woman pick *Prize Fight* by γ and θ.
(3.4a) Find an expression for the man's expected payoff.
(3.4b) Find the first-order condition for the man's choice of strategy.
(3.4c) What are the equilibrium values of γ and θ, and the expected payoffs?
(3.4d) Find the most likely outcome and its probability.
(3.4e) What is the equilibrium payoff in the mixed-strategy equilibrium for Ranked Coordination?
(3.4f) Why is the mixed-strategy equilibrium a better focal point in Battle of the Sexes than in Ranked Coordination?

3.5: A Voting Paradox.

Adam, Charles, and Vladimir are the only three voters in Podunk. Only Adam owns property. There is a proposition on the ballot to tax property-holders 120 dollars and distribute the proceeds equally among all citizens who do not own property. Each citizen dislikes having to go to the polling place and vote (despite the short lines), and would pay 20 dollars to avoid voting. They all must decide whether to vote before going to work. The proposition fails if the vote is tied. Assume that in equilibrium Adam votes with probability θ and Charles and Vladimir each vote with the same probability γ, but they decide to vote independently of each other.
(3.5a) What is the probability that the proposition will pass, as a function of θ and γ?
(3.5b) What are the two possible equilibrium probabilities γ_1 and γ_2 with which Charles might vote? Why, intuitively, are there two symmetric equilibria?
(3.5c) What is the probability θ that Adam will vote in each of the two symmetric equilibria?
(3.5d) What is the probability that the proposition will pass?

3.6: Alba and Rome: Asymmetric information and mixed strategies.

A Roman, Horatius, unwounded, is fighting the three Curatius brothers from Alba, each of whom is wounded. If Horatius continues fighting, he wins with probability 0.1, and the payoffs are (10, −10) for (Horatius, Curatii) if he wins, and (−10, 10) if he loses. With probability $\alpha = 0.5$, Horatius is panic-stricken and runs away. If he runs and the Curatii do not chase him, the payoffs are (−20, 10). If he runs and the Curatius brothers chase and kill him, the payoffs are (−21, 20). If, however, he is not panic-stricken, but he runs anyway and the Curatii give chase, he is able to kill the fastest brother first and thereafter dispose of the other two, for payoffs of (10, −10). Horatius is, in fact, not panic-stricken.

(3.6a) With what probability θ would the Curatii give chase if Horatius were to run?

(3.6b) With what probability γ does Horatius run?

(3.6c) How would θ and γ be affected if the Curatii falsely believed that the probability of Horatius being panic-stricken was 1? What if they believed it was 0.9?

4 Dynamic Games with Symmetric Information

4.1 Subgame Perfectness

In this chapter we will make heavy use of the extensive form to study games with moves that occur in sequence. We start in section 4.1 with a refinement of the Nash equilibrium concept called perfectness that incorporates sensible implications of the order of moves. Perfectness is illustrated in section 4.2 with a game of entry deterrence. Section 4.3 expands on the idea of perfectness using the example of nuisance suits, meritless lawsuits brought in the hopes of obtaining a settlement out of court. Nuisance suits show the importance of a threat being made credible and how sinking costs early or having certain nonmonetary payoffs can benefit a player. This example will also be used to discuss the open-set problem of weak equilibria in games with continuous strategy spaces, in which a player offering a contract chooses its terms to make the other player indifferent about accepting or rejecting. The last perfectness topic will be renegotiation: the idea that when there are multiple perfect equilibria, the players will coordinate on equilibria that are Pareto optimal in subgames but not in the game as a whole. Section 4.5 turns to the completely different topic of discounting payments across time, and section 4.6 is a glimpse into the biological approach to game theory.

The Perfect Equilibrium of Follow the Leader I

Subgame perfectness is an equilibrium concept based on the ordering of moves and the distinction between an equilibrium path and an equilibrium. The **equilibrium path** is the path through the game tree that is followed in equilibrium, but the equilibrium itself is a strategy profile, which includes the players' responses to other players' deviations from the equilibrium path. These off-equilibrium responses are crucial to decisions on the equilibrium path. A threat, for example, is a promise to carry out a certain action if another player deviates from his equilibrium actions, and it has an influence even if it is never used.

Perfectness is best introduced with an example. In section 2.1, a flaw of Nash equilibrium was revealed in the game Follow the Leader I, which has three pure strategy Nash equilibria of which only one is reasonable. The players are Smith and Jones, who choose disk sizes. Both their payoffs are greater if they choose the same size and greatest if they coordinate on *Large*. Smith moves first, so his strategy set is (*Small*, *Large*). Jones' strategy is more complicated, because it must specify an action for each information set, and Jones' information set depends on what Smith chose. A typical element of Jones' strategy set is (*Large*, *Small*), which specifies that he chooses *Large* if Smith chose *Large*, and *Small* if Smith chose *Small*. From the strategic form we found the following three Nash equilibria.

Equilibrium	Strategies	Outcome
E_1	{*Large*, (*Large*, *Large*)}	Both pick *Large*
E_2	{*Large*, (*Large*, *Small*)}	Both pick *Large*
E_3	{*Small*, (*Small*, *Small*)}	Both pick *Small*

Only Equilibrium E_2 is reasonable, because the order of the moves should matter to the decisions which the players make. The problem with the strategic form, and, thus, with simple Nash equilibrium, is that it ignores who moves first. Smith moves first, and it seems reasonable that Jones should be allowed—in fact should be required—to rethink his strategy after Smith moves.

Consider Jones' strategy of (*Small*, *Small*) in equilibrium E_3. If Smith deviated from equilibrium by choosing *Large*, it would be unreasonable for Jones to stick to the response *Small*. Instead, he should also choose *Large*. But if Smith expected a response of *Large*, he would have chosen *Large* in the first place, and E_3 would not be an equilibrium. A similar argument shows that it would be irrational for Jones to choose (*Large*, *Large*), and we are left with E_2 as the unique equilibrium.

We say that equilibria E_1 and E_3 are Nash equilibria but not "perfect" Nash equilibria. A strategy profile is a perfect equilibrium if it remains an equilibrium on all possible paths, includes not only the equilibrium path but all the other paths, which branch off into different "subgames."

A **subgame** *is a game consisting of a node which is a singleton in every player's information partition, that node's successors, and the payoffs at the associated end nodes.*[1]

A strategy profile is a **subgame perfect Nash equilibrium** *if (a) it is a Nash equilibrium for the entire game; and (b) its relevant action rules are a Nash equilibrium for every subgame.*

The extensive form of Follow the Leader I in figure 4.1 (a reprise of figure 2.1) has three subgames: (1) the entire game, (2) the subgame starting at node J_1, and (3) the subgame starting at node J_2. Strategy profile E_1 is not a subgame perfect equilibrium, because it is only Nash in subgames (1) and (3), not in subgame (2). Strategy profile E_3 is not a subgame perfect equilibrium, because it is only Nash

[1] Technically, this is a *proper* subgame because of the information qualifier, but no economist is so ill-bred as to use any other kind of subgame.

Figure 4.1 Follow the Leader I

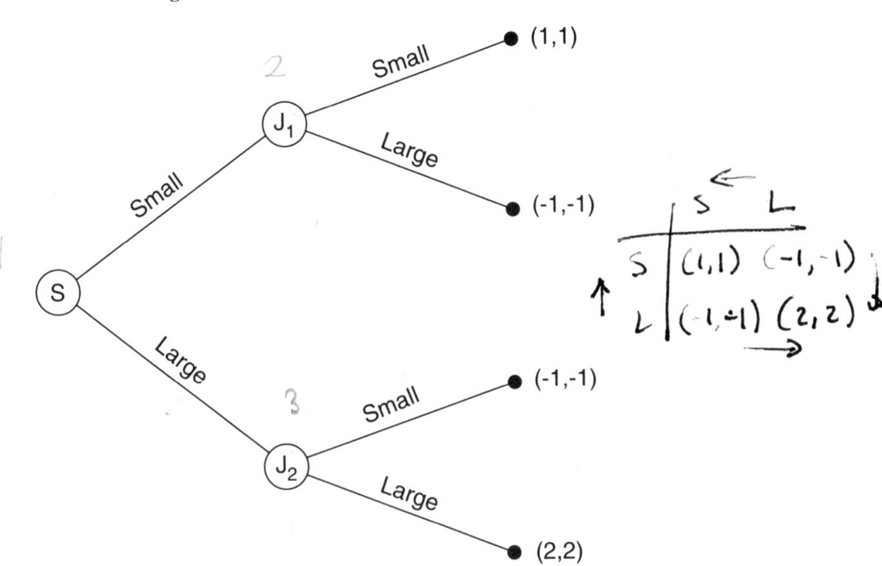

in subgames (1) and (2), not in subgame (3). But strategy profile E_2 is Nash in all three subgames.

The term **sequential rationality** is often used to denote the idea that a player should maximize his payoffs at each point in the game, re-optimizing his decisions at each point and taking into account the fact that he will re-optimize in the future. This is a blend of the economic ideas of ignoring sunk costs and rational expectations. Sequential rationality is so standard a criterion for equilibrium now that often I will speak of "equilibrium" without the qualifier when I wish to refer to an equilibrium that satisfies sequential rationality in the sense of being a "subgame perfect equilibrium" or, in a game of asymmetric information, a "perfect Bayesian equilibrium."

One reason why perfectness (the word "subgame" is usually left off) is a good equilibrium concept is because it represents the idea of sequential rationality. A second reason is that a weak Nash equilibrium is not robust to small changes in the game. So long as he is certain that Smith will not choose *Large*, Jones is indifferent between the never-to-be-used responses (*Small* if *Large*) and (*Large* if *Large*). Equilibria E_1, E_2, and E_3 are all weak Nash equilibria because of this. But if there is even a small probability that Smith will choose *Large*—perhaps by mistake—then Jones would prefer the response (*Large* if *Large*), and equilibria E_1 and E_3 are no longer valid. Perfectness is a way to eliminate some of these less robust weak equilibria. The small probability of a mistake is called a **tremble**, and section 6.1 returns to this **trembling hand** approach as one way to extend the notion of perfectness to games of asymmetric information.

For the moment, however, the reader should note that the tremble approach is distinct from sequential rationality. Consider The Tremble Game in figure 4.2. This game has three Nash equilibria, all weak: (*Out*, *Down*), (*Out*, *Up*), and (*In*, *Up*). Only (*Out*, *Up*) and (*In*, *Up*) are subgame perfect, because although *Down*

Figure 4.2 The Tremble Game: Trembling Hand versus Subgame Perfectness

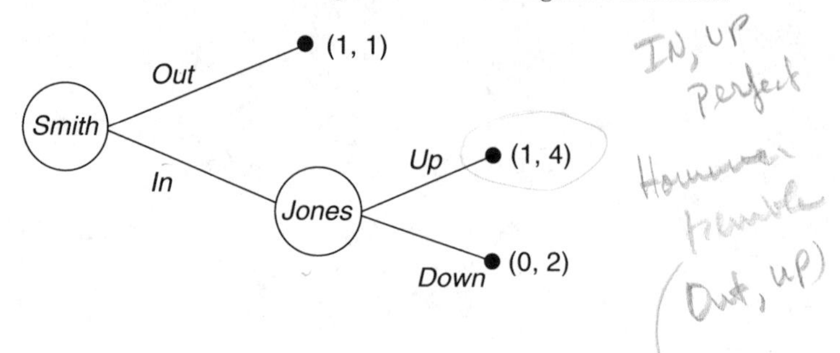

is weakly Jones' best response to Smith's *Out*, it is inferior if Smith chooses *In*. In the subgame starting with Jones' move, the only subgame perfect equilibrium is for Jones to choose *Up*. The possibility of trembles, however, rules out (*In*, *Up*) as an equilibrium. If Jones has even an infinitesimal chance of trembling and choosing *Down*, Smith will choose *Out* instead of *In*. Also, Jones will choose *Up*, not *Down*, because if Smith trembles and chooses *In*, Jones prefers *Up* to *Down*. This leaves only (*Out*, *Up*) as an equilibrium, despite the fact that it is weakly Pareto dominated by (*In*, *Up*).

4.2 An Example of Perfectness: Entry Deterrence I

We turn now to a game in which perfectness plays a role just as important as in Follow the Leader I but in which the players are in conflict. An old question in industrial organization is whether an incumbent monopolist can maintain his position by threatening to wage a price war against any new firm that enters the market. This idea was heavily attacked by Chicago School economists such as McGee (1958) on the grounds that a price war would hurt the incumbent more than collusion with the entrant. Game theory can present this reasoning very cleanly. Let us consider a single episode of possible entry and price warfare, which nobody expects to be repeated. We will assume that even if the incumbent chooses to collude with the entrant, maintaining a duopoly is difficult enough so that market revenue drops considerably.

Entry Deterrence I

Players

Two firms, the entrant and the incumbent.

Order of Play

(1) The entrant decides whether to *Enter* or *Stay Out*.
(2) If the entrant enters, the incumbent can *Collude* with him, or *Fight* by cutting the price drastically.

Payoffs

Market profits are 300 at the monopoly price and 0 at the fighting price. Entry costs are 10. Duopoly competition reduces market revenue to 100, which is split evenly.

The strategy sets can be discovered from the order of play. They are {*Enter*, *Stay Out*} for the entrant, and {*Collude* if entry occurs, *Fight* if entry occurs} for the incumbent. The game has the two Nash equilibria indicated in boldface in table 4.1, (*Enter*, *Collude*) and (*Stay Out*, *Fight*). The equilibrium (*Stay Out*, *Fight*) is weak, because the incumbent would just as soon *Collude* given that the entrant is staying out.

Table 4.1 Entry Deterrence I

		Incumbent		
		Collude		*Fight*
Entrant:	*Enter*	**40, 50**	←	−10, 0
		↑		↓
	Stay Out	0,300	↔	**0,300**

Payoffs to: (Entrant, Incumbent)

A piece of information has been lost by condensing from the extensive form, figure 4.3, to the strategic form, table 4.1, i.e., the fact that the entrant gets to move first. Once he has chosen *Enter*, the incumbent's best response is *Collude*. The threat to fight is not credible and would be employed only if the incumbent

Figure 4.3 Entry Deterrence I

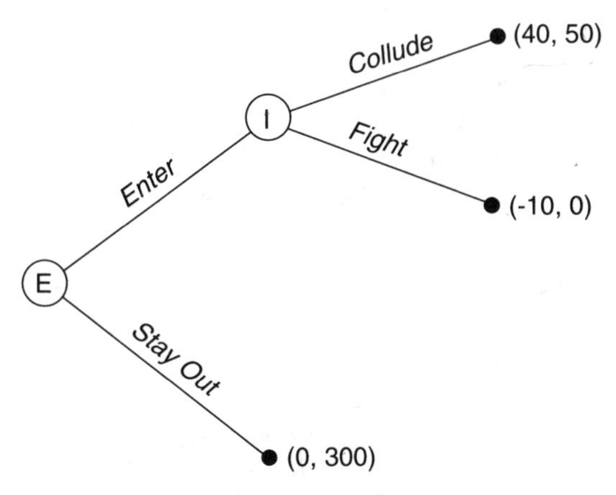

Payoffs to: (Entrant, Incumbent)

could bind himself to fight, in which case he never does fight, because the entrant chooses to stay out. The equilibrium (*Stay Out*, *Fight*), is Nash but not subgame perfect, because if the game is started after the entrant has already entered, the incumbent's best response is *Collude*. This does not prove that collusion is inevitable in duopoly, but it is the equilibrium for Entry Deterrence I.

The trembling hand interpretation of perfect equilibrium can be used here. So long as it is certain that the entrant will not enter, the incumbent is indifferent between *Fight* and *Collude*, but if there were even a small probability of entry—perhaps because of a lapse of good judgement by the entrant—the incumbent would prefer *Collude* and the Nash equilibrium would be broken.

Perfectness rules out threats that are not credible. Entry Deterrence I is a good example because if a communication move were added to the game tree, the incumbent might tell the entrant that entry would be followed by fighting, but the entrant would ignore this noncredible threat. If, however, some means existed by which the incumbent could precommit himself to fight entry, the threat would become credible. The next section will look at one context, nuisance lawsuits, in which such precommitment might be possible.

Should the Modeller Ever Use Nonperfect Equilibria?

A game in which a player can commit himself to a strategy can be modelled in two ways:

(1) As a game in which nonperfect equilibria are acceptable, or
(2) By changing the game to replace the action *Do X* with *Commit to Do X* at an earlier node.

An example of (2) in Entry Deterrence I is to reformulate the game so that the incumbent moves first, deciding in advance whether or not to choose *Fight* before the entrant moves. Approach (2) is better than (1) because if the modeller wants to let players commit to some actions and not to others, he can do this by carefully specifying the order of play. Allowing equilibria to be nonperfect forbids such discrimination and multiplies the number of equilibria. Indeed, the problem with subgame perfectness is not that it is too restrictive but that it still allows too many strategy profiles to be equilibria in games of asymmetric information. A subgame must start at a single node and not cut across any player's information set, so often the only subgame will be the whole game and subgame perfectness does not restrict equilibrium at all. Section 6.1 discusses perfect Bayesian equilibrium and other ways to extend the perfectness concept to games of asymmetric information.

4.3 Credible Threats, Sunk Costs, and the Open-Set Problem in the Game of Nuisance Suits

Like the related concepts of sunks costs and rational expectations, sequential rationality is a simple idea with tremendous power. This section will show that power in another simple game, one which models nuisance suits. We have already come across one application of game theory to law, the Png (1983) model of section 2.5. In some ways, law is particularly well suited to analysis by game theory because the legal process is so concerned with conflict and the provision of definite rules to regulate that conflict. In what other field could an article be titled: "An Economic Analysis of Rule 68"? (Miller [1986], a discussion of a federal rule of procedure that can penalize a litigant if he refuses to accept a settlement offer.) The growth in the area can be seen by comparing the overview in the Ayres (1990) review of the first edition of the present book with the entire book by Baird, Gertner & Picker (forthcoming). In law, even more clearly than in business, a major objective is to avoid inefficient outcomes by restructuring the rules, and nuisance suits are one of the inefficiencies that a good policymaker hopes to eliminate.

Nuisance suits are lawsuits with little chance of success, whose only possible purpose seems to be the hope of a settlement out of court. In the context of entry deterrence, people commonly think that large size is an advantage, and that the large incumbent will threaten the small entrant, but in the contex of nuisance suits people commonly think that large size is a disadvantage and a wealthy individual or corporation is vulnerable to extortionary litigation. Nuisance Suits I models the essentials of the situation: bringing suit is costly and has little chance of success, but because defending the suit is also costly, the defendant might pay a generous amount to settle it out of court. The model is similar to the Png Settlement Game of chapter 2 in many respects, but here the model will be one of symmetric information and we will make explicit the sequential rationality requirement that was implicit the discussion in chapter 2.

Nuisance Suits I: Simple Extortion

Players

A plaintiff and a defendant.

Order of Play

(1) The plaintiff decides whether to bring suit against the defendant, at cost c.
(2) The plaintiff makes a take-it-or-leave-it settlement offer of $s>0$.
(3) The defendant accepts or rejects the settlement offer.
(4) If the defendant rejects the offer, the plaintiff decides whether to give up or go to trial at a cost p to himself and a cost d to the defendant.
(5) If the case goes to trial, the plaintiff wins amount x with probability γ and otherwise wins nothing.

Payoffs

Figure 4.4 shows the payoffs. Let $\gamma x < p$, so the plaintiff's expected winnings are less than his marginal cost of going to trial.

Figure 4.4 The Extensive Form for the Game of Nuisance Suits

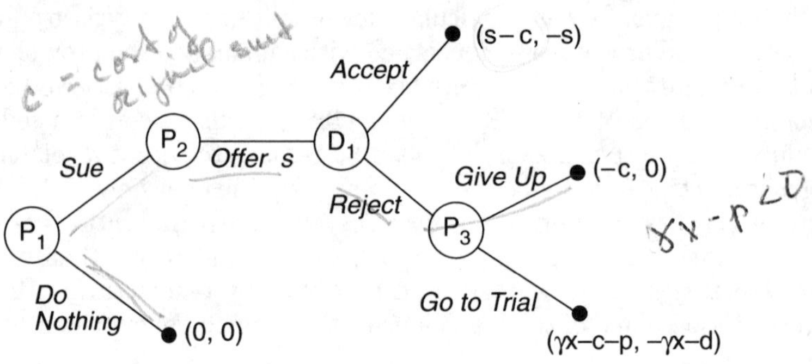

The perfect equilibrium is

> Plaintiff: *Do nothing*, *Offer s*, *Give up*
> Defendant: *Reject*
> Outcome: The plaintiff does not bring a suit.

The equilibrium settlement offer s can be any positive amount. Note that the equilibrium specifies actions at all four nodes of the game, even though only the first is reached in equilibrium.

To find a perfect equilibrium the modeller starts at the end of the game tree, following the advice of Dixit & Nalebuff (1991, p. 34) to "Look ahead and reason back." At node P_3, the plaintiff will choose *Give up*, since by assumption $\gamma x - c - p < -c$. This is because the suit is brought only in the hope of settlement, not in the hope of winning at trial. At node D_1, the defendant, foreseeing that the plaintiff will give up, rejects any positive settlement offer. This makes the plaintiff's offer at P_2 irrelevant, and, looking ahead to a payoff of $-c$ from choosing *Sue* at P_1, the plaintiff chooses *Do nothing*.

Thus, if nuisance suits are brought, it must be for some reason other than the obvious one, the plaintiff's hope of extracting a settlement offer from a defendant who wants to avoid trial costs. This is fallacious because the plaintiff himself bears trial costs and hence cannot credibly make the threat. It is fallacious even if the defendant's legal costs would be much higher than the plaintiff's (d much bigger than p), because the relative size of the costs does not enter into the argument.

One might wonder how risk aversion affects this conclusion. Might not the defendant settle because he is more risk averse than the plaintiff? That is a good question, but Nuisance Suits I can be adapted to risk-averse players with very little change. Risk would enter at the trial stage, as a final move by Nature to decide who wins. In Nuisance Suits I, γx represented the expected value of the award. If both the defendant and the plaintiff are equally risk averse, γx can still represent the expected payoff from the award—one simply interprets x and 0 as the utility of the cash award and the utility of an award of 0, rather than as the actual cash amounts. If the players have different degrees of risk aversion, the expected loss to the defendant is not the same as the expected gain to the plaintiff, and the payoffs must be adjusted. If the defendant is more risk averse, the payoffs from *Go to trial* would change to $(-c - p + \gamma x, -\gamma x - y - d)$, where y represents the extra disutility of risk to the defendant. This, however, makes no difference to the equilibrium. The crux of the game is that the plaintiff is unwilling to go to trial because of the cost to himself, and the cost to the defendant, including the cost of bearing risk, is irrelevant.

If nuisance suits are brought, it must therefore be for some more complicated reason. Already, in chapter 2, we looked at one reason for litigation to reach trial in the Png Settlement Game: incomplete information. That is probably the most important explanation and it has been much studied, as can be seen from the surveys by Cooter & Rubinfeld (1989) and Kennan and R. Wilson (1993). In this section, though, let us confine ourselves to explanations where the probability of a suit's success is common knowledge. Even then, costly threats might be credible because of sinking costs strategically (Nuisance Suits II), or because of the nonmonetary payoffs resulting from going to trial (Nuisance Suits III).

Nuisance Suits II: Using Sunk Costs Strategically[2]

Let us now modify the game so that the plaintiff can pay his lawyer the amount p in advance, with no refund if the case settles. This inability to obtain a refund actually helps the plaintiff, by changing the payoffs from the game so his payoff from *Give up* is $-c - p$, compared to $-c - p + \gamma x$ from *Go to trial*. Having sunk the legal costs, he will go to trial if $\gamma x > 0$—that is, if he has any chance of success at all.

This, in turn, means that the plaintiff would only prefer settlement to trial if $s > \gamma x$. The defendant would prefer settlement to trial if $s < \gamma x + d$, so there is a positive **settlement range** of $[\gamma x, \gamma x + d]$ within which both players are willing to settle. The exact amount of the settlement depends on the bargaining power of the parties, something to be examined in chapter 11. Here, allowing the plaintiff to make a take-it-or-leave-it offer means that $s = \gamma x + d$ in equilibrium, and if $\gamma x + d > p + c$, the nuisance suit will be brought even though $\gamma x < p + c$. Thus, the plaintiff is bringing the suit only because he can extort d, the amount of the defendant's legal costs.

Even though the plaintiff can now extort a settlement, he does it at some cost to himself, so an equilibrium with nuisance suits will require that

$$-c - p + \gamma x + d \geq 0 \tag{1}$$

If inequality (1) is false, then, even if the plaintiff could extract the maximum possible settlement of $s = \gamma x + d$, he would not do so, because he would then have to pay $c + p$ before reaching the settlement stage. This implies that a totally meritless suit (with $\gamma = 0$), would not be brought unless the defendant had higher legal costs than the plaintiff ($d > p$). If inequality (1) is satisfied, however, the following strategy profile is a perfect equilibrium:

[2] The inspiration for this model is Rosenberg & Shavell (1985).

Plaintiff: *Sue, Offer* $s = \gamma x + d$, *Go to trial*
Defendant: *Accept* $s \leq \gamma x + d$
Outcome: Plaintiff sues and offers to settle, to which the defendant agrees.

An obvious counter to the plaintiff's ploy would be for the defendant to also sink his costs, by paying d before the settlement negotiations, or even before the plaintiff decides to file suit. Perhaps this is one reason why large corporations use in-house counsel, who are paid a salary regardless of how many hours they work, as well as outside counsel, hired by the hour. If so, nuisance suits cause a social loss—the wasted time of the lawyers, d—even if nuisance suits are never brought, just as aggressor nations cause social loss in the form of world military expenditure even if they never start a war.[3]

Two problems, however, face the defendant who tries to sink the cost d. First, although it saves him γx if it deters the plaintiff from filing suit, it also means the defendant must pay the full amount d. This is worthwhile if the plaintiff has all the bargaining power, as in Nuisance Suits II, but it might not be if s lay in the middle of the settlement range because the plaintiff was not able to make a take-it-or-leave-it offer. If settlement negotiations resulted in s lying exactly in the middle of the settlement range, so $s = \gamma x + d/2$, then it might not be worthwhile for the defendant to sink d to deter nuisance suits that would settle for $\gamma x + d/2$.

Second, there is an asymmetry in litigation: the plaintiff has the choice of whether to bring suit or not. Since it is the plaintiff who has the initiative, he can sink p and make the settlement offer before the defendant has the chance to sink d. The only way for the defendant to avoid this is to pay d well in advance, in which case the expenditure is wasted if no possible suits arise. What the defendant would like best would be to buy legal insurance which, for a small premium, would pay all defense costs in future suits that might occur. As we will see in chapters 7 and 9, however, insurance of any kind faces problems arising from asymmetric information. In this context, there is the "moral hazard" problem, in that once the defendant is insured he has less incentive to avoid causing harm to the plaintiff and provoking a lawsuit.

The Open-Set Problem in Nuisance Suits II

Nuisance Suits II illustrates a technical point that arises in a great many games with continuous strategy spaces and causes great distress to novices in game theory. The equilibrium in Nuisance Suits II is only a weak Nash equilibrium. The plaintiff proposes $s = \gamma x + d$, and the defendant has the same payoff from accepting or rejecting, but in equilibrium the defendant accepts the offer with probability one, despite his indifference. This seems arbitrary, or even silly. Should not the plaintiff propose a slightly lower settlement to give the defendant

[3] Nonrefundable lawyer's fees, paid in advance, have traditionally been acceptable, but a New York court recently ruled they were unethical. The court thought that such fees unfairly restricted the client's ability to fire his lawyer, an example of how ignorance of game theory can lead to confused rule-making. See "Nonrefundable Lawyers' Fees, Paid in Advance, are Unethical, Court Rules," *Wall Street Journal*, January 29, 1993, p. B3, citing *In the matter of Edward M. Cooperman, Appellate Division of the Supreme Court, Second Judicial Department, Brooklyn, 90-00429.*

a strong incentive to accept it and avoid the risk of having to go to trial? If the parameters are such that $s = \gamma x + d = 60$, for example, why does the plaintiff risk holding out for 60 when he might be rejected and most likely receive 0 at trial, when he could offer 59 and give the defendant a strong incentive to accept?

One answer is that no other equilibrium exists besides $s = 60$. Offering 59 cannot be part of an equilibrium, because it is dominated by offering 59.9; offering 59.9 is dominated by offering 59.99, and so forth. This is known as the **open-set problem**, because the set of offers that the defendant strongly wishes to accept is open and has no maximum—it is bounded at 60, but a set must be bounded *and closed* to guarantee that a maximum exists.

A second answer is that under the assumptions of rationality and Nash equilibrium the objection's premise is false because the plaintiff bears no risk whatsoever in offering $s = 60$. It is fundamental to Nash equilibrium that each player believe that the others will follow equilibrium behavior. Thus, if the equilibrium strategy profile says that the defendant will accept $s \leq 60$, the plaintiff can offer 60 and believe it will be accepted. This is really just to say that a weak Nash equilibrium is still a Nash equilibrium, a point emphasized in chapter 3 in connection with mixed strategies.

A third answer is that the problem is an artifact of using a model with a continuous strategy space, and it disappears if the strategy space is made discrete. Assume that s can only take values in multiples of 0.01, so it could be 59.0, 59.01, 59.02, and so forth, but not 59.001 or 59.002. The settlement part of the game will now have two perfect equilibria. In the strong equilibrium E1, $s = 59.99$ and the defendant accepts any offer $s < 60$. In the weak equilibrium E2, $s = 60$ and the defendant accepts any offer $s \leq 60$. The difference is trivial, so the discrete strategy space has made the model more complicated without any extra insight.[4]

One can also specify a more complicated bargaining game to avoid the issue of how exactly the settlement is determined. Here one could say that the settlement is not proposed by the plaintiff, but simply emerges with a value halfway through the settlement range, so $s = \gamma x + d/2$. This seems reasonable enough, and it adds a little extra realism to the model at the cost of a little extra complexity. It avoids the open-set problem, but only by avoiding being clear about how s is determined. I call this kind of modelling **blackboxing**, because it is as if at some point in the game, variables with certain values go into a black box and come out the other side with values determined by an exogenous process. Blackboxing is perfectly acceptable as long as it neither drives nor obscures the point the model is making. Nuisance Suits III will illustrate this method.

Fundamentally, however, the point to keep in mind is that games are models, not reality. They are meant to clear away the unimportant details of a real situation and simplify it down to the essentials. Since a model is trying to answer a question, it should focus on what answers that question. Here, the question is why nuisance suits might be brought, so it is proper to exclude details of the bargaining if they are irrelevant to the answer. Whether a plaintiff offers 59.99

[4] A good example of the ideas of discrete money values and sequential rationality is in Robert Louis Stevenson's story, "The Bottle Imp" (Stevenson [1987]). The imp grants the wishes of the bottle's owner but will seize his soul if he dies in possession of it. Although the bottle cannot be given away, it can be sold, but only at a price less than that for which it was purchased.

or 60, and whether a rational person accepts an offer with probability 0.99 or 1.00, is part of the unimportant detail, and whatever approach is simplest should be used. If the modeller really thinks that these are important matters, they can indeed be modelled, but they are not important in this context.

One source of concern over the open-set problem, I think, is that perhaps the payoffs are not quite realistic, because the players should derive utility from hurting "unfair" players. If the plaintiff makes a settlement offer of 60, keeping the entire savings made from avoiding the trial for himself, everyday experience tells us that the defendant will indignantly refuse the offer. Guth *et al.* (1982) have found in experiments that people turn down bargaining offers they perceive as unfair, as one might expect. If indignation is truly important, it can be explicitly incorporated into the payoffs, and if that is done, the open-set problem returns. Indignation is not boundless, whatever people may say. Suppose that accepting a settlement offer that benefits the plaintiff more than the defendant gives a disutility of x to the defendant because of his indignation at his unjust treatment. The plaintiff will then offer to settle for exactly $60 - x$, so the equilibrium is still weak and the defendant is still indifferent between accepting and rejecting the offer. The open-set problem persists, even after realistic emotions are added to the model.

I have spent so much time on the open-set problem not because it is important but because it arises so often and is a sticking point for people unfamiliar with modelling. It is not a problem that disturbs experienced modellers, unlike other basic issues we have already encountered—for example, the issue of how a Nash equilibrium comes to be common knowledge among the players— but it is important to understand why it is not important.

Nuisance Suits III: Malicious Emotions

One of the most common misconceptions about game theory, as about economics in general, is that it ignores non-rational and non-monetary motivations. Game theory does take the basic motivations of the players to be exogenous to the model, but those motivations are crucial to the outcome and they often are not monetary, although payoffs are always given numerical values. Game theory does not call somebody irrational who prefers leisure to money or who is motivated by the desire to be world dictator. It does require the players' emotions to be carefully gauged to determine exactly how the actions and outcomes affect the players' utilities.

Emotions are often important to lawsuits, and law professors tell their students that when the cases they study seem to involve disputes too trivial to be worth taking to court, they can guess that the real motivations are emotional. Emotions could enter in a variety of distinct ways. The plaintiff might simply like going to trial, which can be expressed as a value of $p < 0$. This would be true of many criminal cases, because prosecutors like news coverage and want credit with the public for prosecuting certain kinds of crime. The Rodney King trials of 1992 and 1993 were of this variety; regardless of the merits of the case against the policemen who beat Rodney King, the prosecutors needed to go to trial to satisfy the public outrage, and when the state prosecutors failed in the first trial, the federal government was happy to accept the cost of bringing suit in the second trial. A different motivation is that the plaintiff might derive utility from the fact of winning the case quite separately from the monetary award, because he wants a public statement declaring that he is in the right. This is often a motivation in bringing libel suits, or for a criminal defendant who wants to clear his good name.

A different emotional motivation for going to trial is the desire to inflict losses on the defendant, a motivation we will call "malice," although it might as inaccurately be called "righteous anger." In this case, d enters as a positive argument in the plaintiff's utility function. We will construct a model of this kind, called Nuisance Suits III, and assume that $\gamma = 0.1$, $c = 3$, $p = 14$, $d = 50$, and $x = 100$, and that the plaintiff receives additional utility of 0.1 times the defendant's disutility. Let us also adopt the blackboxing technique discussed earlier and assume that the settlement s is in the middle of the settlement range. The payoffs conditional on suit being brought are

$$\pi_{plaintiff}(Defendant\ accepts) = s - c + 0.1s = 1.1s - 3 \tag{2}$$

and

$$\begin{aligned} \pi_{plaintiff}(Go\ to\ trial) &= \gamma x - c - p + 0.1\,(d + \gamma x) \\ &= 10 - 3 - 14 + 6 = -1. \end{aligned} \tag{3}$$

Now, working back from the end in accordance with sequential rationality, note that since the plaintiff's payoff from *Give Up* is -3, he will go to trial if the defendant rejects the settlement offer. The overall payoff from bringing a suit that eventually goes to trial is still -1, which is worse than the payoff of 0 from not bringing suit in the first place, but if s is high enough, the payoff from bringing suit and settling is higher still. If s is greater than 1.82 ($= [-1 + 3]/1.1$, rounded), the plaintiff prefers settlement to trial, and if s is greater than about 2.73 ($= [0 + 3]/1.1$, rounded), he prefers settlement to not bringing the suit at all.

In determining the settlement range, the relevant payoff is the expected incremental payoff since the suit has been brought. The plaintiff will settle for any $s \geq 1.82$, and the plaintiff will settle for any $s \leq \gamma x + d = 60$, as before. The settlement range is [1.82, 60], and $s = 30.91$. The settlement offer is no longer the maximizing choice of a player, and hence is moved to the outcome in the equilibrium description below.

> Plaintiff: *Sue, Go to Trial*
> Defendant: *Accept any* $s \leq 60$
> Outcome: The plaintiff sues and offers $s = 30.91$, and the defendant accepts the settlement.

Perfectness is important here because the defendant would like to threaten never to settle and be believed. The plaintiff would not bring suit given his expected payoff of -1 from bringing a suit that goes to trial, so a believable threat would be effective. But such a threat is not believable. Once the plaintiff does bring suit, the only Nash equilibrium in the remaining subgame is for the defendant to accept his settlement offer. This is interesting because the plaintiff, despite his willingness to go to trial, ends up settling out of court. When information is symmetric, as it is here, there is a tendency for equilibria to be efficient. Although the plaintiff wants to hurt the defendant, he also wants to keep his expenses low. Thus, he is willing to hurt the defendant less if it enables him to save on his own legal costs.

One final point before leaving these models is that much of the value of modelling comes simply from setting up the rules of a game, which helps to show what is important in a situation. One problem that arises in setting up a model of nuisance suits is deciding what a "nuisance suit" really is. In the game of Nuisance Suits, it has been defined as a suit whose expected damages do not repay the plaintiff's costs of going to trial. But having to formulate a definition brings to mind another problem that might be called the problem of nuisance suits: that the plaintiff brings suits he knows will not win unless the court makes a mistake. Since the court might make a mistake with very high probability, the games above would not be appropriate models—γ would be high, and the problem is not that the plaintiff's expected gain from trial is low, but that it is high. This, too, is an important problem, but having to construct a model shows that it is different.

4.4 Recoordination to Pareto-Dominant Equilibria in Subgames: Pareto Perfection

One simple refinement of equilibrium that was mentioned in chapter 1 is to rule out any strategy profiles that are Pareto dominated by Nash equilibria. Thus, in the game of Ranked Coordination, the inferior Nash equilibrium would be ruled out as an acceptable equilibrium. The idea behind this is that in some unmodelled way the players discuss their situation and coordinate to avoid the bad equilibria. Since only Nash equilibria are discussed, the players' agreements are self-enforcing which is a more limited suggestion than the approach in cooperative game theory according to which the players make binding agreements.

The coordination idea can be taken further in various two ways. One is to think about coalitions of players coordinating on favorable equilibria, so that two players might coordinate on an equilibrium even if a third player dislikes it. Bernheim, Peleg, & Whinston (1987) and Bernheim & Whinston (1987) define a Nash strategy profile as a **coalition-proof Nash equilibrium** if no coalition of players could form a self-enforcing agreement to deviate from it. They take the idea further by subordinating it to the idea of sequential rationality. The natural way to do this is to require that no coalition would deviate in future subgames, a notion called by various names, including **renegotiation proofness**, **recoordination** (e.g., Laffont & Tirole (1993), p. 460), and **Pareto perfection** (e.g., Fudenberg & Tirole (1991a), p. 175). The idea has been used extensively in the analysis of infinitely repeated games, which are particularly subject to the problem of multiple equilibria; Abreu, Pearce & Stachetti (1986) is an example of this literature. Whichever name is used, the idea is distinct from the renegotiation problem in the principal-agent models to be studied in chapter 8, which involves the rewriting of earlier binding contracts to make new binding contracts.

The best way to demonstrate the idea of Pareto perfection is by an illustration, the Pareto Perfection Puzzle, whose extensive form is shown in figure 4.5. In this game Smith chooses *In* or *Outside Option 1,* which yields payoffs of 10 to each player. Jones then chooses *Outside Option 2,* which yields 20 to each player, or initiates either a coordination game or a prisoner's dilemma. Rather than draw the full subgames in extensive form, figure 4.5 inserts the payoff matrix for the subgames.

The Pareto Perfection Puzzle illustrates the complicated interplay between perfectness and Pareto dominance. The Pareto-dominant strategy profile is (*In, Prisoner's Dilemma|In, any actions in the coordination subgame, actions yielding (50, 50) in the Prisoner's Dilemma subgame*). Nobody expects this strategy profile to be an equilibrium, since it is neither perfect nor Nash. Perfectness tells us that if the Prisoner's Dilemma subgame is reached, the payoffs will be (0, 0), and if the coordination subgame is reached, they will be either (1, 1) or (2, 30). In light of this, the perfect equilibria of The Pareto Perfection Puzzle are

> *E1:* (*In, outside option 2|In, the actions yielding (1, 1) in the coordination subgame, the actions yielding (0, 0) in the Prisoner's Dilemma subgame*). The payoffs are (20, 20).
>
> E2: (*outside option 1, coordination game|In, the actions yielding (2, 30) in the coordination subgame, the actions yielding (0, 0) in the Prisoner's Dilemma subgame*). The payoffs are (10, 10).

If one applies Pareto dominance without perfection, E1 will be the equilibrium, since both players prefer it. If the players can recoordinate at any point and change their expectations, however, then if play of the game reaches the coordination subgame, the players will recoordinate on actions yielding (2, 30). Pareto perfection thus knocks out E1 as an equilibrium. Not only does it rule out the Pareto-dominant strategy profile that yields (50, 50) as an equilibrium, it also

Figure 4.5 The Pareto Perfection Puzzle

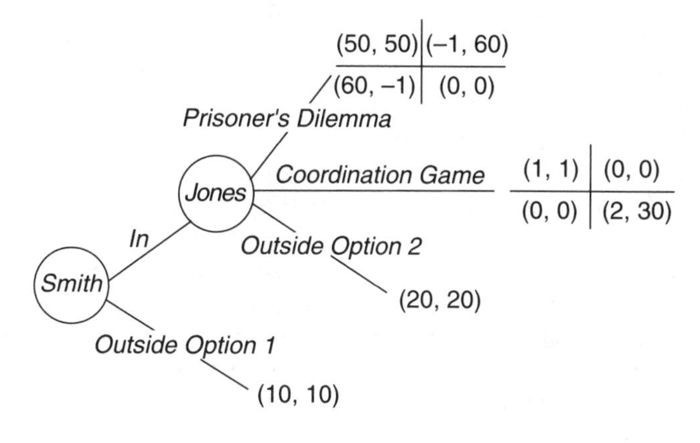

rules out the Pareto-dominant perfect strategy profile that yields (20, 20) as an equilibrium. Rather, the payoffs are (10, 10). Thus, Pareto perfection is not the same thing as simply picking the Pareto-dominant perfect strategy profile.

It is difficult to say which equilibrium is best here, since this is an abstract game and we cannot call upon details from the real world to refine the model. The approach of applying an equilibrium refinement is not as likely to yield results as using the intuition behind the refinement. The intuition here is that the players will somehow coordinate on Pareto-dominant equilibria, perhaps finding open discussion helpful. If we ran an experiment on student players using The Pareto Perfection Puzzle, I would expect to reach different equilibria depending on what communication is allowed. If the players are allowed to talk only before the game starts, it seems more likely that E1 would be the equilibrium, since players could agree to play it and would have no chance to explicitly recoordinate later. If the players could talk at any time as the game proceeded, E2 becomes more plausible. Real-world situations arise with many different communications technologies, so there is no one right answer.

4.5 Discounting

The remaining sections of chapter 4 put aside the issue of perfectness and credible threats. This section introduces the important modelling tool of discounting.

A model in which the action takes place in real time must specify whether payments and receipts are valued less if they are made later, i.e., whether they are **discounted**. Discounting is measured by the discount rate or the discount factor.

> *The* **discount rate**, r, *is the extra fraction of a payoff unit needed to compensate for delaying receipt by one period.*

> *The* **discount factor**, δ, *is the value in present payoff units of one payoff unit to be received one period from the present.*

The discount rate is analogous to the interest rate, and in some models the interest rate determines the discount rate. The discount factor represents exactly the same idea as the discount rate, and $\delta = 1/(1 + r)$. Models use r or δ depending on notational convenience. Zero discounting is equivalent to $r = 0$ and $\delta = 1$, so the notation includes zero discounting as a special case.

Whether to put discounting into a model involves two questions. The first is whether the added complexity will be accompanied by a change in the results or by a surprising demonstration of no change in the results. A second, more specific question is whether the events of the model occur in real time, so that discounting is appropriate. The bargaining game of Alternating Offers from section 11.3 can be interpreted in two ways. One way is that the players make all their offers and counteroffers between dawn and dusk of a single day, so essentially no real time has passed. The other way is that each offer consumes a week of time, so that the delay before the bargain is reached is important to the players. Discounting is appropriate only in the second interpretation.

Discounting has two important sources: time preference and a probability that the game might end, represented by the rate of time preference, ρ, and the probability each period that the game ends, θ. It is usually assumed that ρ and θ are constant. If they both take the value zero, the player does not care whether his payments are scheduled now or ten years from now. Otherwise, a player is indifferent between $x/(1 + \rho)$ now and x guaranteed to be paid one period later. With probability $(1 - \theta)$ the game continues and the later payment is actually made, so the player is indifferent between $(1 - \theta)x/(1 + \rho)$ now and the promise of x to be paid one period later contingent upon the game still continuing. The discount factor is therefore

$$\delta = \frac{1}{1 + r} = \frac{(1 - \theta)}{(1 + \rho)}. \quad (4)$$

Table 4.2 summarizes the implications of discounting for the value of payment streams of various kinds. We will not go into how these are derived, but they all stem from the basic fact that a dollar paid in the future is worth δ dollars now. Continuous time models usually refer to rates of payment rather than lump sums, so the discount factor is not so useful a concept, but discounting works the same way as in discrete time except that payments are continuously compounded. For a full explanation, see a finance text (e.g., Appendix A of Copeland & Weston [1988]).

Table 4.2 Discounting

Payoff Stream	**Discounted Value** r-notation (discount rate)	**Discounted Value** δ-notation (discount factor)
x at the end of one period	$\frac{x}{1+r}$	δx
x at the end of each period in perpetuity	$\frac{x}{r}$	$\frac{\delta x}{1-\delta}$
x at the start of each period in perpetuity	$x + \frac{x}{r}$	$\frac{x}{1-\delta}$
x at the end of each period up through T (first formula)	$\sum_{t=1}^{T} \frac{x}{(1+r)^t}$	$\sum_{t=1}^{T} \delta^t x$
x at the end of each period up through T (second formula)	$\frac{x}{r}\left(1-\frac{1}{(1+r)^T}\right)$	$\frac{\delta x}{1-\delta}\left(1-\delta^T\right)$
x at time t in continuous time	xe^{-rt}	
Flow of x per period up to time T in continuous time	$\int_0^T xe^{-rt}dt$	
Flow of x per period in perpetuity in continuous time	$\frac{x}{r}$	

Figure 4.6 Discounting

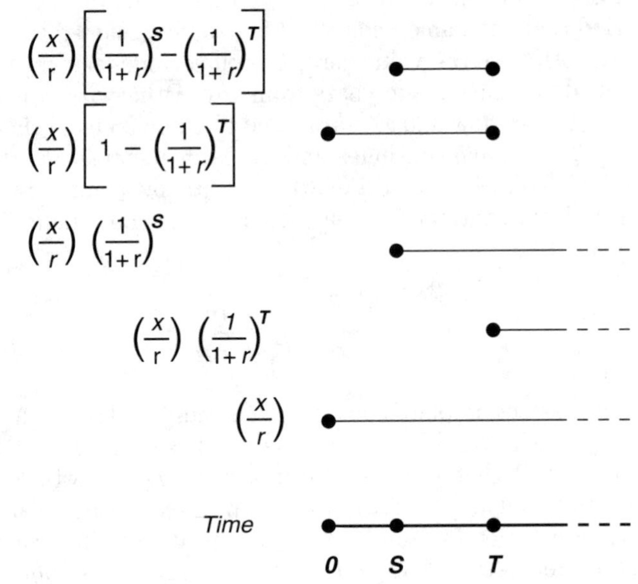

$\left(\frac{x}{r}\right)\left[\left(\frac{1}{1+r}\right)^S - \left(\frac{1}{1+r}\right)^T\right]$

$\left(\frac{x}{r}\right)\left[1 - \left(\frac{1}{1+r}\right)^T\right]$

$\left(\frac{x}{r}\right)\left(\frac{1}{1+r}\right)^S$

$\left(\frac{x}{r}\right)\left(\frac{1}{1+r}\right)^T$

$\left(\frac{x}{r}\right)$

Time

0 S T

The way to remember the formula for an annuity over a period of time is to use the formulas for a payment at a certain time in the future and for a perpetuity. A stream of x paid at the end of each year is worth x/r. A payment of Y at the end of period T has a present value of $-Y/(1+r)^T$. Thus, if at the start of period T you must pay out a perpetuity of x at the end of each year, the present value of that payment is $\left(\frac{x}{r}\right)\left(\frac{1}{1+r}\right)^T$. One may also view a stream of payments each year from the present until period T as the same thing as owning a perpetuity but having to give away a perpetuity in period T. This leaves a present value of $\left(\frac{x}{r}\right)\left(1 - \left(\frac{1}{1+r}\right)^T\right)$, which is the second formula for an annuity given in table 4.2. Figure 4.6 illustrates this approach to annuities and shows how it can also be used to value a stream of income that starts at period S and ends at period T.

Discounting will be left out of most dynamic games in this book, but it is an important issue in infinitely repeated games, and will be discussed further in section 5.2.

4.6 Evolutionary Equilibrium: Hawk-Dove

For most of this book we have been using the Nash equilibrium concept or refinements of it based on information and sequentiality, but in biology such concepts are often inappropriate. The lower animals are less likely than humans to think about the strategies of their opponents at each stage of a game. Their strategies are more likely to be preprogrammed and their strategy sets more restricted than the businessman's, if perhaps not more so than his customer's. In addition, behavior evolves, and any equilibrium must take account of the possibility of odd behavior caused by the occasional mutation. That the equilibrium is common knowledge, or that players cannot precommit to strategies, are not compelling assumptions. Thus, the ideas of Nash equilibrium and sequential rationality are much less useful than when game theory is modelling rational players.

Game theory has grown to some importance in biology, but the style is different than in economics. The goal is not to explain how players would rationally pick actions in a given situation, but to explain how behavior evolves or persists over time under exogenous shocks. Both approaches end up defining equilibria to be strategy profiles that are best responses in some sense, but biologists care much more about the stability of the equilibrium and how strategies interact over time. In section 3.4, we touched briefly on the stability of the Cournot equilibrium, but economists view stability as a pleasing by-product of the equilibrium rather than its justification. For biologists, stability is the point of the analysis.

Consider a game with identical players who engage in pairwise contests. In this special context, it is useful to think of an equilibrium as a strategy profile such that no player with a new strategy can enter the environment (**invade**) and receive a higher expected payoff than the old players. Moreover, the invading strategy should continue to do well even if it plays against itself with finite probability, or its invasion could never grow to significance. In the commonest model in biology, all the players adopt the same strategy in equilibrium, called an **evolutionarily stable strategy**. John Maynard Smith originated this idea, which is somewhat confusing because it really aims at an equilibrium concept, which involves a strategy profile, not just one player's strategy. For games with pairwise interactions and identical players, however, the evolutionarily stable strategy can be used to define an equilibrium concept.

> *A strategy s^* is an* **evolutionarily stable strategy**, *or* **ESS**, *if, using the notation $\pi(s_i, s_{-i})$ for player i's payoff when his opponent uses strategy s_{-i}, for every other strategy s' either*

$$\pi(s^*, s^*) > \pi(s', s^*) \tag{5}$$

or

$$\pi(s^*, s^*) = \pi(s', s^*) \text{ and } \pi(s^*, s') > \pi(s', s'). \tag{6}$$

If Condition (5) holds, then a population of players using s^* cannot be invaded by a deviant using s'. If Condition (6) holds, then s' does well against s^*, but badly against itself, so that if more than one player tried to use s' to invade a population using s^*, the invaders would fail.

We can interpret ESS in terms of Nash equilibrium. Condition (5) says that s^* is a strong Nash equilibrium (although not every strong Nash strategy is ESS). Condition (6) says that if s^* is only a weak Nash strategy, the weak alternative s' is not a best response to itself. ESS is a refinement of Nash, narrowed by the requirement that ESS not only be a best response, but that it (1) have the highest payoff of any strategy (which rules out equilibria with asymmetric payoffs) and (2) be a best response to itself. The motivations behind the two equilibrium concepts are quite different, but the similarities are useful because even if the modeller prefers ESS to Nash, he can start with the Nash strategies in his efforts to find an ESS.

An Example of ESS: Hawk-Dove

The best-known illustration of the ESS is the game of Hawk-Dove. Imagine that we have a population of birds, each of whom can behave as an aggressive Hawk or as a pacific Dove. We will focus on two randomly chosen birds, Bird One and Bird Two. Each bird has a choice of what behavior to choose on meeting another bird. A resource worth $V = 2$ "fitness units" is at stake when the two birds meet. If they both fight, the loser incurs a cost of $C = 4$, which means that the expected payoff when two Hawks meet is -1 ($= 0.5[2] + 0.5[-4]$) for each of them. When two Doves meet, they split the resource, for a payoff of 1 apiece. When a Hawk meets a Dove, the Dove flees for a payoff of 0, leaving the Hawk with a payoff of 2. Table 4.3 summarizes this.

Table 4.3 Hawk-Dove: Economics Notation

		Bird Two		
		Hawk		*Dove*
	Hawk	**−1, −1**	→	**2,0**
Bird One:		↓		↑
	Dove	**0, 2**	←	**1,1**

Payoffs to: (Bird One, Bird Two)

These payoffs are often depicted differently in biology games. Since the two players are identical, one can depict the payoffs by using a table showing the payoffs only of the row player. Applying this to Hawk-Dove generates table 4.4.

Table 4.4 Hawk-Dove: Biology Notation

		Bird Two	
		Hawk	*Dove*
Bird One:	*Hawk*	−1	2
	Dove	0	1

Payoffs to: (Bird One)

Hawk-Dove is Chicken with new feathers. The two games have the same ordinal ranking of payoffs, as can be seen by comparing table 4.3 with table 3.2, and their equilibria are the same except for the mixing parameters. Hawk-Dove has no symmetric pure-strategy Nash equilibrium, and hence no pure-strategy ESS, since in the two asymmetric Nash equilibria, *Hawk* gives a bigger payoff than *Dove*, and the doves would disappear from the population. In the ESS for this game, neither hawks nor doves completely take over the environment. If the population consisted entirely of hawks, a dove could invade and obtain a one-round payoff of 0 against a hawk, compared to the -1 that a hawk obtains against a hawk. If the population consisted entirely of doves, a hawk could invade and obtain a one-round payoff of 2 against a dove, compared to the 1 that a dove obtains against a dove.

In the mixed-strategy ESS, the equilibrium strategy is to be a hawk with probability 0.5 and a dove with probability 0.5, which can be interpreted as a population 50 percent hawks and 50 percent doves. As in the mixed-strategy equilibria in chapter 3, the players are indifferent as to their strategies. The expected payoff from being a hawk is the 0.5(2) from meeting a dove plus the $0.5(-1)$ from meeting another hawk, a sum of 0.5. The expected payoff from being a dove is the 0.5(1) from meeting another dove plus the 0.5(0) from meeting a hawk, also a sum of 0.5. Moreover, the equilibrium is stable in a sense similar to the Cournot equilibrium. If 60 percent of the population were hawks, a bird would have a higher fitness level as a dove. If "higher fitness" means being able to reproduce faster, the number of doves increases and the proportion returns to 50 percent over time.

ESS depends on the strategy sets allowed the players. If two birds can base their behavior on commonly observed random events, such as which bird arrives at the resource first, and $V > C$ (as specified above), then a strategy called the **bourgeois strategy** is an ESS. Under this strategy, the bird respects property rights like a good bourgeois; it behaves as a hawk if it arrives first, and as a dove if it arrives second, where we assume the order of arrival is random. The bourgeois strategy has an expected payoff of 1 from meeting itself, and behaves exactly like a 50:50 randomizer when it meets a strategy that ignores the order of arrival, so it can successfully invade a population of 50:50 randomizers. But the bourgeois strategy is a correlated strategy (see section 3.3), and requires something like the order of arrival to decide which of two identical players will play *Hawk*.

ESS is suited to games in which all the players are identical and interacting in pairs. It does not apply to games with non-identical players—wolves who can be wily or big and deer who can be fast or strong—although other equilibrium concepts of the same flavor can be constructed. The approach follows three steps, specifying (1) the initial population proportions and the probabilities of interactions, (2) the pairwise interactions, and (3) the dynamics by which players with higher payoffs increase in number in the population. Economics games generally use only the second step, which describes the strategies and payoffs from a single interaction.

The third step, the evolutionary dynamics, is especially foreign to economics. In specifying dynamics, the modeller must specify a difference equation (for discrete time) or differential equation (for continuous time) that describes how the strategies employed change over iterations, whether because players differ in the number of their descendants or because they learn to change their strategies over time. In economics games, the adjustment process is usually degenerate: the players jump instantly to the equilibrium. In biology games, the adjustment process is slower and cannot be derived from theory. How quickly the population of hawks increases relative to doves depends on the metabolism of the bird and the length of a generation.

Slow dynamics also make the starting point of the game important, unlike the case when adjustment is instantaneous. Figure 4.7, taken from D. Friedman (1991), shows a way to graphically depict evolution in a game in which all three strategies of *Hawk*, *Dove*, and *Bourgeois* are used. A point in the triangle represents a proportion of the three strategies in the population. At point E_3, for example, half the birds play *Hawk*, half play *Dove*, and none play *Bourgeois*, while at E_4 all the birds play *Bourgeois*.

Figure 4.7 shows the result of dynamics based on a function specified by Friedman that gives the rate of change of a strategy's proportion based on its payoff relative to the other two strategies. Points E_1, E_2, E_3, and E_4 are all fixed points in the sense that the proportions do not change no matter which of these points the game starts from. Only point E_4 represents an evolutionarily stable equilibrium, however, and if the game starts with any positive proportion of birds playing *Bourgeois*, the proportions tend towards E_4. The original Hawk-Dove which excluded the bourgeois strategy can be viewed as the HD line at the bottom of the triangle, and E_3 is evolutionarily stable in that restricted game.

Figure 4.7 also shows the importance of mutation in biological games. If the population of birds is one hundred percent dove, as at E_2, it stays that way in the absence of mutation, since if there are no hawks to begin with, the fact that they would reproduce at a faster rate than doves becomes irrelevant. If, however, a bird could mutate to play *Hawk* and then pass this behavior on to his offspring, then eventually some bird would do so and the mutant strategy would be successful. The technology of mutations can be important to the ultimate equilibrium. In more complicated games than Hawk-Dove, it can matter whether mutations happen to be small, accidental shifts to strategies similar to those that are currently being played, or can be of arbitrary size, so that a superior strategy quite different from the existing strategies might be reached.

Figure 4.7 Evolutionary Dynamics in The Hawk-Dove-Bourgeois Game

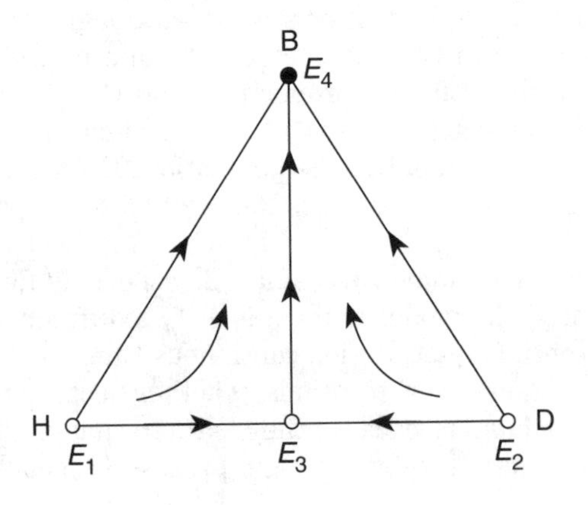

The idea of mutation is distinct from the idea of evolutionary dynamics, and it is possible to use one without the other. In economics models, a mutation would correspond to the appearance of a new action in the action set of one of the players in a game. This is one way to model innovation: not as research followed by stochastic discoveries, but as accidental learning. The modeller might specify that the discovered action becomes available to players slowly through evolutionary dynamics, or instantly, in the usual style of economics. This style of research has promise for economics, but since the technologies of dynamics and mutation are important there is a danger of simply multiplying models without reliable results unless the modeller limits himself to a narrow context and bases his technology on empirical measurements.

Notes

N4.1 Subgame Perfectness

- The terms "perfectness" and "perfection" are used synonymously. Selten (1965) proposed the equilibrium concept in an article written in German. "Perfectness" is used in Selten (1975) and conveys an impression of completeness more appropriate to the concept than the goodness implied by "perfection." "Perfection," however, is more common.
- It is debatable whether the definition of subgame ought to include the original game. Gibbons (1992, p. 122) does not, for example, and modellers usually do not in their conversation.
- Perfectness is not the only way to eliminate weak Nash equilibria like (*Stay Out, Collude*). In Entry Deterrence I, (*Enter, Collude*) is the only iterated dominance equilibrium, because *Fight* is weakly dominated for the incumbent.
- The distinction between perfect and non-perfect Nash equilibria is like the distinction between **closed loop** and **open loop** trajectories in dynamic programming. Closed loop (or **feedback**) trajectories can be revised after they start, like perfect equilibrium strategies, while open loop trajectories are completely prespecified (though they may depend on state variables). In dynamic programming the distinction is not so important, because prespecified strategies do not change the behavior of the other players. No threat, for example, is going to alter the pull of the moon's gravity on a rocket.
- A subgame can be infinite in length, and infinite games can have non-perfect equilibria. The infinitely repeated Prisoner's Dilemma is an example; here every subgame looks exactly like the original game, but begins at a different point in time.
- **Sequential Rationality in Macroeconomics.** In macroeconomics the requirement of **dynamic consistency** or **time consistency** is similar to perfectness. These terms are less precisely defined than perfectness, but they usually require that strategies need only be best responses in subgames starting from nodes on the equilibrium path, instead of all subgames. Under this interpretation, time consistency is a less stringent condition than perfectness.

  The Federal Reserve, for example, might like to induce inflation to stimulate the economy, but the economy is stimulated only if the inflation is unexpected. If the inflation is expected, its effects are purely bad. Since members of the public know that the Fed would like to fool them, they disbelieve its claims that it will not generate inflation (see Kydland & Prescott [1977]). Likewise, the government would like to issue nominal debt, and promises lenders that it will keep inflation low, but once the debt is issued, the government has an incentive to inflate its real value to zero. One reason the US Federal Reserve Board was established to be independent of Congress was to diminish this problem.

  The amount of game theory used in macroeconomics has been increasing at a fast rate. For references see Canzoneri & Henderson's 1991 book, which focuses on international coordination and pays particular attention to trigger strategies.
- Often, irrationality—behavior that is automatic rather than strategic—is an advantage. The Doomsday Machine in the movie *Dr Strangelove* is one example. The Soviet Union decides that it cannot win a rational arms race against the richer United States, so it creates a bomb which automatically blows up the entire world if anyone explodes a nuclear bomb. The movie also

illustrates a crucial detail without which such irrationality is worse than useless: you have to tell the other side that you have the Doomsday Machine.

President Nixon reportedly told his aide H.R. Haldeman that he followed a more complicated version of this strategy: "I call it the Madman Theory, Bob. I want the North Vietnamese to believe that I've reached the point where I might do *anything* to stop the war. We'll just slip the word to them that 'for God's sake, you know Nixon is obsessed about Communism. We can't restrain him when he's angry—and he has his hand on the nuclear button'—and Ho Chi Minh himself will be in Paris in two days begging for peace"(Haldeman & DiMona [1978] p. 83). The Gang of Four model in section 6.4 tries to model a situation like this.

- The "lock-up agreement" is an example of a credible threat: in a takeover defense, the threat to destroy the firm is made legally binding. See Macey & McChesney (1985) p. 33.

N4.3 An Example of Perfectness: Entry Deterrence I

- The Stackelberg equilibrium of a duopoly game (section 3.4) can be viewed as the perfect equilibrium of a Cournot game modified so that one player moves first, a game similar to Entry Deterrence I. The player moving first is the Stackelberg leader and the player moving second the Stackelberg follower. The follower could threaten to produce a high output, but he will not carry out his threat if the leader produces a high output first.
- Perfectness is not so desirable a property of equilibrium in biological games. The reason the order of moves matters is because the rational best reply depends on the node at which the game has arrived. In many biological games the players act by instinct and unthinking behavior is not unrealistic.
- Reinganum & Stokey (1985) is a clear presentation of the implications of perfectness and commitment illustrated with the example of natural resource extraction.

N4.6 Evolutionary Equilibrium: The Hawk-Dove Game

- Dawkins (1989) is a good verbal introduction to evolutionary conflict. See also Axelrod & Hamilton (1981) for a short article on biological applications of the Prisoner's Dilemma, Hines (1987) for a survey, and Maynard Smith (1982) for a book. J. Hirshleifer (1982) compares the approaches of economists and biologists. Among economists, Cornell & Roll (1981) and Riley (1979a) have used ESS. Jacquemin's 1985 industrial organization text is sympathetic to the evolutionary approach, and Boyd & Richerson (1985) uses it to examine cultural transmission, which has important differences from purely genetic transmission. Sugden (1986) has written a book using game theory from the biological approach to look at social theory.

Problems

4.1: Repeated Entry Deterrence.

Consider two repetitions without discounting of the game Entry Deterrence I from section 4.2. Assume that there is one entrant, who sequentially decides whether to enter two markets that have the same incumbent.

(4.1a) Draw the extensive form of this game.
(4.1b) What are the 16 elements of the strategy sets of the entrant?
(4.1c) What is the subgame perfect equilibrium?
(4.1d) What is one of the nonperfect Nash equilibria?

4.2: Evolutionarily Stable Strategies.

A population of scholars are playing the following coordination game over their two possible conversation topics over lunch, football and economics. Let $N_t(F)$ and $N_t(E)$ be the numbers who talk football and economics in period t, and let θ

be the percentage who talk football, so $\theta = \frac{N(football)}{N(football) + N(economics)}$. Government regulations requiring lunchtime attendance and stipulating the topics of conversation have maintained the values $\theta = 0.5$, $N_t(F) = 50{,}000$ and $N_t(E) = 50{,}000$ up to this year's deregulatory reform. In the future, some people may decide to go home for lunch instead, or change their conversation. Table 4.5 shows the payoffs.

Table 4.5 Evolutionarily Stable Strategies

		Scholar 2	
		Football (θ)	*Economics* ($1-\theta$)
Scholar 1	*Football* (θ)	1,1	0,0
	Economics ($1-\theta$)	0,0	5,5

Payoffs to: (Scholar 1, Scholar 2)

(4.2a) There are three Nash equilibria: (*Football, Football*), (*Economics, Economics*), and a mixed-strategy equilibrium. What are the evolutionarily stable strategies?

(4.2b) Let $N_t(s)$ be the number of scholars playing a particular strategy in period t and let $\pi_t(s)$ be the payoff. Devise a Markov difference equation to express the population dynamics from period to the next: $N_{t+1}(s) = f(N_t(s), \pi_t(s))$. Start the system with a population of 100,000, half the scholars talking football and half talking economics. Use your dynamics to finish Table 4.6.

Table 4.6 Conversation Dynamics

t	$N_t(F)$	$N_t(E)$	θ	$\pi_t(F)$	$\pi_t(E)$
−1	50,000	50,000	0.5	0.5	2.5
0					
1					
2					

(4.2c) Repeat part (b), but specifying non-Markov dynamics, in which $N_{t+1}(s) = f(N_t(s), \pi_t(s), \pi_{t-1}(s))$.

4.3: Pliny and the Freedmens' Trial.

(Pliny, 1963, pp. 221–4, Riker, 1986, pp. 78–88). Afranius Dexter died mysteriously, perhaps dead by his own hand, perhaps killed by his freedmen (servants a step above slaves), or perhaps killed by his freedmen on his own orders. The freedmen went on trial before the Roman Senate. Assume that 45 percent of the senators favor acquittal, 35 percent favor banishment, and 20 percent favor execution, and that the preference rankings in the three groups are $A > B > E$, $B > A > E$, and $E > B > A$. Also assume that each group has a leader and votes as a bloc.

(4.3a) Modern legal procedure requires the court to decide guilt first and then assign a penalty if the accused is found guilty. Draw a tree to represent the sequence of events (this will not be a game tree, since it will represent the actions of groups of players, not of individuals). What is the outcome in a perfect equilibrium?

(4.3b) Suppose that the acquittal bloc can pre-commit to how they will vote in the second round if guilt wins in the first round. What will they do, and what will happen? What would the execution bloc do if they could control the second-period vote of the acquittal bloc?

(4.3c) The normal Roman procedure began with a vote on execution versus no execution, and then voted on the alternatives in a second round if execution failed to gain a majority. Draw a tree to represent this. What would happen in this case?

(4.3d) Pliny proposed that the Senators divide into three groups, depending on whether they supported acquittal, banishment, or execution, and that the outcome with the most votes should win. This proposal caused a roar of protest. Why did he propose it?

(4.3e) Pliny did not get the result he wanted with his voting procedure. Why not?

(4.3f) Suppose that personal considerations made it most important to a senator that he show his stand by his vote, even if he had to sacrifice his preference for a particular outcome. If there were a vote over whether to use the traditional Roman procedure or Pliny's procedure, who would vote with Pliny, and what would happen to the freedmen?

4.4: Grab the Dollar.

Table 4.7 shows the payoffs for the simultaneous-move game of Grab the Dollar. A silver dollar is put on the table between Smith and Jones. If one grabs it, he keeps the dollar, for a payoff of 4 utils. If both grab, then neither gets the dollar, and both feel bitter. If neither grabs, each gets to keep something.

Table 4.7 Grab the Dollar

		Jones	
		Grab (θ)	*Wait* ($1-\theta$)
Smith:	*Grab* (θ)	$-1,-1$	4,0
	Wait (θ)	0,4	1,1

Payoffs to: (Smith, Jones)

(4.4a) What are the evolutionarily stable strategies?

(4.4b) Suppose each player in the population is a point on a continuum, and that the initial amount of players is 1, evenly divided between *Grab* and *Wait*. Let $N_t(s)$ be the number of players playing a particular strategy in period t and let $\pi_t(s)$ be the payoff. Let the population dynamics be $N_{t+1}(i) = (2N_t(i))\left(\frac{\pi_t(i)}{\sum_j \pi_t(j)}\right)$. Find the missing entries in table 4.8.

Table 4.8 Grab the Dollar: Dynamics

t	$N_t(G)$	$N_t(W)$	$N_t(total)$	θ	$\pi_t(G)$	$\pi_t(w)$
0	0.5	0.5	1	0.5	1.5	0.5
1						
2						

(4.4c) Repeat part (b), but with the dynamics $N_{t+t}(s) = \left[1 + \frac{\pi_t(s)}{\sum_j \pi_t(j)}\right][2N_t(s)]$.

enter — collude — (40,50)

enter — fight — (-10,0)

stay out — (0,300)

perfect

	collude	fight
enter	(40,50)	(-10,0)
stay out	(0,300)	(0,300)

5 Reputation and Repeated Games with Symmetric Information

5.1 Finitely Repeated Games and the Chainstore Paradox

Chapter 4 showed how to refine the concept of Nash equilibrium to find sensible equilibria in games with moves in sequence over time, so-called dynamic games. An important class of dynamic games are repeated games, in which players repeatedly make the same decision in the same environment. Section 4.6 came close to this in its discussion of biological games, where a simple game like Hawk-Dove was repeated many times, but a crucial part of the biological games is that the population playing the game changes over time. The equilibrium, in fact, is a stable population mix. Chapter 5 will look at repeated games in which the rules of the game remain unchanged with each repetition. All that changes is the "history" which grows as time passes, and, if the number of repetitions is finite, the approach of the end of the game. It is also possible for the information structure to change over time, since players' moves may convey their private information, but this chapter will confine itself to games of symmetric information.

Section 5.1 will show the perverse unimportance of repetition for the games of Entry Deterrence and The Prisoner's Dilemma, a phenomenon known as the Chainstore Paradox. Neither discounting, probabilistic end dates, infinite repetitions, nor precommitment are satisfactory escapes from the Chainstore Paradox. This is summarized in the Folk Theorem of section 5.2. Section 5.2 will also discuss strategies which punish players who fail to cooperate in a repeated game—strategies such as the Grim Strategy, Tit-for-Tat, and Minimax. Section 5.3 builds a framework for reputation models based on The Prisoner's Dilemma, and section 5.4 presents one particular reputation model, the Klein-Leffler model of product quality. Section 5.5 concludes the chapter with an overlapping generations model of consumer switching costs which uses the idea of Markov strategies to narrow down the number of equilibria.

Chainstore Paradox (Selten [1978])

Suppose that we repeat Entry Deterrence I (from section 4.2) 20 times in the context of a chainstore that is trying to deter entry into 20 markets where it has outlets. We have seen that entry into just one market would not be deterred, but perhaps with 20 markets the outcome is different because the chainstore would fight the first entrant to deter the next 19.

The repeated game is much more complicated than the **one-shot game**, as the unrepeated version is called. A player's action is still to *Enter* or *Stay Out*, to *Fight* or *Collude*, but his strategy is a potentially very complicated rule telling him what action to choose depending on what actions both players took in each of the previous periods. Even the five-round repeated Prisoner's Dilemma has a strategy set for each player with over two billion strategies, and the number of strategy profiles is even greater (Sugden [1986], p. 108).

The obvious way to solve the game is from the beginning, where there is the least past history on which to condition a strategy, but that is not the easy way. We have to follow Kierkegaard, who said, "Life can only be understood backwards, but it must be lived forwards" (Kierkegaard [1938], p. 465). In picking his first action, a player looks ahead to its implications for all the future periods, so it is easiest to start by understanding the end of a multi-period game, where the future is shortest.

Consider the situation in which 19 markets have already been invaded (and maybe the chainstore fought, or maybe not). In the last market, the subgame in which the two players find themselves is identical to the one-shot Entry Deterrence I, so the entrant will *Enter* and the chainstore will *Collude*, regardless of the past history of the game. Next, consider the next-to-last market. The chainstore can gain nothing from building a reputation for ferocity, because it is common knowledge that he will *Collude* with the last entrant anyway. So he might as well *Collude* in the 19th market. But we can say the same of the 18th market and—by continuing backward induction—of every market, including the first. This result is called the **Chainstore Paradox**.

Backward induction ensures that the strategy profile is a subgame perfect equilibrium. There are other Nash equilibria—(*Always Fight*, *Never Enter*), for example—but because of the Chainstore Paradox they are not perfect.

The Repeated Prisoner's Dilemma

The Prisoner's Dilemma is similar to Entry Deterrence I. Here the prisoners would like to commit themselves to *Deny*, but, in the absence of commitment, they *Confess*. The Chainstore Paradox can be applied to show that repetition does not induce cooperative behavior. Both prisoners know that in the last repetition, both will *Confess*. After 18 repetitions, they know that no matter what happens in the 19th, both will *Confess* in the 20th, so they might as well *Confess* in the 19th too. Building a reputation is pointless, because in the 20th period it is not going to matter. Proceeding inductively, both players *Confess* in every period, the unique perfect equilibrium outcome.

In fact, as a consequence of the fact that the one-shot Prisoner's Dilemma has a dominant strategy equilibrium, confessing is the only Nash outcome for the repeated Prisoner's Dilemma, not just the only perfect outcome. The argument of the previous paragraph did not show that confessing was the unique Nash outcome. To show subgame perfectness, we worked back from the end using longer and longer subgames. To show that confessing is the only Nash outcome, we do not look at subgames, but instead rule out successive classes of strategies from being Nash. Consider the portions of the strategy which apply to the equilibrium path (that is, the portions directly relevant to the payoffs). No strategy in the class that calls for *Deny* in the last period can be a Nash strategy, because the same strategy with *Confess* replacing *Deny* would dominate it. But if both players have strategies calling for confessing in the last period, then no strategy that does not call for confessing in the next-to-last period is Nash, because a player should deviate by replacing *Deny* with *Confess* in the next-to-last period. The argument can be carried back to the first period, ruling out any class of strategies that does not call for confessing everywhere along the equilibrium path.

The strategy of always confessing is not a dominant strategy, as it is in the one-shot game, because it is not the best response to various suboptimal strategies such as (*Deny until the other player Confesses, then Deny for the rest of the game*). Moreover, the uniqueness is only on the equilibrium path. Non-perfect Nash strategies could call for cooperation at nodes far away from the equilibrium path, since that action would never have to be taken. If Row has chosen (*Always Confess*), one of Column's best responses is (*Always Confess unless Row has chosen Deny ten times; then always Deny*).

5.2 Infinitely Repeated Games, Minimax Punishments, and the Folk Theorem

The contradiction between the Chainstore Paradox and what many people think of as real-world behavior has been most successfully resolved by adding incomplete information to the model, as in section 6.4. Before we turn to incomplete information, however, we will explore certain other modifications. One idea is to repeat The Prisoner's Dilemma an infinite number of times instead of a finite number (after all, few economies have a known end date). Without a last period, the inductive argument in the Chainstore Paradox fails.

In fact, we can find a simple perfect equilibrium for the infinitely repeated Prisoner's Dilemma in which both players cooperate—a game in which both players adopt the Grim Strategy.

Grim Strategy

(*1*) *Start by choosing Deny.*

(*2*) *Continue to choose Deny unless some player has chosen Confess, in which case choose Confess forever.*

Notice that the Grim Strategy says that even if a player is himself the first to deviate and choose *Confess*, he continues to choose *Confess* thereafter.

If Column uses the Grim Strategy, the Grim Strategy is weakly Row's best response. If Row cooperates, he will continue to receive the high (*Deny*, *Deny*) payoff forever. If he confesses, he will receive the higher (*Confess*, *Deny*) payoff once, but the best he can hope for thereafter is the (*Confess*, *Confess*) payoff.

Even in the infinitely repeated game, cooperation is not immediate, and not every strategy that punishes confessing is perfect. A notable example is the strategy of Tit-for-Tat.

Tit-for-Tat

(*1*) *Start by choosing Deny.*

(*2*) *Thereafter, in period n choose the action that the other player chose in period (n − 1).*

If Column uses Tit-for-Tat, Row does not have an incentive to *Confess* first, because if Row cooperates he will continue to receive the high (*Deny, Deny*) payoff, but if he confesses and then returns to Tit-for-Tat, the players alternate (*Confess, Deny*) with (*Deny, Confess*) forever. Row's average payoff from this alternation would be lower than if he had stuck to (*Deny, Deny*), and would swamp the one-time gain. But Tit-for-Tat is not perfect in the infinitely repeated Prisoner's Dilemma without discounting, because it is not rational for Column to punish Row's initial *Confess*. Adhering to Tit-for-Tat's punishments results in a miserable alternation of *Confess* and *Deny*, so Column would rather ignore Row's first *Confess*. The deviation is not from the equilibrium path action of *Deny*, but from the off-equilibrium action rule of *Confess in response to a Confess*. Tit-for-Tat, unlike the Grim Strategy, cannot enforce cooperation.[1]

Unfortunately, although eternal cooperation is a perfect equilibrium outcome in the infinite game under at least one strategy, so is practically anything else, including eternal confessing. The multiplicity of equilibria is summarized by the Folk Theorem, so called because its origins are hazy.

Theorem 5.1 (The Folk Theorem)

In an infinitely repeated n-person game with finite action sets at each repetition, any combination of actions observed in any finite number of repetitions is the unique outcome of some subgame perfect equilibrium given

Condition 1: *The rate of time preference is zero, or positive and sufficiently small;*

Condition 2: *The probability that the game ends at any repetition is zero, or positive and sufficiently small;*

and

Condition 3: *The set of payoff profiles that strictly Pareto dominate the minimax payoff profiles in the mixed extension of the one-shot game is n-dimensional.*

[1] See problem 5.5 for elaboration of this point.

What the Folk Theorem tells us is that claiming that particular behavior arises in a perfect equilibrium is meaningless in an infinitely repeated game. This applies to any game that meets Conditions 1 to 3, not just to The Prisoner's Dilemma. If an infinite amount of time always remains in the game, a way can always be found to make one player willing to punish some other player for the sake of a better future, even if the punishment currently hurts the punisher as well as the punished. Any finite interval of time is insignificant compared to eternity, so the threat of future reprisal makes the players willing to carry out the punishments needed.

We will next discuss conditions 1 to 3.

Condition 1: Discounting

The Folk Theorem helps answer the question of whether discounting future payments lessens the influence of the troublesome Last Period. Quite to the contrary, with discounting, the present gain from confessing is weighted more heavily and future gains from cooperation more lightly. If the discount rate is very high, the game almost returns to being one-shot. When the real interest rate is 1,000 percent, a payment next year is little better than a payment a hundred years hence, so next year is practically irrelevant. Any model that relies on a large number of repetitions also assumes that the discount rate is not too high.

Allowing a little discounting is nonetheless important to show that there is no discontinuity at a discount rate of zero. If we come across an undiscounted, infinitely repeated game with many equilibria, the Folk Theorem tells us that adding a low discount rate will not reduce the number of equilibria. This contrasts with the effect of changing the model by having a large but finite number of repetitions, a change which often eliminates all but one outcome by inducing the Chainstore Paradox.

A discount rate of zero supports many perfect equilibria, but if the rate is high enough, the only equilibrium outcome is eternal confessing. We can calculate the critical value for given parameters. The Grim Strategy imposes the heaviest possible punishment for deviant behavior. Using the payoffs for the Prisoner's Dilemma from table 5.2a in the next section, the equilibrium payoff from the Grim Strategy is the current payoff of 5 plus the value of the rest of the game, which from table 4.2 is $5/r$. If Row deviated by confessing, he would receive a current payoff of 10, but the value of the rest of the game would fall to 0. The critical value of the discount rate is found by solving the equation $5 + 5/r = 10 + 0$, which yields $r = 1$, for a discount rate of 100 percent or a discount factor of $\delta = 0.5$. Unless the players are extremely impatient, confessing is not much of a temptation.

Condition 2: A Probability of the Game Ending

Time preference is fairly straightforward, but what is surprising is that assuming that the game ends in each period with probability θ does not make a drastic difference. In fact, we could even allow θ to vary over time, so long as it never became too large. If $\theta > 0$, the game ends in finite time with probability one; or, put less dramatically, the expected number of repetitions is finite, but it still behaves like a discounted infinite game, because the expected number of future repetitions is always large, no matter how many repetitions have already occurred. The game still has no Last Period, and it is still true that imposing a last period, no matter how far beyond the expected number of repetitions, would radically change the results.

I find it interesting that the claim (a) "the game will end at some uncertain date before *T*," is different from the claim (b) "there is a constant probability of the game ending." Under (a), the game is like a finite game, because, as time passes, the maximum length of time still to run shrinks to zero. Under (b), even if the game will end by *T* with high probability, if it actually lasts until *T* the game looks exactly the same as at time zero. The fourth verse from the hymn *Amazing Grace* puts this stationarity very nicely (though I expect the hymn is supposed to apply to a game with $\theta = 0$).

When we've been there ten thousand years,
Bright shining as the sun,
We've no less days to sing God's praise
Than when we'd first begun.

Condition 3: Dimensionality

The "minimax payoff" mentioned in theorem 5.1 is the payoff that results if all the other players pick strategies solely to punish player i, and he protects himself as best he can.

The set of strategies s^{i*}_{-i} *is the set of* $(n-1)$ **minimax strategies** *chosen by all the players except i to keep i's payoff as low as possible, no matter how he responds.* s^{i*}_{-i} *solves*

$$\underset{s_{-i}}{Minimize} \quad \underset{s_i}{Maximum}\ \pi_i(s_i, s_{-i}). \tag{1}$$

Player i's **minimax payoff, minimax value,** *or* **security value** *is his payoff from the solution of (1).*

The dimensionality condition is needed only for games with three or more players. It is satisfied if there is some payoff profile for each player in which his payoff is greater than his minimax payoff but still different from the payoff of every other player. Figure 5.1 shows how this condition is satisfied for the two-person Prisoner's Dilemma of table 5.2a below, but not for the two-person Ranked Coordination game. It is also satisfied by the n-person Prisoner's Dilemma in which a solitary confesser gets a higher payoff than his cooperating fellow-prisoners, but not by the n-person Ranked Coordination game, in which all the players have the same payoff. The condition is necessary because establishing the desired behavior requires some way for the other players to punish a deviator without punishing themselves.

Figure 5.1 The Dimensionality Condition

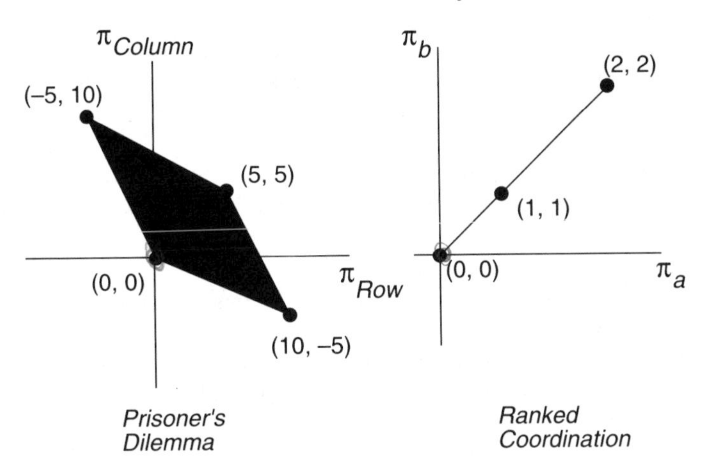

Minimax and Maximin

In discussions of strategies which enforce cooperation, the question of deciding on the maximum severity of punishment strategies frequently arises. The idea of the minimax strategy is useful for this in that the minimax strategy is defined as the most severe sanction possible if the offender does not cooperate in his own punishment. The corresponding strategy for the offender, trying to protect himself from punishment, is the maximin strategy:

> *The strategy s_i^* is a* **maximin strategy** *for player i if, given that the other players pick strategies to make i's payoff as low as possible, s_i^* gives i the highest possible payoff. In our notation, s_i^* solves*
>
> $$\underset{s_i}{Maximize} \quad \underset{s_{-i}}{Minimum} \; \pi_i(s_i, s_{-i}). \tag{2}$$

The following formulae show how to calculate the minimax and maximin strategies for a two-player game with player 1 as i.

$$\text{Maximin:} \quad \underset{s_1}{Maximum} \quad \underset{s_2}{Minimum} \quad \pi_1$$

$$\text{Minimax:} \quad \underset{s_2}{Minimum} \quad \underset{s_1}{Maximum} \quad \pi_1$$

In The Prisoner's Dilemma, the minimax and maximin strategies are both *Confess*. Although The Welfare Game (table 3.1) has only a mixed-strategy Nash equilibrium, if we restrict ourselves to the pure strategies the Pauper's maximin strategy is *Try to Work*, which guarantees him at least 1, and his strategy for minimaxing the Government is *Be Idle*, which prevents the Government from getting more than zero.

Under minimax, player 2 is purely malicious but must move first (at least in choosing a mixing probability) in his attempt to cause player 1 the maximum pain. Under maximin, player 1 moves first, in the belief that player 2 is out to get him. In variable-sum games, minimax is for sadists and maximin for paranoids. In zero-sum games, the players are merely neurotic: minimax is for optimists and maximin for pessimists.

The maximin strategy need not be unique, and can be in mixed strategies. Since maximin behavior can also be viewed as minimizing the maximum loss that might be suffered, decision theorists refer to such a policy as a **minimax criterion,** a catchier phrase (Luce & Raiffa [1957], p. 279).

It is tempting to use maximin strategies as the basis of an equilibrium concept. A **maximin equilibrium** is made up of a maximin strategy for each player. Such a strategy might seem reasonable because each player then has protected himself from the worst harm possible. Maximin strategies have very little justification, however, for a rational player. They are not simply the optimal strategies for risk-averse players, because risk aversion is accounted for in the utility payoffs. The players' implicit beliefs can be inconsistent in a maximin equilibrium, and a player must believe that his opponent would choose the most harmful strategy out of spite rather than self-interest if maximin behavior is to be rational.

The usefulness of minimax and maximin strategies is not in directly predicting the best strategies of the players, but in setting the bounds of how their strategies affect their payoffs, as in condition 3 of theorem 5.1.

It is important to remember that minimax and maximin strategies are not always pure strategies. In The Minimax Illustration Game of table 5.1, Row can guarantee himself a payoff of 0 by choosing *Down*, so that this is his maximin strategy. Column cannot hold Row's payoff down to 0, however, by using a pure minimax strategy. If Column chooses *Left*, Row can choose *Middle* and get a payoff of 1; if Column chooses *Right*, Row can choose *Up* and get a payoff of 1. Column can, however hold Row's payoff down to 0 by choosing a mixed minimax strategy of (*Probability 0.5 of Left, Probability 0.5 of Right*). Row would then respond with *Down*, for a minimax payoff of 0, since either *Up*, *Middle*, or a mixture of the two would give him a payoff of -0.5 ($= 0.5(-2) + 0.5(1)$).[2]

Table 5.1 The Minimax Illustration Game[3]

		Column	
		Left	*Right*
	Up	−2, [2]	[1], −2
Row:	*Middle*	[1], −2	−2, [2]
	Down	0, [1]	0, [1]

Payoffs to: (Row, Column)

2 Column's maximin and minimax strategies can also be computed. The strategy for minimaxing Column is (*Probability* 0.5 *of Up, Probability* 0.5 *of Middle*), his maximin strategy is (*Probability* 0.5 *of Left, Probability* 0.5 *of Right*), and his minimax payoff is 0.

3 This example is from p. 150 of Fudenberg & Tirole (1991a).

In two-person zero-sum games, minimax and maximin strategies are more directly useful, because when player 1 reduces player 2's payoff, he increases his own payoff. Punishing the other player is equivalent to rewarding yourself. This is the origin of the celebrated **Minimax Theorem** (von Neumann [1928]), which says that a minimax equilibrium exists in pure or mixed strategies for every two-person zero-sum game and is identical to the maximin equilibrium. Unfortunately, the games that come up in applications are almost never zero-sum games, so the Minimax Theorem is of limited applicability.

Precommitment

What if we use metastrategies, abandoning the idea of perfectness by allowing players to commit at the start to a strategy for the rest of the game? We would still want to keep the game noncooperative by disallowing binding promises, but we could model it as a game with simultaneous choices by both players, or with one move each in sequence.

If precommitted strategies are chosen simultaneously, the equilibrium outcome of the finitely repeated Prisoner's Dilemma calls for always confessing, because allowing commitment is the same as allowing equilibria to be non-perfect, in which case, as was shown earlier, the unique Nash outcome is always confessing.

A different result is achieved if the players precommit to strategies in sequence. The outcome depends on the particular values of the parameters, but one possible equilibrium is the following: Row moves first and chooses the strategy (*Deny* until Column *Confesses*; thereafter always *Confess*), and Column chooses (*Deny* until the last period; then *Confess*). The observed outcome would be for both players to deny until the last period, and then for Row to again deny, but for Column to confess. Row would submit to this because if he chose a strategy that initiated confessing earlier, Column would choose a strategy of starting to confess earlier too. The game has a second-mover advantage.

5.3 Reputation: One-sided Prisoner's Dilemma

Part II of this book will analyze moral hazard and adverse selection. Under moral hazard, a player wants to commit to high effort, but he cannot credibly do so. Under adverse selection, a player wants to prove he is high ability, but he cannot. In both, the problem is that the penalties for lying are insufficient. Reputation seems to offer a way out of the problem. If the relationship is repeated, perhaps a player is willing to be honest in early periods in order to establish a reputation for honesty which will be valuable to himself later.

Reputation plays a similar role in making threats to punish credible. Usually punishment is costly to the punisher as well as the punished, and it is not clear why the punisher should not let bygones be bygones. Yet in 1988 the Soviet Union paid off a 70-year-old debt to dissuade the Swiss authorities from blocking a mutually beneficial new bond issue ("Soviets Agree to Pay Off Czarist Debt to Switzerland," *Wall Street Journal*, January 19, 1988, p. 60). Why were the Swiss so vindictive towards Lenin?

The questions of why players do punish and do not cheat are really the same questions that arise in the repeated Prisoner's Dilemma, where only the fact of an infinite number of repetitions allows cooperation. That is the great problem of reputation. Since everyone knows that a player will *Confess*, choose low effort, or default on debt in the last period, why do they suppose he will bother to build up a reputation in the present? Why should past behavior be any guide to future behavior?

Not all reputation problems are quite the same as The Prisoner's Dilemma, but they have much the same flavor. Some games, like duopoly or the original Prisoner's Dilemma, are **two-sided** in the sense that each player has the same strategy set and the payoffs are symmetric. Others, such as the game of Product Quality (see below), are what we might call **one-sided Prisoner's Dilemmas**, which have properties similar to the Prisoner's Dilemma, but do not fit the usual definition because they are asymmetric. Table 5.2 shows the normal forms for both the original Prisoner's Dilemma and the one-sided version.[4] The important difference is that in the one-sided Prisoner's Dilemma at least one player really does prefer the profile equivalent to (*Deny, Deny*), which is (*High Quality, Buy*) in table 5.2b, to anything else. He confesses defensively, rather than both offensively and defensively. The payoff (0, 0) can often be interpreted as the refusal of one player to interact with the other, for example, the motorist who refuses to buy cars from Chrysler because he knows they once falsified odometers. Table 5.2 lists examples of both one-sided and two-sided games. Versions of the Prisoner's Dilemma with three or more players can also be classified as two-sided or one-sided, depending on whether or not all players find *Confess* a dominant strategy.

Table 5.2 Prisoner's Dilemmas

(a) Two-sided (conventional)

		Column		
		Deny		*Confess*
Row:	*Deny*	5, 5	→	−5,10
		↓		↓
	Confess	10, −5	→	**0, 0**

Payoffs to: (Row, Column)

(b) One-sided

		Consumer (Column)		
		Buy		*Boycott*
Seller (Row):	*High Quality*	5, 5	←	0,0
		↓		↕
	Low Quality	10, −5	→	**0,0**

Payoffs to: (Seller, Consumer)

[4] The exact numbers are different from the Prisoner's Dilemma in table 1.1, but the ordinal rankings are the same. Numbers such as those in table 5.2 are more commonly used, because it is convenient to normalize the (*Confess, Confess*) payoffs to (0, 0) and to make most of the numbers positive rather than negative.

Table 5.3 Repeated Games in Which Reputation is Important

Application	Sidedness	Players	Actions
Prisoner's Dilemma	two-sided	Row	*Deny/Confess*
		Column	*Deny/Confess*
Duopoly	two-sided	Firm	*High price/Low price*
		Firm	*High price/Low price*
Employment	two-sided	Employer	*Bonus/No bonus*
		Employee	*Work/Shirk*
Product Quality	one-sided	Consumer	*Buy/Boycott*
		Seller	*High quality/low quality*
Entry Deterrence	one-sided	Incumbent	*Low price/High price*
		Entrant	*Enter/Stay out*
Financial Disclosure	one-sided	Corporation	*Truth/Lies*
		Investor	*Invest/Refrain*
Borrowing	one-sided	Lender	*Lend/Refuse*
		Borrower	*Repay/Default*

The Nash and iterated dominance equilibria in the one-sided Prisoner's Dilemma are still (*Confess*, *Confess*), but it is not a dominant-strategy equilibrium. Column does not have a dominant strategy, because if Row were to choose *Deny*, Column would also choose *Deny*, to obtain a payoff of 5; but if Row chooses *Confess*, Column would choose *Confess*, for a payoff of 0. *Confess* is however, weakly dominant for Row, which makes (*Confess*, *Confess*) the iterated dominant-strategy equilibrium. In both games, the players would like to persuade each other that they will cooperate, and devices that induce cooperation in the one-sided game will usually obtain the same result in the two-sided game.

5.4 Product Quality in an Infinitely Repeated Game

The Folk Theorem tells us that some perfect equilibrium of an infinitely repeated game—sometimes called an **infinite-horizon model**—can generate any pattern of behavior observed over a finite number of periods. But since the Folk Theorem is no more than a mathematical result, the strategies that generate particular patterns of behavior may be unreasonable. The theorem's value is in provoking close scrutiny of infinite horizon models and forcing the modeller to show why his equilibrium is better than a host of others. He must go beyond satisfaction of the technical criterion of perfectness and justify the strategies on other grounds.

In the simplest model of product quality, a seller can choose between producing costly high quality or costless low quality, and the buyer cannot determine quality before he purchases. If the seller would produce high quality under symmetric information, we have a one-sided Prisoner's Dilemma, as in table 5.2b. Both players are better off when the seller produces high quality and the buyer purchases the product, but the seller's weakly dominant strategy is to produce low quality, so the buyer will not purchase. This is also an example of moral hazard, the topic of chapter 7.

A potential solution is to repeat the game, allowing the firm to choose quality at each repetition. If the number of repetitions is finite, however, the outcome stays the same because of the Chainstore Paradox. In the last repetition, the subgame is identical to the one-shot game, so the firm chooses low quality. In the next-to-last repetition, it is foreseen that the last period's outcome is independent of the current actions, so the firm also chooses low quality, an argument that can be carried back to the first repetition.

If the game is repeated an infinite number of times, the Chainstore Paradox is inapplicable and the Folk Theorem says that a wide range of outcomes can be observed in equilibrium. Klein & Leffler (1981) construct a plausible equilibrium for an infinite-period model. Their original article, in the traditional verbal style of UCLA, does not phrase the result in terms of game theory, but we will recast it here.[5] In equilibrium, the firm is willing to produce a high quality product because it can sell at a high price for many periods, but consumers refuse to ever buy again from a firm that has once produced low quality. The equilibrium price is high enough that the firm is unwilling to sacrifice its future profits for a one-time windfall from deceitfully producing low quality and selling it at a high price. Although this is only one of a large number of subgame perfect equilibria, the consumers' behavior is simple and rational: no consumer can benefit by deviating from the equilibrium.

Product Quality

Players

An infinite number of potential firms and a continuum of consumers.

Order of Play

(1) An endogenous number n of firms decide to enter the market at cost F.
(2) A firm that has entered chooses its quality to be *High* or *Low*, incurring the constant marginal cost c if it picks *High* and zero if it picks *Low*. The choice is unobserved by consumers. The firm also picks a price p.
(3) Consumers decide which firms (if any) to buy from, choosing firms randomly if they are indifferent. The amount bought from firm i is denoted q_i.
(4) All consumers observe the quality of all goods purchased in that period.
(5) The game returns to (2) and repeats.

Payoffs

The consumer benefit from a product of low quality is zero, but consumers are willing to buy quantity $q(p) = \Sigma_{i=1}^{n} q_i$ for a product believed to be high quality, where $dq/dp < 0$.

[5] This is my formalization in Rasmusen (1989b). An earlier formalization is Dybvig & Spatt (unpublished).

If a firm stays out of the market, its payoff is zero.

If firm i enters, it receives $-F$ immediately. Its current end-of-period payoff is $q_i p$ if it produces *Low* quality and $q_i(p - c)$ if it produces *High* quality. The discount rate is $r \geq 0$.

That the firm can produce low quality items at zero marginal cost is unrealistic, but it is only a simplifying assumption. By normalizing the cost of producing low quality to zero, we avoid having to carry an extra variable through the analysis without affecting the result.

The Folk Theorem tells us that this game has a wide range of perfect outcomes, including a large number with erratic quality patterns like (*High*, *High*, *Low*, *High*, *Low*, *Low*, . . .). If we confine ourselves to pure-strategy equilibria with a stationary outcome of constant quality and identical behavior by all firms in the market, then the two outcomes are low quality and high quality. Low quality is always an equilibrium outcome, since it is an equilibrium of the one-shot game. If the discount rate is low enough, high quality is also an equilibrium outcome, and this will be the focus of our attention. Consider the following strategy profile:

Firms. $\tilde{n}$ firms enter. Each produces high quality and sells at price $\tilde{p}$. If a firm ever deviates from this, it thereafter produces low quality (and sells at the same price $\tilde{p}$). The values of $\tilde{p}$ and $\tilde{n}$ are given by equations (4) and (8) below.

Buyers. Buyers start by choosing randomly among the firms charging $\tilde{p}$. Thereafter, they remain with their initial firm unless it changes its price or quality, in which case they switch randomly to a firm that has not changed its price or quality.

This strategy profile is a perfect equilibrium. Each firm is willing to produce high quality and refrain from price-cutting because otherwise it would lose all its customers. If it has deviated, it is willing to produce low quality because quality is unimportant, given the absence of customers. Buyers stay away from a firm that has produced low quality because they know it will continue to do so, and they stay away from a firm that has cut prices because they know it will produce low quality. For this story to work, however, the equilibrium must satisfy three constraints that will be explained in more depth in section 7.3: incentive compatibility, competition, and market clearing.

The **incentive compatibility** constraint says that the individual firm must be willing to produce high quality. Given the buyers' strategy, if the firm ever produces low quality it receives a one-time windfall profit, but loses its future profits. The tradeoff is represented by constraint (3), which is satisfied if the discount rate is low enough.

$$\frac{q_i p}{1 + r} \leq \frac{q_i(p - c)}{r} \qquad \text{(incentive compatibility)} \qquad (3)$$

Inequality (3) determines a lower bound for the price, which must satisfy

$$\tilde{p} \geq (1 + r)c \tag{4}$$

Condition (4) will be satisfied as an equality, because any firm trying to charge a price higher than the quality-guaranteeing $\tilde{p}$ would lose all its customers.

The second constraint is that competition drives profits to zero, so firms are indifferent between entering and staying out of the market.

$$\frac{q_i(p - c)}{r} = F \quad \text{(competition)} \tag{5}$$

Treating (3) as an equation and using it to replace p in equation (5) gives

$$q_i = \frac{F}{c} \tag{6}$$

We have now determined p and q_i, and only n remains, which is determined by the equality of supply and demand. The market does not always clear in models of asymmetric information (see Stiglitz [1987]), and in this model each firm would like to sell more than its equilibrium output at the equilibrium price, but the market output must equal the quantity demanded by the market.

$$nq_i = q(p) \quad \text{(market clearing)} \tag{7}$$

Combining equations (3), (6), and (7) yields

$$\tilde{n} = \frac{cq([1 + r]c)}{F}. \tag{8}$$

We have now determined the equilibrium values, the only difficulty being the standard existence problem caused by the requirement that the number of firms be an integer (see note N5.4).

The equilibrium price is fixed because F is exogenous and demand is not perfectly inelastic, which pins down the size of firms. If there were no entry cost, but demand were still elastic, the equilibrium price would still be the unique p that satisfied constraint (3), and the market quantity would be determined by $q(p)$, but F and q_i would be undetermined. If consumers believed that any firm which might possibly produce high quality paid an exogenous dissipation cost F, the result would be a continuum of equilibria. The firms' best response would be for $\tilde{n}$ of them to pay F and produce high quality at price $\tilde{p}$, where $\tilde{n}$ is determined by the zero profit condition as a function of F. Klein & Leffler note this indeterminacy and suggest that the profits might be dissipated by some sort of brand-specific capital. The history of the industry may also explain the number of firms. Schmalensee (1982) shows how a pioneering brand can retain a large market share because consumers are unwilling to investigate the quality of new brands.

This idea can be extended to labor contracts also, and is very similar to the notion of the "efficiency wage" in section 8.4.

5.5 Markov Equilibria and Overlapping Generations in the Game of Customer Switching Costs

The next model demonstrates a general modelling technique, the **overlapping generations model**, in which different cohorts of otherwise identical players enter and leave the game with overlapping "lifetimes," and presents a new equilibrium concept, "Markov equilibrium." The best-known example of an overlapping-generations model is the original consumption-loans model of Samuelson (1958). The models are most often used in macroeconomics, but they can also be useful in microeconomics. Klemperer (1987) has stimulated considerable interest in customers who incur costs in moving from one seller to another. The model used here will be that of Farrell & C. Shapiro (1988).

Customer Switching Costs

(Farrell & C. Shapiro [1988])

Players

Firms Apex and Brydox, and a series of customers, each of whom is first called a youngster and then an oldster.

Order of Play

(1a) Brydox, the initial incumbent, picks the incumbent price p_1^i.
(1b) Apex, the initial entrant, picks the entrant price p_1^e.
(1c) The oldster picks a firm.
(1d) The youngster picks a firm.
(1e) Whichever firm attracted the youngster becomes the incumbent.
(1f) The oldster dies and the youngster becomes an oldster.
(2a) Return to (1a), possibly with new identities for entrant and incumbent.

Payoffs

The discount factor is δ. The customer reservation price is R and the switching cost is c. The per period payoffs in period t are, for $j = (i, e)$,

$$\pi_{firm\ j} = \begin{cases} 0 & \text{if no customers are attracted} \\ p_t^j & \text{if just oldsters or just youngsters are attracted} \\ 2p_t^j & \text{if both oldsters and youngsters are attracted} \end{cases}$$

$$\pi_{oldster} = \begin{cases} R - p_t^i & \text{if he buys from the incumbent} \\ R - p_t^e - c & \text{if he switches to the entrant} \end{cases}$$

$$\pi_{youngster} = \begin{cases} R - p_t^i & \text{if he buys from the incumbent} \\ R - p_t^e & \text{if he buys from the entrant} \end{cases}$$

Finding all the perfect equilibria of an infinite game like this one is difficult, so we will follow Farrell and Shapiro in limiting ourselves to the much easier task of finding the perfect Markov equilibrium, which is unique.

A **Markov strategy** *is a strategy that, at each node, chooses the action independently of the history of the game except for the immediately preceding action (or actions, if they were simultaneous).*

Here, a firm's Markov strategy is its price as a function of whether the particular firm is the incumbent or the entrant, and not a function of the entire past history of the game.

There are two ways to use Markov strategies: (1) just look for equilibria that use Markov strategies, and (2) disallow non-Markov strategies and then look for equilibria. Because the first approach does not disallow non-Markov strategies, the equilibrium must be such that no player wants to deviate by using any other strategy, whether Markov or not. This is just a way of eliminating possible multiple equilibria by discarding ones that use non-Markov strategies. The second way is much more dubious, because it requires the players not to use non-Markov strategies, even if they are best responses. A **perfect Markov equilibrium** uses the first approach: it is a perfect equilibrium that happens to use only Markov strategies.

Brydox, the initial incumbent, moves first and chooses p^i low enough that Apex is not tempted to choose $p^e < p^i - c$ and steal away the oldsters. Apex's profit is p^i if it chooses $p^e = p^i$ and serves just youngsters, and $2(p^i - c)$ if it chooses $p^e = p^i - c$ and serves both oldsters and youngsters. Brydox chooses p^i to make Apex indifferent between these alternatives, so

$$p^i = 2(p^i - c), \tag{9}$$

and

$$p^i = p^e = 2c. \tag{10}$$

In equilibrium, Apex and Brydox take turns being the incumbent and charge the same price.

Because the game lasts forever and the equilibrium strategies are Markov, we can use a trick from dynamic programming to calculate the payoffs from being the entrant versus being the incumbent. The equilibrium payoff of the current entrant is the immediate payment of p^e plus the discounted value of being the incumbent in the next period:

$$\pi_e^* = p^e + \delta\pi_i^*. \tag{11}$$

The incumbent's payoff can be similarly stated as the immediate payment of p^i plus the discounted value of being the entrant in the next period:

$$\pi_i^* = p^i + \delta\pi_e^*. \tag{12}$$

We could use equation (10) to substitute for p^e and p^i, which would leave us with the two equations (11) and (12) for the two unknowns π_i^* and π_e^*, but an easier way to compute the payoff is to realize that in equilibrium the incumbent and the entrant sell the same amount at the same price, so $\pi_i^* = \pi_e^*$ and equation (12) becomes

$$\pi_i^* = 2c + \delta \pi_i^*. \tag{13}$$

It follows that

$$\pi_i^* = \pi_e^* = \frac{2c}{1 - \delta}. \tag{14}$$

Prices and total payoffs are increasing in the switching cost c, because that is what gives the incumbent market power and prevents ordinary competition of the ordinary Bertrand kind analyzed in section 13.2. The total payoffs are increasing in δ for the usual reason that future payments increase in value as δ approaches one.

Notes

N5.1 Finitely Repeated Games and the Chainstore Paradox

- The Chainstore Paradox does not apply to all games as neatly as to Entry Deterrence and The Prisoner's Dilemma. If the one-shot game has only one Nash equilibrium, the perfect equilibrium of the finitely repeated game is unique and has that same outcome. But if the one-shot game has multiple Nash equilibria, the perfect equilibrium of the finitely repeated game can have not only the one-shot outcomes, but others besides. See Benoit & Krishna (1985), Harrington (1987), and Moreaux (1985).
- John Heywood is Bartlett's source for the term "tit-for-tat," from the French "tant pour tant."
- A realistic expansion of a game's strategy space may eliminate the Chainstore Paradox. D. Hirshleifer & Rasmusen (1989), for example, show that allowing the players in a multi-person finitely repeated Prisoner's Dilemma to ostracize offenders can enforce cooperation even if there are economies of scale in the number of players who cooperate and are not ostracized.
- The peculiarity of the unique Nash equilibrium for the repeated Prisoner's Dilemma was noticed long before Selten (1978) (see Luce & Raiffa [1957] p. 99), but the term Chainstore Paradox is now generally used for all unravelling games of this kind.
- *An* **epsilon-equilibrium** *is a strategy profile s^* such that no player has more than an ε incentive to deviate from his strategy given that the other players do not deviate. Formally,*

$$\forall i, \; \pi_i(s_i^*, s_{-i}^*) \geq \pi_i(s_i', s_{-i}^*) - \varepsilon, \; \forall s_i' \in S_i. \tag{15}$$

  Radner (1980) has shown that cooperation can arise as an ε-equilibrium of the finitely repeated Prisoner's Dilemma. Fudenberg & Levine (1986) compare the ε-equilibria of finite games with the Nash equilibria of infinite games. Other concepts besides Nash can also use the ε-equilibrium idea.
- A general way to decide whether a mathematical result is a trick of infinity is to see if the same result is obtained as the limit of results for longer and longer finite models. Applied to games, a good criterion for picking among equilibria of an infinite game is to select one which is the limit of the equilibria for finite games as the number of periods gets longer. Fudenberg & Levine (1986) show under what conditions one can find the equilibria of infinite-horizon games by this process. For the Prisoner's Dilemma, (*Always Confess*) is the only equilibrium in all finite games, so it uniquely satisfies the criterion.

- Defining payoffs in games that last an infinite number of periods presents the problem that the total payoff is infinite for any positive payment per period. Ways to distinguish one infinite amount from another include the following.

  (1) Use an **overtaking criterion**. Payoff stream π is preferred to $\tilde{\pi}$ if there is some time T^* such that for every $T \geq T^*$,

$$\sum_{t=1}^{T} \delta^t \pi_t > \sum_{t=1}^{T} \delta^t \tilde{\pi}_t.$$

  (2) Specify that the discount rate is strictly positive, and use the present value. Since payments in distant periods count for less, the discounted value is finite unless the payments are growing faster than the discount rate.

  (3) Use the average payment per period, a tricky method since some sort of limit needs to be taken as the number of periods averaged approaches infinity.

  Whatever the approach, game theorists assume that the payoff function is **additively separable** over time, which means that the total payoff is based on the sum or average, possibly discounted, of the one-shot payoffs. Macroeconomists worry about this assumption, which rules out, for example, a player whose payoff is very low if any of his one-shot payoffs dips below a certain subsistence level. The issue of separability will arise again in section 12.5 when we discuss durable monopoly.
- Ending in finite time with probability one means that the limit of the probability that the game has ended by date t approaches one as t tends to infinity; the probability that the game lasts till infinity is zero. Equivalently, the expectation of the end date is finite, which it could not be were there a positive probability of infinite length.

N5.2 Infinitely Repeated Games, Minimax Punishments, and the Folk Theorem

- References on the Folk Theorem include Aumann (1981), Fudenberg & Maskin (1986), Fudenberg & Tirole (1991a, pp. 152-62), and Rasmusen (1992a). The most commonly cited version of the Folk Theorem says that if Conditions 1 to 3 are satisfied, then:

  Any payoff profile that strictly Pareto dominates the minimax payoff profiles in the mixed extension of an n-person one-shot game with finite action sets is the average payoff in some perfect equilibrium of the infinitely repeated game.

- The evolutionary approach can also be applied to the repeated Prisoner's Dilemma. Boyd & Lorberbaum (1987) show that no pure strategy, including Tit-for-Tat, is evolutionarily stable in a population-interaction version of the Prisoner's Dilemma. J. Hirshleifer & Martinez-Coll (1988) have found that Tit-for-Tat is no longer part of an ESS in an evolutionary Prisoner's Dilemma if (1) more complicated strategies have higher computation costs; or (2) sometimes a *Deny* is observed to be a *Confess* by the other player.
- **Trigger strategies** or *trigger-price strategies* are an important kind of strategies for repeated games. Consider the oligopolist facing uncertain demand (as in Stigler [1964]). He cannot tell whether the low demand he observes facing him is due to Nature or to price cutting by his fellow oligopolists. Two things that could trigger him to cut his own price in retaliation are a series of periods with low demand or a single period of especially low demand. Finding an optimal trigger strategy is a difficult problem (see Porter [1983a]). Trigger strategies are usually not subgame perfect unless the game is infinitely repeated, in which case they are a subset of the equilibrium strategies. Recent work has looked carefully at what trigger strategies are possible and optimal for players in infinitely repeated games; see Abreu, Pearce & Staccheti (1990).

  Empirical work on trigger strategies includes Porter (1983b), who examines price wars between railroads in the 19th century, and Slade (1987), who concluded that price wars among gas stations in Vancouver used small punishments for small deviations rather than big punishments for big deviations.

- A macroeconomist's technical note related to the similarity of infinite games and games with a constant probability of ending is Blanchard (1979), which discusses speculative bubbles.
- In the repeated Prisoner's Dilemma, if the end date is infinite with positive probability and only one player knows it, cooperation is possible by reasoning similar to that of the Gang of Four Theorem in section 6.4.
- An alternative to Condition 3 (dimensionality) in the Folk Theorem is

Condition 3′: *The repeated game has a "desirable" subgame-perfect equilibrium in which the strategy profile $\bar{s}$ played each period gives player i a payoff that exceeds his payoff from some other "punishment" subgame-perfect equilibrium in which the strategy profile $\underline{s}^i$ is played each period:*

$$\exists \bar{s} : \forall i, \exists \underline{s}^i : \pi_i(\underline{s}^i) < \pi_i(\bar{s}). \tag{16}$$

Condition 3′ is useful because sometimes it is easy to find a few perfect equilibria. To enforce the desired pattern of behavior, use the "desirable" equilibrium as a carrot and the "punishment" equilibrium as a self-enforcing stick (See Rasmusen [1992a]).

- Any Nash equilibrium of the one-shot game is also a perfect equilibrium of the finitely or infinitely repeated game.

N5.3 Reputation: The One-sided Prisoner's Dilemma

- *A game that is repeated an infinite number of times without discounting is called a* **supergame**. There is no connection between the terms "supergame" and "subgame."
- The terms, "one-sided" and "two-sided" Prisoner's Dilemma, are my inventions. Only the two-sided version is a true Prisoner's Dilemma according to the definition of note N1.2.
- Empirical work on reputation is scarce. One worthwhile effort is Jarrell & Peltzman (1985), which finds that product recalls inflict costs greatly in excess of the measurable direct costs of the operations. The investigations into actual business practice of Macaulay (1963) are much cited and little imitated. He notes that reputation seems to be more important than the written details of business contracts.
- **Vengeance and Gratitude.** Most models have excluded these feelings (although see J. Hirshleifer [1987]), which can be modelled in two ways.
  (1) A player's current utility from *Confess* or *Deny* depends on what the other player has played in the past; or
  (2) A player's current utility depends on current actions and the other players' current utility in a way that changes with the past actions of the other player.
  The two approaches are subtly different in interpretation. In (1), the joy of revenge is in the action of confessing. In (2), the joy of revenge is in the discomfiture of the other player. Especially if the players have different payoff functions, these two approaches can lead to different results.

N5.4 Product Quality in an Infinitely Repeated Game

- The game of Product Quality may also be viewed as a principal agent model of moral hazard (see chapter 7). The seller (an agent), takes the action of choosing quality that is unobserved by the buyer (the principal), but which affects the principal's payoff, an interpretation used in much of the Stiglitz (1987) survey of the links between quality and price.

  The intuition behind the Klein & Leffler model is similar to the explanation for high wages in the Shapiro & Stiglitz (1984) model of involuntary unemployment (section 8.4). Consumers, seeing a low price, realize that with a price that low the firm cannot resist lowering quality to make short-term profits. A large margin of profit is needed for the firm to decide on continuing to produce high quality.

- A paper related to Klein & Leffler (1981) is Shapiro (1983), which reconciles a high price with free entry by requiring that firms price under cost during the early periods in order to build up a reputation. If consumers believe, for example, that any firm charging a high price for any of the first five periods has produced a low-quality product, but any firm charging a high price thereafter has produced high quality, then firms behave accordingly and the beliefs are confirmed. That the beliefs are self-confirming does not make them irrational; it only means that many different beliefs are rational in the many different equilibria.
- An equilibrium exists in the Product Quality model only if the entry cost F is just the right size to make n an integer in equation (8). Any of the usual assumptions to get around the integer problem could be used: allowing potential sellers to randomize between entering and staying out; assuming that, for historical reasons, n firms have already entered; or assuming that firms lie on a continuum and that the fixed cost is a uniform density across firms that have entered.

N5.5 Markov Equilibria and Overlapping Generations in the Game of Customer Switching Costs

- We assumed that the incumbent chooses its price first, but the alternation of incumbency remains even if we make the opposite assumption. The natural assumption is that prices are chosen simultaneously, but because of the discontinuity in the payoff function, that subgame has no equilibrium in pure strategies.

Problems

5.1: Overlapping Generations.[6]

There is a long sequence of players. One player is born in each period t, and he lives throughout periods t and $t + 1$. Thus, two players are alive in any one period, a youngster and an oldster. Each player is born with one unit of chocolate, which cannot be stored. Utility is increasing in chocolate consumption, and a player is very unhappy if he consumes less than 0.3 units of chocolate in a period: the per-period utility functions are $U(C) = -1$ for $C < 0.3$ and $U(C) = C$ for $C \geq 0.3$, where C is consumption. Players can give away their chocolate, but, since chocolate is the only good, they cannot sell it. A player's action is to consume X units of chocolate as a youngster and give away $1 - X$ to some oldster.

(5.1a) If there is finite number of generations, what is the unique Nash equilibrium?

(5.1b) If there are an infinite number of generations, what are the two Pareto-ranked perfect equilibria?

(5.1c) If there is a probability θ at the end of each period (after consumption takes place) that barbarians will invade and steal all the chocolate (leaving the civilized people with payoffs of -1 for any X), what is the highest value of θ that still allows for an equilibrium with $X = 0.5$?

[6] See Samuelson (1958).

5.2: Product Quality with Lawsuits.

Modify the game of Product Quality (section 5.8) by assuming that if the seller misrepresents his quality he must, as a result of a class-action suit, pay damages of x per unit sold, where $x \in (0, c]$.

(5.2a) What is $\tilde{p}$ as a function of x, F, c, and r? Is $\tilde{p}$ greater than when $x = 0$?
(5.2b) What is the equilibrium output per firm? Is it greater than when $x = 0$?
(5.2c) What is the equilibrium number of firms? Is it greater than when $x = 0$?
(5.2d) If, instead of x per unit, the seller pays X to a law firm to successfully defend him, what is the incentive compatibility constraint?

5.3: Repeated Games.[7]

Players Benoit and Krishna repeat the following game three times, with discounting:

Table 5.4 A Benoit-Krishna Game

		Krishna		
		Deny	*Waffle*	*Confess*
	Deny	10, 10	−1, −12	−1, 15
Benoit:	*Waffle*	−12, −1	8, 8	−1, −1
	Confess	15, −1	8, −1	0, 0

Payoffs to: (Benoit, Krishna)

(5.3a) Why is there no equilibrium in which the players play *Deny* in all three periods?
(5.3b) Describe a perfect equilibrium in which both players pick *Deny* in the first two periods.
(5.3c) Adapt your equilibrium to the twice-repeated game.
(5.3d) Adapt your equilibrium to the T-repeated game.
(5.3e) What is the greatest discount rate for which your equilibrium still works in the 3-period game?

5.4: Repeated Entry Deterrence.

Assume that Entry Deterrence I is repeated an infinite number of times, with a tiny discount rate and with payoffs received at the start of each period. In each period, the entrant chooses *Enter* or *Stay out*, even if he entered previously.

(5.4a) What is a perfect equilibrium in which the entrant enters each period?
(5.4b) Why is (*Stay out*, *Fight*) not a perfect equilibrium?
(5.4c) What is a perfect equilibrium in which the entrant never enters?

[7] See Benoit & Krishna (1985).

(5.4d) What is the maximum discount rate for which your strategy profile in part (c) is still an equilibrium?

5.5: Repeated Prisoner's Dilemma.

Set $P = 0$ in the general Prisoner's Dilemma in table 1.9, and assume that $2R > S + T$.

(5.5a) Show that the Grim Strategy, when played by both players, is a perfect equilibrium for the infinitely repeated game. What is the maximum discount rate for which the Grim Strategy remains an equilibrium?

(5.5b) Show that Tit-for-Tat is not a perfect equilibrium in the infinitely repeated Prisoner's Dilemma with no discounting.

6 Dynamic Games with Asymmetric Information

6.1 Perfect Bayesian Equilibrium: Entry Deterrence II and III

Asymmetric information, and, in particular, incomplete information, is enormously important in game theory. This is particularly true for dynamic games, since when the players have several moves in sequence, their earlier moves may convey private information that is relevant to the decisions of players moving later on. Revealing and concealing information are the basis of much of strategic behavior and are especially useful as ways of explaining actions that would be irrational in a nonstrategic world.

Chapter 4 showed that even if there is symmetric information in a dynamic game, Nash equilibrium may need to be refined using subgame perfectness if the modeller is to make sensible predictions. Asymmetric information requires a somewhat different refinement to capture the idea of sunk costs and credible threats, and section 6.1 sets out the standard refinement of perfect Bayesian equilibrium. Section 6.2 shows that even this may not be enough refinement to guarantee uniqueness and discusses further refinements based on out-of-equilibrium beliefs. Section 6.3 uses the idea to show that a player's ignorance may work to his advantage, and to explain how even when all players know something, lack of common knowledge still affects the game. Section 6.4 introduces incomplete information into the repeated Prisoner's Dilemma and shows the Gang of Four solution to the Chainstore Paradox of Chapter 5. Section 6.5 describes the celebrated Axelrod tournament, an experimental approach to the same paradox.

Subgame Perfectness is Not Enough

In games of asymmetric information, we will still require that an equilibrium be subgame perfect, but the mere forking of the game tree might not be relevant to a player's decision, because with asymmetric information he does not know which fork the game has taken. Smith might know he is at one of two different nodes depending on whether Jones has high or low production costs, but if he does not know the exact node, the "subgames" starting at each node are irrelevant to his decisions. In fact, they are not even subgames as we have defined them, because they cut across Smith's information sets. This can be seen in an asymmetric information version of Entry Deterrence I (section 4.3). In Entry Deterrence I, the incumbent colluded with the entrant because fighting him was more costly than colluding once the entrant had entered. Now, let us set up the game to allow some entrants to be *Strong* and some *Weak* in the sense that it is more costly for the incumbent to choose *Fight* against a *Strong* entrant than a *Weak* one. The incumbent's payoff from *Fight|Strong* will be 0, as before, but his payoff from *Fight|Weak* will be X, where X will take values ranging from 0 (Entry Deterrence I) to 300 (Entry Deterrence IV and V) in different versions of the game.

Entry Deterrence II, III, and IV will all have the extensive form shown in figure 6.1. With 50 percent probability, the incumbent's payoff from *Fight* is X rather than the 0 in Entry Deterrence I, but the incumbent does not know which payoff is the correct one in the particular realization of the game. This is modelled as an initial move by Nature, who chooses between the entrant being *Weak* or *Strong*, unobserved by the incumbent.

Entry Deterrence II: Fighting is Never Profitable

In Entry Deterrence II, $X = 1$, so information is not very asymmetric. It is common knowledge that the incumbent never benefits from *Fight*, even though his exact payoff might be zero or it might be one. Unlike in the case of Entry Deterrence I, however, subgame perfectness does not rule out any Nash equilibria, because the only subgame is the subgame starting at node N, which is the entire game. A subgame cannot start at nodes E_1 or E_2, because neither of those nodes are singletons in the information partitions. Thus, the implausible Nash equilibrium, (*Stay Out*, *Fight*), escapes elimination by a technicality.

Figure 6.1 Entry Deterrence II, III, and IV

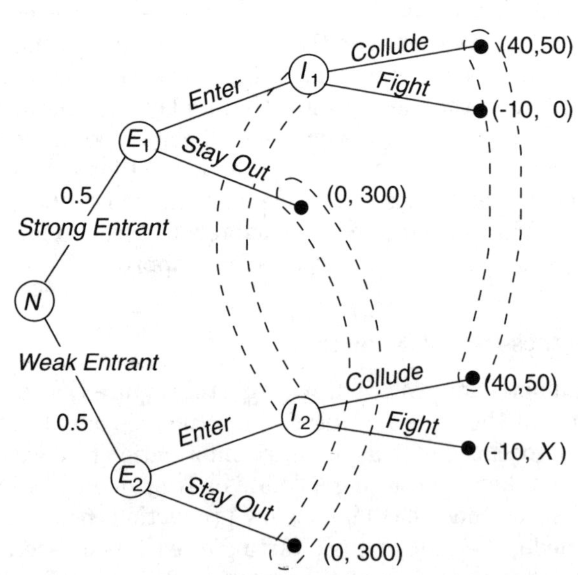

The equilibrium concept needs to be refined in order to eliminate the implausible equilibrium. Two general approaches can be taken: either introduce small "trembles" into the game, or require that strategies be best responses given rational beliefs. The first approach takes us to the "trembling hand-perfect" equilibrium, while the second takes us to the "perfect Bayesian" and "sequential" equilibrium. The results are similar whichever approach is taken.

Trembling-Hand Perfectness

Trembling-hand perfectness is an equilibrium concept introduced by Selten (1975) according to which a strategy that is to be part of an equilibrium must continue to be optimal for the player even if there is a small chance that the other player will pick an out-of-equilibrium action (i.e., that the other player's hand will "tremble").

Trembling-hand perfectness is defined for games with finite action sets as follows.

> *The strategy profile s^* is a* **trembling-hand perfect** *equilibrium if for any ε there is a vector of positive numbers $\delta_1, \ldots, \delta_n \in [0,1]$ and a vector of completely mixed strategies $\sigma_1, \ldots \sigma_n$ such that the perturbed game where every strategy is replaced by $(1 - \delta_i)s_i + \delta_i \sigma_i$ has a Nash equilibrium in which every strategy is within distance ε of s^*.*

Every trembling-hand perfect equilibrium is subgame perfect; indeed, section 4.1 justified subgame perfectness using a tremble argument. Unfortunately, it is often hard to tell whether a strategy profile is trembling-hand perfect, and the concept is undefined for games with continuous strategy spaces because it is hard to work with mixtures of a continuum (see note N3.1). Moreover, the equilibrium depends on which trembles are chosen, and deciding why one tremble should be more common than another may be difficult.

Perfect Bayesian Equilibrium and Sequential Equilibrium

The second approach to asymmetric information, introduced by Kreps & Wilson (1982b) in the spirit of Harsanyi (1967), is to start with prior beliefs, common to all players, that specify the probabilities with which Nature chooses the types of the players at the beginning of the game. Some of the players observe Nature's move and update their beliefs, while other players can update their beliefs only by deductions they make from observing the actions of the informed players.

The deductions used to update beliefs are based on the actions specified by the equilibrium. When players update their beliefs, they assume that the other players are following the equilibrium strategies, but since the strategies themselves depend on the beliefs, an equilibrium can no longer be defined based on strategies alone. Under asymmetric information, an equilibrium is a strategy profile and a set of beliefs such that the strategies are best responses. The combination of beliefs and strategies is called an **assessment** by Kreps and Wilson.

On the equilibrium path, all that the players need to update their beliefs are their priors and Bayes' Rule, but off the equilibrium path this is not enough.

Suppose that in equilibrium the entrant always enters. If for whatever reason the impossible happens and the entrant stays out, what is the incumbent to think about the probability that the entrant is weak? Bayes' Rule does not help, because when *Prob(data)* = 0, which is the case for data such as *Stay Out* which is never observed in equilibrium, the posterior belief cannot be calculated using Bayes' Rule. From section 2.4,

$$Prob(Weak|Stay\ Out) = \frac{Prob(Stay\ Out|Weak)Prob(Weak)}{Prob(Stay\ Out)}. \tag{1}$$

The posterior *Prob(Weak|Stay Out)* is undefined, because (1) requires dividing by zero.

A natural way to define equilibrium is as a strategy profile consisting of best responses given that equilibrium beliefs follow Bayes' Rule and that out-of-equilibrium beliefs follow a specified pattern that does not contradict Bayes' Rule.

> *A* **perfect Bayesian equilibrium** *is a strategy profile s and a set of beliefs μ such that at each node of the game:*
>
> *(1) The strategies for the remainder of the game are Nash given the beliefs and strategies of the other players.*
>
> *(2) The beliefs at each information set are rational given the evidence appearing thus far in the game (meaning that they are based, if possible, on priors updated by Bayes' Rule, given the observed actions of the other players under the hypothesis that they are in equilibrium).*

Kreps & Wilson (1982b) use this idea to form their equilibrium concept of sequential equilibrium, but they impose a third condition, defined only for games with discrete strategies, to restrict beliefs a little further:

> *(3) The beliefs are the limit of a sequence of rational beliefs, i.e., if (μ^*, s^*) is the equilibrium assessment, then some sequence of rational beliefs and completely mixed strategies converges to it:*
>
> $(\mu^*, s^*) = \text{Lim}_{n\to\infty}\ (\mu^n, s^n)$ for some sequence (μ^n, s^n) in $\{\mu, s\}$.

Condition (3) is quite reasonable and makes sequential equilibrium close to trembling-hand perfect equilibrium, but it adds more to the concept's difficulty than to its usefulness. If players are using the sequence of completely mixed strategies s^n, then every action is taken with some positive probability, so Bayes' Rule can be applied to form the beliefs μ^n after any action is observed. Condition (3) says that the equilibrium assessment has to be the limit of some such sequence (though not of every such sequence). For the rest of the book we will use perfect Bayesian equilibrium and dispense with Condition (3), although it usually can be satisfied.

Sequential equilibria are always subgame perfect (Condition (1) takes care of that). Every trembling-hand perfect equilibrium is a sequential equilibrium, and "almost every" sequential equilibrium is trembling-hand perfect. Every sequential equilibrium is perfect Bayesian, but not every perfect Bayesian equilibrium is sequential.

Back to Entry Deterrence II

Armed with the concept of the perfect Bayesian equilibrium, we can find a sensible equilibrium for Entry Deterrence II.

> Entrant: *Enter|Weak, Enter|Strong*
> Incumbent: *Collude*
> Beliefs: *Prob(Strong|Stay Out)* = 0.4

In this equilibrium the entrant enters whether he is *Weak* or *Strong*. The incumbent's strategy is *Collude*, which is not conditioned on Nature's move, since he does not observe it. Because the entrant enters regardless of Nature's move, an out-of-equilibrium belief for the incumbent if he should observe *Stay Out* must be specified, and this belief is arbitrarily chosen to be that the incumbent's subjective probability that the entrant is *Strong* is 0.4 given his observation that the entrant deviated by choosing *Stay Out*. Given this strategy profile and out-of-equilibrium belief, neither player has incentive to change his strategy.

There is no perfect Bayesian equilibrium in which the entrant chooses *Stay Out*. *Fight* is a bad response even under the most optimistic possible belief, that the entrant is *Weak* with probability 1. Notice that perfect Bayesian equilibrium is not defined structurally, like subgame perfectness, but rather in terms of optimal responses. This enables it to come closer to the economic intuition which we wish to capture by an equilibrium refinement.

Finding the perfect Bayesian equilibrium of a game, like finding the Nash equilibrium, requires intelligence. Algorithms are not useful. To find a Nash equilibrium, the modeller thinks about his game, picks a plausible strategy profile, and tests whether the strategies are best responses to each other. To make it a perfect Bayesian equilibrium, he notes which actions are never taken in equilibrium and specifies the beliefs that players use to interpret those actions. He then tests whether each player's strategies are best responses given his beliefs at each node, checking in particular whether any player would like to take an out-of-equilibrium action in order to set in motion the other players' out-of-equilibrium beliefs and strategies. This process does not involve testing whether a player's beliefs are beneficial to the player, because players do not choose their own beliefs; the priors and out-of-equilibrium beliefs are exogenously specified by the modeller.

One might wonder why the beliefs have to be specified in Entry Deterrence II. Does not the game tree specify the probability that the entrant is *Weak*? What difference does it make if the entrant stays out? Admittedly, Nature does choose each type with probability 0.5, so if the incumbent had no other information than this prior, that would be his belief. But the entrant's action might convey additional information. The concept of perfect Bayesian equilibrium leaves the modeller free to specify how the players form beliefs from that additional information, so long as the beliefs do not violate Bayes' Rule. (A technically valid choice of beliefs by the modeller might still be met with scorn, though, as with any silly assumption.) Here, the equilibrium says that if the entrant stays out, the incumbent believes he is *Strong* with probability 0.4 and *Weak* with probability 0.6, beliefs that are arbitrary but do not contradict Bayes' Rule.

In Entry Deterrence II the out-of-equilibrium beliefs do not and should not matter. If the entrant chooses *Stay Out*, the game ends, so the incumbent's beliefs are irrelevant. Perfect Bayesian equilibrium was only introduced as a way out of a technical problem. In the next section, however, the precise out-of-equilibrium beliefs will be crucial to which strategy profiles are equilibria.

6.2 Refining Perfect Bayesian Equilibrium: The PhD Admissions Game

Entry Deterrence III: Fighting is Sometimes Profitable

In Entry Deterrence III, assume that $X = 60$, not $X = 1$. This means that fighting is more profitable for the incumbent than collusion if the entrant is *Weak*. As before, the entrant knows if he is *Weak*, but the incumbent does not. Retaining the prior after observing out-of-equilibrium actions, which in this game is *Prob(Strong)* = 0.5, is a convenient way to form beliefs that is called **passive conjectures.** The following is a perfect Bayesian equilibrium which uses passive conjectures.

> **A Plausible Equilibrium for Entry Deterrence III**
> Entrant: *Enter|Weak, Enter|Strong*
> Incumbent: *Collude*
> Beliefs: *Prob(Strong|Stay Out)* = 0.5

In choosing whether to enter, the entrant must predict the incumbent's behavior. If the probability that the entrant is *Weak* is 0.5, the expected payoff to the incumbent from choosing *Fight* is 30 (= 0.5[0] + 0.5[60]), which is less than the payoff of 50 from *Collude.* The incumbent will collude, so the entrant enters. The entrant may know that the incumbent's payoff is actually 60, but that is irrelevant to the incumbent's behavior.

The out-of-equilibrium belief does not matter to this first equilibrium, although it will in other equilibria of the same game. Although beliefs in a perfect Bayesian equilibrium must follow Bayes' Rule, that puts very little restriction on how players interpret out-of-equilibrium behavior. Out-of-equilibrium behavior is "impossible," so when it does occur there is no obvious way the player should react. Some beliefs may seem more reasonable than others, however, and Entry Deterrence III has another equilibrium that requires less plausible beliefs off the equilibrium path.

> **An Implausible Equilibrium for Entry Deterrence III**
> Entrant: *Stay Out|Weak, Stay Out|Strong*
> Incumbent: *Fight*
> Beliefs: *Prob(Strong|Enter)* = 0.1

This is an equilibrium because if the entrant were to deviate and enter, the incumbent would calculate his payoff from fighting to be 54 (= 0.1[0] + 0.9[60]), which is greater than the *Collude* payoff of 50. The entrant would therefore stay out.

The beliefs in the implausible equilibrium are different and less reasonable than in the plausible equilibrium. Why should the incumbent believe that weak entrants would enter mistakenly nine times as often as strong entrants? The beliefs do not violate Bayes' Rule, but they have no justification.

The reasonableness of the beliefs is important because if the incumbent uses passive conjectures, the implausible equilibrium breaks down. With passive conjectures, the incumbent would want to change his strategy to *Collude*, because the expected payoff from *Fight* would be less than 50. The implausible equilibrium is less robust with respect to beliefs than the plausible equilibrium, and it requires beliefs that are harder to justify.

Even though dubious outcomes may be perfect Bayesian equilibria, the concept does have some bite, ruling out other dubious outcomes. There does not, for example, exist an equilibrium in which the entrant enters only if he is *Strong* and stays out if he is *Weak* (called a "separating equilibrium" because it separates out different types of players). Such an equilibrium would have to look like this:

A Conjectured Separating Equilibrium for Entry Deterrence III
Entrant: *Stay Out|Weak*, *Enter|Strong*
Incumbent: *Collude*

No out-of-equilibrium beliefs are specified for the conjectures in the separating equilibrium because there is no out-of-equilibrium behavior about which to specify them. Since the incumbent might observe either *Stay out* or *Enter* in equilibrium, the incumbent will always use Bayes' Rule to form his beliefs. He will believe that an entrant who stays out must be weak and an entrant who enters must be strong. This conforms to the idea behind Nash equilibrium that each player assumes that the other follows the equilibrium strategy, and then decides how to reply. Here the incumbent's best response, given his beliefs, is *Collude|Enter*, so that is the second part of the proposed equilibrium. But this cannot be an equilibrium, because the entrant would want to deviate. Knowing that entry would be followed by collusion, even the weak entrant would enter. So there cannot be an equilibrium in which the entrant enters only when strong.

PhD Admissions Game

Passive conjectures may not always be the most satisfactory belief, as the next example shows. Suppose that a university knows that 90 percent of the population hate economics and would be unhappy in its PhD program, and that 10 percent love economics and would do well. In addition, it cannot observe the applicant's type. If the university rejects an application, its payoff is 0 and the applicant's is -1 because of the trouble needed to apply. If the university accepts the application of someone who hates economics, the payoffs of both university and student are -10, but if the applicant loves economics, the payoffs are $+20$ for each player. Figure 6.2 shows this game in extensive form. The population proportions are represented by a node at which Nature chooses the student to be a *Lover* or *Hater* of economics.

Figure 6.2 PhD Admissions

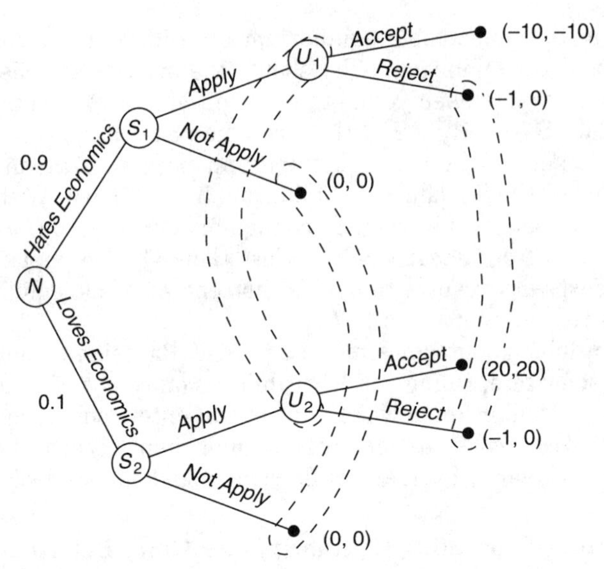

PhD Admissions is a signalling game of the kind we will look at in chapter 10. It has various perfect Bayesian equilibria that differ in their out-of-equilibrium beliefs, but the equilibria can be divided into two distinct categories, depending on the outcome: the **separating equilibrium**, in which the lovers of economics apply and the haters do not, and the **pooling equilibrium**, in which neither type of student applies.

A Separating Equilibrium for PhD Admissions
Student: *Apply|Lover, Do Not Apply|Hater*
University: *Admit*

The separating equilibrium does not need to specify out-of-equilibrium beliefs, because Bayes' Rule can always be applied whenever both of the two possible actions *Apply* and *Do Not Apply* can occur in equilibrium.

A Pooling Equilibrium for PhD Admissions
Student: *Do Not Apply|Lover, Do Not Apply|Hater*
University: *Reject*
Beliefs: *Prob(Hater|Apply)* = 0.9 (passive conjectures)

The pooling equilibrium is supported by passive conjectures. Both types of students refrain from applying because they believe correctly that they would be rejected and receive a payoff of −1; and the university is willing to reject any student who foolishly applied, believing that he is a *Hater* with 90 percent probability.

Because the perfect Bayesian equilibrium concept imposes no restrictions on out-of-equilibrium beliefs, researchers starting with McLennan (1985) have come up with a variety of exotic refinements of the equilibrium concept, a number of which are listed in note N6.2. Let us consider whether various alternatives to passive conjectures would support the pooling equilibrium in PhD Admissions.

Passive Conjectures.

$$Prob(Hater|Apply = 0.9.$$

This is the belief specified above, under which out-of-equilibrium behavior leaves beliefs unchanged from the prior. The argument for passive conjectures is that the student's application is a mistake, and that both types are equally likely to make mistakes, although *Haters* are more common in the population. This supports the pooling equilibrium.

The Intuitive Criterion.

$$Prob(Hater|Apply) = 0.$$

Under the Intuitive Criterion of Cho & Kreps (1987), if there is a type of informed player who could not benefit from the out-of-equilibrium action no matter what beliefs were held by the uninformed player, the uninformed player's belief must put zero probability on that type. Here, the *Hater* could not benefit from applying under any possible beliefs of the university, so the university puts zero probability on an applicant being a *Hater*. This argument will not support the pooling equilibrium, because if the university holds this belief, it will want to admit anyone who applies.

Complete Robustness.

$$Prob(Hater|Apply) = m, 0 \leq m \leq 1.$$

Under this approach, the equilibrium strategy profile must consist of responses that are best, given any and all out-of-equilibrium beliefs. Our equilibrium for Entry Deterrence II satisfied this requirement. Complete robustness rules out a pooling equilibrium in PhD Admissions, because a belief like $m = 0$ makes accepting applicants a best response, in which case only the *Lover* will apply. A useful first step in analyzing conjectured pooling equilibria is to test whether they can be supported by extreme beliefs such as $m = 0$ and $m = 1$.

An ad hoc Specification.

$$Prob(Hater|Apply) = 1.$$

Sometimes the modeller can justify beliefs by the circumstances of the particular game. Here, one could argue that anyone so foolish as to apply knowing that the university would reject them could not possibly have the good taste to love economics. This supports the pooling equilibrium also.

An alternative approach to the problem of out-of-equilibrium beliefs is to remove its origin by building a model in which every outcome is possible in equilibrium because different types of players take different equilibrium actions. In the PhD Admissions Game, we could assume that there are a few students who both love economics and actually enjoy writing applications. Those students would always apply in equilibrium, so there would never be a pure pooling equilibrium in which nobody applied, and Bayes' Rule could always be used. In equilibrium, the university would always accept someone who applied, because applying is never out-of-equilibrium behavior and it always indicates that the applicant is a *Lover*. This approach is especially attractive if the modeller takes the possibility of trembles literally, instead of just using it as a technical tool.

The arguments for different kinds of beliefs can also be applied to Entry Deterrence III, which had two different pooling equilibria and no separating equilibrium. We used passive conjectures in the "plausible" equilibrium. The intuitive criterion would not restrict beliefs at all, because both types would enter if the incumbent's beliefs were such as to make him collude, and both would stay out if they made him fight. Complete robustness would rule out as an equilibrium the strategy profile in which the entrant stays out regardless of type, because the optimality of staying out depends on the beliefs. It would support the strategy profile in which the entrant enters and out-of-equilibrium beliefs do not matter.

6.3 The Importance of Common Knowledge: Entry Deterrence IV and V

To demonstrate the importance of common knowledge, let us consider two more versions of the game of Entry Deterrence. We will use passive conjectures in both. In Entry Deterrence III, the incumbent was hurt by his ignorance. Entry Deterrence IV will show how he can benefit from ignorance, and Entry Deterrence V will show what can happen when the incumbent has the same information as the entrant but the information is not common knowledge.

Entry Deterrence IV: Incumbent Benefits from Ignorance

To construct Entry Deterrence IV, let $X = 300$ in figure 6.1, so fighting is even more profitable than in Entry Deterrence III but the game is otherwise the same: the entrant knows his type, but the incumbent does not. The following is the unique perfect Bayesian equilibrium.

> **Equilibrium for Entry Deterrence IV**
> Entrant: *Stay Out|Weak, Stay Out|Strong*
> Incumbent: *Fight*
> Beliefs: *Prob(Strong|Enter)* $= 0.5$ (passive conjectures)

This equilibrium can be supported by other out-of-equilibrium beliefs, but no equilibrium is possible in which the entrant enters. There is no pooling equilibrium in which both types of entrant enter, because then the incumbent's expected payoff from *Fight* would be 150 ($= 0.5[0] + 0.5[300]$), which is greater than the

Collude payoff of 50. There is no separating equilibrium, because if only the strong entrant entered and the incumbent always colluded, the weak entrant would be tempted to imitate him and enter as well.

In Entry Deterrence IV, unlike Entry Deterrence III, the incumbent benefits from his own ignorance, because he would always fight entry, even if the payoff were (unknown to himself) just zero. The entrant would very much like to communicate the costliness of fighting, but the incumbent would not believe him, so entry never occurs.

Entry Deterrence V: Lack of Common Knowledge of Ignorance

In Entry Deterrence V, it may happen that both the entrant and the incumbent know the payoff from (*Enter*, *Fight*), but the entrant does not know whether the incumbent knows. The information is known to both players, but is not common knowledge.

Figure 6.3 depicts this somewhat complicated situation. The game begins with Nature assigning the entrant a type, *Strong* or *Weak*, as before. This is observed by the entrant but not by the incumbent. Next, Nature moves again and either tells the incumbent the entrant's type or remains silent. This is observed by the incumbent, but not by the entrant. The four games starting at nodes G_1 to G_4

Figure 6.3 Entry Deterrence V

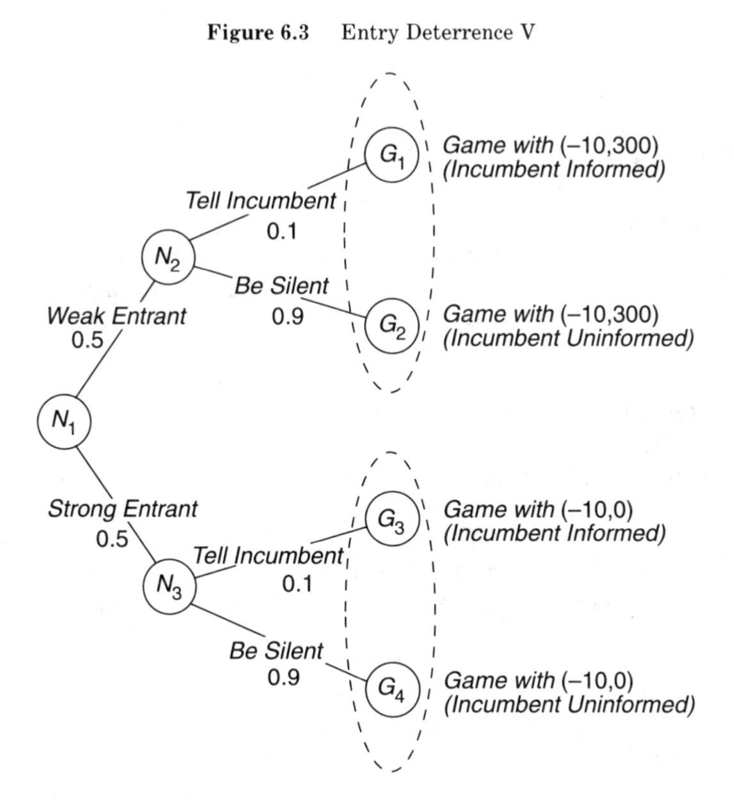

represent different combinations of payoffs from (*Enter*, *Fight*) and knowledge of the incumbent. The entrant does not know how well informed the incumbent is, so the entrant's information partition is ($\{G_1, G_2\}, \{G_3, G_4\}$).

> **Equilibrium for Entry Deterrence V**
> Entrant: *Stay Out|Weak*, *Stay Out|Strong*
> Incumbent: *Fight|Nature said "Weak"*, *Collude|Nature said "Strong"*, *Fight|Nature said nothing*
> Beliefs: *Prob(Strong|Enter, Nature said nothing)* = 0.5 (passive conjectures).

Since the entrant puts a high probability on the incumbent not knowing, the entrant should stay out, because the incumbent will fight for either of two reasons. With probability 0.9, Nature has said nothing and the incumbent calculates his expected payoff from *Fight* to be 150, and with probability 0.05 (= 0.1[0.5]) Nature has told the incumbent that the entrant is weak and the payoff from *Fight* is 300. Even if the entrant is weak and Nature tells this to the incumbent, the entrant would choose *Stay Out*, because he does not know that the incumbent knows, and his expected payoff from *Enter* would be −7.5 (= [0.9 + 0.05] [−10] + 0.05[40]).

If it were common knowledge that the entrant was strong, the entrant would enter and the incumbent would collude. If it is known by both players, but not common knowledge, the entrant stays out, even though the incumbent would collude if he entered. Such is the importance of common knowledge.

6.4 Incomplete Information in the Repeated Prisoner's Dilemma: The Gang of Four Model

Chapter 5 explored various ways to steer between the Scylla of the Chainstore Paradox and the Charybdis of the Folk Theorem to find a resolution to the problem of repeated games. In the end, uncertainty turned out to make little difference to the problem, but incomplete information was left unexamined in chapter 5. One might imagine that if the players did not know each others' types, the resulting confusion might allow cooperation. Let us investigate this by adding incomplete information to the finitely repeated Prisoner's Dilemma and finding the perfect Bayesian equilibria.

One way to incorporate incomplete information would be to assume that a large number of players are irrational, but that a given player does not know whether any other player is of the irrational type or not. In this vein, one might assume that with high probability Row is a player who blindly follows the strategy of Tit-for-Tat. If Column thinks he is playing against a Tit-for-Tat player, his optimal strategy is to *Deny* until near the last period (how near depending on the parameters), and then *Confess*. If he were not certain of this, but the probability were high that he faced a Tit-for-Tat player, Row would choose that same strategy. Such a model begs the question, because it is not the incompleteness of the information that drives the model, but the high probability that one player blindly uses Tit-for-Tat. Tit-for-Tat is not a rational strategy, and to assume that many players use it is to assume away the problem. A more surprising result is that a small amount of incomplete information can make a big difference to the outcome.[1]

The Gang of Four Model

One of the most important explanations of reputation is that of Kreps, Milgrom, Roberts & Wilson (1982), hereafter referred to as the Gang of Four. In their model, a few players are genuinely unable to play any strategy but Tit-for-Tat, and many players pretend to be of that type. The beauty of the model is that it requires only a small amount of incomplete information, and a low probability γ that player Row is a Tit-for-Tat player. It is not unreasonable to suppose that a world that contains neo-Ricardians and McGovernites contains a few mildly irrational Tit-for-Tat players, and such behavior is especially plausible among consumers, who are subject to less evolutionary pressure than firms.

It may even be misleading to call the Tit-for-Tat players "irrational", because they may just have unusual payoffs, particularly since we will assume that they are rare. The unusual players have a small direct influence, but they matter because other players imitate them. Even if Column knows that with high probability Row is just pretending to be Tit-for-Tat, Column does not care what the truth is so long as Row keeps on pretending. Hypocrisy is not only the tribute vice pays to virtue; it can be just as good for deterring misbehavior.

Theorem 6.1: The Gang of Four Theorem

Consider a T-*stage, repeated prisoner's dilemma, without discounting but with a probability γ of a Tit-for-Tat player. In any perfect Bayesian equilibrium, the number of stages in which either player chooses* Confess *is less than some number* M *that depends on γ but not on* T.

The significance of the Gang of Four Theorem is that while the players do resort to *Confess* as the last period approaches, the number of periods during which they *Confess* is independent of the total number of periods. Suppose M = 2,500. If T = 2,500, there might be a *Confess* every period. But if T = 10,000, there are 7,500 periods without a *Confess*. For reasonable probabilities of the unusual type, the number of periods of cooperation can be much larger. Wilson (unpublished) has set up an entry deterrence model in which the incumbent fights entry (the equivalent of *Deny* above) up to seven periods from the end, although the probability the entrant is of the unusual type is only 0.008.

The Gang of Four Theorem characterizes the equilibrium outcome rather than the equilibrium. Finding perfect Bayesian equilibria is difficult and tedious, since the modeller must check all the out-of-equilibrium subgames, as well as the equilibrium path. Modellers usually content themselves with describing important characteristics of the equilibrium strategies and payoffs.[2]

[1] Begging the question is not as illegitimate in modelling as in rhetoric, however, because it may indicate that the question is a vacuous one in the first place. If the payoffs of Prisoner's Dilemma are not those of most of the people one is trying to model, the Chainstore Paradox becomes irrelevant.

[2] Section 14.1 does contain a somewhat more detailed description of what happens in a model of repeated entry deterrence with incomplete information.

To get a feeling for why Theorem 6.1 is correct, consider what would happen in a 10,001-period game with a probability of 0.01 that Row is playing the Grim Strategy of *Deny* until the first *Confess*, and *Confess* every period thereafter. If the payoffs are as in table 5.2a, a best response for Column to a known grim player is (*Confess* only in the last period, unless Row chooses *Confess* first, in which case respond with *Confess*). Both players will choose *Deny* until the last period, and Column's payoff will be 50,010 (= (10,000)(5) + 10). Suppose for the moment that if Row is not grim, he is highly aggressive, and will choose *Confess* every period. If Column follows the strategy just described, the outcome will be (*Confess, Deny*) in the first period and (*Confess, Confess*) thereafter, for a payoff to Column of −5 (= −5 + (10,000)(0)). If the probabilities of the two outcomes are 0.01 and 0.99, Column's expected payoff from the strategy described is 495.15. If instead he follows a strategy of (*Confess* every period), his expected payoff is just 1 (= 0.01 (10) + 0.99 (0)). It is clearly in Column's advantage to take a chance by cooperating with Row, even if Row has a 0.99 probability of following a very aggressive strategy.

The aggressive strategy, however, is not Row's best response to Column's strategy. A better response is for Row to choose *Deny* until the second-to-last period, and then to choose *Confess*. Given that Column is cooperating in the early periods, Row will cooperate also. This argument has not described what the Nash equilibrium actually is, since the iteration back and forth between Row and Column can be continued, but it does show why Column chooses *Deny* in the first period, which is the leverage the argument needs: the payoff is so great if Row is actually the grim player that it is worthwhile for Column to risk a low payoff for one period.

The Gang of Four Theorem provides a way out of the Chainstore Paradox, but it creates a problem of multiple equilibria in much the same way as the infinitely repeated game. For one thing, if the asymmetry is two-sided, so both players might be unusual types, it becomes much less clear what happens in threat games such as Entry Deterrence. Also, what happens depends on which unusual behaviors have positive, if small, probability. Theorem 6.2 says that the modeller can make the average payoffs take any particular values by making the game last long enough and choosing the form of the irrationality carefully.

Theorem 6.2: The Incomplete Information Folk Theorem (Fudenberg & Maskin [1986] p. 547)

For any two-person repeated game without discounting, the modeller can choose a form of irrationality so that for any probability $\varepsilon > 0$ there is some finite number of repetitions such that with probability $(1 - \varepsilon)$ a player is rational and the average payoffs in some sequential equilibrium are closer than ε to any desired payoffs greater than the minimax payoffs.

6.5 The Axelrod Tournament

Another way to approach the repeated Prisoner's Dilemma is through experiments, such as the round robin tournament described by political scientist Robert

Axelrod in his 1984 book. Contestants submitted strategies for a 200 repetition Prisoner's Dilemma. Since the strategies could not be updated during play, players could precommit, but the strategies could be as complicated as they wished. If a player wanted to specify a strategy which simulated subgame perfectness by adapting to past history just as a noncommitted player would, he was free to do so, but he could also submit a non-perfect strategy such as Tit-for-Tat or the slightly more forgiving Tit-for-Two-Tats. Strategies were submitted in the form of computer programs that were matched with each other and played automatically. In Axelrod's first tournament, 14 programs were submitted as entries. Every program played every other program, and the winner was the one with the greatest sum of payoffs over all the plays. The winner was Anatol Rapoport, whose strategy was Tit-for-Tat.

The tournament helps to show which strategies are robust against a variety of other strategies in a game with given parameters. It is quite different from trying to find a Nash equilibrium, because it is not common knowledge what the equilibrium is in such a tournament. The situation could be viewed as a game of incomplete information in which Nature chooses the number and cognitive abilities of the players and their priors regarding each other.

After the results of the first tournament were announced, Axelrod ran a second tournament, adding the probability $\theta = 0.00346$ that the game would end each round so as to avoid the Chainstore Paradox. The winner among the 62 entrants was again Anatol Rapoport, and again he used Tit-for-Tat.

Before choosing his tournament strategy, Rapoport had written an entire book on the Prisoner's Dilemma in analysis, experiment, and simulation (Rapoport & Chammah [1965]). Why did he choose such a simple strategy as Tit-for-Tat? Axelrod points out that Tit-for-Tat has three strong points.

(1) It never initiates confessing (**niceness**).
(2) It retaliates instantly against confessing (**provokability**).
(3) It forgives a confesser who goes on to cooperate (it is **forgiving**).

Despite these advantages, care must be taken in interpreting the results of the tournament. It does not follow that Tit-for-Tat is the best strategy, or that cooperative behavior should always be expected in repeated games.

First, Tit-for-Tat never beats any other strategy in a one-on-one contest. It won this tournament by piling up points through cooperation, having lots of high score plays and very few low score plays. In an elimination tournament, Tit-for-Tat would be eliminated very early, because it scores *high* payoffs but never the *highest* payoff.

Second, the other players' strategies matter to the success of Tit-for-Tat. In neither tournament were the strategies submitted a Nash equilibrium. If a player knew what strategies he was facing, he would want to revise his own. Some of the strategies submitted in the second tournament would have won the first, but they did poorly because the environment had changed. Other programs, designed to try to probe the strategies of their opposition, wasted too many (*Confess*, *Confess*) episodes on the learning process, but if the games had lasted a thousand repetitions they would have done better.

Third, in a game in which players occasionally confessed because of trembles, two Tit-for-Tat players facing each other would do very badly. The strategy instantly punishes a confessing player, and it has no provision for ending the punishment phase.

Optimality depends on the environment. When information is complete and the payoffs are all common knowledge, confessing is the only equilibrium outcome, but in practically any imaginable situation, information is slightly incomplete, so cooperation becomes more plausible. Tit-for-Tat is suboptimal for any given environment, but it is robust across environments, and that is its advantage.

Notes

N6.1 Perfect Bayesian Equilibrium: Entry Deterrence I and II

- Section 4.1 showed that even in games of perfect information, not every subgame perfect equilibrium is trembling-hand perfect. In games of perfect information, however, every subgame perfect equilibrium is a perfect Bayesian equilibrium, since no out-of-equilibrium beliefs need to be specified.

N6.2 Refining Perfect Bayesian Equilibrium: The PhD Admissions Game

- Fudenberg & Tirole (1991b) is a careful analysis of the issues involved in defining perfect Bayesian equilibrium.
- Section 6.2 is about debatable ways of restricting beliefs such as passive conjectures or equilibrium dominance, but less controversial restrictions are sometimes useful. In a three-player game, consider what happens when Smith and Jones have incomplete information about Brown, and then Jones deviates. Should Smith and Jones update their priors on Brown's type? Passive conjectures seems much more reasonable. If Brown deviates, can the out-of-equilibrium beliefs specify that Smith and Jones update their beliefs about Brown in different ways? This seems dubious in light of the Harsanyi doctrine that everyone begins with the same priors.
- For discussions of the appropriateness of different equilibrium concepts in actual economic models see Rubinstein (1985b) on bargaining, Shleifer & Vishny (1986) on greenmail and D. Hirshleifer & Titman (1990) on tender offers.
- **Exotic Refinements.** Binmore (1990) and Kreps (1990b) are booklength treatments of rationality and equilibrium concepts. New and improved equilibrium concepts refining perfect Bayesian equilibrium have been suggested by McClennan (1985), Cho & Kreps (1987) (**intuitive criterion**), Banks & Sobel (1987) (**divinity**), and Grossman & Perry (1986) (**perfect sequential equilibrium**).

  Two refinements somewhat different in flavor are stability and properness.

  Stability. (Kohlberg & Mertens [1986]) This concept is axiomatic, rather than based on beliefs or trembles. Its main requirements are that an equilibrium exist, that it be subgame perfect, that game trees that are essentially the same generate the same equilibria, and that deleting dominated strategies does not change the equilibrium. These requirements are so stringent that for many games no stable equilibrium exists. (The use of the term "stability" is unrelated to its use by von Neumann-Morgenstern or in connection with the Cournot equilibrium.)

  Proper Equilibrium. (Myerson [1978]) Properness is different from most refinements in being tremble-based rather than belief-based. The idea is that "worse" trembles are less likely. Take the trembling-hand perfect equilibrium concept and set the trembles so that if pure strategy s_i' is a worse reply than pure strategy s_i'', the probability weight put by the tremble's mixed strategy on s_i' is less than ε times the weight on s_i''. Take the limit as ε tends to zero, and a proper equilibrium results.
- **Beer-Quiche Game** (Cho & Kreps [1987]). To illustrate their "intuitive criterion," Cho and Kreps use the Beer-Quiche Game. In this game, Player I might be either weak or strong in his duelling ability, but he wishes to avoid a duel even if he thinks he can win. Player II wishes to fight a duel only if Player I is weak, which has a probability of 0.1. Player II does not know Player I's type, but he observes what Player I has for breakfast. He knows that weak players

Figure 6.4 The Beer-Quiche Game

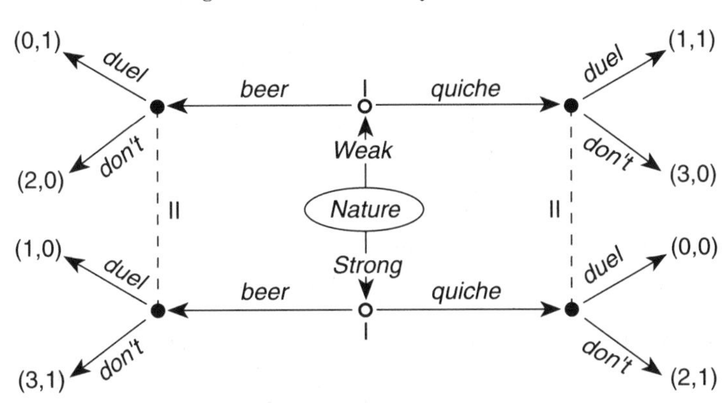

prefer quiche for breakast, while strong players prefer beer. The payoffs are shown in figure 6.4.

Figure 6.4 illustrates a few twists on how to draw an extensive form. It begins with Nature's choice of *Strong* or *Weak* in the middle of the diagram. Player I then chooses whether to breakfast on *beer* or *quiche*. Player II's nodes are connected by a dotted line if they are in the same information set. Player II chooses *Duel* or *Don't*, and payoffs are then received.

This game has two perfect Bayesian equilibrium outcomes, both of which are pooling. In E_1, Player I has beer for breakfast regardless of type, and Player II chooses not to duel. This is supported by the out-of-equilibrium belief that a quiche-eating Player I is weak with probability over 0.5, in which case Player II would choose to duel on observing quiche. In E_2, Player I has quiche for breakfast regardless of type, and Player II chooses not to duel. This is supported by the out-of-equilibrium belief that a beer-drinking Player I is weak with probability greater than 0.5, in which case Player II would choose to duel on observing beer.

Passive conjectures and the intuitive criterion both rule out equilibrium E_2. According to the reasoning of the intuitive criterion, Player I could deviate without fear of a duel by giving the following convincing speech,

> I am having beer for breakfast, which ought to convince you I am strong. The only conceivable benefit to me of breakfasting on beer comes if I am strong. I would never wish to have beer for breakfast if I were weak, but if I am strong and this message is convincing, then I benefit from having beer for breakfast.

N6.5 The Axelrod Tournament

- Hofstadter (1983) is a nice discussion of the Prisoner's Dilemma and the Axelrod tournament by an intelligent computer scientist who came to the subject untouched by the preconceptions or training of economics. It is useful for elementary economics classes. Axelrod's 1984 book provides a fuller treatment.

Problems

6.1: Cournot Duopoly Under Incomplete Information about Costs.

This problem introduces incomplete information into the Cournot model of Chapter 3 and allows for a continuum of player types.

(6.1a) Modify "The Cournot Game" of Chapter 3 by specifying that Apex' average cost of production is c per unit, while Brydox' remains zero. What are the outputs of each firm if the costs are common knowledge? What are the numerical values if $c = 10$?

(6.1b) Let Apex' cost c be c_{max} with probability θ and 0 with probability $1 - \theta$, so Apex is one of two types. Brydox does not know Apex' type. What are the outputs of each firm?

(6.1c) Let Apex' cost c be drawn from the interval $[0, c_{max}]$ using the uniform distribution, so there is a continuum of types. Brydox does not know Apex' type. What are the outputs of each firm?

(6.1d) Outputs were 40 for each firm in the zero-cost game in chapter 3. Check your answers in parts (b) and (c) by seeing what happens if $c_{max} = 0$.

(6.1e) Let $c_{max} = 20$ and $\theta = 0.5$, so the expectation of Apex' average cost is 10 in parts (a), (b), and (c). What are the average outputs for Apex in each case?

(6.1f) Modify the model of part (b) so that $c_{max} = 20$ and $\theta = 0.5$, but somehow $c = 30$. What outputs do your formulas from part (b) generate? Is there anything this could sensibly model?

Problem 6.2: Limit Pricing.[3]

An incumbent firm operates in the local computer market, which is a natural monopoly in which only one firm can survive. The incumbent can price *Low*, losing 40 in profits, or *High*, losing nothing. It knows its own operating cost C, which is 20 with probability 0.75 and 30 with probability 0.25. A potential entrant knows those probabilities, but not the incumbent's exact cost. The entrant can enter at a cost of 100, and its operating cost of 25 is common knowledge. The firm with the highest operating cost immediately drops out if it has a competitor, and the survivor earns the monopoly revenue of 150.

(6.2a) Draw the extensive form for this game.

(6.2b) Why is there no perfect equilibrium in which the incumbent prices *Low* only when its costs are 20? (no separating equilibrium)

(6.2c) In a perfect Bayesian equilibrium in which the incumbent prices *Low* regardless of its costs (a pooling equilibrium), what do the out-of-equilibrium beliefs have to specify?

(6.2d) What are two different perfect Bayesian equilibria for this game?

(6.2e) What is a set of out-of-equilibrium beliefs that do not support a pooling equilibrium at a *Low* price?

6.3: Symmetric Information and Prior Beliefs.

In The Expensive-Talk Game of table 6.1, Battle of the Sexes is preceded by a communication move in which the man chooses *Silence* or *Talk*. *Talk* costs 1 payoff unit and consists of a declaration by the man that he is going to the prize fight. This declaration is just talk; it is not binding on him.

[3] See Milgrom and Roberts (1982a).

Table 6.1 Subgame Payoffs in The Expensive-Talk Game

		Woman	
		Fight	*Ballet*
Man:	*Fight*	3,1	0,0
	Ballet	0,0	1,3

Payoffs to: (Man, Woman)

(6.3a) Draw the extensive form for this game, putting the man's move first in the simultaneous-move subgame.
(6.3b) What are the strategy sets for the game? (start with the woman's)
(6.3c) What are the three perfect pure-strategy equilibrium outcomes in terms of observed actions? (remember: strategies are not the same thing as outcomes)
(6.3d) Describe the equilibrium strategies for a perfect equilibrium in which the man chooses to talk.
(6.3e) The idea of "forward induction" says that an equilibrium should remain an equilibrium even if strategies that are dominated in that equilibrium are removed from the game and the procedure is iterated. Show that this procedure rules out SBB (*Silence, Ballet, Ballet*) as an equilibrium outcome.[4]

6.4: Lack of Common Knowledge.

This problem looks at what happens if the parameter values in Entry Deterrence V are changed.
(6.4a) Why does *Pr(Strong|Enter, Nature said nothing)* = 0.95 not support the equilibrium in section 6.3?
(6.4b) Why is the equilibrium in section 6.3 not an equilibrium if 0.7 is the probability that Nature tells the incumbent?
(6.4c) Describe the equilibrium if 0.7 is the probability that Nature tells the incumbent. For what out-of-equilibrium beliefs does this remain the equilibrium?

[4] See Van Damme (1989). In fact, this procedure rules out TFF (*Talk, Fight, Fight*) also.

Part II

Asymmetric Information

7 Moral Hazard: Hidden Actions

7.1 Categories of Asymmetric Information Models

It used to be that the economist's generic answer to someone who brought up peculiar behavior which seemed to contradict basic theory was, "It must be some kind of price discrimination." Today, we have a new answer: "It must be some kind of asymmetric information." In a game of asymmetric information, player Smith knows something that player Jones does not. This covers a broad range of models (including price discrimination nowadays), so perhaps it is not surprising that so many situations come under its rubric. We will divide games of asymmetric information into five categories, to be studied in four chapters.

(1) **Moral hazard with hidden actions** (chapter 7)

Smith and Jones begin with symmetric information and agree to a contract, but then Smith takes an action unobserved by Jones. Information is complete.

(2) **Moral hazard with hidden knowledge** (or **hidden information**) (chapter 8)

Smith and Jones begin with symmetric information and agree to a contract. Nature then makes a move observed by Smith but not by Jones, and Smith takes some action, which may be simply a report of Nature's move. Information is complete.

(3) **Adverse selection** (chapter 9)

Nature begins the game by choosing Smith's type (his payoff and strategies), unobserved by Jones. Smith and Jones then agree to a contract. Information is incomplete.

(4, 5) **Signalling** and **Screening** (chapter 10)

Nature begins the game by choosing Smith's type, unobserved by Jones. To demonstrate his type, Smith takes actions that Jones can observe. If Smith takes the action before they agree to a contract, he is signalling; if he takes it afterwards, he is being screened. Information is incomplete.

Signalling and screening are special cases of adverse selection, which is itself a situation of hidden knowledge. Information is complete in either kind of moral hazard, and incomplete in adverse selection, signalling, and screening.

Note that some people may say that information *becomes* incomplete in a model of moral hazard with hidden knowledge, even though it is complete at the start of the game. That statement runs contrary to the definition of complete information in chapter 2, however. The most important distinctions to keep in mind are whether or not the players agree to a contract before or after information becomes asymmetric and whether or not their own actions are common knowledge.

We will make heavy use of the "principal-agent model" to analyze asymmetric information. Usually this term is applied to moral hazard models, since the problems studied in the law of agency usually involve an employee who disobeys orders by choosing the wrong actions, but the paradigm will be useful in all of these contexts. The two players are the principal and the agent, who are usually representative individuals. The principal hires an agent to perform a task, and the agent acquires an informational advantage about his type, his actions, or the outside world at some point in the game. It is usually assumed that the players can make a binding **contract** at some point in the game, which is to say that the principal can commit to paying the agent an agreed sum if he observes a certain outcome. In the implicit background of such models are courts which will punish any player who breaks a contract in a way that can be proven with public information.

The **principal** *(or* **uninformed player***) is the player who has the coarser information partition.*

The **agent** *(or* **informed player***) is the player who has the finer information partition.*

Figure 7.1 shows the game trees for five principal-agent problems corresponding to the categories listed above. In each model the principal (P) offers the agent (A) a contract, which he accepts or rejects. In some, Nature (N) makes a move or the agent chooses an effort level, message, or signal. The moral hazard models (figures 7.1a and 7.1b), are games of complete information with uncertainty. The principal offers a contract, and after the agent accepts, Nature adds noise to the task being performed. In moral hazard with hidden actions (figure 7.1a), the agent moves before Nature, and in moral hazard with hidden knowledge (figure 7.1b), the agent moves after Nature and conveys a "message" to the principal about Nature's move.

Figure 7.1 Categories of Asymmetric Information Models

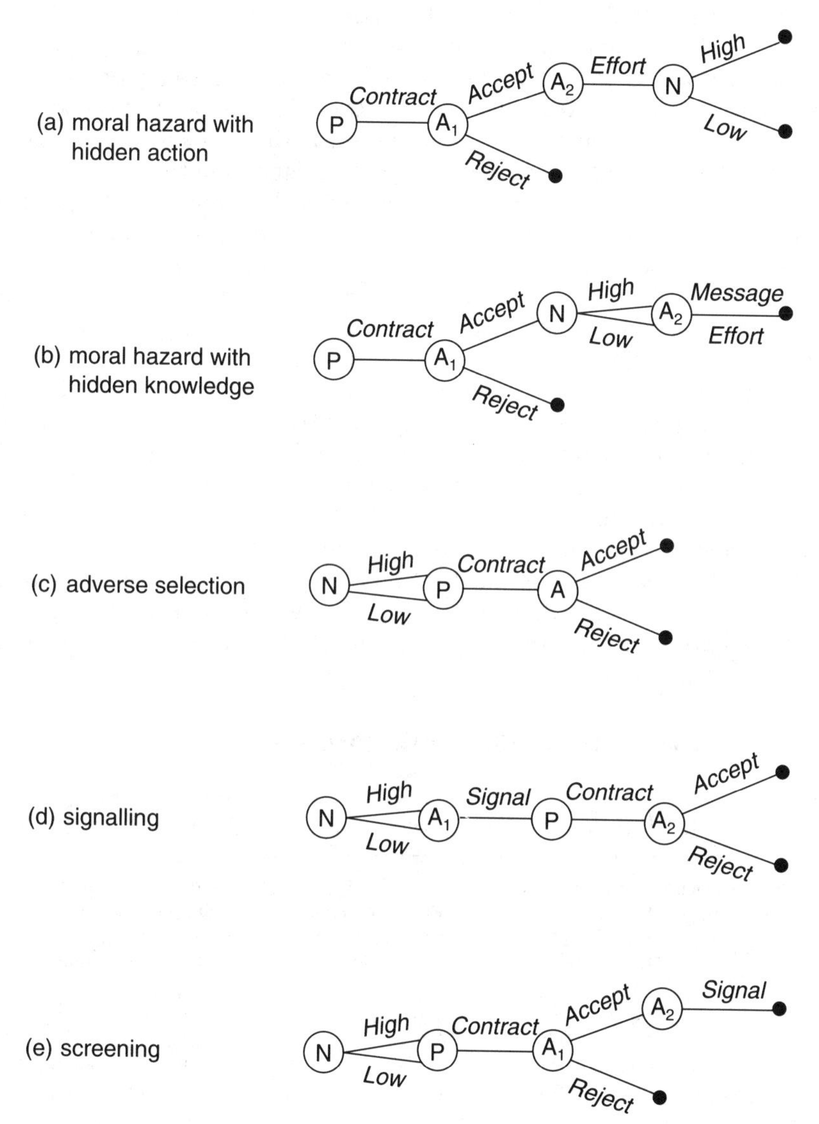

Adverse selection models have incomplete information, so Nature moves first and picks the type of the agent, generally on the basis of his ability to perform the task. In the simplest model, figure 7.1c, the agent simply accepts or rejects the contract. If the agent can send a "signal" to the principal, as in figures 7.1d and 7.1e, the model is signalling if he sends the signal before the principal offers a contract, and is screening otherwise. A "signal" is different from a "message" because it is not a costless statement, but a costly action. Some adverse selection models contain uncertainty and some do not.

A problem we will consider in detail is that of an employer (the principal) hiring a worker (the agent). If the employer knows the worker's ability but not his effort level, the problem is moral hazard with hidden actions. If neither player knows the worker's ability at first, but the worker discovers it once he starts working, the problem is moral hazard with hidden knowledge. If the worker knows his ability from the start, but the employer does not, the problem is adverse selection. If, in addition to the worker knowing his ability from the start, he can acquire credentials before he makes a contract with the employer, the problem is signalling. If the worker acquires his credentials in response to a wage offer made by the employer, the problem is screening.

The five categories are only gradually rising from the swirl of the literature on agency models, and the definitions are not well established. In particular, some would argue that what I have called moral hazard with hidden knowledge and screening are essentially the same as adverse selection. Myerson (1991, p. 263), for example, suggests calling the problem of players taking the wrong action "moral hazard" and the problem of misreporting information, "adverse selection." Many economists do not realize that screening and signalling are different and use the terms interchangeably. "Signal" is such a useful word that it is often used simply to indicate any variable conveying information. Most people have not thought very hard about any of the definitions, but the importance of the distinctions will become clear as we explore the properties of the models. For readers whose minds are more synthetic than analytic, table 7.1 may be as helpful as anything in clarifying the categories.

Table 7.1 Applications of the Principal-Agent Model

	Principal	Agent	Effort or Type and Signal
Moral hazard with hidden actions	Insurance company	Policyholder	Care to avoid theft
	Insurance company	Policyholder	Drinking and Smoking
	Plantation owner	Sharecropper	Farming Effort
	Bondholders	Stockholders	Riskiness of corporate projects
	Tenant	Landlord	Upkeep of building
	Landlord	Tenant	Upkeep of building
	Society	Criminal	Number of robberies
Moral hazard with hidden knowledge	Shareholders	Company president	Investment decision
	FDIC	Bank	Safety of loans
Adverse selection	Insurance Company	Policyholder	Infection with HIV virus
	Employer	Worker	Skill
Signalling and screening	Employer	Worker	Skill and education
	Buyer	Seller	Durability and warranty
	Investor	Stock-issuer	Stock value and percentage retained

Section 7.2 discusses the roles of uncertainty and asymmetric information in a principal-agent model of moral hazard with hidden actions, called the Production Game, and section 7.3 shows how various constraints are satisfied in equilibrium.

Section 7.4 collects several unusual contracts produced under moral hazard and discusses the properties of optimal contracts using the example of the Broadway Game. Section 7.5 uses diagrams and the Insurance Game to approach the classical problem of moral hazard: lack of effort by the insured party to avoid accidents.

7.2 A Principal-Agent Model: The Production Game

In the archetypal principal-agent model, the principal is a manager and the agent a worker. In this section we will devise a series of these types of games, the last of which will be the standard principal-agent model.

Denote the monetary value of output by $q(e)$, which is increasing in effort, e. The agent's utility function, $U(e, w)$, is decreasing in effort and increasing in the wage, w, while the principal's utility, $V(q - w)$, is increasing in the difference between output and the wage.

The Production Game

Players

The principal and the agent.

Order of Play

(1) The principal offers the agent a wage w.
(2) The agent decides whether to accept or reject the contract.
(3) If the agent accepts, he exerts effort e.
(4) Output equals $q(e)$, where $q' > 0$.

Payoffs

If the agent rejects the contract, then $\pi_{agent} = \bar{U}$ and $\pi_{principal} = 0$.
If the agent accepts the contract, then $\pi_{agent} = U(e, w)$ and $\pi_{principal} = V(q - w)$.

An assumption common to most principal-agent models is that either the principal or the agent is one of many perfect competitors. In the background, other principals compete to employ the agent, so the principal's equilibrium profit equals zero; or many agents compete to work for the principal, so the agent's equilibrium utility equals the minimum for which he will accept the job, called the **reservation utility**, $\bar{U}$. There is some reservation utility level even if the principal is a monopolist, however, because the agent has the option of remaining unemployed if the wage is too low.

One way of viewing the assumption in the Production Game that the principal moves first is that many agents compete for one principal. The order of moves allows the principal to make a take-it-or-leave-it offer, leaving the agent with as little bargaining room as if he had to compete with a multitude of other agents. This is really just a modelling convenience, however, since the agent's reservation utility, $\bar{U}$, can be set at the level a principal would have to pay the agent in competition with other principals. This level of $\bar{U}$ can even be calculated, since it is the level at which the principal's payoff from profit maximization using the optimal contract is driven down to the principal's reservation utility by competition with other principals. Here the principal's reservation utility is zero, but that too can be chosen to fit the situation being modelled. As in the game of Nuisance Suits in section 4.3, the main concern in choosing who makes the offer is to avoid getting caught up in a bargaining subgame.

Refinements of the equilibrium concept will not be important in this chapter; Nash equilibrium will be sufficient for our purposes, because information is complete and the concerns of perfect Bayesian equilibrium will not arise. Subgame perfectness will be required, since otherwise the agent might commit to reject any contract that does not give him all of the gains from trade, but it will not drive the important results.

We will go through a series of five versions of the Production Game in this chapter.

Production Game I: Full Information

In the first version of the game, every move is common knowledge and the contract is a function $w(e)$.

Finding the equilibrium involves finding the best possible contract from the point of view of the principal, given that he must make the contract acceptable to the agent and that he foresees how the agent will react to the contract's incentives. The principal must decide what he wants the agent to do and how to give him incentives to do it as cheaply as possible.

The agent must be paid some amount $\tilde{w}(e)$ to exert effort e, where $\tilde{w}(e)$ is defined to be the w that solves the participation constraint

$$U(e, w(e)) = \bar{U}. \tag{1}$$

Thus, the principal's problem is

$$\underset{e}{Maximize}\ V(q(e) - \tilde{w}(e)). \tag{2}$$

The first-order condition for this problem is

$$V'(q(e) - \tilde{w}(e))\left(\frac{\partial q}{\partial e} - \frac{\partial \tilde{w}}{\partial e}\right) = 0, \tag{3}$$

which implies that

$$\frac{\partial q}{\partial e} = \frac{\partial \tilde{w}}{\partial e}. \tag{4}$$

From the implicit function theorem (see section 13.4) and the participation constraint,

$$\frac{\partial \tilde{w}}{\partial e} = -\left(\frac{\partial U}{\partial e} \Big/ \frac{\partial U}{\partial \tilde{w}}\right). \tag{5}$$

Combining equations (4) and (5) yields

$$\frac{\partial U}{\partial \tilde{w}} \frac{\partial q}{\partial e} = -\frac{\partial U}{\partial e}. \tag{6}$$

Equation (6) says that at the optimal effort level e^*, the marginal utility to the agent which would result if he kept all the marginal output from extra effort equals the marginal disutility to him of that effort. As usual, the outcome can be efficient even if the agent does not actually keep the extra output, since who keeps the output is a distributional question.

Figure 7.2 shows this graphically. The agent has indifference curves in effort-wage space that slope upwards, since if effort is increased the wage must be increased to keep utility the same. The principal's indifference curves also slope upwards, because although he does not care about effort directly, he does care about output, which rises with effort. The principal might be either risk averse or risk neutral; his indifference curve is concave rather than linear in either case because figure 7.2 shows a technology with diminishing returns to effort. If effort starts out being higher, extra effort yields less additional output so the wage cannot rise as much without reducing profits.

Under perfect competition among the principals the profits are zero, so the reservation utility $\bar{U}$ is chosen so that at the profit-maximizing effort e^*, $\tilde{w}(e^*) = q(e^*)$, or

$$U(e^*, q(e^*)) = \bar{U}. \tag{7}$$

Figure 7.2 The Efficient Effort Level in Production Game I

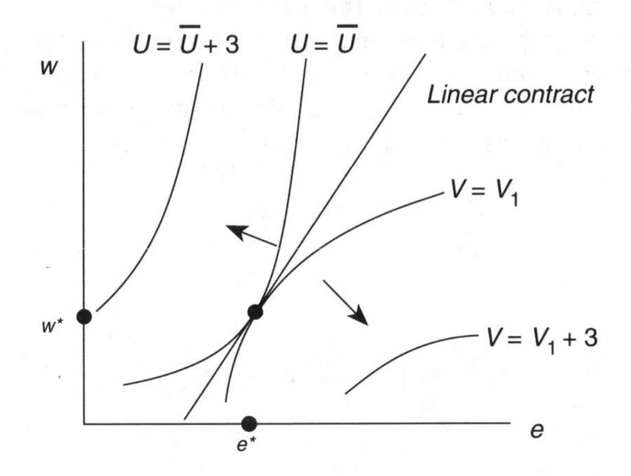

The principal selects the point on the $U = \bar{U}$ indifference curve that maximizes his profits, which is at the effort e^* and wage w^*. He must then design a contract that will induce the agent to choose this effort level. The following three contracts are equally effective under full information.

(1) The **forcing contract** sets $w(e^*) = w^*$ and $w(e \neq e^*) = 0$. This is certainly a strong incentive for the agent to choose exactly $e = e^*$.

(2) The **threshold contract** sets $w(e \geq e^*) = w^*$ and $w(e < e^*) = 0$. This can be viewed as a flat wage for low effort levels, equal to 0 in this contract, plus a bonus if effort reaches e^*. Since the agent dislikes effort, the agent will choose exactly $e = e^*$.

(3) The **linear contract** shown in figure 7.2 sets $w(e) = \alpha + \beta e$, where α and β are chosen so that $w^* = \alpha + \beta e^*$ and the contract line is tangent to the indifference curve $U = \bar{U}$ at e^*. The most northwesterly of the agent's indifference curves that touch this contract line touches it at e^*.

Before going on to versions of the game with asymmetric information, it will be useful to look at one other version of the game with full information, in which the agent, not the principal, proposes the contract. This will be called Production Game II.

Production Game II: Full Information. Agent Moves First

In this version every move is common knowledge and the contract is a function $w(e)$. The order of play, however, is now as follows.

Order of Play

(1) The agent offers the principal a contract $w(e)$.
(2) The principal decides whether to accept or reject the contract.
(3) If the principal accepts, the agent exerts effort e.
(4) Output equals $q(e)$, where $q' > 0$.

In this game, the agent has all the bargaining power, not the principal. The participation constraint is now that the principal must earn zero profits, so $q(e) - w(e) \geq 0$. The agent will maximize his own payoff by driving the principal to exactly zero profits, so $w(e) = q(e)$. Substituting $q(e)$ for $w(e)$ to account for the participation constraint, the maximization problem for the agent in proposing an effort level e at a wage $w(e)$ can therefore be written as

$$\underset{e}{Maximize}\ U(e, q(e)). \tag{8}$$

The first-order condition is

$$\frac{\partial U}{\partial e} + \left(\frac{\partial U}{\partial q}\right)\left(\frac{\partial q}{\partial e}\right) = 0. \tag{9}$$

Since $\frac{\partial U}{\partial q} = \frac{\partial U}{\partial w}$ when wages equal output, equation (9) implies that

$$\frac{\partial U}{\partial w}\frac{\partial q}{\partial e} = -\frac{\partial U}{\partial e}. \tag{10}$$

Comparing this to equation (6), the equation when the principal had the bargaining power, it is clear that e^* is identical in Production Game I and Production Game II. It does not matter who has the bargaining power; the efficient effort level stays the same.

Figure 7.2 can be used to illustrate this game as well. Suppose that $V_1 = 0$. The agent must choose a point on the $V_1 = 0$ indifference curve that maximizes his own utility, and then provide himself with contract incentives to choose that point. The agent's payoff is highest at effort e^* given that he must make $V_1 = 0$, and all three contracts described in Production Game I provide him with the correct incentives.

The efficient-effort level is independent of which side has the bargaining power because the gains from efficient production are independent of how those gains are distributed so long as each party has no incentive to abandon the relationship. This as the same lesson as that of the Coase Theorem, which says that under general conditions the activities undertaken will be efficient and independent of the distribution of property rights (Coase [1960]). This property of the efficient-effort level means that the modeller is free to make the assumptions on bargaining power that help to focus attention on the information problems he is studying.

Production Game III: Flat Wage under Certainty

In this version of the game, the principal can condition the wage neither on effort nor on output. This is modelled as a principal who observes neither effort nor output, so information is asymmetric.

It is easy to imagine a principal who cannot observe effort, but it seems very strange that he cannot observe output, especially since he can deduce the output from the value of his payoff. It is not ridiculous that he cannot base wages on output, however, because a contract must be enforceable by some third party such as a court. Law professors complain about economists who speak of "unenforceable contracts." In law school, a contract is defined as an enforceable agreement, and most of a contracts class is devoted to discovering which agreements are contracts. If, for example, a teacher does a poor job of inspiring his students, that may be clear to his school but very costly to prove in court, so his wage cannot be based on his output of inspiration. For such situations, Production Game III is appropriate. Output is not **contractible** or **verifiable**, which leads to the same outcome as when it is unobservable in a contracting model.

The outcome of Production Game III is simple and inefficient. If the wage is nonnegative, the agent accepts the job and exerts zero effort, so the principal offers a wage of zero.

If there is nothing on which to condition the wage, the agency problem cannot be solved by designing the contract carefully. If it is to be solved at all, it will be by some other means such as reputation or repetition of the game, the solutions of chapter 5. Typically, however, there is some contractible variable such as output upon which the principal can condition the wage. Such is the case in Production Game IV.

Production Game IV: Output-Based Wage under Certainty

In this version, the principal cannot observe effort but can observe output and specify the contract to be $w(q)$.

Now the principal picks not a number w but a function $w(q)$. His problem is not quite so straightforward as in Production Game I, where he picked the function $w(e)$, but here, too, it is possible to achieve the efficient effort level e^* despite the unobservability of effort. The principal starts by finding the optimal effort level e^*, as in Production Game I. That effort yields the efficient output level $q^* = q(e^*)$. To give the agent the proper incentives, the contract must reward him when output is q^*. Again, a variety of contracts could be used. The forcing contract, for example, would be any wage function such that $U(e^*, w(q^*)) = \bar{U}$ and $U(e, w(q)) < \bar{U}$ for $e \neq e^*$.

Production Game IV shows that the unobservability of effort is not a problem in itself, if the contract can be conditioned on something which is observable and perfectly correlated with effort. The true agency problem occurs when that perfect correlation breaks down, as in Production Game V.

Production Game V: Output-Based Wage under Uncertainty

In this version, the principal cannot observe effort but can observe output and specify the contract to be $w(q)$. Output, however, is a function $q(e, \theta)$ both of effort and the state of the world $\theta \in \mathbf{R}$, which is chosen by Nature according to the probability density $f(\theta)$ as the new move (5) of the game. Move (5) comes just after the agent chooses effort, so the agent cannot choose a low effort knowing that Nature will take up the slack. (If the agent can observe Nature's move before his own, the game becomes moral hazard with both hidden knowledge and hidden actions.)

Because of the uncertainty about the state of the world, effort does not map cleanly onto the observed output in Production Game V. A given output might have been produced by any of several different effort levels, so a forcing contract will not necessarily achieve the desired effort. Unlike the case in Production Game IV, here the principal cannot deduce that $e = e^*$ from the fact that $q = q^*$. Moreover, even if the contract does induce the agent to choose e^*, if it does so by penalizing him heavily when $q \neq q^*$ it will be expensive for the principal. The agent's expected utility must be kept equal to $\bar{U}$ because of the participation constraint, and if the agent is sometimes paid a low wage because output happens not to equal q^*, he must be paid more when output does equal q^* to make up for it. If the agent is risk averse, this variability in his wage requires that his expected wage be higher than the w^* found earlier, because he must be compensated for the extra risk. There is a tradeoff between incentives and insurance against risk.

Moral hazard becomes a problem when $q(e)$ is not a one-to-one function because a single value of e might result in any of a number of values of q, depending on the value of θ. In this case the output function is not invertible; knowing q, the principal cannot deduce the value of e perfectly without assuming equilibrium behavior on the part of the agent.

The combination of unobservable effort and lack of invertibility in Production Game V means that no contract can induce the agent to put forth the efficient effort level without incurring extra costs, which usually take the form of an extra risk imposed on the agent. We will still try to find a contract that is efficient in the sense of maximizing welfare given the informational constraints. The terms "first-best" and "second-best" are used to distinguish these two kinds of optimality.

> A **first-best contract** *achieves the same allocation as the contract that is optimal when the principal and the agent have the same information set and all variables are contractible.*

> A **second-best contract** *is Pareto optimal given information asymmetry and constraints on writing contracts.*

The difference in welfare between the first-best world and the second-best world is the cost of the agency problem.

The first four production games were easier because the principal could find a first-best contract without searching very far. But even defining the strategy space in a game like Production Game V is tricky, because the principal may wish to choose a very complicated function $w(q)$. Finding the optimal contract when a forcing contract cannot be used becomes a difficult problem without general answers, because of the tremendous variety of possible contracts. The rest of the chapter will show how the problem may at least be approached, if not actually solved.

7.3 Finding Optimal Contracts: The Three-Step Procedure and the Incentive Compatibility and Participation Constraints

The principal's objective in Production Game V is to maximize his utility knowing that the agent is free to reject the contract entirely and that the contract must give the agent an incentive to choose the desired effort. These two constraints arise in every moral hazard problem, and they are named the **participation constraint** and the **incentive compatibility constraint**. Mathematically, the principal's problem is

$$\underset{w(\cdot)}{Maximize}\ EV(q(\tilde{e},\theta) - w(q(\tilde{e}, \theta))) \tag{11}$$

subject to

$$\tilde{e} = \underset{e}{argmax}\ EU(e,\ w(q(e,\ \theta))) \quad \text{(incentive compatibility constraint)} \tag{11a}$$

$$EU(\tilde{e},\ w(q(\tilde{e},\ \theta))) \geq \bar{U} \quad \text{(participation constraint)} \tag{11b}$$

The incentive-compatibility constraint takes account of the fact that the agent moves second, so the contract must induce him to voluntarily pick the desired effort. The participation constraint, also called the **reservation utility** or **individual rationality** constraint, requires that the worker prefer the contract to leisure, home production, or alternative jobs.

Expression (11) is the way an economist instinctively sets up the problem, but setting it up is often as far as he can get with the **first-order condition approach**. The difficulty is not just that the maximizer is choosing a wage function instead of a number, because control theory or the calculus of variations can solve such problems. Rather, it is that the constraints are nonconvex—they do not rule out a nice convex set of points in the space of wage functions as would the constraint "$w \geq 4$," but rather rule out a very complicated set of possible wage functions.

A different approach, developed by Grossman & Hart (1983) and called the **three-step procedure** by Fudenberg & Tirole (1990), is to focus on contracts that induce the agent to pick a particular action rather than to directly attack the problem of maximizing profits. The first step is to find for each possible effort level the set of wage contracts that induce the agent to choose that effort level. The second step is to find the contract which supports that effort level at the lowest cost to the principal. The third step is to choose the effort level that maximizes profits, given the necessity to support that effort with the costly wage contract from the second step.

To support the effort level e, the wage contract $w(\cdot)$ must satisfy the incentive compatibility and participation constraints. Mathematically, the problem of finding the least cost $C(\tilde{e})$ of supporting the effort level $\tilde{e}$ combines steps one and two.

$$C(\tilde{e}) = \underset{w(\cdot)}{Minimum}\ Ew(q(\tilde{e},\ \theta)) \tag{12}$$

subject to constraints (11a) and (11b).

Step three takes the principal's problem of maximizing his payoff, (11), and restates it as

$$\underset{\tilde{e}}{Maximize}\ EV(q(\tilde{e},\ \theta) - C(\tilde{e})). \tag{13}$$

After finding out which contract most cheaply induces each effort, the principal discovers the optimal effort by solving problem (13).

Breaking the problem into parts makes it easier to solve. Perhaps the most important lesson of the three-step procedure, however, is to reinforce the points that the goal of the contract is to induce the agent to choose a particular effort level and that asymmetric information increases the cost of the inducements.

7.4 Optimal Contracts: The Broadway Game

Relation Between Output and Compensation

The next game, inspired by Mel Brooks's offbeat film *The Producers*, illustrates a peculiarity of optimal contracts: sometimes the agent's reward should not increase with his output. Investors advance funds to the producer of a Broadway show that might succeed or might fail. The producer has the choice of embezzling or not embezzling the funds advanced to him, with a direct gain to himself of 50 if he embezzles. If the show is a success, the revenue is 500 if he did not embezzle and 100 if he did. If the show is a failure, revenue is −100 in either case, because extra expenditure on a fundamentally flawed show is useless.

Broadway Game I

Players

Producer and investors.

Order of Play

(1) The investors offer a wage contract $w(q)$ as a function of revenue q.
(2) The producer accepts or rejects the contract.
(3) The producer chooses *Embezzle* or *Do not embezzle*.
(4) Nature picks the state of the world to be *Success* or *Failure* with equal probability. Table 7.2 shows the resulting revenue q.

Payoffs

The producer is risk averse and the investors are risk neutral. The producer's payoff is $U(100)$ if he rejects the contract, where $U' > 0$ and $U'' < 0$, and the investors' payoff is 0. Otherwise,

$$\pi_{producer} = \begin{cases} U(w(q) + 50) & \textit{if he embezzles} \\ U(w(q)) & \textit{if he does not embezzle} \end{cases}$$

$$\pi_{investors} = q - w(q)$$

Table 7.2 Broadway Game I: Profits

		State of the World	
		Failure (0.5)	*Success* (0.5)
Effort	*Embezzle*	−100	+100
	Do not embezzle	−100	+500

Another way to tabulate outputs, shown in table 7.3, is to put the probabilities of the outcomes in the boxes, with effort in the rows and output in the columns.

Table 7.3 Broadway Game I: Probabilities of Profits

		Profit			
		−100	**+100**	**+500**	**Total**
Effort	*Embezzle*	0.5	0.5	0	1
	Do not embezzle	0.5	0	0.5	1

The investors will observe q to equal either -100, $+100$, or $+500$, so the producer's contract will specify at most three different wages: $w(-100)$, $w(+100)$, and $w(+500)$. The producer's expected payoffs from his two possible actions are

$$\pi(\textit{Do not embezzle}) = 0.5U\,(w(-100)) + 0.5U\,(w(+500)) \tag{14}$$

and

$$\pi(\textit{Embezzle}) = 0.5U\,(w(-100) + 50) + 0.5U(w(+100) + 50). \tag{15}$$

The incentive compatibility constraint is $\pi(\textit{Do not embezzle}) \geq \pi(\textit{Embezzle})$, so

$$\begin{aligned} &0.5U(w(-100)) + 0.5U(w(+500)) \\ &\qquad \geq 0.5U(w(-100) + 50) + 0.5U(w(+100) + 50), \end{aligned} \tag{16}$$

and the participation constraint is

$$\pi(\textit{Do not embezzle}) = 0.5U(w(-100)) + 0.5U(w(+500)) \geq U(100). \tag{17}$$

The investors want the participation constraint (17) to be satisfied at as low a dollar cost as possible. This means they want to impose as little risk on the producer as possible, since he requires a higher expected value for his wage if the risk is higher. Ideally, $w(-100) = w(+500)$, which provides full insurance. The usual agency tradeoff is between smoothing out the agent's wage and providing him with incentives. Here, no tradeoff is required, because of a special feature of the problem: there exists an outcome that could not occur unless the producer chooses the undesirable action. That outcome is $q = +100$, and it means that the following **boiling-in-oil contract** provides both riskless wages and effective incentives.

$$\begin{aligned} w(+500) &= 100 \\ w(-100) &= 100 \\ w(+100) &= -\infty. \end{aligned}$$

Under this contract, the producer's wage is a flat 100 when he does not embezzle, so the participation constraint is satisfied. It is also binding, because it is satisfied as an equality, and the investors would have a higher payoff if the constraint were relaxed. If the producer does embezzle, he faces a payoff of $-\infty$ with probability 0.5, so the incentive compatibility constraint is satisfied. It is nonbinding, because it is satisfied as a strong inequality and the investors' equilibrium payoff does not fall if the constraint is tightened a little by making the producer's earnings from embezzlement slightly higher. Note that the cost of the contract to the investors is 100 in equilibrium, so that their overall expected payoff is $0.5(-100) + 0.5(+500) - 100 = 100$, which is greater than zero and thus gives the investors enough return to be willing to back the show.

The boiling-in-oil contract is an application of the **sufficient statistic condition**, which states that for incentive purposes, if the agent's utility function is separable in effort and money, wages should be based on whatever evidence best indicates effort, and only incidentally on output (see Holmstrom [1979] and note N7.2). In the spirit of the three-step procedure, what the principal wants is to induce the agent to choose the appropriate effort, *Do not embezzle*, and his data on what the agent chose is the output. In equilibrium (though not out of it), the datum $q = +500$ contains exactly the same information as the datum $q = -100$. Both lead to the same posterior probability that the agent chose *Do not embezzle*, so the wages conditioned on each datum should be the same. We need to insert the qualifier, "in equilibrium," because to form the posterior probabilities the principal needs to have some beliefs as to the agent's behavior. Otherwise, he could not interpret $q = -100$ at all.

Milder contracts than this would also be effective. Two wages will be used in equilibrium, a low wage $\underline{w}$ for an output of $q = 100$ and a high wage $\bar{w}$ for any other output. The participation and incentive compatibility constraints provide two equations to solve for these two unknowns. To find the mildest possible contract, the modeller must also specify a function for $U(w)$ which, interestingly enough, was unnecessary for finding the first boiling-in-oil contract. Let us specify that

$$U(w) = 100w - 0.1w^2. \tag{18}$$

A quadratic utility function like this is only increasing if its argument is not too large, but since the wage will not exceed $w = 1000$, it is a reasonable utility function for this model. Substituting (18) into the participation constraint (17) and solving for the full-insurance high wage $\bar{w} = w(-100) = w(+500)$ yields $\bar{w} = 100$ and a reservation utility of 9000. Substituting into the incentive compatibility constraint, (16), yields

$$9000 \geq 0.5U(100 + 50) + 0.5U(\underline{w} + 50). \tag{19}$$

When (19) is solved using the quadratic equation, it yields (with rounding error), $\underline{w} \leq 5.6$. A low wage of $-\infty$ is far more severe than what is needed.

If both the producer and the investors were risk averse, risk sharing would change the part of the contract that applied in equilibrium. The optimal contract would then provide for $w(-100) < w(+500)$ to share the risk. The principal would have a lower marginal utility of wealth when output was $+500$, so he would be better able to pay an extra dollar of wages in that state than when output was -100.

One of the oddities of Broadway Game I is that the wage is higher for an output of -100 than for an output of $+100$. This illustrates the idea that the principal's aim is to reward not output, but input. If the principal pays more simply because output is higher, he is rewarding Nature, not the agent. People usually believe that higher pay for higher output is "fair," but Broadway Game I shows that this ethical view is too simple. Higher effort usually leads to higher output, so higher pay is usually a good incentive, but this is not invariably true.

The decoupling of reward and result has broad applications. Becker (1968) in criminal law and Polinsky & Che (1991) in tort law note that if society's objective is to keep the amount of enforcement costs and harmful behavior low, the penalty applied should not simply be matched to the harm. Very high penalties that are seldom inflicted will provide the proper incentives and keep enforcement costs low, even though a few unlucky offenders will receive penalties all out of proportion to the harm they have caused.

A less gaudy name for a boiling-in-oil contract is the alliterative "**shifting support scheme**," so named because the contract depends on the support of the output distribution being different when effort is optimal than when effort is other than optimal. Put more simply, the set of possible outcomes under optimal effort must be different from the set of possible outcomes under any other effort level. As a result, certain outputs show without doubt that the producer embezzled. Very heavy punishments inflicted only for those outputs achieve the first-best because a non-embezzling producer has nothing to fear.

Figure 7.3 shows shifting supports in a model where output can take not three, but a continuum of values. If the agent shirks instead of working, certain low outputs become possible and certain high outputs become impossible. In a case like this, where the support of the output shifts when behavior changes, boiling-in-oil contracts are useful: the wage is $-\infty$ for the low outputs possible only under shirking. When there is a limit to the amount the agent can be punished, or the support is the same under all actions, the threat of boiling-in-oil might not achieve the first-best contract, but similar contracts can still be used. The conditions favoring such contracts are

Figure 7.3 Shifting Supports in an Agency Model

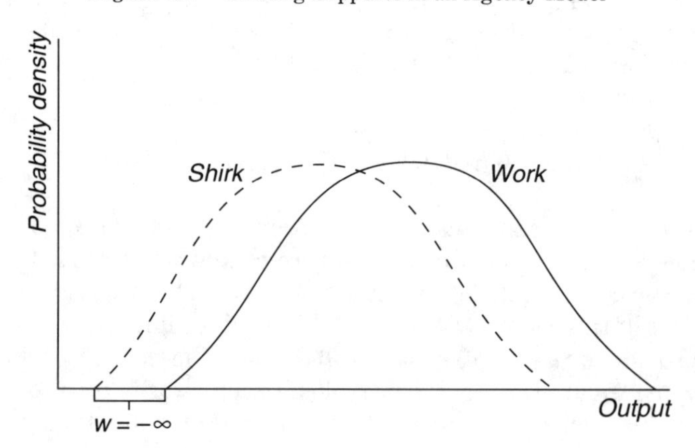

(1) The agent is not very risk averse.
(2) There are outcomes with high probability under shirking that have low probability under optimal effort.
(3) The agent can be severely punished.
(4) It is credible that the principal will carry out the severe punishment.

Selling the Store

Another first-best contract that can sometimes be used is **selling the store**. Under this arrangement, the agent buys the entire output for a flat fee paid to the principal, becoming the **residual claimant**, since he keeps every additional dollar of output that his extra effort produces. This is equivalent to fully insuring the principal, since his payoff becomes independent of the moves of the agent and of Nature.

In Broadway Game I, selling the store takes the form of the producer paying the investors 100 (= 0.5[−100] + 0.5[+500] − 100) and keeping all the profits for himself. The drawbacks are that (1) the producer might not be able to afford to pay the investors the flat price of 100; and (2) the producer might be risk averse and incur a heavy utility cost in bearing the entire risk. These two drawbacks are why producers go to investors in the first place.

Public Information that Hurts the Principal and the Agent

We can modify The Broadway Game to show how having more public information available can hurt both players. This will also provide a little practice in using information sets. Let us split *Success* into two states of the world, *Minor Success* and *Major Success*, which have probabilities 0.3 and 0.2, as shown in Table 7.4.

Table 7.4 Broadway Game II: Profits

		State of the World		
		Failure (0.5)	*Minor Success* (0.3)	*Major Success* (0.2)
Effort	*Embezzle*	−100	−100	+400
	Do not embezzle	−100	+450	+575

Under the optimal contract,

$$w(-100) = w(+450) = w(+575) > w(+400) + 50. \tag{3}$$

This is so because the producer is risk averse and only the datum $q = +400$ is proof that the producer embezzled. The optimal contract must do two things: it must deter embezzlement and it must pay the producer as predictable a wage as possible. For predictability, the wage is made constant unless $q = +400$. To deter embezzlement, the producer must be punished if $q = +400$. As in Broadway Game I, the punishment would not have to be infinitely severe, and the minimum effective punishment could be calculated in the same way as in that game. The investors would pay the producer a wage of 100 in equilibrium and their expected payoff would be $100(= 0.5(-100) + 0.3(450) + 0.2(575) - 100)$. Thus, a contract can be found for Broadway Game II such that the agent would not embezzle.

But consider what happens when the information set is refined so that before the agent takes his action both he and the principal can tell whether the show will be a major success or not. Let us call this game Broadway Game III. Under the refinement, each player's initial information partition is

$$(\{Failure, \ Minor \ Success\}, \{Major \ Success\}),$$

instead of the original coarse partition

$$(\{Failure, \ Minor \ Success, \ Major \ Success\}).$$

If the information sets were refined all the way to singletons, this would be very useful to the investors because they could abstain from investing in a failure and they could easily determine whether the producer embezzled or not. As it is, however, the refinement does not help the investors decide when to finance the show. If they can still hire the producer and prevent him from embezzling at a cost of 100, the payoff from investing in a major success is 475 $(=575 - 100)$. But the payoff from investing in a show given the information set {*Failure, Minor Success*} would be about 6.25 $\left(= \left(\frac{0.5}{0.5 + 0.3}\right)(-100) + \left(\frac{0.3}{0.5 + 0.3}\right)(450) - 100\right)$, which is still positive. So the improvement in information is no help with respect to the decision of when to invest.

Although the refinement has no direct effect on the efficiency of investment, it ruins the producer's incentives. If he observes {*Failure, Minor Success*}, he is free to embezzle without fear of the oil-boiling output of $+400$. He would still refrain from embezzling if he observed {*Major Success*}, but no contract that does not impose risk on a nonembezzling producer can stop him from embezzling if he observes {*Failure, Minor Success*}. Whether a risky contract can be found that would prevent the producer from embezzling at a cost of less than 6.25 to the investors depends on the producer's risk aversion. If he is very risk averse, the cost of the incentive is more than 6.25, and the investors will give up investing in shows that might be minor successes. Better information reduces welfare, because it increases the producer's temptation to misbehave.

7.5 State-Space Diagrams: Insurance Games I and II

The principal-agent models discussed so far in this chapter have been presented in terms of algebraic equations or outcome matrices. Another approach, especially useful when the strategy space is continuous, is to use diagrams. The term

Figure 7.4 Insurance Game I

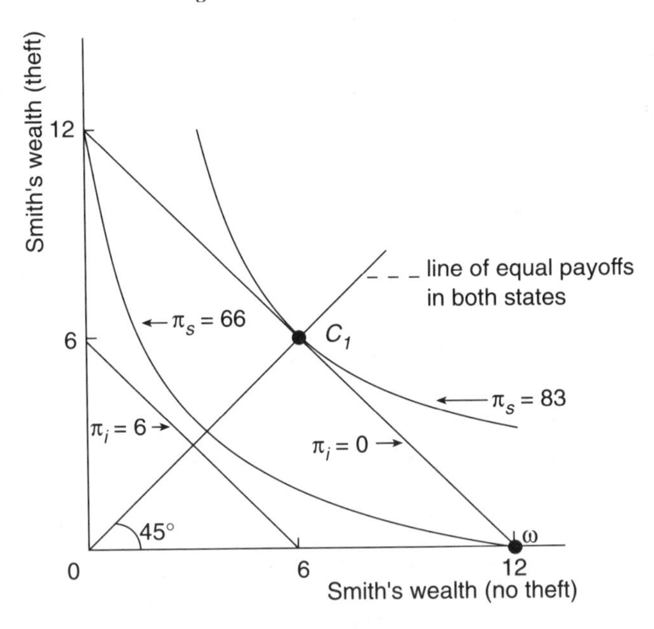

"moral hazard," comes from the insurance industry. Suppose Mr Smith (the agent) is considering buying theft insurance for a car with a value of 12. Figure 7.4, which illustrates his situation, is an example of a **state-space diagram**, a diagram whose axes measure the values of one variable in two different states of the world. Before Smith buys insurance, his dollar wealth is 0 if there is a theft and 12 otherwise, depicted as his endowment, $\omega = (12, 0)$. The point (12, 0) indicates a wealth of 12 in one state and 0 in the other, while the point (6, 6) indicates a wealth of 6 in each state.

One cannot tell the probabilities of each state just by looking at the state-space diagram. Let us specify that if Smith is careful where he parks, the state *Theft* occurs with probability 0.5, but if he is careless the probability rises to 0.75. He is risk averse, and, other things equal, he has a mild preference to be careless, a preference worth only ε to him. Other things are not equal, however, and he would choose to be careful were he uninsured because of the high correlation of carlessness with carelessness.

The insurance company (the principal) is risk neutral, perhaps because it is owned by diversified shareholders. We assume that no transaction costs are incurred in providing insurance and that the market is competitive, a switch from Production Game V, where the principal collected all the gains from trade. If the insurance company can require Smith to park carefully, it offers him insurance at a premium of 6, with a payout of 12 if theft occurs, leaving him with an allocation $C_1 = (6, 6)$. This satisfies the competition constraint because it is the most attractive contract any company can offer without making losses. Smith, whose allocation is 6 no matter what happens, is **fully insured**. In state-space diagrams, allocations like C_1 which fully insure one player are on the 45° line through the origin, the line along which his allocations in the two states are equal.

The game is described below in a specification that includes two insurance companies to simulate a competitive market. For Smith, who is risk averse, we must distinguish between dollar *allocations* such as (12, 0) and utility *payoffs* such as $0.5U(12) + 0.5U(0)$. The curves in figure 7.4 are labelled in units of utility for Smith and dollars for the insurance company.

Insurance Game I: Observable Care

Players

Smith and two insurance companies.

Order of Play

(1) Smith chooses to be either *Careful* or *Careless*, observed by the insurance company.
(2) Insurance company 1 offers a contract (x, y), in which Smith pays premium x and receives compensation y if there is a theft.
(3) Insurance company 2 also offers a contract of the form (x, y).
(4) Smith picks a contract.
(5) Nature chooses whether there is a theft, with probability 0.5 if Smith is *Careful* or 0.75 if Smith is *Careless*.

Payoffs

Smith is risk averse and the insurance companies are risk neutral. The insurance company not picked by Smith has a payoff of zero.
Smith's utility function U is such that $U' > 0$ and $U'' < 0$. If Smith picks contract (x, y), the payoffs are:
If Smith chooses *Careful*,

$$\pi_{Smith} = 0.5\,U(12 - x) + 0.5U(0 + y - x)$$
$$\pi_{company} = 0.5x + 0.5(x - y), \quad \text{for his insurer.}$$

If Smith chooses *Careless*,

$$\pi_{Smith} = 0.25U(12 - x) + 0.75U(0 + y - x) + \varepsilon$$
$$\pi_{company} = 0.25x + 0.75(x - y), \quad \text{for his insurer}$$

In the equilibrium of Insurance Game I Smith chooses to be *Careful* because he foresees that otherwise his insurance will be more expensive. Figure 7.4 is the corner of an Edgeworth box which shows the indifference curves of Smith and his insurance company given that Smith's care keeps the probability of a theft down to 0.5. The company is risk neutral, so its indifference curve, $\pi_i = 0$, is a straight line with slope $-1/1$. Its payoffs are higher on indifference curves such as $\pi_i = 6$ that are closer to the origin and thus have smaller expected payouts to Smith. The insurance company is indifferent between ω and C_1, at both of which its profits are zero. Smith is risk averse, so if he is *Careful* his indifference curves are closest to the origin on the 45° line, where his wealth in the two states is equal. Picking the numbers 66 and 83 for concreteness, I have labelled his original indifference curve $\pi_s = 66$ and drawn the preferred indifference curve $\pi_s = 83$ through the equilibrium contract C_1. The equilibrium contract is C_1, which satisfies the competition constraint by generating the highest expected utility for Smith that allows nonnegative profits to the company.

Insurance Game I is a game of symmetric information. Insurance Game II changes that. Suppose that

(1) The company cannot observe Smith's action,
or
(2) The state insurance commission does not allow contracts to require Smith to be careful,
or
(3) A contract requiring Smith to be careful is impossible to enforce because of the cost of proving carelessness.

In each case Smith's action is a noncontractible variable, so we model all three the same way by putting Smith's move second. The new game is like Production Game V, with uncertainty, unobservability, and two levels of output, *Theft* and *No Theft*. The insurance company may not be able to directly observe Smith's action, but his dominant strategy is to be *Careless*, so the company knows the probability of a theft is 0.75. Insurance Game II is the same as Insurance Game I except for the following.

Insurance Game II: Unobservable Care

Order of Play

(1) Insurance company 1 offers a contract of form (x, y), under which Smith pays premium x and receives compensation y if there is a theft.
(2) Insurance company 2 offers a contract of form (x, y).
(3) Smith picks a contract.
(4) Smith chooses either *Careful* or *Careless*.
(5) Nature chooses whether there is a theft, with probability 0.5 if Smith is *Careful* or 0.75 if Smith is *Careless*.

Smith's dominant strategy is *Careless*, so in contrast to Insurance Game I, the insurance company must offer a contract with a premium of 9 and a payout of 12 to prevent losses, which leaves Smith with an allocation $C_2 = (3, 3)$. Making thefts more probable reduces the slopes of both players' indifference curves, because it decreases the utility of points to the southeast of the 45° line and increases utility to the northwest. In figure 7.5, the insurance company's isoprofit curve swivels from the solid line $\pi_i = 0$ to the dotted line $\tilde{\pi}_i = 0$. It swivels around ω because that is the point at which the company's profit is independent of how probable it is that Smith's car will be stolen, since the company is not insuring him at point ω. Smith's indifference curve also swivels, from the solid curve $\pi_s = 66$ to the dotted curve $\tilde{\pi}_s = 66 + \varepsilon$. It swivels around the intersection of the $\pi_s = 66$ curve with the 45° line, because on that line the probability of theft does not affect Smith's payoff. The ε difference appears because Smith gets to choose the action *Careless*, which he slightly prefers.

Figure 7.5 shows that no full insurance contract will be offered. The contract C_1 is acceptable to Smith, but not to the insurance company, because it earns negative profits, and the contract C_2 is acceptable to the insurance company, but not to Smith, who prefers ω. Smith would like to commit himself to being careful, but he cannot make his commitment credible. If the means existed to prove his honesty, he would use them even if they were costly. He might, for example, agree to buy off-street parking even though locking his car would be cheaper, if verifiable.

Although no full insurance contract such as C_1 or C_2 is mutually agreeable, other contracts can be used. Consider the partial insurance contract C_3 in figure 7.5, which has a premium of 6 and a payout of 8. Smith would prefer C_3 to his endowment of $\omega = (12, 0)$ whether he chooses *Careless* or *Careful*. We can think of C_3 in two ways:

(1) Full insurance except for a **deductible** of four. The insurance company pays for all losses in excess of four.

(2) Insurance with a **co-insurance** rate of one-third. The insurance company pays two-thirds of all losses.

Figure 7.5 Insurance Game II with Full and Partial Insurance

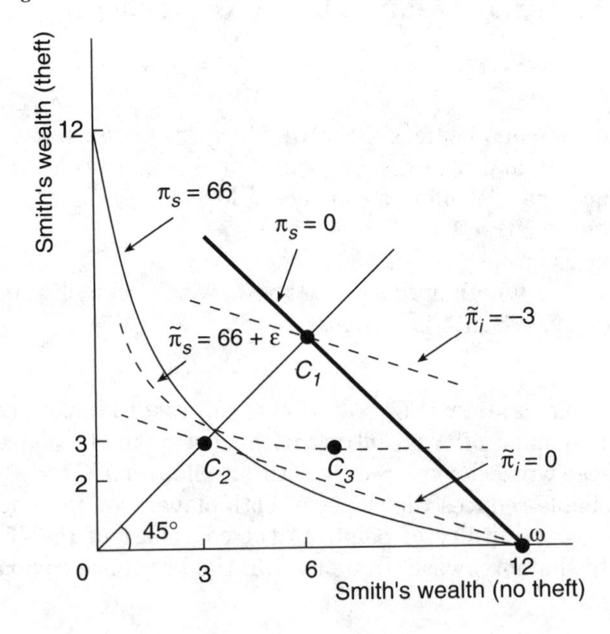

Figure 7.6 More on Partial Insurance in Insurance Game II

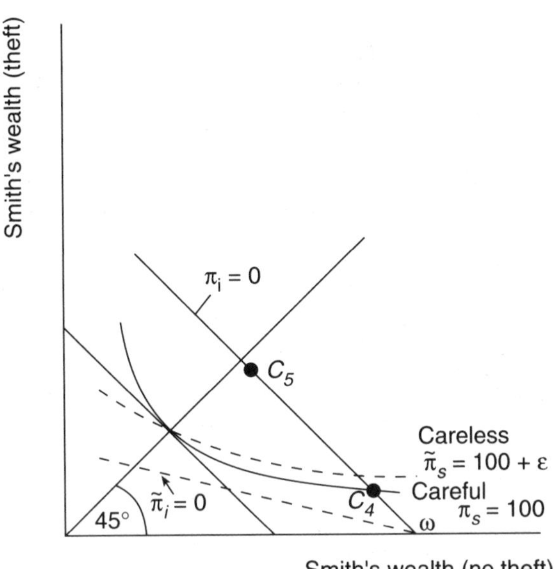

The outlook is bright, because Smith chooses *Careful* under a partial insurance contract like C_3. The moral hazard is "small" in the sense that Smith barely prefers *Careless*. With even a small deductible, Smith would choose *Careful* and the probability of theft would fall to 0.5, allowing the company to provide much more generous insurance. The solution of full insurance is "almost" reached. In reality, we rarely observe truly full insurance, because insurance contracts repay only the price of the car and not the bother of replacing it, which is great enough to deter owners from leaving their cars unlocked.

Figure 7.6 illustrates effort choice under partial insurance. Smith has a choice between dashed indifference curves (*Careless*) and solid ones (*Careful*). To the southeast of the 45° line, the dashed indifference curve for a particular utility level is always above that utility's solid indifference curve. Offered contract C_4, Smith chooses *Careful*, remaining on the solid indifference curve, so C_4 yields zero profit to the insurance company. In fact, the competing insurance companies will offer contract C_5 in equilibrium, which is almost full insurance, but just almost, so that Smith will choose *Careful* to avoid the small amount of risk he still bears.

Thus, as in the principal-agent model there is a tradeoff between efficient effort and efficient risk allocation. Even when the ideal of full insurance and efficient effort cannot be reached, there exists some best choice like C_5 in the set of feasible contracts, a second-best insurance contract that recognizes the constraints of informational asymmetry.

Notes

N7.1 Categories of Asymmetric Information Models

- The separation of asymmetric information into hidden actions and hidden knowledge is suggested in Arrow (1985) and commented upon in Hart & Holmstrom (1987). The term "hidden knowledge" seems to have become more popular than "hidden information," which I used in the first edition.

- Surveys of agency theory include Baiman (1982) and Hart & Holmstrom (1987). Hess' 1983 book on organization covers some of the earlier literature. Sonnenschein (1983) is a short and relatively nontechnical essay on hidden knowledge. Contracting theory also has widespread applications to finance; see Harris & Raviv (1992) for a survey.
- Empirical work on agency problems includes Joskow (1985, 1987) on coal mining, Masten & Crocker (1985) on natural gas contracts, Monteverde & Teece (1982) on auto components, Murphy (1986) on executive compensation, Rasmusen (1988b) on the mutual organization in banking, Staten & Umbeck (1982) on air traffic controllers and disability payments, and Wolfson (1985) on the reputation of partners in oil-drilling.
- A large literature of nonmathematical theoretical papers looks at organizational structure in the light of the agency problem. See Alchian & Demsetz (1972), Fama (1980), and Klein, Crawford, & Alchian (1978). Milgrom & Roberts (1992) have written a book on organization theory that describes what has been learned about the principal-agent problem at a technical level that M.B.A. students can understand.
- For examples of agency problems, see "Many Companies Now Base Workers' Raises on Their Productivity," *Wall Street Journal*, November 15, 1985, pp. 1, 15; "Big Executive Bonuses Now Come with a Catch: Lots of Criticism," *Wall Street Journal*, May 15, 1985, p. 33; "Bribery of Retail Buyers is Called Pervasive," *Wall Street Journal*, April 1, 1985, p. 6; "Some Employers Get Tough on Use of Air-Travel Prizes," *Wall Street Journal*, March 22, 1985, p. 27.
- We have lots of "prinsipuls" in economics. I find this paradigm helpful for remembering spelling: "The principal's principal principle was to preserve his principal."
- "Principal" and "Agent" are legal terms, and agency is an important area of the law. Economists have focussed on quite different questions than lawyers. Economists focus on effort: how the principal induces the agent to do things. Lawyers focus on malfeasance and third parties: how the principal stops the agent from doing the wrong things and who bears the burden if he fails. If, for example, the manager of a tavern enters into a supply contract against the express command of the owner, who must be disappointed—the owner or the third-party supplier?
- *Double-sided moral hazard.* The text described one-sided moral hazard. Moral hazard can also be double-sided, as when each player takes actions unobservable by the other that affect the payoffs of both of them. An example is tort negligence by both plaintiff and defendant: if a careless auto driver hits a careless pedestrian, and they go to law, the court must try to allocate blame, and the legislature must try to set up laws to induce the proper amount of care. Landlords and tenants also face double moral hazard, as implied in table 7.1.
- A common convention in principal-agent models is to make one player male and the other female, so that "his" and "her" can be used to distinguish between them. I find this distracting, since gender is irrelevant to most models and adds one more detail for the reader to keep track of. If readers naturally thought "male" when they saw "principal," this would not be a problem—but they do not.

N7.2 A Principal-Agent Model: The Production Game

- In Production Game III, we could make the agent's utility depend on the state of the world as well as on effort and wages. Little would change from the simpler model.
- The model in the text uses "effort" as the action taken by the agent, but effort is used to represent a variety of real-world actions. The cost of pilferage by employees is an estimated $8 billion a year in the USA. Employers have offered rewards for detection, one even offering the option of a year of twice-weekly lottery tickets instead of a lump sum. The Chicago department store Marshall Field's, with 14,000 workers, in one year gave out 170 rewards of $500 each, catching almost 500 dishonest employees. ("Hotlines and Hefty Rewards: Retailers Step Up Efforts to Curb Employee Theft," *Wall Street Journal*, September 17, 1987, p. 37.)

  For an illustration of the variety of kinds of "low effort," see "Hermann Hospital Estate, Founded for the Poor, has Benefited the Wealthy, Investigators Allege," *Wall Street Journal*, March 13, 1985, p. 4, which describes such forms of misbehavior as pleasure trips on company funds, high salaries, contracts for redecorating awarded to girlfriends, phony checks, kicking back real-estate commissions, and investing in friendly companies. Nonprofit enterprises, often lacking both principles and principals, are especially vulnerable, as are governments, for the same reason.

- The Production Game assumes that the agent dislikes effort. Is this realistic? People differ. My father tells of his experience in the navy when the sailors were kept busy by being ordered to scrape loose paint. My father found it a way to pass the time but says that other sailors would stop chipping when they were not watched, preferring to stare into space. *De gustibus non est disputandum.*[1] But even if effort has positive marginal utility at low levels, it has negative marginal utility at high enough levels—including, perhaps, at the efficient level. This is as true for professors as for sailors.
- Suppose that the principal does not observe the variable θ (which might be effort), but he does observe t and x (which might be output and profits). From Holmstrom (1979) and Shavell (1979) we have, restated in my words,

The Sufficient Statistic Condition. *If t is a sufficient statistic for θ relative to x, then the optimal contract needs to be based only on t if both principal and agent have separable utility functions.*

The variable t is a **sufficient statistic** *for θ relative to x if, for all t and x,*

$$Prob(\theta|t, x) = Prob(\theta|t). \tag{21}$$

This implies, from Bayes' Rule, that $Prob(t, x \mid \theta) = Prob(x \mid t)Prob(t \mid \theta)$; that is, x depends on θ only because x depends on t and t depends on θ.

The Sufficient Statistic Condition is closely related to the Rao-Blackwell Theorem (see Cox & Hinkley [1974] p. 258), which says that the decision rule for nonstrategic decisions ought not to be random.

Gjesdal (1982) notes that if the utility functions are not separable, the theorem does not apply and randomized contracts may be optimal. Suppose there are two actions the agent might take. The principal prefers action X, which reduces the agent's risk aversion, to action Y, which increases it. The principal could offer a randomized wage contract, so the agent would choose action X and make himself less risk averse. This randomization is not a mixed strategy. The principal is not indifferent between high and low wages; he prefers to pay a low wage, but we allow him to commit to a random wage earlier in the game.

N7.3 The Incentive Compatibility, Participation, and Competition Constraints

- Discussions of the first-order condition approach can be found in Grossman & Hart (1983) and Hart & Holmstrom (1987).
- The term, "individual rationality constraint," is more common, but "participation constraint" is more sensible. Since in modern modelling every constraint requires individuals to be rational, the name is singularly ill-chosen.
- **Paying the Agent More than His Reservation Wage.** If agents compete to work for principals, the participation constraint is binding whenever there are only two possible outcomes or whenever the agent's utility function is separable in effort and wages. Otherwise, it might happen that the principal picks a contract giving the agent more expected utility than is necessary to keep him from quitting. The reason is that the principal not only wants to keep the agent working, but to choose a high effort.
- If the distribution of output satisfies the **monotone likelihood ratio property** (MLRP), the optimal contract specifies higher pay for higher output. Let $f(q \mid e)$ be the probability density of output. The MLRP is satisfied if

$$\forall e' > e \text{ and } q' > q, \quad f(q'|e')f(q|e) - f(q'|e)f(q|e') > 0, \tag{22}$$

or, in other words, if when $e' > e$, the ratio $f(q \mid e')/f(q \mid e)$ is increasing in q. Alternatively, f satisfies the MLRP if $q' > q$ implies that q' is a more favorable message than q in the sense of Milgrom (1981b). Less formally, the MLRP is satisfied if the ratio of the likelihood of a high

[1] "About tastes there can be no arguing."

effort to a low effort rises with observed output. The distributions in the Broadway Game of section 7.5 violate the MLRP, but the normal, exponential, Poisson, uniform, and chi-square distributions all satisfy it. Stochastic dominance does not imply the MLRP. If effort of 0 produces outputs of 10 or 12 with equal probability, and effort of 1 produces outputs of 11 or 13 also with equal probability, the second distribution is stochastically dominant, but the MLRP is not satisfied.

- Finding general conditions that allow the modeller to characterize optimal contracts is difficult. Much of Grossman & Hart (1983) is devoted to the rather obscure Spanning Condition or Linear Distribution Function Condition (LDFC), under which the first-order condition approach is valid. The survey by Hart & Holmstrom (1987) makes a valiant attempt at explaining the LDFC.

N7.4 Optimal Contracts: The Broadway Game

- Daniel Asquith suggested the idea behind Broadway Game II.
- Franchising is one compromise between selling the store and paying a flat wage. See Mathewson & Winter (1985), Rubin (1978), and Klein & Saft (1985).
- Mirrlees (1974) is an early reference on the idea of the boiling-in-oil contract.
- Broadway Game II shows that improved information could reduce welfare by increasing a player's incentive to misbehave. This is distinct from the nonstrategic insurance reason why improved information can be harmful. Suppose that Smith is insuring Jones against hail ruining Jones' wheat crop during the next year, increasing Jones' expected utility and giving a profit to Smith. If someone comes up with a way to forecast the weather before the insurance contract is agreed upon, both players will be hurt. Insurance will break down, because if it is known that hail will ruin the crop, Smith will not agree to share the loss, and if it is known there will be no hail, Jones will not pay a premium for insurance. Both players prefer not knowing the outcome in advance.

N7.5 State-Space Diagrams: Insurance Game I and II

- The utility function of the agent in Insurance Game I is **separable** in effort and money; if effort changes from *Careful* to *Careless*, that does not change the marginal utility of a dollar for a given wealth level. Separability matters to the relative slopes of the dashed and solid indifference curves in the diagrams, because it means the slopes differ at a particular point only because the probabilities of an accident differ because of effort, not because the value of money differs.
- The text uses premiums and payouts to describe insurance. Another way to describe insurance is as the sale of an asset which provides different returns in different states. In Insurance Game I, the insurance company sells a security, C_1, with return $(-6, 6)$ and an expected value of 0, for a price of 0. Although Smith is risk averse, he is eager to buy this risky security because it cancels the risk of his initial asset, the car, which has return (12, 0).

  In Insurance Game II, which has a higher probability of theft in equilibrium, the insurance company offers a security, C_2, with return $(-9, 3)$ and expected value of 0 $(= 0.25[-9] + 0.75[3])$ if Smith chooses *Careless*. The partial insurance contract C_3 is a $(-6, 2)$ security with expected value 0 $(= 0.25[-6] + 0.75[2])$, which when combined with Smith's original asset leaves him with allocation [0.75,(6,2)], where 0.75 is the chance of a theft and (6, 2) is his wealth under the contract.
- When the sale of insurance is monopolized, the market behaves differently. Not only is the quantity of insurance restricted, but the contracts offered may specify both the price and the quantity of insurance which an individual can buy.
- Gary Schwartz pointed out to me that one reason why liability insurance rarely has coinsurance is that it covers legal disputes, where the need for the victim's side to have a single voice makes it especially useful to have a single residual claimant. With coinsurance, the benefit from each extra dollar awarded would be split between the insurer and the insured, making litigation less efficient.
- Defining optimality is not always straightforward in models of asymmetric information. Holmstrom & Myerson (1983) distinguish three kinds of efficiency, depending on when expected payoffs are evaluated: **ex ante**, before any player has private information; **interim**, when each player has received private information; and **ex post**, after all information has become common knowledge.

Problems

7.1: First-Best Solutions in a Principal-Agent Model.

Suppose an agent has the utility function $U = \sqrt{w} - e$, where e can assume the levels 0 or 1. Let the reservation utility level be $\bar{U} = 3$. The principal is risk neutral. Denote the agent's wage, conditioned on output, as $\underline{w}$ if output is 0 and $\bar{w}$ if output is 100. Table 7.5 shows the outputs.

Table 7.5 A Moral Hazard Game

Effort	Probability of Output 0	100	Total
Low ($e = 0$)	0.3	0.7	1
High ($e = 1$)	0.1	0.9	1

(7.1a) What would the agent's effort choice and utility be if he owned the firm?

(7.1b) If agents are scarce and principals compete for them, what will the agent's contract be under full information? His utility?

(7.1c) If principals are scarce and agents compete to work for them, what would the contract be under full information? What will the agent's utility and the principal's profit be in this situation?

(7.1d) Suppose that $U = w - e$. If principals are the scarce factor and agents compete to work for principals, what would the contract be when the principal cannot observe effort? (Negative wages are allowed.) What will the agent's utility and the principal's profit be in this situation?

7.2: The Principal-Agent Problem.

Suppose the agent has a utility function of $U = \sqrt{w} - e$, where e can assume the levels 0 or 7, and a reservation utility of $\bar{U} = 4$. The principal is risk neutral. Denote the agent's wage, conditioned on output, as $\underline{w}$ if output is 0 and $\bar{w}$ if output is 1,000. Only the agent observes his effort. Principals compete for agents. Table 7.6 shows the output.

Table 7.6 Output from Low and High Effort

Effort	Probability of Output 0	1000	Total
Low ($e = 0$)	0.9	0.1	1
High ($e = 7$)	0.2	0.8	1

(7.2a) What are the incentive compatibility, participation, and zero-profit constraints for obtaining high effort?
(7.2b) What would utility be if wages were fixed and could not depend on output or effort?
(7.2c) What is the optimal contract? What is the agent's utility?
(7.2d) What would the agent's utility be under full information? Under asymmetric information, what is the agency cost (the lost utility) as a percentage of the utility the agent receives?

7.3: Why Entrepreneurs Sell Out.

Suppose an agent has a utility function of $U = \sqrt{w} - e$, where e can assume the levels 0 or 2.4, and his reservation utility is $\bar{U} = 7$. The principal is risk neutral. Denote the agent's wage, conditioned on output, as $w(0)$, $w(49)$, $w(100)$, or $w(225)$. Table 7.7 shows the output.

Table 7.7 Entrepreneurs Selling Out

Method	Probability of Output 0	49	100	225	Total
Safe ($e = 0$)	0.1	0.1	0.8	0	1
Risky ($e = 2.4$)	0	0.5	0	0.5	1

(7.3a) What would the agent's effort choice and utility be if he owned the firm?
(7.3b) If agents are scarce and principals compete for them, what will the agent's contract be under full information? His utility?
(7.3c) If principals are scarce and agents compete to work for principals, what will the contract be under full information? What will the agent's utility and the principal's profit be in this situation?
(7.3d) If agents are the scarce factor and principals compete for them, what will the contract be when the principal cannot observe effort? What will the agent's utility and the principal's profit be in this situation?

7.4: Bankruptcy Constraints.

A risk-neutral principal hires an agent with utility function $U = \mathrm{w} - e$ and reservation utility $\bar{U} = 5$. Effort is either 0 or 10. There is a bankruptcy constraint: $w \geq 0$. Output is given by Table 7.8.
(7.4a) What would the agent's effort choice and utility be if he owned the firm?
(7.4b) If agents are scarce and principals compete for them, what will the agent's contract be under full information? His utility?
(7.4c) If principals are scarce and agents compete to work for them, what will the contract be under full information? What will the agent's utility be?

Table 7.8 Bankruptcy

Effort	Probability of Output 0	400	Total
Low ($e = 0$)	0.5	0.5	1
High ($e = 10$)	0.1	0.9	1

(7.4d) If principals are scarce and agents compete to work for them, what will the contract be when the principal cannot observe effort? What will the payoffs be for each player?

(7.4e) Suppose there is no bankruptcy constraint. If principals are the scarce factor and agents compete to work for them, what will the contract be when the principal cannot observe effort? What will the payoffs be for principal and agent?

7.5: Worker Effort.

A worker can be *Careful* or *Careless*, efforts which generate mistakes with probabilities 0.25 and 0.75. His utility function is $U = 100 - 10/w - x$, where w is his wage and x takes the value 2 if he is careful and 0 otherwise. Whether a mistake is made is contractible, but effort is not. Risk-neutral employers compete for the worker, and his output is worth 0 if a mistake is made and 20 otherwise. No computation is needed for any part of this problem.

(7.5a) Will the worker be paid anything if he makes a mistake?

(7.5b) Will the worker be paid more if he does not make a mistake?

(7.5c) How would the contract be affected if employers were also risk averse?

(7.5d) What would the contract look like if a third category, "slight mistake," with an output of 19, occurs with probability 0.1 after *Careless* effort and with probability zero after *Careful* effort?

8 Topics in Moral Hazard

8.1 Pooling versus Separating Equilibria and the Revelation Principle

In chapter 8 we will continue with moral hazard by taking up hidden knowledge and a variety of special cases that apply to both kinds of moral hazard. Section 8.1 introduces hidden knowledge and distinguishes between pooling and separating equilibria. It also discusses a modelling simplification called the revelation principle. Section 8.2 uses diagrams to apply the model to the selection of a sales strategy. Section 8.3 returns to moral hazard with hidden actions and addresses the problem of contract renegotiation. We then turn to remedies for moral hazard in sections on efficiency wages (section 8.4), tournaments (section 8.5), and various other remedies (section 8.6). The chapter concludes with section 8.7, which discusses the teams model of joint production by a group of agents.

Information is complete in moral hazard games, but in moral hazard with hidden knowledge, the agent, but not the principal, observes a move of Nature after the game begins. Information is symmetric at the time of contracting, but becomes asymmetric later. From the principal's point of view, agents are identical at the beginning of the game but develop private types midway through it, depending on what they have seen. His chief concern is to give them incentives to disclose their types later, which gives games with hidden knowledge a flavor close to that of the adverse-selection models to be studied in chapter 9. The agent may exert effort, but effort's contractibility is less important when the principal does not know which effort is appropriate because he is ignorant of the state of the world chosen by Nature. The main difference in technical analysis between moral hazard with hidden knowledge and adverse selection is that if the game begins with symmetric information and only becomes asymmetric after a contract has been agreed upon, the contract must satisfy a participation constraint which takes into account the fact that the agent's type is not yet known to him.

There is more hope for obtaining efficient outcomes in moral hazard with hidden knowledge than in the other two kinds of games of asymmetric information. The advantage over adverse selection is that information is symmetric at the time of contracting, so neither player can use private information to extract surplus from the other by choosing inefficient contract terms. The advantage over hidden actions is that the post-contractual asymmetry is with respect to knowledge only, which is neutral in itself, rather than over whether the agent exerted high effort, which causes direct disutility to him.

For a comparison between the two types of moral hazard, let us modify Production Game V from section 7.2 to turn it into a game of hidden knowledge.

Production Game VI: Hidden Knowledge

Players

The principal and the agent.

Order of Play

(1) The principal offers the worker a wage contract of the form $w(q, m)$.
(2) The agent accepts or rejects the principal's offer.
(3) Nature chooses the state of the world θ according to probability distribution $F(\theta)$. The agent observes θ, but the principal does not.
(4) If the agent accepts, he exerts effort e and sends a message m, both observed by the principal.
(5) Output is $q(e, \theta)$.

Payoffs

If the agent rejects the contract, $\pi_{agent} = \bar{U}$ and $\pi_{principal} = 0$.
If the agent accepts the contract, $\pi_{agent} = U(e, w, \theta)$ and $\pi_{principal} = V(q - w)$.

The principal would like to know θ so he can tell which effort level is appropriate. In an ideal world he would employ an honest agent who always chose $m = \theta$, but in noncooperative games, talk is cheap. Since the agent's words are worthless, the principal must try to design a contract that either provides incentives for truthfulness or takes lying into account. The **mechanism design** literature addresses this question. It says that the principal **implements** a **mechanism** to extract the agent's information.

Pooling and Separating Equilibria

In hidden-action models, the principal tries to construct a contract which will induce the agent to take the single appropriate action. In hidden-knowledge models, the principal tries to make different actions attractive under different states of the world, so the agent's choice depends on the hidden state.

> *If all types of agents pick the same strategy in all states, the equilibrium is* **pooling**. *Otherwise, it is* **separating**.

The distinction between pooling and separating is different from the distinction between equilibrium concepts. A model might have multiple Nash equilibria, some pooling and some separating. Moreover, a single equilibrium—even a pooling one—can include several contracts, but if it is pooling the agent always uses the same strategy, regardless of type. If the agent's equilibrium strategy is mixed, the equilibrium is pooling if the agent always picks the same mixed strategy, even though the messages and efforts would differ across realizations of the game.

These two terms came up in section 6.2 in the game of PhD Admissions. Neither type of student applied in the pooling equilibrium, but one type did in the separating equilibrium. In a principal-agent model, the principal tries to design the contract to achieve separation unless the incentives turn out to be too costly.

A separating contract need not be fully separating. If agents who observe $\theta \leq 4$ accept contract C_1 but other agents accept C_2, the equilibrium is separating but it does not separate out every type. We say that the equilibrium is **fully revealing** if the agent's choice of contract always conveys his private information to the principal. Between pooling and fully revealing equilibria are the **imperfectly separating** equilibria, also called **semi-separating**, **partially separating**, **partially revealing**, or **partially pooling** equilibria.

The principal's problem, as in Production Game V, is to maximize his profits subject to

(1) **Incentive compatibility** (the agent picks the desired contract and actions).
(2) **Participation** (the agent prefers the contract to his reservation utility).

In a model with hidden knowledge, the incentive compatibility constraint is customarily called the **self-selection constraint**, because it induces the different types of agents to pick different contracts. As with hidden actions, if principals compete in offering contracts, a **competition constraint** is added: the equilibrium contract must be as attractive as possible to the agent, since otherwise another principal could profitably lure him away. An equilibrium may also need to satisfy a part of the competition constraint not found in hidden actions models: either a **nonpooling constraint** or a **nonseparating constraint**. If one of several competing principals wishes to construct a pair of separating contracts C_1 and C_2, he must construct it so that not only do agents choose C_1 and C_2 depending on the state of the world (to satisfy incentive compatibility), but they also prefer (C_1, C_2) to a pooling contract C_3 (to satisfy nonpooling).

Unravelling the Truth when Silence is the Only Alternative

Before going on to look at a self-selection contract, let us look at a special case in which hidden knowledge paradoxically makes no difference. The usual hidden-knowledge model has no penalty for lying, but let us briefly consider what happens if the agent cannot lie but he can be silent or tell half-truths. Suppose that Nature uses the uniform distribution to assign the variable θ some value in the interval [0, 10], and the agent's payoff is increasing in the principal's estimate of θ. Usually we assume that the agent can lie freely, sending a message m that takes any value in [0, 10], but let us assume instead that he cannot lie, although he is free to conceal information. Thus, if $\theta = 2$, he can send the uninformative message $m \geq 0$ (equivalent to no message), or $m \geq 1$, or $m = 2$, but not the lie $m \geq 4$.

When $\theta = 2$ the agent might as well send a message that is the exact truth: "$m = 2$." If he were to choose "$m \geq 1$," for example, the principal's first thought might be to estimate θ as the average value of the interval [1, 10], which is 5.5. But the principal would realize that no agent with a value of θ greater than 5.5 would want to send that message in a Nash equilibrium. This realization restricts the possible interval to [1, 5.5], which in turn has an average of 3.25. But then no agent with $\theta > 3.25$ would send the message "$m \geq 1$." The principal can continue this process of logical **unravelling** to conclude that $\theta = 1$. The message "$m \geq 0$" would be even worse, making the principal believe that $\theta = 0$. In this model, "No news is bad news." The agent would therefore not send the message "$m \geq 1$" and he would be indifferent between "$m = 2$" and "$m \geq 2$" because the principal would make the same deduction from either message.

Perfect revelation is paradoxical, but that is because the assumptions just described are rarely satisfied in the real world. In particular, unpunishable lying and genuine ignorance allow information to be concealed. If the seller is free to lie without punishment, then, in the absence of other incentives, he always pretends that his information is extremely favorable, so nothing he says conveys any information, favorable or unfavorable. If he really is ignorant in some states of the world, then his silence could mean either that he has nothing to say or that he has nothing he wants to say. The unravelling argument fails because if he sends an uninformative message the buyers will attach some probability to "no news" instead of "bad news." Problem 8.3 at the end of this chapter explores unravelling further.

The Revelation Principle

The principal might choose to offer a contract that induces the agent to lie in equilibrium, since he can take lying into account when he designs the contract, but this complicates the analysis. Each state of the world has a single truth, but a continuum of lies: generically speaking, almost everything is false. The revelation principle helps us simplify.

> **The Revelation Principle.** *For every contract $w(q,m)$ that leads to lying (that is, to $m \neq \theta$), there is a contract $w^*(q, m)$ with the same outcome for every θ but no incentive for the agent to lie.*

In many possible contracts, sending false messages is profitable for the agent in that when the state of the world is a he receives a reward of x_1 for the true report of a and $x_2 > x_1$ for the false report of b. A contract which gives the agent a reward of x_2 regardless of whether he reported a or b would lead to exactly the same payoffs for each player while giving the agent no incentive to lie. The revelation principle notes that a contract with no lying can always be found by imitating the relation between states of the world and payoffs in the equilibrium of the contract with lying. This idea can also be applied to games in which both players must make reports to each other.

Applied to concrete examples, the revelation principle may seem obvious. Suppose we are concerned with the effect on the moral climate of cheating on income taxes, but anyone who makes $70,000 a year can claim he makes $50,000 and the government does not have the resources to catch him. The revelation principle says that we can rewrite the tax code to set the tax to be the same for taxpayers earning $70,000 and for those earning $50,000, and the same amount of taxes will be collected without anyone having the incentive to lie. Applied to moral education, the principle says that the mother who agrees never to punish her daughter if she tells her all her escapades will never hear any untruths. Clearly, the principle's usefulness is not so much to improve outcomes as to simplify contracts. The principal (and the modeller) need only look at contracts which induce truth-telling, so the relevant strategy space is shrunk and we can add a third constraint to the incentive compatibility and participation constraints to help calculate the equilibrium:

(3) **Truth-telling.** The equilibrium contract makes the agent willing to choose $m = \theta$.

The revelation principle says that a truth-telling equilibrium exists, but not that it is unique. It may well happen that the equilibrium is a weak Nash equilibrium in which the optimal contract gives the agent no incentive to lie but also no incentive to tell the truth. This is similar to the open-set problem discussed in section 4.3; the optimal contract may satisfy the agent's participation constraint but makes him indifferent between accepting and rejecting the contract. If agents derive the slightest utility from telling the truth, of course, then truth-telling becomes a strong equilibrium, but if their utility from telling the truth is really significant, it should be made an explicit part of the model. If the utility of truth-telling is strong enough, in fact, agency problems and the costs associated with them disappear. This is one reason why morality is useful to business.

8.2 An Example of Moral Hazard with Hidden Knowledge: The Salesman Game

The next game illustrates the differences between pooling and separating equilibria. The manager of a company has told his salesman to investigate a potential customer, who is either a *Pushover* or a *Bonanza*. If he is a *Pushover*, the efficient sales effort is low and sales should be moderate. If he is a *Bonanza*, the effort and sales should be higher.

The Salesman Game

Players

A manager and a salesman.

Order of Play

(1) The manager offers the salesman a contract of the form $w(q, m)$, where q is sales and m is a message.
(2) The salesman decides whether or not to accept the contract.
(3) Nature chooses whether the customer is a *Bonanza* or a *Pushover* with probabilities 0.2 and 0.8. Denote the state variable "customer status" by θ. The salesman observes the state, but the manager does not.
(4) If the salesman has accepted the contract, he chooses his sales level q, which implicitly measures his effort.

Payoffs

The manager is risk neutral and the salesman is risk averse. If the salesman rejects the contract, his payoff is $\bar{U} = 8$ and the manager's is zero. If he accepts the contract, then

$$\pi_{manager} = q - w.$$

$$\pi_{salesman} = U(q, w, \theta), \text{ where } \frac{\partial U}{\partial q} < 0, \frac{\partial^2 U}{\partial q^2} < 0, \frac{\partial U}{\partial w} > 0, \frac{\partial^2 U}{\partial w^2} < 0.$$

Figure 8.1 shows the indifference curves of manager and salesman, labelled with numerical values for exposition. The manager's indifference curves are straight lines with slope 1 because he is acting on behalf of a risk-neutral company. If the wage and the quantity both rise by a dollar, profits are unchanged, and the profits do not depend directly on whether θ takes the value *Pushover* or the value *Bonanza.*

The salesman's indifference curves also slope upwards, because he must receive a higher wage to compensate for the extra effort that makes q greater. They are convex because the marginal utility of dollars is decreasing and the marginal disutility of effort is increasing. As figure 8.1 shows, the salesman has two sets of indifference curves, solid for *Pushovers* and dashed for *Bonanzas*, since the effort that secures a given level of sales depends on the state.

Because of the participation constraint, the manager must provide the salesman with a contract giving him at least his reservation utility of 8, which is the same in both states. If the true state is that the customer is a *Bonanza*, the manager would like to offer a contract that leaves the salesman on the dashed indifference curve $\tilde{U}_s = 8$, and the efficient outcome is (q_2, w_2), the point at which the salesman's indifference curve is tangent to one of the manager's indifference curves. At that point, if the salesman sells an extra dollar, he requires an extra dollar of compensation.

If it were common knowledge that the customer was a *Bonanza*, the principal could choose w_2 so that $U(q_2, w_2, Bonanza) = 8$ and offer the forcing contract

$$w = \begin{cases} 0 & \text{if } q < q_2 \\ w_2 & \text{if } q \geq q_2 \end{cases} \qquad (1)$$

Figure 8.1 The Salesman Game with Curves for Pooling Equilibrium

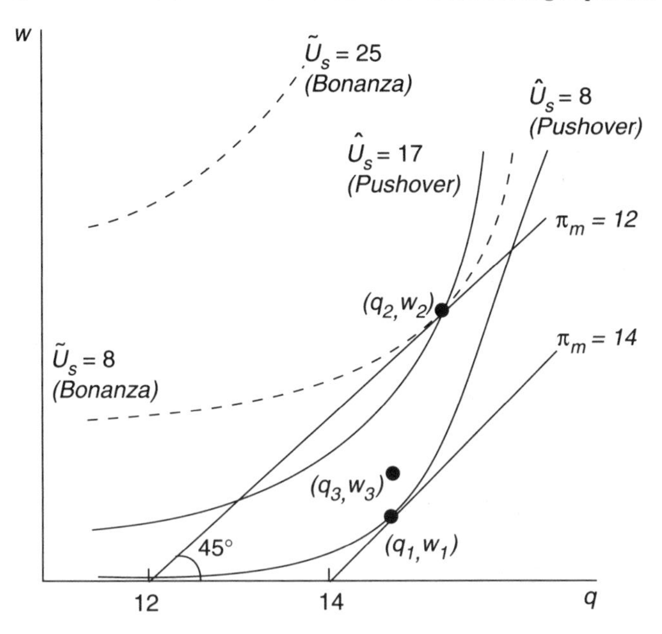

The salesman would accept the contract and choose $q = q_2$. But if the customer were actually a *Pushover*, the salesman would still choose $q = q_2$, an inefficient outcome that does not maximize profits. High sales would be inefficient because the salesman would be willing to give up more than a dollar of wages to escape having to make his last dollar of sales. Profits would not be maximized, because the salesman achieves a utility of 17, and he would have been willing to work for less.

The revelation principle says that in searching for the optimal contract we need only look at contracts that induce the agent to truthfully reveal what kind of customer he faces. If it required more effort to sell any quantity to the *Bonanza*, as shown in figure 8.1, the salesman would always want the manager to believe that he faced a *Bonanza*, so he could extract the extra pay necessary to achieve a utility of 8 selling to *Bonanzas*. The only optimal truth-telling contract is the pooling contract that pays the intermediate wage of w_3 for the intermediate quantity of q_3, and zero for any other quantity, regardless of the message. The pooling contract is a second-best contract, a compromise between the optimum for *Pushovers* and the optimum for *Bonanzas*. The point (q_3, w_3) is closer to (q_1, w_1) than to (q_2, w_2), because the probability of a *Pushover* is higher and the contract must satisfy the participation constraint

$$0.8U(q_3, w_3, Pushover) + 0.2U(q_3, w_3, Bonanza) \geq 8. \tag{2}$$

The nature of the equilibrium depends on the shapes of the indifference curves. If they are shaped as in figure 8.2, the equilibrium is separating, not pooling, and there does exist a first-best, fully revealing contract.

Figure 8.2 Indifference Curves for a Separating Equilibrium

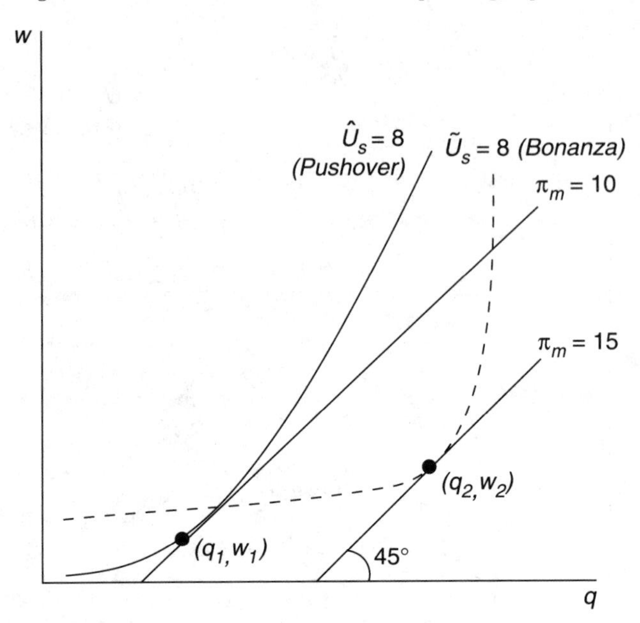

$$\text{Separating Contract} \begin{cases} \text{Agent announces } \textit{Pushover}\text{: } w = \begin{cases} 0 \text{ if } q < q_1 \\ w_1 \text{ if } q \geq q_1 \end{cases} \\ \\ \text{Agent announces } \textit{Bonanza}\text{: } w = \begin{cases} 0 \text{ if } q < q_2 \\ w_2 \text{ if } q \geq q_2 \end{cases} \end{cases} \quad (3)$$

Again, we know from the revelation principle that we can narrow attention to contracts that induce the salesman to tell the truth. With the indifference curves of figure 8.2, contract (3) induces the salesman to be truthful and the incentive compatibility constraint is satisfied. If the customer is a *Bonanza*, but the salesman claims to observe a *Pushover* and chooses q_1, his utility is less than 8 because the point (q_1, w_1) lies below the $\tilde{U}_s = 8$ indifference curve. If the customer is a *Pushover* and the salesman claims to observe a *Bonanza*, then although (q_2, w_2) does yield the salesman a higher wage than (q_1, w_1), the extra income is not worth the extra effort, because (q_2, w_2) is far below the indifference curve $\hat{U}_s = 8$.

Another way to phrase the description of a separating equilibrium is to say that it gives the salesman a choice of contracts, rather than saying that it gives him a single contract that specifies different wages for different outputs. He agrees to work with the manager, and after he discovers what type the customer is he chooses either the contract (q_1, w_1) or the contract (q_2, w_2), where each is a forcing contract and he receives 0 if after choosing the contract (q_i, w_i) he produces output $q \neq q_i$. In this interpretation, we say that the manager offers a **menu of contracts** and the salesman selects one of them after learning his type. This is simply a different way of describing the same equilibrium.

Sales contracts in the real world are often complicated, because it is easy to measure the major component of output, sales, and hard to measure the inputs of workers who are out in the field away from direct supervision. The Salesman Game is a real problem. Gonik (1978) describes hidden knowledge contracts used by IBM's subsidiary in Brazil. Salesmen were first assigned quotas. They then announced their own sales forecast as a percentage of quota and chose from among a set of contracts, one for each possible forecast. Inventing some numbers for illustration, if Smith were assigned a quota of 400 and he announced 100 percent, he would get $w = 70$ if he sold 400 and $w = 80$ if he sold 450; but if he had announced 120 percent, he would have gotten $w = 60$ for 400 and $w = 90$ for 450. The contract encourages extra effort when the extra effort is worth the extra sales. The idea here, as in the Salesman Game, is to reward salesmen not just for high effort, but for appropriate effort.

The Salesman Game illustrates a number of ideas. It can have either a pooling or a separating equilibrium, depending on the utility function of the salesman. The revelation principle can be applied to avoid having to consider contracts in which the manager must interpret the salesman's lies. It also shows how to use diagrams when the algebraic functions are intractable or unspecified, a problem that does not arise in most of the two-valued numerical examples in this book.

8.3 Renegotiation of Contracts: The Repossession Game

Renegotiation comes up in two very different contexts in game theory. Chapter 4 looked at the situation where players can coordinate on Pareto-superior subgame equilibria that might be Pareto inferior for the entire game, an idea linked to the problem of selecting among multiple equilibria. This section looks at a completely different context, one in which the players have signed a binding contract, but, in a subsequent subgame, both players might agree to scrap the old contract and write a new one using the old contract as a starting point in their negotiations. Here, the questions are not about equilibrium selection, but instead concern which strategies should be allowed in the game. This is an issue that frequently arises in principal-agent models, and it was first pointed out in the context of hidden knowledge by Dewatripont (1989). Here we will use a model of hidden actions to illustrate renegotiation, a model in which a bank wants to lend money to a consumer so that he can buy a car, and must worry whether the consumer will work hard enough to repay the loan.

The Repossession Game

Players

A bank and a consumer.

Order of Play

(1) The bank can do nothing or it can offer the consumer an auto loan which allows him to buy a car that costs 11, but requires him to pay back L or lose possession of the car to the bank.

(2) The consumer accepts or rejects the loan.
(3) The consumer chooses to *Work*, for an income of 15, or *Play*, for an income of 8. The disutility of work is 5.
(4) The consumer repays the loan or defaults.
(4a) In one version of the game, the bank offers to settle for an amount S and leave possession of the car to the consumer.
(4b) The consumer accepts or rejects the settlement S.
(5) If the bank has not been paid L or S, it repossesses the car.

Payoffs

If the bank does not make any loan or the consumer rejects it, both players' payoffs are zero. The value of the car is 12 to the consumer and 7 to the bank, so the bank's payoff if a loan is made is

$$\pi_{bank} = \begin{cases} L - 11 & \text{if the original loan is repaid} \\ S - 11 & \text{if a settlement is made} \\ 7 - 11 & \text{if the car is repossessed.} \end{cases}$$

If the consumer chooses *Work* his income W is 15 and his disutility of effort D is -5. If he chooses *Play*, then $W = 8$ and $D = 0$. His payoff is

$$\pi_{consumer} = \begin{cases} W + 12 - L - D & \text{if the original loan is repaid} \\ W + 12 - S - D & \text{if a settlement is made} \\ W - D & \text{if the car is repossessed.} \end{cases}$$

We will consider two versions of the game, both of which allow commitment in the sense of legally binding agreements over transfers of money and wealth but do not allow the consumer to commit directly to *Work*. If the consumer does not repay the loan, the bank has the legal right to repossess the car, but the bank cannot have the consumer thrown into prison for breaking a promise to choose *Work*. Where the two versions of the game will differ is in whether they allow the renegotiation moves (4a) and (4b).

Repossession Game I

The first version of the game does not allow renegotiation, so moves (4a) and (4b) are dropped from the game. In equilibrium, the bank will make the loan at a rate of $L = 12$, and the consumer will choose *Work* and repay the loan. Working back from the end of the game in accordance with sequential rationality, the consumer is willing to repay because by repaying 12 he receives a car worth 12.[1] He will choose *Work* because he can then repay the loan and his payoff will be 10 ($= 15 +$

[1] As usual, we could change the model slightly to make the consumer strongly desire to repay the loan, by substituting a bargaining subgame that splits the gains from trade between bank and consumer rather than specifying that the bank make a take-it-or-leave-it offer. See section 4.3.

12 − 12 − 5), but if he chooses *Play* he will not be able to repay the loan and the bank will repossess the car, reducing his payoff to 8 (= 8 − 0). The bank will offer a loan at $L = 12$ because the consumer will repay it and that is the maximum repayment to which the consumer will agree. The bank's equilibrium payoff is 1 (= 12 − 11). This is an efficient outcome because the consumer does buy the car, which he values at more than its cost to the car dealer, although it is the bank rather than the consumer that gains the surplus, because of the bank's bargaining power over the terms of the loan.

Repossession Game II

The second version of the game does allow renegotiation, so moves (4a) and (4b) are added back into the game. Renegotiation turns out to be harmful, because it results in an equilibrium in which the bank refuses to make a loan, reducing the payoffs of bank and consumer to (0, 10) instead of (1, 10); the gains from trade are lost.

The equilibrium in Repossession Game I breaks down because the consumer would deviate by choosing *Play*. In Repossession Game I, this would result in the bank repossessing the car, and in Repossession Game II, the bank still has the right to do this, for a payoff of −4 (= 7 − 11). If the bank chooses to renegotiate and offer $S = 8$, however, this settlement will be accepted by the consumer, since in exchange he gets to keep a car worth 12, and the payoffs of bank and consumer are −3 (= 8 − 11) and 12 (= 8 + 12 − 8). Thus, the bank will renegotiate, and the consumer will have increased his payoff from 10 to 12 by choosing *Play*. Looking ahead to this from move (1), however, the bank will see that it can do better by refusing to make the loan, resulting in the payoffs (0, 10). The bank cannot even break even by raising the loan rate L. If $L = 30$, for instance, the consumer will still happily accept, knowing that when he chooses *Play* and defaults the ultimate amount he will pay will be just $S = 8$.

Renegotiation has a paradoxical effect. In the subgame starting with consumer default it increases efficiency, by allowing the players to make a Pareto improvement over an inefficient punishment. In the game as a whole, however, it reduces efficiency by preventing players from using punishments to deter inefficient actions. This is true of any situation in which punishment imposes a deadweight loss instead of being simply a transfer from the punished to the punisher. This may be why American judges are less willing than the general public to impose punishments on criminals. By the time a criminal reaches the courtroom, extra years in jail have no beneficial effect (incapacitation aside) and impose real costs on both criminal and society, and judges are unwilling to impose sentences which in each particular case are inefficient.

The renegotiation problem also comes up in principal-agent models because of risk bearing by a risk-averse agent when the principal is risk neutral. Optimal contracts impose risk on risk-averse agents to provide incentives for high effort or self-selection. If at some point in the game it is common knowledge that the agent has chosen his action or report, but Nature has not yet moved, the agent bears needless risk. The principal knows the agent has already moved, so the two of them are willing to recontract to put the risk from Nature's move back on the principal. But the expected future recontracting makes a joke of the original contract and reduces the agent's incentives for effort or truthfulness (see Fudenberg & Tirole [1990]).

The Repossession Game illustrates other ideas besides renegotiation. It is a game of perfect information but has the feel of a game of moral hazard with hidden actions. This is because the game has an implicit bankruptcy constraint, so that the contract cannot sufficiently punish the consumer for an inefficient choice of effort. Restricting the strategy space has the same effect as restricting the information available to a player. It is another example of the distinction between observability and contractibility—the consumer's effort is observable, but it is not really contractible, because the bankruptcy constraint prevents him from being punished for his low effort.

This game also illustrates the difficulty of deciding what "bargaining power" means. This is a term that is very important to how many people think about law and public policy but which they define hazily. Chapter 11 will analyze bargaining in great detail, using the paradigm of splitting a pie. The natural way to think of bargaining power is to treat it as the ability to get a bigger share of the pie. Here, the pie to be split is the surplus of 1 from the consumer's purchase of a car at cost 11 which will yield him 12 in utility. Both versions of the Repossession Game give all the bargaining power to the bank in the sense that whenever there is a surplus to be split, the bank gets 100 percent of it. But this does not help the bank in Repossession Game II, because the consumer can put himself in a position where the bank ends up a loser from the transaction despite its bargaining power.

8.4 Efficiency Wages

The next three sections are about remedies for moral hazard that can be applied to either hidden actions or hidden knowledge. Most of the illustrations will be of hidden actions, but usually "low effort" can be replaced by "lying about the hidden knowledge."

Shapiro & Stiglitz (1984) show how involuntary unemployment can be explained by a principal-agent model. When all workers are employed at the market wage, a worker who is caught shirking and fired can immediately find another job just as good. Firing is ineffective and effective penalties like boiling-in-oil are excluded from the strategy spaces of legal businesses. Becker & Stigler (1974) have suggested that workers post performance bonds, but if workers are poor this is impractical. Without bonds or boiling-in-oil, the worker chooses low effort and receives a low wage.

To induce a worker not to shirk, the firm can offer to pay him a premium over the market-clearing wage, which he loses if he is caught shirking and fired. If one firm finds it profitable to raise the wage, however, so do all firms. One might think that after the wages are equalized, the incentive not to shirk would disappear. But when a firm raises its wage, its demand for labor falls, and when all firms raise their wages, the market demand for labor falls, creating unemployment. Even if all firms pay the same wage, a worker has an incentive not to shirk, because if he were fired he would stay unemployed, and even if there is a random chance of leaving the unemployment pool, the unemployment rate rises sufficiently high that workers choose not to risk being caught shirking. The equilibrium is not first-best efficient, because even though the marginal revenue of labor equals the wage, it exceeds the marginal disutility of effort, but it is efficient in a second-best sense. By deterring shirking, the hungry workers hanging around the factory gates are performing a socially valuable function (but they mustn't be paid for it!).

The idea of paying high wages to increase the threat of dismissal is old, and can even be found in *The Wealth of Nations* (Smith [1776] p. 207). What is new in Shapiro & Stiglitz is the observation that unemployment is generated by these "efficiency wages." These firms behave paradoxically. They pay workers more than necessary to attract them, and outsiders who offer to work for less are turned away. Can this explain why "overqualified" jobseekers are unsuccessful and mediocre managers are retained? Employers are unwilling to hire someone talented, because he could find another job after being fired for shirking, and trustworthiness matters more than talent in some jobs.

This discussion should remind you of the game of Product Quality of section 5.4. There too, purchasers paid more than the reservation price in order to give the seller an incentive to behave properly, because a seller who misbehaved could be punished by termination of the relationship. The key characteristics of such models are that there is a constraint on the amount of contractual punishment for misbehavior and that the participation constraint is not binding in equilibrium.

8.5 Tournaments

Games in which relative performance is important are called **tournaments**. Tournaments are similar to auctions, the difference being that the actions of the losers matter directly, unlike in auctions. Like auctions, they are especially useful when the principal wants to elicit information from the agents. A principal-designed tournament is sometimes called a **yardstick competition** because the agents provide the measure for their wages.

Farrell (unpublished) uses a tournament to explain how "slack" might be the major source of welfare loss from monopoly, an old idea usually prompted by faulty reasoning. The usual claim is that monopolists are inefficient because, unlike competitive firms, they do not have to maximize profits to survive. This relies on the dubious assumption that firms care about survival, not profits. Farrell makes a subtler point: although the shareholders of a monopoly maximize profit, the managers maximize their own utility, and moral hazard is severe without the benchmark of other firms' performances.

Let firm Apex have two possible production techniques, *Fast* and *Careful*. Independently for each technique, Nature chooses production cost $c = 1$ with probability θ and $c = 2$ with probability $1 - \theta$. The manager can either choose a technique at random or investigate the costs of both techniques at a utility cost to himself of α. The shareholders can observe the resulting production cost, but not whether the manager investigates. If they see the manager pick *Fast* and a cost $c = 2$, they do not know whether he chose it without investigating, or investigated both techniques and found they were both costly. The wage contract is based on what the shareholders can observe, so it takes the form (w_1, w_2), where w_1 is the wage if $c = 1$ and w_2 if $c = 2$. The manager's utility is $\log w$ if he does not investigate, $\log w - \alpha$ if he does, and the reservation utility of $\log \bar{w}$ if he quits.

If the shareholders want the manager to investigate, the contract must satisfy the self-selection constraint

$$U(\textit{not investigate}) \leq U(\textit{investigate}). \tag{4}$$

If the manager investigates, he still fails to find a low-cost technique with probability $(1 - \theta)^2$, so (4) is equivalent to

$$\theta \log w_1 + (1 - \theta)\log w_2 \leq [1 - (1 - \theta)^2]\log w_1 + (1 - \theta)^2 \log w_2 - \alpha. \tag{5}$$

The self-selection constraint is binding, since the shareholders want to keep the manager's compensation to a minimum. Turning inequality (5) into an equality and simplifying yields

$$\theta(1 - \theta) \log \frac{w_1}{w_2} = \alpha. \tag{6}$$

The participation constraint, which is also binding, is $U(\bar{w}) = U(\text{investigate})$, or

$$\log \bar{w} = [1 - (1 - \theta)^2]\log w_1 + (1 - \theta)^2 \log w_2 - \alpha. \tag{7}$$

Solving equations (6) and (7) together for w_1 and w_2, yields

$$\begin{aligned} w_1 &= \bar{w}e^{\alpha/\theta} \\ w_2 &= \bar{w}e^{\alpha/(1-\theta)}. \end{aligned} \tag{8}$$

The expected cost to the firm is

$$[1 - (1 - \theta)^2]\bar{w}e^{\alpha/\theta} + (1 - \theta)^2 \bar{w}e^{-\alpha/(1-0)}. \tag{9}$$

If the parameters are $\theta = 0.1$, $\alpha = 1$, and $\bar{w} = 1$, the rounded values are $w_1 = 22{,}026$ and $w_2 = 0.33$, and the expected cost is 4,185. Quite possibly, the shareholders decide it is not worth making the manager investigate.

But suppose that Apex has a competitor, Brydox, in the same situation. The shareholders of Apex can threaten to boil their manager in oil if Brydox adopts a low-cost technology and Apex does not. If Brydox does the same, the two managers are in a prisoner's dilemma, both wishing not to investigate, but each investigating from fear of the other doing so. The forcing contract for Apex specifies $w_1 = w_2$ to fully insure the manager, and boiling-in-oil if Brydox has lower costs than Apex. The contract need satisfy only the participation constraint that $\log w - \alpha = \log \bar{w}$, so $w = 2.72$ and the cost of learning to Apex is only 2.72, not 4,185. Competition raises efficiency, not through the threat of firms going bankrupt but through the threat of managers being fired.

8.6 Institutions and Agency Problems

Ways to Alleviate Agency Problems

Usually when agents are risk averse, the first-best cannot be achieved, because some tradeoff must be made between providing the agent with incentives and keeping his compensation from varying too much between states of the world, or because it is not possible to punish him sufficiently. We have looked at a number of different ways to solve the problem, and at this point a listing might be useful. Each method is illustrated by application to the particular problem of executive compensation, which is empirically important, and interesting both because explicit incentive contracts are used and because they are not used more often (see Baker, Jensen & Murphy [1988]).

(1) **Reputation** (sections 5.3, 5.4, 6.4, 15.1)

Managers are promoted on the basis of past effort or truthfulness.

(2) **Risk-Sharing Contracts** (sections 7.3, 7.4, 7.5)

The executive receives not only a salary, but call options on the firm's stock. If he reduces the stock value, his options fall in value.

(3) **Boiling-in-Oil** (section 7.4)

If the firm would only become unable to pay dividends if the executive shirked and was unlucky, the threat of firing him when the firm skips a dividend will keep him working hard.

(4) **Selling the Store** (section 7.4)

The managers buy the firm in a leveraged buyout.

(5) **Efficiency Wages** (section 8.4)

To make him fear losing his job, the executive is paid a higher salary than his ability warrants (cf. Rasmusen [1988b] on mutual banks).

(6) **Tournaments** (section 8.5)

Several vice presidents compete and the winner succeeds the president.

(7) **Monitoring** (section 3.4)

The directors hire a consultant to evaluate the executive's performance.

(8) Repetition

Managers are paid less than their marginal products for most of their career, but are rewarded later with higher salaries or generous pensions if their career record has been good.

(9) Changing the Type of the Agent

Older executives encourage the younger by praising ambition and hard work.

We have talked about all but the last two solutions. Repetition enables the contract to come closer to the first-best if the discount rate is low (Radner [1985]). Production Game V failed to attain the first-best in section 7.2 because output depended on both the agent's effort and random noise. If the game were repeated 50 times with independent drawings of the noise, the randomness would average out and the principal could form an accurate estimate of the agent's effort. This is, in a sense, begging the question, by saying that in the long run effort can be deduced after all.

Changing the agent's type by increasing the direct utility from desirable or decreasing that from undesirable behavior is a solution that has received little attention from economists, who have focussed on changing the utility by changing monetary rewards. Akerlof (1983), one of the few papers on the subject of changing type, points out that the moral education of children, not just their intellectual education, affects their productivity and success. The attitude of economics, however, has been that while virtuous agents exist, the rules of an organization need to be designed with the unvirtuous agents in mind. As the Chinese thinker Han Fei Tzu said some two thousand years ago,

> Hardly ten men of true integrity and good faith can be found today, and yet the offices of the state number in the hundreds. If they must be filled by men of integrity and good faith, then there will never be enough men to go around; and if the offices are left unfilled, then those whose business it is to govern will dwindle in number while disorderly men increase. Therefore the way of the enlighted ruler is to unify the laws instead of seeking for wise men, to lay down firm policies instead of longing for men of good faith. (Han Fei Tzu [1964], p. 109 from his chapter, "The Five Vermin")

The number of men of true integrity has probably not increased as fast as the size of government, so Han Fei Tzu's observation remains valid, but it should be kept in mind that honest men do exist and honesty can enter into rational models. There are tradeoffs between spending to foster honesty and spending for other purposes, and there may be tradeoffs between using the second-best contracts designed for agents indifferent about the truth and using the simpler contracts appropriate for honest agents.

Government Institutions and Agency Problems

The field of law is well suited to analysis by principal-agent models. Even in the 19th century, Holmes (1881, p. 31) conjectured in *The Common Law* that the reason why sailors at one time received no wages if their ship was wrecked was to discourage them from taking to the lifeboats too early instead of trying to save it. The reason why such a legal rule may have been suboptimal is not that it was unfair—presumably sailors knew the risk before they set out—but because incentive compatibility and insurance work in opposite directions. If sailors are more risk averse than ship owners, and pecuniary advantage would not add much to their effort during storms, the owner ought to provide insurance to the sailors by guaranteeing them wages whether the voyage succeeds or not.

Another legal question is who should bear the cost of an accident: the victim (for example, a pedestrian hit by a car) or the person who caused it (the driver). The economist's answer depends on who has the most severe moral hazard. If the pedestrian could have prevented the accident at the lowest cost, he should pay; otherwise, the driver. This idea of the **least-cost avoider** is extremely useful in the economic analysis of law, and is a major theme of Posner's classic treatise on law and economics (Posner [1992]). Insurance or wealth transfer may also enter as considerations. If pedestrians are more risk averse, drivers should bear the cost, and, according to some political views, if pedestrians are poorer, drivers should bear the cost. Note that this last consideration—wealth transfer—is not relevant to private contracts. If a principal earning zero profits is required to bear the cost of work accidents, for example, the agent's wage will be lower than if he bore them instead.

Criminal law is also concerned with tradeoffs between incentives and insurance. Holmes (1881, p. 40) notes, approvingly, that Macaulay's draft of the Indian Penal Code made breach of contract for the carriage of passengers a criminal offense. The reason is that the palanquin-bearers were too poor to pay damages for abandoning their passengers in desolate regions, so the power of the state was needed to provide heavier punishments than bankruptcy. In general, however, the legal rules actually used seem to diverge more from optimality in criminal law than civil law. If, for example, there is no chance that an innocent man can be convicted of embezzlement, boiling embezzlers in oil might be good policy, but most countries would not allow this. Taking the example a step further, if the evidence for murder is usually less convincing than for embezzling, our analysis could easily indicate that the penalty for murder should be less, but such reasoning offends the common notion that the severity of punishment should be matched with harm from the crime.

Private Institutions and Agency Problems

While agency theory can be used to explain and perhaps improve government policy, it also helps explain the development of many curious private institutions. Agency problems are an important hindrance to economic development, and may explain a number of apparently irrational practices. Popkin (1979, pp. 66, 73, 157) notes a variety of these. In Vietnam, for example, absentee landlords were more lenient than local landlords, but improved the land less, as one would expect of principals who suffer from informational disadvantages *vis-à-vis* their agents. Along the pathways in the fields, farmers would plant early-harvesting rice that the farmer's family could harvest by itself in advance of the regular crop, so that hired labor could not grab handfuls as they travelled. In 13th century England, beans were seldom grown, despite their nutritional advantages, because they were too easy to steal. Some villages tried to solve the problem by prohibiting anyone from entering the beanfields except during certain hours marked by the priest's ringing the church bell, so everyone could tend and watch their beans at the same official time.

In less exotic settings, moral hazard provides another reason besides tax benefits why employees take some of their wages in fringe benefits. Professors are granted some of their wages in university computer time because this induces them to do more research. Having a zero marginal cost of computer time is a way around the moral hazard of slacking on research, despite being a source of moral hazard in wasting computer time. A less typical but more imaginative example is that of the bank in Minnesota which, concerned about its image, gave each employee $100 in credit at certain clothing stores to upgrade their style of dress. By compromising between paying cash and issuing uniforms the bank could hope to raise both its profits and the utility of its employees. ("The $100 Sounds Good, but What do They Wear on the Second Day?" *Wall Street Journal*, October 16, 1987, p. 17.)

Longterm contracts are an important occasion for moral hazard, since so many variables are unforeseen, and hence noncontractible. The term **opportunism** has been used to describe the behavior of agents who take advantage of noncontractibility to increase their payoff at the expense of the principal (see Williamson [1975] and Tirole [1986]). Smith may be able to extract a greater payment from Jones than was agreed upon in their contract, because when a contract is incomplete, Smith can threaten to harm Jones in some way. This is called **hold-up potential** (Klein, Crawford, & Alchian [1978]). Hold-up potential can even make an agent introduce competing agents into the game, if competition is not so extreme as to drive rents to zero. Michael Granfield tells me that Fairchild once developed a new patent on a component of electronic fuel injection systems that it sought to sell to another firm, TRW. TRW offered a much higher price if Fairchild would license its patent to other producers, fearing the hold-up potential of buying from just one supplier. TRW could have tried writing a contract to prevent hold-up, but knew that it would be difficult to prespecify all the ways that Fairchild could cause harm, including not only slow delivery, poor service, and low quality, but also sins of omission like failing to sufficiently guard the plant from shutdown due to accidents and strikes.

It should be clear from the variety of these examples that moral hazard is a common problem. Now that the first flurry of research on the principal-agent problem has finished, researchers are beginning to use the new theory to study institutions that were formerly relegated to descriptive "soft" scholarly work.

8.7 Teams

To conclude this chapter, let us switch our focus from the individual agent to a group of agents. We have already looked at tournaments, which involve more than one agent, but a tournament still takes place in a situation where each agent's output is distinct. The tournament is a solution to the standard problem, and the principal could always fall back on other solutions such as individual risk-sharing contracts. In this section, the existence of a group of agents results in destroying the effectiveness of the individual risk-sharing contracts, because observed output is a joint function of the unobserved effort of many agents. Even though there is a group, a tournament is impossible, because only one output is observed. The situation has much of the flavor of the Civic Duty Game of chapter 3: the actions of a group of players produce a joint output, and each player wishes that the others would carry out the costly actions. A teams model is defined as follows.

A **team** *is a group of agents who independently choose effort levels that result in a single output for the entire group.*

We will look at teams using the following game.

Teams

(Holmstrom [1982])

Players

A principal and n agents.

Order of Play

(1) The principal offers a contract to each agent i of the form $w_i(q)$, where q is total output.
(2) The agents decide whether or not to accept the contract.
(3) The agents simultaneously pick effort levels e_i, ($i = 1, \ldots, n$).
(4) Output is $q(e_1, \ldots, e_n)$.

Payoffs

If any agent rejects the contract, all payoffs equal 0. Otherwise,

$$\pi_{principal} = q - \sum_{i=1}^{n} w_i;$$

$$\pi_i = w_i - v_i(e_i), \text{ where } v_i' > 0 \text{ and } v_i'' > 0.$$

Despite the risk neutrality of the agents, "selling the store" fails to work here, because the team of agents still has the same problem as the employer had. The team's problem is cooperation between agents, and the principal is peripheral.

Denote the efficient vector of actions by e^*. An efficient contract is

$$w_i(q) = \begin{cases} b_i & \text{if } q \geq q(e^*) \\ 0 & \text{if } q < q(e^*) \end{cases} \tag{10}$$

where $\sum_{i=1}^{n} b_i = q(e^*)$ and $b_i > v_i(e_i^*)$.

Contract (10) gives agent i the wage b_i if all agents pick the efficient effort, and nothing if any of them shirks, in which case the principal keeps the output. The teams model gives one reason to have a principal: he is the residual claimant who keeps the forfeited output. Without him, it is questionable whether the agents would carry out the threat to discard the output if, say, output were 99 instead of the efficient 100. There is a problem of dynamic consistency. The agents would like to commit in advance to throw away output, but only because they never have to do so in equilibrium. If the modeller wishes to disallow discarding output, he imposes the **budget-balancing constraint** that the sum of the wages exactly equal the output, no more and no less. But budget balancing creates a problem for the team that is summarized in proposition 8.1.

Proposition 8.1

If there is a budget-balancing constraint, no differentiable wage contract $w_i(q)$ generates an efficient Nash equilibrium.

Agent i's problem is

$$\underset{e_i}{Maximize}\; w_i(q(e)) - v_i(e_i). \tag{11}$$

His first-order condition is

$$\left(\frac{dw_i}{dq}\right)\left(\frac{dq}{de_i}\right) - \frac{dv_i}{de_i} = 0. \tag{12}$$

With budget balancing and a linear utility function, the Pareto optimum maximizes the sum of utilities (something not generally true), so the optimum solves

$$\underset{e_1, \ldots, e_n}{Maximize}\; q(e) - \textstyle\sum_{i=1}^{n} v_i(e_i). \tag{13}$$

The first-order condition is that the marginal dollar contribution to output equal the marginal disutility of effort:

$$\frac{dq}{de_i} - \frac{dv_i}{de_i} = 0. \tag{14}$$

Equation (14) contradicts (12), the agent's first-order condition, because dw_i/dq is not equal to one. If it were, agent i would be the residual claimant and receive the entire marginal increase in output—but under budget balancing, not every agent can do that. Because each agent bears the entire burden of his marginal effort and only part of the benefit, the contract does not achieve the first-best. Without budget balancing, on the other hand, if the agent shirked a little he would gain the entire leisure benefit from shirking, but he would lose his entire wage under the optimal contract.

Discontinuities in Public Good Payoffs

Ordinarily, there is a free rider problem if several players each pick a level of effort which increases the level of some public good whose benefits they share. Noncooperatively, they choose effort levels lower than if they could make binding promises. Mathematically, let identical risk-neutral players indexed by i choose effort levels e_i to produce amount $q(e_1, \ldots, e_n)$ of the public good, where q is a continuous function. Player i's problem is

$$\underset{e_i}{Maximize}\; q(e_1, \ldots, e_n) - e_i, \tag{15}$$

which has the first-order condition

$$\frac{\partial q}{\partial e_i} - 1 = 0, \tag{16}$$

whereas the greater, first-best effort n-vector e^* is characterized by

$$\sum_{i=1}^{n} \frac{\partial q}{\partial e_i} - 1 = 0. \tag{17}$$

If the function q is discontinuous at e^* (for example, $q = 0$ if $e_i < e_i^*$ for any i), the strategy profile e^* can be a Nash equilibrium. In the game of Teams the same effect is at work. Although the Teams function is not discontinuous, contract (10) is constructed as if it were, in order to obtain the same incentives.

The first-best can be achieved because the discontinuity at e^* makes every player the marginal, decisive player: if he shirks a little, output falls drastically and with certainty. Either of the following two modifications restores the free rider problem and induces shirking:

(1) Let q be a function not only of effort but of random noise—Nature moves after the players. Uncertainty makes the expected output a continuous function of effort.

(2) Let players have incomplete information about the critical value—Nature moves before the players and chooses e^*. Incomplete information makes the estimated output a continuous function of effort.

The discontinuity phenomenon is common. Examples, not all of which note the problem, include:

(1) Effort in teams (Holmstrom [1982], Rasmusen [1987]).
(2) Entry deterrence by an oligopoly (Bernheim [1984b], Waldman [1987]).
(3) Output in oligopolies with trigger strategies (Porter [1983a]).
(4) Patent races (section 14.1).
(5) Tendering shares in a takeover (Grossman & Hart [1980], section 13.5).
(6) Preferences for levels of a public good.

Notes

N8.1 Pooling versus Separating Equilibria, and the Revelation Principle

- The books by Fudenberg & Tirole (1991a), Laffont & Tirole (1993), and Spulber (1989), and Baron's chapter in the *Handbook of Industrial Organization* edited by Schmalensee and Willig are good places to look for more on mechanism design.
- Levmore (1982) discusses hidden knowledge problems in tort damages, corporate freezeouts, and property taxes in a law review article.
- In moral hazard with hidden knowledge, the contract must ordinarily satisfy only one participation constraint, whereas in adverse selection problems there is a different participation constraint for each type of agent. An exception is if there are constraints limiting how much an agent can be punished in different states of the world. If, for example, there are bankruptcy constraints, then, if the agent has different wealths across the N possible states of the world, there will be N constraints for how negative his wage can be, in addition to the single participation constraint. These can be looked at as **interim** participation constraints, since they represent the idea that the agent wants to get out of the contract once he observes the state of the world midway through the game.
- The revelation principle was named by Myerson (1979) and can be traced back to Gibbard (1973). A further reference is Dasgupta, Hammond & Maskin (1979). Myerson's game theory book is, as one might expect, a good place to look for further details (Myerson [1991, pp. 258–63, 294–99]).
- Moral hazard frequently occurs in public policy. Should the doctors who prescribe drugs also be allowed to sell them? The question trades off the likelihood of over-prescription against the potentially lower cost and greater convenience of doctor-dispensed drugs. See "Doctors as Druggists: Good Rx for Consumers?" *Wall Street Journal*, 25 June 1987, p. 24.
- For a careful discussion of the unravelling argument for information revelation, see Milgrom (1981b).
- A hidden knowledge game requires that the state of the world matter to one of the players' payoffs, but not necessarily in the same way as in Production Game VI. The Salesman Game of section 8.2 effectively uses the utility function $U(e, w, \theta)$ for the agent and $V(q - w)$ for the principal. The state of the world matters because the agent's disutility of effort varies across states. In other problems, his utility of money might vary across states.

N8.2 An Example of Moral Hazard with Hidden Knowledge: The Salesman Game

- Sometimes students know more about their class rankings than the professor does. One professor of labor economics used a mechanism of the following kind for grading class discussion. Each student i reports a number evaluating other students in the class. Student i's grade is an increasing function of the evaluations given i by other students and of the correlation between i's evaluations and the other students'. There are many Nash equilibria, but telling the truth is a focal point.

- In dynamic games of moral hazard with hidden knowledge the **ratchet effect** is important: the agent takes into account that his information-revealing choice of contract in this period will affect the principal's offerings in the next period. A principal might allow high prices to a public utility in the first period to discover that its costs are lower than expected, but in the next period the prices would be lowered. The contract is ratcheted irreversibly to be more severe. Hence, the company might not choose a contract which reveals its costs in the first period. This is modelled in Freixas, Guesnerie & Tirole (1985).

  Baron (1989) notes that the principal might purposely design the equilibrium to be pooling in the first period so self selection does not occur. Having learned nothing, he can offer a more effective separating contract in the second period.

N8.4 Efficiency Wages

- For surveys of the efficiency wage literature, see the article by L. Katz (1986), the book of articles edited by Akerlof & Yellen (1986), and the book-length survey by Weiss (1990).
- While the efficiency wage model does explain involuntary unemployment, it does not explain cyclical changes in unemployment.
- The efficiency wage idea is based on the same idea as the Klein & Leffler (1981) model of product quality formalized in section 5.3. If no punishment is available for a player who is tempted to misbehave, a punishment can be created by giving him something to take away. This something can be a high-paying job or a loyal customer. It is also similar to the idea of **co-opting** opponents familiar in politics and university administration. To tame the radical student association, give them an office of their own which can be taken away if they seize the dean's office. Rasmusen (1988b) shows yet another context: when depositors do not know which investments are risky and which are safe, mutual bank managers can be highly paid to deter them from making risky investments that might cost them their jobs.
- Adverse selection can also drive an efficiency wage model. We will see in chapter 9 that a customer might be willing to pay a high price to attract sellers of high-quality cars when he cannot detect quality directly.

N8.5 Tournaments

- An article which stimulated much interest in tournaments is Lazear & Rosen (1981), which discusses in detail the importance of risk aversion and adverse selection.
- One example of a tournament is the two-year, three-man contest for the new chairman of Citicorp. The company appointed three candidates as vice-chairmen: the head of consumer banking, the head of corporate banking, and the legal counsel. Earnings reports were even split into three components, two of which were the corporate and consumer banking (the third was the "investment" bank, irrelevant to the tournament). See "What Made Reed Wriston's Choice at Citicorp," *Business Week*, July 2, 1984, p. 25.
- General Motors has tried a tournament among its production workers. During a depressed year, management credibly threatened to close down the auto plant with the lowest productivity. Reportedly, this did raise productivity. Such a tournament is interesting because it helps explain why a firm's supply curve could be upward sloping even if all its plants are identical, and why it might hold excess capacity. Should information on a plant's current performance have been released to other plants? See "Unions Say Auto Firms Use Interplant Rivalry to Raise Work Quotas," *Wall Street Journal*, November 8, 1983, p. 1.
- Under adverse selection, tournaments must be used differently than under moral hazard because agents cannot control their effort. Instead, tournaments are used to deter agents from accepting contracts in which they must compete for a prize with other agents of higher ability.
- Interfirm management tournaments run into difficulties when shareholders want managers to cooperate in some arenas. If managers collude in setting prices, for example, they can also collude to make life easier for each other.
- Antle & Smith (1986) is an empirical study of tournaments in managers' compensation. Rosen (1986) is a theoretical model of a labor tournament in which the prize is promotion.

- Suppose a firm conducts a tournament in which the best-performing of its vice-presidents becomes the next president. Should the firm fire the most talented vice-president before it starts the tournament? The answer is not obvious. Maybe in the tournament's equilibrium, Mr Talent works less hard because of his initial advantage, so that all of the vice-presidents retain the incentive to work hard.
- A tournament can reward the winner, or shoot the loser. Which is better? Nalebuff & Stiglitz (1983) say to shoot the loser, and Rasmusen (1987) finds a similar result for teams, but for a different reason. Nalebuff & Stiglitz's result depends on uncertainty and a large number of agents in the tournament, while Rasmusen's depends on risk aversion. If a utility function is concave because the agent is risk averse, the agent is hurt more by losing a given sum than he would benefit by gaining it. Hence, for incentive purposes the carrot is inferior to the stick, a result unfortunate for efficiency since penalties are often bounded by bankruptcy or legal constraints.
- Using a tournament, the equilibrium effort might be greater in a second-best contract than in the first-best, even though the second-best is contrived to get around the problem of inducing sufficient effort. Also, a pure tournament, in which the prizes are distributed solely according to the ordinal ranking of output by the agents, is often inferior to a tournament in which an agent must achieve a significant margin of superiority over his fellows in order to win (Nalebuff & Stiglitz [1983]). Companies using sales tournaments sometimes have prizes for record yearly sales besides ordinary prizes, and some long distance athletic races have nonordinal prizes to avoid dull events in which the best racers run "tactical races."
- Organizational slack of the kind described in the Farrell model has important practical implications. In dealing with bureaucrats, one must keep in mind that they are usually less concerned with the organization's prosperity than with their own. In complaining about bureaucratic ineptitude, it may be much more useful to name particular bureaucrats and send them copies of the complaint than to stick to the abstract issues at hand. Private firms, at least, are well aware that customers help monitor agents.

N8.6 Institutions and Agency Problems

- Gaver & Zimmerman (1977) describes how a performance bond of 100 percent was required for contractors building the BART subway system in San Francisco. "Surety companies" generally bond a contractor for five to 20 times his net worth, at a charge of 0.6 percent of the bond per year, and absorption of their bonding capacity is a serious concern for contractors in accepting jobs.
- Even if a product's quality need not meet government standards, the seller may wish to bind himself to them voluntarily. Stroh's *Erlanger* beer proudly announces on every bottle that although it is American, "Erlanger is a special beer brewed to meet the stringent requirements of Reinheitsgebot, a German brewing purity law established in 1516." Inspection of household electrical appliances by an independent lab to get the "U_L" listing is a similarly voluntary adherence to standards.
- The stock price is a way of using outside analysts to monitor an executive's performance. When General Motors bought EDS, they created a special class of stock, GM-E, which varied with EDS performance and could be used to monitor it.

N8.7 Teams

- **Team theory**, as developed by Marschak & Radner (1972) is an older mathematical approach to organization. In the old usage of "team" (different from the current, Holmstrom [1982] usage), several agents who have different information but cannot communicate it must pick decision rules. The payoff is the same for each agent, and their problem is coordination, not motivation.
- The efficient contract (10) supports the efficient Nash equilibrium, but it also supports a continuum of inefficient Nash equilibria. Suppose that in the efficient equilibrium all workers work equally hard. Another Nash equilibrium is for one worker to do no work and the others to work inefficiently hard to make up for him.

- **A Teams contract with hidden knowledge.** In the 1920s, National City Co. assigned 20 percent of profits to compensate management as a group. A management committee decided how to share it, after each officer submitted an unsigned ballot suggesting the share of the fund that Chairman Mitchell should have, and a signed ballot giving his estimate of the worth of each of the other eligible officers, himself excluded (Galbraith [1954] p. 157).
- **A First-Best, Budget-Balancing Contract when Agents are Risk Averse.** Proposition 8.1 can be shown to hold for any contract, not just for differentiable sharing rules, but it does depend on risk neutrality and separability of the utility function. Consider the following contract from Rasmusen (1987):

$$w_i = \begin{cases} b_i & \text{if } q \geq q(e^*). \\ \left.\begin{array}{ll} 0 & \text{with probability } (n-1)/n \\ q & \text{with probability } 1/n \end{array}\right\} & \text{if } q < q(e^*) \end{cases} \quad (18)$$

  If the worker shirks, he enters a lottery. If his risk aversion is strong enough, he prefers the riskless return b_i, so he does not shirk. If agents' wealth is unlimited, then for any positive risk aversion we could construct such a contract, by making the losers in the lottery accept negative pay.
- A teams contract like (10) is not a tournament. Only absolute performance matters, even though the level of absolute performance depends on what all the players do.
- **The budget-balancing constraint.** The legal doctrine of "consideration" makes it difficult to make binding, Pareto-suboptimal promises. An agreement is not a legal contract unless it is more than a promise: both parties have to receive something valuable for the courts to enforce the agreement.
- Adverse selection can be incorporated into a teams model. A team of workers who may differ in ability produce a joint output, and the principal tries to ensure that only high-ability workers join the team (see Rasmusen & Zenger [1990]).

Problems

8.1: Monitoring with Error.

An agent has a utility function $U = \sqrt{w} - \alpha e$, where $\alpha = 1$ and e is either 0 or 5. His reservation utility level is $\bar{U} = 9$, and his output is 100 with low effort and 250 with high effort. Principals are risk neutral and scarce, and agents compete to work for them. The principal cannot condition the wage on effort or output, but he can, if he wishes, spend five minutes of his time, worth 10 dollars, to drop in and watch the agent. If he does that, he observes the agent *Daydreaming* or *Working*, with probabilities that differ depending on the agent's effort. He can condition the wage on those two things, so the contract will be $\{\underline{w}, \bar{w}\}$. The probabilities are given by table 8.1.

Table 8.1 Monitoring with Error

Effort	Probability of Daydreaming	Probability of Working
Low ($e = 0$)	0.6	0.4
High ($e = 5$)	0.1	0.9

(8.1a) What are the profits in the absence of monitoring, if the agent is paid enough to make him willing to work for the principal?
(8.1b) Show that high effort is efficient under full information.
(8.1c) If $\alpha = 1.2$, is high effort still efficient under full information?
(8.1d) Under asymmetric information, with $\alpha = 1$, what are the participation and incentive compatibility constraints?
(8.1e) Under asymmetric information, with $\alpha = 1$, what is the optimal contract?

8.2: Monitoring with Error: Second Offenses.[2]

Individuals who are risk-neutral must decide whether to commit zero, one, or two robberies. The cost to society of robbery is 10, and the benefit to the robber is 5. No robber is ever convicted and jailed, but the police beat up any suspected robber they find. They beat up innocent people mistakenly sometimes, as shown by table 8.2, which shows the probabilities of zero or more beatings for someone who commits zero, one, or two robberies.

Table 8.2 Crime

		Beatings		
		0	1	2
Robberies	0	0.81	0.18	0.01
	1	0.60	0.34	0.06
	2	0.49	0.42	0.09

(8.2a) How big should p^*, the disutility of a beating, be made to deter crime completely while inflicting a minimum of punishment on the innocent?
(8.2b) In equilibrium, what percentage of beatings are of innocent people? What is the payoff of an innocent man?
(8.2c) Now consider a more flexible policy, which inflicts heavier beatings on repeat offenders. If such flexibility is possible, what are the optimal severities for first- and second-time offenders? (call these p_1 and p_2). What is the expected utility of an innocent person under this policy?
(8.2d) Suppose that the probabilities are as given in table 8.3 on the next page. What is an optimal policy for first and second offenders?

8.3: Unravelling.

A prospector owns a gold mine worth an amount θ drawn from the uniform distribution $U[0, 100]$, which nobody knows, including himself. He will certainly sell

[2] See Rubinstein (1979).

Table 8.3 More Crime

Robberies	Beatings 0	Beatings 1	Beatings 2
0	0.9	0.1	0
1	0.6	0.3	0.1
2	0.5	0.3	0.2

the mine, since he is too old to work it and it has no value to him if he does not sell it. The several prospective buyers are all risk neutral. The prospector can, if he desires, dig deeper into the hill and collect a sample of gold ore that will reveal the value of θ. If he shows the ore to the buyers, however, he must show genuine ore, since an unwritten Law of the West says that fraud is punished by hanging offenders from joshua trees as food for buzzards.

(8.3a) For how much can he sell the mine if he is clearly too feeble to have dug into the hill and examined the ore? What is the price in this situation if, in fact, the true value is $\theta = 70$?

(8.3b) For how much can he sell the mine if he can dig a test tunnel at zero cost? Will he show the ore? What is the price in this situation if, in fact, the true value is $\theta = 70$?

(8.3c) For how much can he sell the mine if, after digging a tunnel at zero cost and discovering θ, it costs him an additional 10 to verify the results for the buyers? What is his expected payoff?

(8.3d) What is the prospector's expected payoff if, with probability 0.5, digging a tunnel is costless, but, with probability 0.5, it costs 120?

8.4: Teams.

A team of two workers produces and sells widgets for the principal. Each worker chooses high or low effort. An agent's utility is $U = w - 20$ if his effort is high, and $U = w$ if it is low, with a reservation utility of $\bar{U} = 0$. Nature chooses business conditions to be excellent, good, or bad, with probabilities θ_1, θ_2, and θ_3. The principal observes output but not business conditions, as shown in table 8.4.

Table 8.4 Team Output

	Excellent (θ_1)	**Good (θ_2)**	**Bad (θ_3)**
High, High	100	100	60
High, Low	100	50	20
Low, Low	50	20	0

(8.4a) Suppose $\theta_1 = \theta_2 = \theta_3$. Why is $\{(w(100) = 30, w(not\ 100) = 0), (High, High)\}$ not an equilibrium?

(8.4b) Suppose $\theta_1 = \theta_2 = \theta_3$. Is it optimal to induce high effort? What is an optimal contract with nonnegative wages?

(8.4c) Suppose $\theta_1 = 0.5$, $\theta_2 = 0.5$, and $\theta_3 = 0$. Is it optimal to induce high effort? What is an optimal contract (possibly with negative wages)?

(8.4d) Should the principal stop the agents from talking to each other?

8.5: Efficiency Wages and Risk Aversion.[3]

In each of two periods of work, a worker decides whether to steal amount v, and is detected with probability α and suffers legal penalty p if he, in fact, did steal. A worker who is caught stealing can also be fired, after which he earns the reservation wage w_0. If the worker does not steal, his utility in the period is $U(w)$; if he steals, it is $U(w + v) - \alpha p$, where $U(w_0 + v) - \alpha p > U(w_0)$. The worker's marginal utility of income is diminishing: $U' > 0$, $U'' < 0$, and $\lim_{x \to \infty} U'(x) = 0$. There is no discounting. The firm definitely wants to deter stealing in each period, if at all possible.

(8.5a) Show that the firm can indeed deter theft, even in the second period, and, in fact, do so with a second-period wage w_2^* that is higher than the reservation wage w_0.

(8.5b) Show that the equilibrium second-period wage w_2^* is higher than the first-period wage w_1^*.

[3] See Rasmusen (1992c).

9 Adverse Selection

9.1 Introduction: Production Game VII

In chapter 7, games of asymmetric information were divided into games with moral hazard, in which agents are identical, and games with adverse selection, in which agents differ. In moral hazard with hidden knowledge and adverse selection, the principal tries to sort out agents of different types. In moral hazard with hidden knowledge, the emphasis is on the agent's action rather than his choice of contract, and agents accept contracts before acquiring information. Under adverse selection, the agent has private information about his type or the state of the world before he agrees to a contract, which means that the emphasis is on which contract he will accept.

For comparison with moral hazard, let us consider still another version of the Production Game of chapters 7 and 8.

Production Game VII: Adverse Selection

Players

The principal and the agent.

Order of Play

(0) Nature chooses the agent's ability a, unobserved by the principal, according to distribution $F(a)$.
(1) The principal offers the agent one or more wage contracts $w_1(q)$, $w_2(q)$, . . .
(2) The agent accepts one contract or rejects them all.
(3) Nature chooses a value for the state of the world, θ, according to distribution $G(\theta)$, and $q = q(a, \theta)$.

Payoffs

If the agent rejects all contracts, then $\pi_{agent} = \bar{U}$ and $\pi_{principal} = 0$.
Otherwise, $\pi_{agent} = U(w)$ and $\pi_{principal} = V(q - w)$.

Under adverse selection, it is not the worker's effort but his ability that is noncontractible. Without uncertainty (move (3)), the principal would provide a single contract specifying high wages for high output and low wages for low output, but unlike under moral hazard, either high or low output might be observed in equilibrium if both types of agent accepted the contract. Also, in adverse selection, unlike moral hazard, offering multiple contracts can be an improvement over offering a single contract. The principal might, for example, provide a contract with a flat wage for the low-ability agents and an incentive contract for the high-ability agents.

Production Game VII is a fairly complicated game, so let us begin in sections 9.2 and 9.3 with a certainty game, although we will return to uncertainty in section 9.4. The first game will model a used car market in which the quality of the car is known to the seller but not to the buyer, and the various versions of the game will differ in the types and numbers of the buyers and sellers. Sections 9.4 and 9.5 will return to models with uncertainty, in a model of adverse selection in insurance. One result will be that a Nash equilibrium in pure strategies fails to exist for certain parameter values. Section 9.6 will describe the Groves mechanism, a way to deal with hidden knowledge when a group of players submit reports and the principal takes one action which affects them all. Section 9.7 will describe a wide variety of other applications of adverse selection.

9.2 Adverse Selection under Certainty: Lemons I and II

Akerlof stimulated an entire field of research with his 1970 model of the market for shoddy used cars ("lemons"), in which adverse selection arises because car quality is better known to the seller than to the buyer. In agency terms, the principal contracts to buy from the agent a car whose quality, which might be high or low, is noncontractible despite the lack of uncertainty. Such a model may sound like moral hazard with hidden knowledge, but the difference is that in the used car market the seller has private information about his own type before making any kind of agreement. If, instead, the seller agreed to resell his car where he first bought it, the model would be moral hazard with hidden knowledge, because there would be no asymmetric information at the time of contracting, just an expectation of future asymmetry.

We will spend considerable time adding twists to a model of the market in used cars. The game will have one buyer and one seller, but this will simulate competition between buyers, as discussed in section 7.2, because the seller moves first. If the model had symmetric information there would be no consumer surplus. It will often be convenient to discuss the game as if it had many sellers, interpreting a seller whom Nature randomly assigns a type as a population of sellers of different types, one of whom is drawn by Nature to participate in the game.

Basic Lemons Model

Players

A buyer and a seller.

Order of Play

(0) Nature chooses quality type θ for the seller according to the distribution $F(\theta)$. The seller knows θ, but while the buyer knows F, he does not know the θ of the particular seller he faces.
(1) The buyer offers a price P.
(2) The buyer accepts or rejects.

Payoffs

If the buyer rejects the offer, both players receive payoffs of zero.
Otherwise, $\pi_{buyer} = V(\theta) - P$ and $\pi_{seller} = P - U(\theta)$, where V and U will be defined later.

The payoffs of both players are normalized to zero if no transaction takes place. A normalization is part of the notation of the model rather than a substantive assumption. Here, the model assigns the players' utility a base value of zero when no transaction takes place, and the payoff functions show changes from that base. The seller, for instance, gains P if the sale takes place but loses $U(\theta)$ from giving up the car.

There are various ways to specify $F(\theta)$, $U(\theta)$, and $V(\theta)$. We start with identical tastes and two types (Lemons I), and generalize to a continuum of types (Lemons II). Section 9.3 specifies first that the sellers are identical and value cars more than buyers (Lemons III), next that the sellers have heterogeneous tastes (Lemons IV). We will look less formally at other modifications involving risk aversion and the relative numbers of buyers and sellers.

Lemons I: Identical Tastes, Two Types of Sellers

Let good cars have quality 6,000 and bad cars (lemons) quality 2,000, so $\theta \in \{2{,}000, 6{,}000\}$, and suppose that half the cars in the world are of the first type and the other half of the second type. A payoff profile of (0, 0) will represent the status quo, in which the buyer has \$50,000 and the seller has the car. Assume that both players are risk neutral and they value quality at one dollar per unit, so after a trade the payoffs are $\pi_{buyer} = \theta - P$ and $\pi_{seller} = P - \theta$. The extensive form is shown in figure 9.1.

Figure 9.1 An Extensive Form for Lemons I

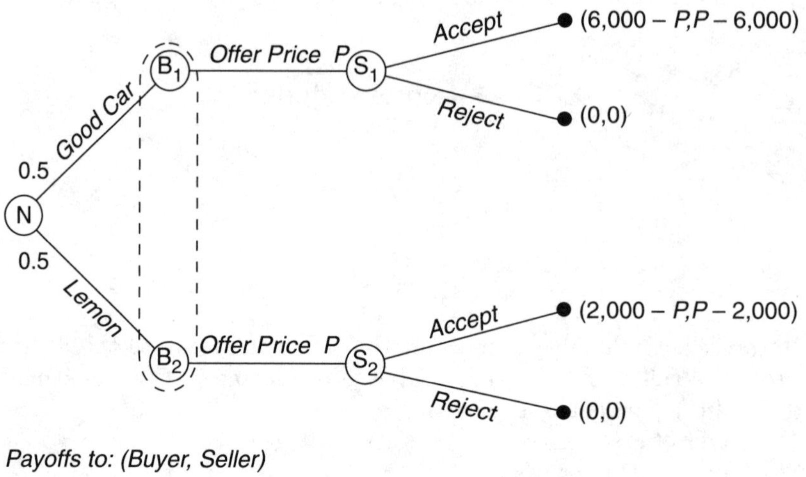

N; 0.5 Good Car; 0.5 Lemon; B_1; B_2; Offer Price P; S_1; S_2; Accept; Reject
(6,000 − P, P − 6,000)
(0,0)
(2,000 − P, P − 2,000)
(0,0)

Payoffs to: (Buyer, Seller)

If he could observe quality at the time of his purchase, the buyer would be willing to accept a contract to pay \$6,000 for a good car and \$2,000 for a lemon. He cannot observe quality, and we assume that he cannot enforce a contract based on his discoveries once the purchase is made. Given these restrictions, if the seller offers \$4,000, a price equal to the average quality, the buyer will deduce that the seller does not have a good car. The very fact that the car is for sale demonstrates its low quality. Knowing that for \$4,000 he would be sold only lemons, the buyer would refuse to pay more than \$2,000. Let us assume that an indifferent seller sells his car, in which case half of the cars are traded in equilibrium, all of them lemons.

A friendly advisor might suggest to the owner of a good car that he wait until all the lemons have been sold and then sell his own car, since everyone knows that only good cars have remained unsold. But allowing for such behavior changes the model by adding a new action. If it were anticipated, the owners of lemons would also hold back and wait for the price to rise. Such a game could be formally analyzed as a war of attrition (section 3.2).

The outcome that half the cars are held off the market is interesting, though not startling, since half the cars do have genuinely higher quality. It is a formalization of Groucho Marx's wisecrack that he would refuse to join any club that would accept him as a member. Lemons II will have a more dramatic outcome.

Lemons II: Identical Tastes, a Continuum of Types of Sellers

One might wonder whether the outcome of Lemons I was an artifact of the assumption of just two types. Lemons II generalizes the game by allowing the seller to be any one of a continuum of types. We will assume that the quality types are uniformly distributed between 2,000 and 6,000. The average quality is $\bar{\theta} = 4{,}000$, which is therefore the price the buyer would be willing to pay for a car of unknown quality if all cars were on the market. The probability density is zero except on the support [2,000, 6,000], where it is $f(\theta) = 1/(6{,}000 - 2{,}000)$, and the cumulative density is

$$F(\theta) = \int_{2{,}000}^{\theta} f(x)dx. \tag{1}$$

After substituting the uniform density for $f(\theta)$ and integrating (1) we obtain

$$F(\theta) = \frac{\theta}{4{,}000} - 0.5. \tag{2}$$

The payoff functions are the same as in Lemons I.

The equilibrium price must be less than \$4,000 in Lemons II because, as in Lemons I, not all cars are put on the market at that price. Owners are willing to sell only if the quality of their cars is less than 4,000, so while the average quality of all used cars is 4,000, the average quality offered for sale is 3,000. The price cannot be \$4,000 when the average quality is 3,000, so the price must drop at least to \$3,000. If that happens, the owners of cars with values from 3,000 to 4,000 pull their cars off the market and the average of those remaining is 2,500. The acceptable price falls to \$2,500, and the unravelling continues, just as in section 8.1, until the price reaches its equilibrium level of \$2,000. But at $P = 2{,}000$ the number of cars on the market is infinitesimal. The market has completely collapsed!

Figure 9.2 puts the price of used cars on one axis and the average quality of cars offered for sale on the other. Each price leads to a different average quality,

Figure 9.2 Lemons II: Identical Tastes

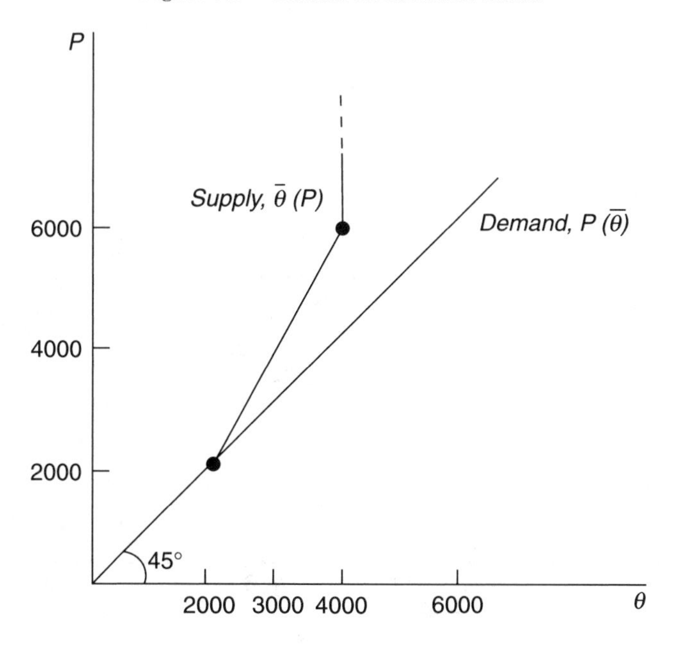

$\bar{\theta}(P)$, and the slope of $\bar{\theta}(P)$ is greater than one because average quality does not rise proportionately with price. If the price rises, the quality of the *marginal* car offered for sale equals the new price, but the quality of the *average* car offered for sale is much lower. In equilibrium, the average quality must equal the price, so the equilibrium lies on the 45° line through the origin. That line is a demand schedule of sorts, just as $\bar{\theta}(P)$ is a supply schedule. The only intersection is the point ($2,000, 2,000).

9.3 Heterogeneous Tastes: Lemons III and IV

The outcome that no cars are traded is extreme, but there is no efficiency loss in either Lemons I or Lemons II. Since all the players have identical tastes, it does not matter who ends up owning the cars. But the players of this section, whose tastes differ, have real need of a market.

Lemons III: Buyers Value Cars More than Sellers

Assume that sellers value their cars at exactly their qualities θ, but that buyers have valuations 20 percent greater, and, moreover, outnumber the sellers. The payoffs if a trade occurs are $\pi_{buyer} = 1.2\theta - P$ and $\pi_{seller} = P - \theta$. In equilibrium, the sellers will capture the gains from trade.

In figure 9.3, the curve $\bar{\theta}(P)$ is much the same as in Lemons II, but the equilibrium condition is no longer that price and average quality lie on the 45° line, but that they lie on the demand schedule $P(\bar{\theta})$, which has a slope of 1.2 instead of 1.0. The demand and supply schedules intersect only at (P = $3,000, $\bar{\theta}(P)$ = 2,500). Because buyers are willing to pay a premium, we only see **partial adverse**

Figure 9.3 Adverse Selection When Buyers Value Cars More than Sellers: Lemons III

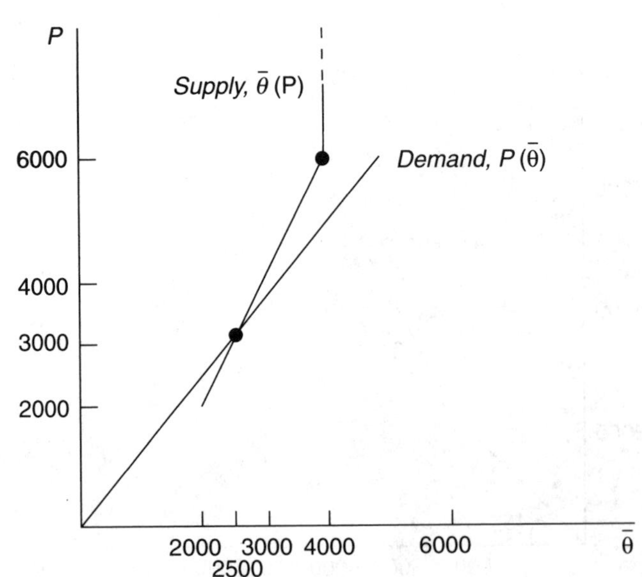

selection; the equilibrium is partially pooling. The outcome is inefficient, because in a world of perfect information all the cars would be owned by the "buyers," who value them more, but under adverse selection they only end up owning the low-quality cars.

Lemons IV: Sellers' Valuations Differ

In Lemons IV, we dig a little deeper to explain why trade occurs, and we model sellers as consumers whose valuations of quality have changed since they bought their cars. For a particular seller, the valuation of one unit of quality is $1 + \varepsilon$, where the random disturbance ε can be either positive or negative and has an expected value of zero. The disturbance could arise because of the seller's mistake—he did not realize how much he would enjoy driving when he bought the car—or because conditions changed—he switched to a job closer to home. Payoffs if a trade occurs are $\pi_{buyer} = \theta - P$ and $\pi_{seller} = P - (1 + \varepsilon)\theta$.

If $\varepsilon = -0.15$ and $\theta = 2{,}000$, then \$1,700 is the lowest price at which the player would resell his car. The average quality of cars offered for sale at price P is the expected quality of cars valued by their owners at less than P, i.e.,

$$\bar{\theta}(P) = E(\theta \mid (1 + \varepsilon)\theta \leq P). \tag{3}$$

Suppose that a large number of new buyers, greater in number than the sellers, appear in the market, and let their valuation of one unit of quality be \$1. The demand schedule, shown in figure 9.4, is the 45° line through the origin. Figure 9.4 shows one possible shape for the supply schedule $\bar{\theta}(P)$, although to specify it precisely we would have to specify the distribution of the disturbances.

Figure 9.4 Lemons IV: Sellers' Valuations Differ

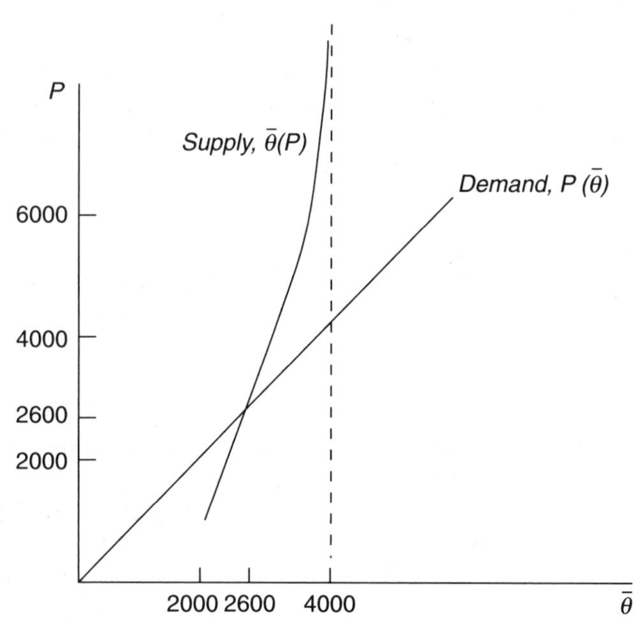

In contrast to Lemons I, II, and III, here if $P \geq \$6{,}000$ some car owners would be reluctant to sell, because they received positive disturbances to their valuations. The average quality of cars on the market is less than 4,000 even at $P = \$6{,}000$. On the other hand, even if $P = \$2{,}000$ some sellers with low quality cars *and* negative realizations of the disturbance still sell, so the average quality remains above 2,000. Under some distributions of ε, a few sellers hate their cars so much they would pay to have them taken away.

The equilibrium drawn in figure 9.4 is ($P = \$2{,}600$, $\bar{\theta} = 2{,}600$). Some used cars are sold, but the number is inefficiently low. Some of the sellers have high quality cars but negative disturbances, and although they would like to sell their cars to someone who values them more, they will not sell at a price of \$2,600.

A theme running through all four Lemons models is that when quality is unknown to the buyer, less trade occurs. Lemons I and II show how trade diminishes, while Lemons III and IV show that the disappearance can be inefficient because some sellers value cars less than some buyers. Next we will use Lemons III, the simplest model with gains from trade, to look at various markets with more sellers than buyers, excess supply, and risk-averse buyers.

More Sellers than Buyers

In analyzing Lemons III, we assumed that buyers outnumbered sellers. As a result, the sellers earned producer surplus. In the original equilibrium, all the sellers with quality less than 3,000 offered a price of \$3,000 and earned a surplus of up to \$1,000. There were more buyers than sellers, so every seller who wished to sell was able to do so, but the price equalled the buyers' expected utility, so no buyer who failed to purchase was dissatisfied. The market cleared.

If, instead, sellers outnumber buyers, what price should a seller offer? At \$3,000, not all would-be sellers can find buyers. A seller who proposed a lower price would find willing buyers despite the somewhat lower expected quality. The buyer's tradeoff between lower price and lower quality is shown in figure 9.3, in which the expected consumer surplus is the vertical distance between the price (the height of the supply schedule) and the demand schedule. When the price is \$3,000 and the average quality is 2,500, the buyer expects a consumer surplus of zero, which is $\$3{,}000 - \$1.2 \cdot 2{,}500$. The combination of price and quality that buyers like best is (\$2,000, 2,000), because if there were enough sellers with quality $\theta = 2{,}000$ to satisfy the demand, each buyer would pay $P = \$2{,}000$ for a car worth \$2,400 to him, acquiring a surplus of \$400. If there were fewer sellers, the equilibrium price would be higher and some sellers would receive producer surplus.

Heterogeneous Buyers: Excess Supply

If buyers have different valuations for quality, the market might not clear, as C. Wilson (1980) points out. Assume that the number of buyers willing to pay \$1.2 per unit of quality exceeds the number of sellers, but that buyer Smith is an eccentric whose demand for high quality is unusually strong. He would pay \$100,000 for a car of quality 5,000 or greater, and \$0 for a car of any lower quality.

In Lemons III without Smith, the outcome is a price of $3,000, an average market quality of 2,500, and a market quality range between 2,000 and 3,000. Smith would be unhappy with this, since he has zero probability of finding a car he likes. In fact, he would be willing to accept a price of $6,000, so that all the cars, from quality 2,000 to 6,000, would be offered for sale and the probability that he is able to buy a satisfactory car would rise from 0 to 0.25. But Smith would not want to buy all the cars offered to him, so the equilibrium has two prices, $3,000 and $6,000, with excess supply at the higher price.

Strangely enough, Smith's demand function is upward sloping. At a price of $3,000, he is unwilling to buy; at a price of $6,000, he is willing, because expected quality rises with price. This does not contradict basic price theory, for the standard assumption of *ceteris paribus* is violated. As the price increases, the quantity demanded would fall if all else stayed the same, but all else does not—quality rises.

Risk Aversion

We have implicitly assumed, by the choice of payoff functions, that the buyers and sellers are both risk neutral. What happens if they are risk averse—that is, if the marginal utilities of wealth and car quality are diminishing? Again we will use Lemons III and the assumption of many buyers.

On the seller's side, risk aversion changes nothing. The seller runs no risk because he knows exactly the price he receives and the quality he surrenders. But the buyer does bear risk, because he buys a car of uncertain quality. Although he would pay $3,600 for a car he knows has quality 3,000, if he is risk averse he will not pay that much for a car with expected quality 3,000 but actual quality of possibly 2,500 or 3,500: he would obtain less utility from adding 500 quality units than from subtracting 500. The buyer would pay perhaps $2,900 for a car whose expected quality is 3,000 where the demand schedule is nonlinear, lying everywhere below the demand schedule of the risk-neutral buyer. As a result, the equilibrium has a lower price and average quality.

9.4 Adverse Selection under Uncertainty: Insurance Game III

The term "adverse selection," like "moral hazard," comes from insurance. Insurance pays more if there is an accident than otherwise, so it benefits accident-prone customers more than safe ones and a firm's customers are "adversely selected" to be accident-prone. The classic article on adverse selection in insurance markets is Rothschild & Stiglitz (1976), which begins, "Economic theorists traditionally banish discussions of information to footnotes." How things have changed! Within ten years, information problems came to dominate research in both microeconomics and macroeconomics.

We will follow Rothschild & Stiglitz in using state-space diagrams, and we will use a version of the Insurance Game of section 7.5. Under moral hazard, Smith chose whether to be *Careful* or *Careless*. Under adverse selection, Smith cannot affect the probability of a theft, which is chosen by Nature. Rather, Smith is either *Safe* or *Unsafe*, and while he cannot affect the probability that his car will be stolen, he does know what the probability is.

Insurance Game III

Players

Smith and two insurance companies.

Order of Play

(0) Nature chooses Smith to be either *Safe*, with probability 0.6, or *Unsafe*, with probability 0.4. Smith knows what type he is, but the insurance companies do not.
(1) Each insurance company offers its own contract (x, y) under which Smith pays premium x unconditionally and receives compensation y if there is a theft.
(2) Smith picks a contract.
(3) Nature chooses whether there is a theft, using probability 0.5 if Smith is *Safe* and 0.75 if he is *Unsafe*.

Payoffs

Smith's payoff depends on his type and the contract (x, y) that he accepts. Let $U' > 0$ and $U'' < 0$.

$$\pi_{Smith}\,(Safe) \quad = 0.5U(12 - x) + 0.5U(0 + y - x).$$
$$\pi_{Smith}\,(Unsafe) = 0.25U(12 - x) + 0.75(0 + y - x).$$

The companies' payoffs depend on what types of customers accept their contracts, as shown in table 9.1.

Table 9.1 Insurance Game III: Payoffs

Company payoff	Type of customers
0	no customers
$0.5x + 0.5(x - y)$	just *Safe*
$0.25x + 0.75(x - y)$	just *Unsafe*
$0.6[0.5x + 0.5(x - y)] + 0.4[0.25x + 0.75(x - y)]$	*Unsafe* and *Safe*

Smith is *Safe* with probability 0.6 and *Unsafe* with probability 0.4. Without insurance, Smith's dollar wealth is 12 if there is no theft and 0 if there is, depicted in figure 9.5 as his endowment in state space, $\omega = (12, 0)$. If Smith is *Safe*, a

Figure 9.5 Insurance Game III: Nonexistence of a Pooling Equilibrium

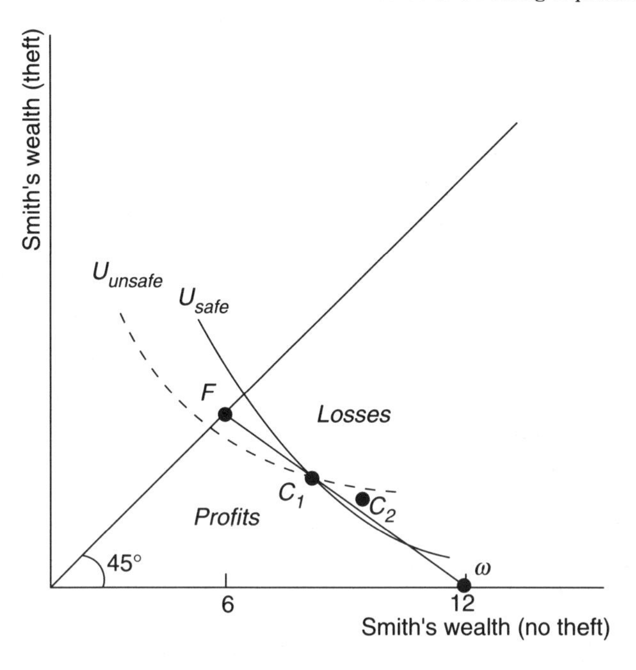

theft occurs with probability 0.5, but if he is *Unsafe* the probability is 0.75. Smith is risk averse (because $U'' < 0$) and the insurance companies are risk neutral.

If an insurance company knew that Smith was *Safe*, it could offer him insurance at a premium of 6 with a payout of 12 after a theft, leaving Smith with an allocation of (6, 6). This is the most attractive contract that is not unprofitable, because it fully insures Smith. Whatever the state, his allocation is 6.

Figure 9.5 shows the indifference curves of Smith and an insurance company. The insurance company is risk neutral, so its indifference curve is the straight line ωF if Smith is a customer regardless of his type. The insurance company is indifferent between ω and C_1, at both of which its expected profits are zero. Smith is risk averse, so his indifference curves are convex, and closest to the origin along the 45 degree line if the probability of *Theft* is 0.5. He has two sets of indifference curves, solid if he is *Safe* and dashed if he is *Unsafe*.

Figure 9.5 shows why no Nash pooling equilibrium exists. To make zero profits, the equilibrium must lie on the line ωF. It is easiest to think about these problems by imagining an entire population of Smiths, whom we will call "customers." Pick a contract C_1 anywhere on ωF and think about drawing the indifference curves for the *Unsafe* and *Safe* customers that pass through C_1. *Safe* customers are always willing to trade *Theft* wealth for *No Theft* wealth at a higher rate than *Unsafe* customers. At any point, therefore, the slope of the solid (*Safe*) indifference curve is steeper than that of the dashed (*Unsafe*) curve. Since the slopes of the dashed and solid indifference curves differ, we can insert another contract, C_2, between them and just barely to the right of ωF. The *Safe* customers prefer contract C_2 to C_1, but the *Unsafe* customers stay with C_1, so C_2 is profitable—since C_2 only attracts *Safes*, it need not be to the left of ωF to avoid losses. But then the original contract C_1 was not a Nash equilibrium, and since our argument holds for any pooling contract, no pooling equilibrium exists.

The attraction of the *Safe* customers away from pooling is referred to as **cream skimming**, although profits are still zero when there is competition for the cream. We next consider whether a separating equilibrium exists, using figure 9.6. The zero profit condition requires that the *Safe* customers take contracts on ωC_4 and the *Unsafes* on ωC_3.

The *Unsafes* will be completely insured in any equilibrium, although at a high price. On the zero-profit line ωC_3, the contract they like best is C_3, which the *Safes* are not tempted to take. The *Safes* would prefer contract C_4, but C_4 uniformly dominates C_3, so it would attract *Unsafes* too, and generate losses. To avoid attracting *Unsafes*, the *Safe* contract must be below the *Unsafe* indifference curve. Contract C_5 is the fullest insurance the *Safes* can get without attracting *Unsafes*: it satisfies the self-selection and competition constraints.

Contract C_5, however, might not be an equilibrium either. Figure 9.7 is the same as figure 9.6 with a few additional points marked. If one firm offered C_6, it would attract both types, *Unsafe* and *Safe*, away from C_3 and C_5, because it is to the right of the indifference curves passing through those points. Would C_6 be profitable? That depends on the proportions of the different types. The assumption on which the equilibrium of figure 9.6 is based is that the proportion of *Safes* is 0.6, so that the zero-profit line for pooling contracts is ωF and C_6 would be unprofitable. In figure 9.7 it is assumed that the proportion of *Safes* is higher, so the zero-profit line for pooling contracts would be $\omega F'$ and C_6, lying to its left, is profitable. But we already showed that no pooling contract is Nash, so C_6 cannot

Figure 9.6 A Separating Equilibrium for Insurance Game III

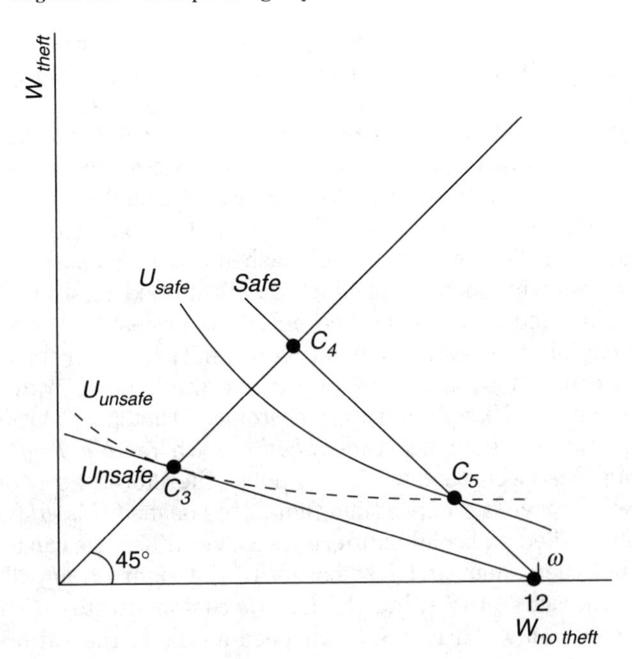

Figure 9.7 Curves for Which There is No Equilibrium in Insurance Game III

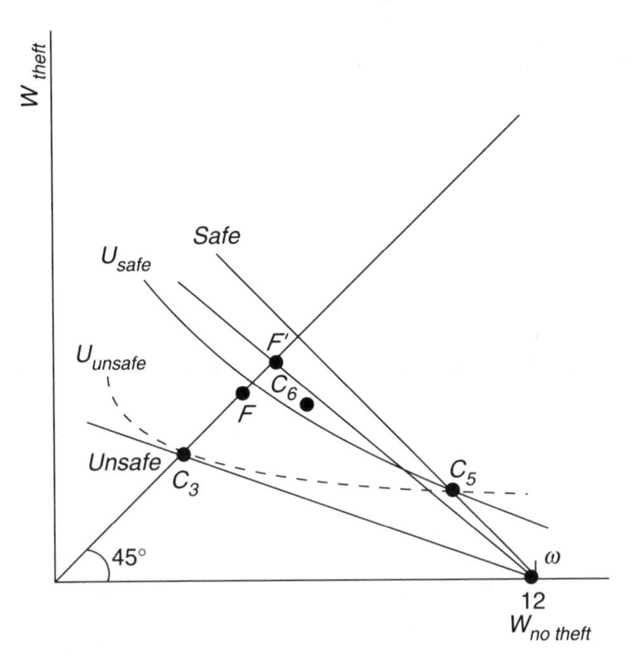

be an equilibrium. Since neither a separating pair like (C_3, C_5) nor a pooling contract like C_6 is an equilibrium, no equilibrium whatsoever exists.

The essence of nonexistence here is that if separating contracts are offered, some company is willing to offer a superior pooling contract; but if a pooling contract is offered, some company is willing to offer a separating contract that makes it unprofitable. A monopoly would have a pure-strategy equilibrium, but in a competitive market only a mixed-strategy Nash equilibrium exists (see Dasgupta & Maskin [1986b]).

9.5 Other Equilibrium Concepts: Wilson Equilibrium and Reactive Equilibrium

In Insurance Game III, any pooling contract is vulnerable to a cream-skimming contract that draws away the *Safes*, but this is a little strange, because it seems that after that happens the now unprofitable old pooling contract (which was soaking up the *Unsafes*) would be withdrawn. The game tree does not reflect this, nor does the Nash equilibrium concept.

One way to obtain a pure-strategy equilibrium is to redefine the equilibrium concept. C. Wilson (1980) suggests that the pooling equilibrium is legitimate because a principal (an uninformed player) who was thinking about introducing the new contract would realize that it would be unprofitable once the old contract was withdrawn.

A **Wilson equilibrium** *is a set of contracts such that when the agents (informed players) choose among them so as to maximize profits,*

(1) All contracts make nonnegative profits;
and

(2) No new contract (or set of contracts) could be offered that would make positive profits even after all contracts that would make negative profits as a result of its entry were withdrawn.

The Wilson equilibrium is the same as the Nash separating equilibrium if that exists, and otherwise it is the pooling contract most preferred by the *Safes*. In figure 9.6, the Wilson equilibrium is the same as the Nash equilibrium, the separating pair (C_3, C_5). In figure 9.7, where no Nash equilibrium exists, the Wilson equilibrium is the zero-profit pooling contract, F'. It is on the line $\omega F'$, so it satisfies part (a) of the definition. It provides the fullest insurance of any zero-profit pooling contract, so that no new pooling contract would be more attractive, and while some new separating contract might be profitable if the *Unsafes* stayed with F', any such contract would cause F' to be withdrawn and would be unprofitable thereafter.

The idea of Wilson equilibrium can also be incorporated into the game by modifying the game tree instead of redefining the equilibrium concept, as suggested by Fernandez & Rasmusen (unpublished), to obtain the Wilson outcome as the perfect equilibrium of the modified game.

Wilson Equilibrium

(1) Principals simultaneously offer contracts, called "old contracts."

(2) Principals may simultaneously offer other contracts, called "new contracts."

(3) Principals may simultaneously withdraw any old contracts.

(4) Agents choose from among the remaining old and new contracts, and trading occurs.

In the perfect Bayesian equilibrium of this game, the principals offer the contracts that form a Wilson equilibrium in move (1). The approach of changing the game tree may seem more complicated than changing the equilibrium concept, but that is because it clearly delineates the somewhat vague intuition behind the equilibrium concept. Making use of the Wilson concept is not just a technical assumption: it is assuming that the market has a particular structure.

Riley (1979b) uses reasoning similar to Wilson's to justify his concept of "reactive equilibrium." Under this concept, an equilibrium is a set of contracts such that though some new contract might be profitable, that new contract would itself become unprofitable if a second new contract were introduced. More formally, following Engers & Fernandez (1987),

A **reactive equilibrium** *is a set of contracts S yielding nonnegative profits such that for any nonempty set of contracts S′ (the defection), where $S \cup S'$ is closed, there exists a closed set of contracts S″ (the reaction) such that:*

(*1*) *S' incurs losses when only these three sets are tendered, and*

(*2*) *S'' does not incur losses when these three sets are tendered, whether or not other contracts are also offered.*

In both figure 9.6 and 9.7, the reactive equilibrium is the separating pair (C_3, C_5). That pair yields zero profits, and while in figure 9.7 there is a profitable deviation (C_6), that deviation would become unprofitable if a cream-skimming contract were added as a reaction. Moreover, condition (2) of the definition is met, because if the reactive cream-skimming contract is chosen carefully, no additional contracts can be added which make it unprofitable, given that C_6 continues to be offered.

A separating reactive equilibrium always exists because any pooling contract disrupting it could be reacted against: reactive equilibrium makes constructive use of the nonexistence of a pooling equilibrium. The Wilson concept is based on withdrawing contracts in response to deviation, whereas the reactive concept is based on adding them. As a result, when Nash equilibrium does not exist, the Wilson concept favors a pooling equilibrium, while the reactive concept favors a separating equilibrium.

9.6 The Groves Mechanism

Hidden knowledge is particularly important in public economics, the study of government spending and taxation. Government policy involves moral hazard (recall The Welfare Game and The Auditing Game), but often the government's task is simply to extract information from the citizens in order to maximize welfare. The optimal taxation literature starting with Mirrlees (1971) is an example: citizens differ in their income-producing ability, and the government wishes to demand higher taxes from the more able citizens. An even purer problem of hidden knowledge is the problem of public goods with private preferences. The government must decide whether it is worthwhile to buy a public good based on the combined preferences of all the citizens, but first it needs to discover those preferences. Unlike in the previous games in this chapter, a group of agents is involved, not just a single agent. Moreover, the government is an altruistic principal who cares directly about the utility of the agents, rather than a car buyer or an insurance seller who cares about the agents' utility only in order to satisfy self-selection and participation constraints.

The next example is adapted from p. 426 of Varian (1992). The mayor of a town is considering installing a streetlight costing $100. Each of the five houses near the light would be taxed exactly $20, but the mayor will only install it if he decides that the sum of the residents' valuations for it is greater than the cost. The problem is to discover the valuations. If the mayor simply asks them, householder Smith could say that his valuation is $5,000, and Brown says he likes the dark and would pay $5,000 to keep the street dark, but all the mayor could conclude would be that Smith's valuation exceeds $20 and Brown's does not. Talk is cheap, and the dominant strategy is to overreport or underreport.

The flawed mechanism just described can be denoted by

$$M_1: \left(20, \sum_{i=1}^{5} m_i \geq 100\right), \tag{4}$$

which means that each resident pays 20, and the light is installed if the sum of the messages exceeds 100.

An alternative mechanism is to make resident i pay the amount of his message, or pay zero if it is negative. This mechanism is

$$M_2: \left(Max\{m_i, 0\}, \sum_{j=1}^{5} m_j \geq 100\right), \tag{5}$$

in which case there is no dominant strategy. Player i would announce $m_i = 0$ if he thought the project would go through without his support, but he would announce up to his valuation if necessary. There is a continuum of Nash equilibria that attain the efficient result. Most of these are asymmetric, and there is a problem of how the equilibrium to be played out becomes common knowledge. This is a simple mechanism, however, and it already teaches a lesson: that people are more likely to report their true political preferences if they must bear part of the costs themselves.

Instead of just ensuring that the correct decision is made in a Nash equilibrium, it may be possible to design a mechanism that makes truthfulness a **dominant-strategy mechanism**. Consider the mechanism

$$M_3: \left(100 - \sum_{j \neq i} m_j, \sum_{j=1}^{5} m_j \geq 100\right). \tag{6}$$

Under mechanism (6), player i's message does not affect his tax bill except by its effect on whether or not the streetlight is installed. If player i's valuation is v_i, his full payoff is $v_i - 100 + \Sigma_{j \neq i} m_j$ if $m_i + \Sigma_{j \neq i} m_j \geq 100$, and zero otherwise. It is not hard to see that he will be truthful in a Nash equilibrium in which the other players are truthful, but we can go further. Truthfulness is weakly dominant. Moreover, the players will tell the truth whenever lying would alter the mayor's decision.

Consider a numerical example. Suppose that Smith's valuation is 40 and that the sum of the valuations is 110, so the project is indeed efficient. If the other players report their truthful sum of 70, Smith's payoff from truthful reporting is his valuation of 40 minus his tax of 30. Reporting more would not change his payoff, while reporting less than 30 would reduce it to 0.

If we are wondering whether Smith's strategy is dominant, we must also consider his best response when the other players lie. If they underreported, announcing 50 instead of the truthful 70, Smith could make up the difference by overreporting 60, but his payoff would be -10 ($= 40 + 50 - 100$) so he would do better to report the truthful 40, killing the project and leaving him with a payoff of 0. If the other players overreported, announcing 80 instead of the truthful 70, then Smith benefits if the project goes through, and he should report at least 20 to obtain his payoff of 40 minus 20. He is willing to report exactly 40, so there is an equilibrium with truth-telling.

The problem with a dominant-strategy mechanism like the one facing Smith is that it is not budget balancing. The government raises less in taxes than it spends on the project (in fact, the taxes would be negative). Lack of budget balancing is a crucial feature of dominant-strategy mechanisms. While the government deficit can be made either positive or negative, it cannot be made zero, unlike in the case of Nash mechanisms.

9.7 A Variety of Applications

Price Dispersion

Usually the best model for explaining price dispersion is a search model—Salop & Stiglitz (1977), for example, which is based on the assumption of buyers whose search costs differ. But although we passed over it quickly in section 9.3, the Lemons model with Smith, the quality-conscious consumer, generated not only excess supply, but price dispersion as well. Cars of the same average quality were sold for $3,000 and $6,000.

Similarly, while the most obvious explanation for why brands of stereo amplifiers sell at different prices is that customers are willing to pay more for higher quality, adverse selection contributes another explanation. Consumers might be willing to pay high prices because they know that high-priced brands could include both high-quality and low-quality amplifiers, whereas low-priced brands are invariably low-quality. The low-quality amplifier ends up selling at two prices: a high price in competition with high-quality amplifiers, and, in different stores or under a different name, a low price aimed at customers less willing to trade dollars for quality.

This explanation does depend on sellers of amplifiers incurring a large enough fixed set-up or operating cost. Otherwise, too many low-quality brands would crowd into the market, and the proportion of high-quality brands would be too small for consumers to be willing to pay the high price. The low-quality brands would benefit as a group from entry restrictions: too many of them spoil the market, not through price competition but through degrading the average quality.

Health Insurance and Medicare

Medical insurance is subject to adverse selection because some people are healthier than others. The variance in health is particularly high among old people, who sometimes have difficulty obtaining any insurance at all. Under basic economic theory this is a puzzle: the price should rise until supply equals demand. The problem is pooling: when the price of insurance is appropriate for the average old person, healthier ones stop buying. The price must rise to keep profits nonnegative, and the market disappears, just as in Lemons II.

If the facts indeed fit this story, adverse selection is an argument for government-enforced pooling. If all old people are required to purchase government insurance (Medicare in the United States), then while the healthier of them may be worse off, the vast majority could be helped.

Using adverse selection to justify Medicare, however, points out how dangerous many of the models in this book can be. For policy questions, the best default opinion is that markets are efficient. On closer examination, we have found that many markets are inefficient because of strategic behavior or information asymmetry. It is dangerous, however, to immediately conclude that the government should intervene, because the same arguments applied to government show that the cure might be worse than the disease. The analyst of health care needs to take seriously the moral hazard and rent-seeking that arise from government insurance. Doctors and hospitals will increase the cost and amount of treatment if the government pays for it, and the transfer of wealth from young people to the elderly, which is likely to swamp the gains in efficiency, might distort the shape of the government program from the economist's ideal.

Henry Ford's Five-Dollar Day

In 1914 Henry Ford made a much-publicized decision to raise the wage of his auto workers to $5 a day, considerably above the market wage. This pay hike occurred without pressure from the workers, who were nonunionized. Why did Ford do it?

The pay hike could be explained by either moral hazard or adverse selection. In accordance with the idea of efficiency wages (section 8.4), Ford might have wanted workers who would be worried about losing their premium job at his factory, because they would work harder and refrain from shirking. Adverse selection could also explain the pay hike: by raising his wage Ford attracted a mixture of low- and high-quality workers, rather than low-quality alone (see Raff & Summers [1987]).

Bank Loans

Suppose that two people come to you for an unsecured loan of $10,000. One offers to pay an interest rate of 10 percent and the other offers 200 percent. Who do you accept? Like the car buyer who chooses to buy at a high price, you may choose to lend at a low interest rate.

If a lender raises his interest rate, both his pool of loan applicants and their behavior changes because adverse selection and moral hazard contribute to a rise in default rates. Borrowers who expect to default are less concerned about the high interest rate than dependable borrowers, so the number of loans shrinks and the default rate rises (see Stiglitz & Weiss [1981]). In addition, some borrowers shift to higher-risk projects with greater chance of default but higher yields when they are successful. In section 15.1 we will go through the model of D. Diamond (1989) which looks at the evolution of this problem as firms age.

Whether because of moral hazard or adverse selection, asymmetric information can also result in excess demand for bank loans. The savers who own the bank do not save enough at the equilibrium interest rate to provide loans to all the borrowers who want loans. Thus, the bank makes a loan to John, while denying one to Joe, his observationally equivalent twin. Policymakers should carefully consider any laws that rule out arbitrary loan criteria or require banks to treat all customers equally. A bank might wish to restrict its loans to left-handed people, neither from prejudice nor because they really are better credit risks, but because it is useful to ration loans according to some criterion arbitrary enough to avoid the moral hazard of favoritism by loan officers.

Bernanke (1983) suggests adverse selection in bank loans as an explanation for the Great Depression in the United States. The difficulty in explaining the Depression is not so much the initial stock market crash as the persistence of the unemployment that followed. Bernanke notes that the crash wiped out local banks and dispersed the expertise of the loan officers. After the loss of this expertise, the remaining banks were less willing to lend because of adverse selection, which made it difficult for the economy to recover.

Solutions to Adverse Selection

Even in markets where it apparently does not occur, the threat of adverse selection, like the threat of moral hazard, can be an important influence on market institutions. Adverse selection can be circumvented in a number of ways besides the contractual solutions we have been analyzing. I will mention some of them in the context of the used car market.

One set of solutions consists of ways to make car quality contractible. Buyers who find that their car is defective may have recourse to the legal system if the sellers were fraudulent, although in the United States the courts are too slow and costly to be fully effective. Other government bodies such as the Federal Trade Commission may do better by issuing regulations particular to the industry. Even without regulation, private warranties—promises to repair the car if it breaks down—may be easier to enforce than oral claims, by dispelling ambiguity about what level of quality is guaranteed.

Testing (the equivalent of moral hazard's monitoring) is always used to some extent. The prospective driver tries the car on the road, inspects the body, and otherwise tries to reduce information asymmetry. At a cost, he could even reverse the asymmetry by hiring mechanics, learning more about the car than the owner himself. The rule is not always *caveat emptor*; what should one's response be to an antique dealer who offers to pay $500 for an apparently worthless old chair?

Reputation can solve adverse selection, just as it can solve moral hazard, but only if the transaction is repeated and the other conditions of the models in chapters 5 and 6 are met. An almost opposite solution is to show that there are innocent motives for a sale; that the owner of the car has gone bankrupt, for example, and his creditor is selling the car cheaply to avoid the holding cost.

Penalties not strictly economic are also important. One example is the social ostracism inflicted by the friend to whom a lemon has been sold; the seller is no longer invited to dinner. Or, the seller might have moral principles that prevent him from defrauding buyers. Such principles, provided they are common knowledge, would help him obtain a higher price in the used-car market. Akerlof himself has worked on the interaction between social custom and markets in his 1980 and 1983 articles. The second of these looks directly at the value of inculcating moral principles, using theoretical examples to show that parents might wish to teach their children principles, and that society might wish to give hiring preference to students from elite schools.

It is by violating the assumptions needed for perfect competition that asymmetric information enables government and social institutions to raise efficiency. This points to a major reason for studying asymmetric information: where it is important, noneconomic interference can be helpful instead of harmful. I find the social solutions particularly interesting since, as mentioned earlier in connection with health care, government solutions introduce agency problems as severe as the information problems they solve. Noneconomic behavior is important under adverse selection, in contrast to perfect competition, which allows an "Invisible Hand" to guide the market to efficiency, regardless of the moral beliefs of the traders. If everyone were honest, the lemons problem would disappear because the sellers would truthfully disclose quality. If some fraction of the sellers were honest, but buyers could not distinguish them from the dishonest sellers, the outcome would presumably be somewhere between the outcomes of complete honesty and complete dishonesty. The subject of market ethics is important, and would profit from investigation by scholars trained in economic analysis.

Notes

N9.1 Introduction: Production Game VII

- For an example of an adverse selection model in which workers also choose effort level, see Akerlof (1976) on the "rat race." The model is not moral hazard, because while the employer observes effort, the worker's types—their utility costs of hard work—are known only to themselves.
- Gresham's Law ("Bad money drives out good") is a statement of adverse selection. Only debased money will be circulated if the payer knows the quality of his money better than the receiver. The same result occurs if quality is common knowledge, but for legal reasons the receiver is obligated to take the money, whatever its quality. An example of the first is adulterated Roman silver coins with low silver content; and of the second, Zambian currency with an overvalued exchange rate.
- Most adverse selection models have types that could be called "good" and "bad," because one type of agent would like to pool with the other, who would rather be separate. It is also possible to have a model in which both types would rather separate—types of workers who prefer night shifts versus those who prefer day shifts, for example—or two types who both prefer pooling—male and female college students.
- Two curious features of labor markets is that workers of widely differing outputs seem to be paid identical wages and that tests are not used more in hiring decisions. Schmidt and Judiesch have found that in jobs requiring only unskilled and semi-skilled blue-collar workers, the top 1 percent of workers, as defined by performance on ability tests not directly related to output, were 50 percent more productive than the average. In jobs defined as "high complexity" the difference was 127 percent (cited in Seligman [1992], p. 145).

  At about the same time as Akerlof (1970), another seminal paper appeared on adverse selection, Mirrlees (1971), although the relation only became clear later. Mirrlees looked at optimal taxation, and the problem of how the government chooses a tax schedule given that it cannot observe the abilities of its citizens to earn income, and this began the literature on mechanism design. Used cars and income taxes do not appear similar, but in both situations an uninformed player must decide how to behave with respect to another player whose type he does not know. Section 15.4 sets out a descendant of Mirrlees (1971) in a model of government procurement: much of government policy is motivated by the desire to create incentives for efficiency at minimum cost while eliciting information from individuals with superior information.

N9.2 Adverse Selection under Certainty: Lemons I and II

- Dealers in new cars and other durables have begun offering "extended-service contracts" in recent years. These contracts, offered either by the manufacturers or by independent companies, pay for repairs after the initial warranty expires. For reasons of moral hazard or adverse selection, the contracts usually do not cover damage from accidents. Oddly enough, they also do not cover items like oil changes despite their usefulness in prolonging engine life. Such contracts have their own problems, as shown by the fact that several of the independent companies went bankrupt in the late 1970s and early 1980s, making their contracts worthless.[1]
- Suppose that the cars of Lemons II lasted two periods and did not physically depreciate. A naive economist looking at the market would see new cars selling for $6,000 (twice $3,000) and old cars selling for $2,000 and conclude that the service stream had depreciated by 33 percent. Depreciation and adverse selection are hard to untangle using market data.
- Bond (1982) is an empirical paper on prices of used trucks.
- Lemons II uses a uniform distribution. For a general distribution F, the average quality $\bar{\theta}(P)$ of cars with quality P or less is

$$\bar{\theta}(P) = E(\theta \mid \theta \leq P) = \int_{-\infty}^{P} \frac{xF'(x)dx}{F(P)}. \tag{7}$$

Equation (7) also arises in physics (equation for a center of gravity) and nonlinear econometrics (the likelihood equation). Think of $\bar{\theta}(P)$ as a weighted average of the values of θ up to P, the weights being densities. Having multiplied by all these weights in the numerator, we have to divide by their "sum," $F(P) = \int_{-\infty}^{P} F'(x)dx$, in the denominator, giving rise to equation (7).

N9.3 Heterogeneous Tastes: Lemons III and IV

- You might object to a model in which the buyers of used cars value quality more than the sellers, since the sellers are often richer people. Remember that quality here is "quality of used cars," which is different from "quality of cars." The utility functions could be made more complicated without abandoning the basic model. We could specify something like $\pi_{buyer} = \theta + k/\theta - P$, where $\theta^2 > k$. Such a specification implies that the lower the quality of the car, the greater the difference between the valuations of buyer and seller.
- In Akerlof (1970) the quality of new cars is uniformly distributed between zero and 2, and the model is set up differently, with the demand and supply curves offered by different types of traders and net supply and gross supply presented rather confusingly. Usually the best way to model a situation in which traders sell some of their endowment and consume the rest is to use only gross supplies and demands. Each old owner supplies his car to the market, but in equilibrium he might buy it back, having better information about his car than the other consumers. Otherwise, it is easy to count a given unit of demand twice, once in the demand curve and once in the net supply curve.
- See Stiglitz (1987) for a good survey of the relation between price and quality. Leibenstein (1950) uses diagrams to analyze the implications of individual demand being linked to the market price of quantity in markets for "bandwagon," "snob," and "Veblen" goods. See also "Pricing of Products is Still an Art, Often Having Little Link to Costs," *Wall Street Journal*, November 25, 1981, p. 29.
- Risk aversion is concerned only with variability of outcomes, not their level. If the quality of used cars ranges from 2,000 to 6,000, buying a used car is risky. If all used cars are of quality 2,000, buying a used car is riskless, because the buyer knows exactly what he is getting.

  In Insurance Game III in section 9.4, the separating contract for the *Unsafe* consumer fully insures him: he bears no risk. But in constructing the equilibrium, we had to be very careful to keep the *Unsafes* from being tempted by the risky contract designed for the *Safes*. Risk is a bad thing, but, as with old age, the alternative is worse. If Smith were certain his car would be stolen, he would bear no risk, because he would be certain to have low utility.

[1] "Extended-Service Contracts for New Cars Shed Bad Reputation as Repair Bills Grow," *Wall Street Journal*, June 10, 1985, p. 25.

Figure 9.8 Lemons III: Buyers Value Cars More and Minimum Quality is Zero

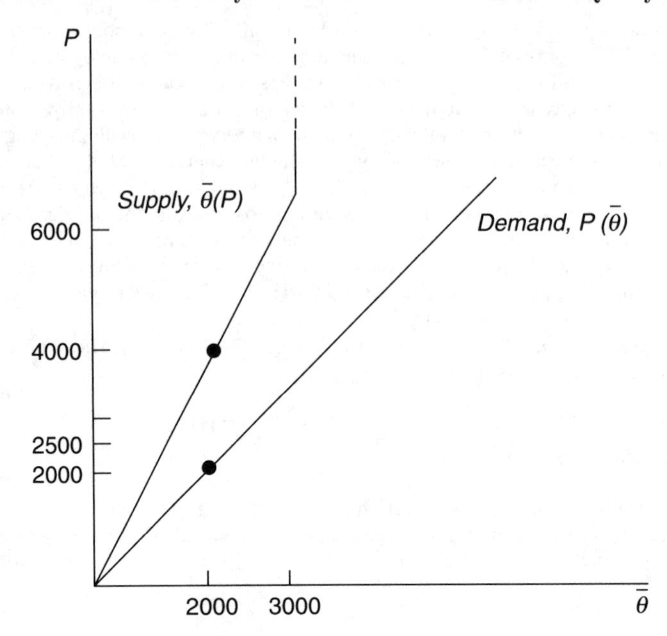

- To the buyers in Lemons IV, the average quality of cars for a given price is stochastic because they do not know which values of ε were realized. To them, the curve $\bar{\theta}(P)$ is only the *expectation* of the average quality.
- **Lemons III′: Minimum Quality of 0.** If the minimum quality of car in Lemons III were zero, not 2,000, the resulting game (Lemons III′) would be close to the original Akerlof (1970) specification. As figure 9.8 shows, the supply schedule and the demand schedule intersect at the origin, so that the equilibrium price is zero and no cars are traded. The market has shut down entirely, because of the unravelling effect described in Lemons II. Even though the buyers are willing to accept a quality lower than the dollar price, the price that buyers are willing to pay does not rise with quality as fast as the price needed to extract that average quality from the sellers, and a car of minimum quality is valued exactly the same by buyers and sellers. A 20 percent premium on zero is still zero. The efficiency implications are even stronger than before, because at the optimum all the old cars are sold to new buyers, but in equilibrium, none are.

N9.4 Adverse Selection under Uncertainty: Insurance Game III

- Markets with two types of customers are very common in insurance, because it is easy to distinguish male from female, both those types are numerous, and the difference between them is important. Males under age 25 pay almost twice the auto insurance premiums of females, and females pay 10 to 30 percent less for life insurance. The difference goes both ways, however: Aetna charges a 35-year old woman 30 to 50 percent more than a man for medical insurance. One market in which rates do not differ much is disability insurance. Women do make more claims, but the rates are the same because relatively few women buy the product (*Wall Street Journal*, 27 August 1987, p. 21).

N9.5 Other Equilibrium Concepts: Wilson Equilibrium and Reactive Equilibrium

- Engers & Fernandez (1987) show how to transform a simultaneous move game into a sequential move game such that the reactive equilibrium of the original game is one of the perfect Bayesian equilibria of the transformed game.

- In adverse-selection games it often matters whether the informed player or the uninformed player offers the contract. Wilson and reactive equilibrium are important when the uninformed player offers the contract, since it is only he who runs the risk of receiving something unexpected in the transaction and he might then wish to withdraw an offer. The issues involved are the same as in the difference between screening and signalling, which will be discussed at length in chapter 10.

N9.6 The Groves Mechanism

- Vickrey (1961) first suggested the non-budget-balancing mechanism for revelation of preferences, but it was rediscovered later (Groves [1973]) and became known as the Groves Mechanism.
- Roth (1984) is an interesting analysis of the system of matching hospitals with interns after the order of their preferences has been announced.
- An article using moral hazard to look at the problems of risk-sharing in optimal taxation, insurance, and price discrimination is Stiglitz (1977). Preference revelation is at the heart of the price-discrimination problem, the standard reference for which is Phlips's 1983 book.

N9.7 A Variety of Applications

- Bagehot (1971) is the earliest reference on the adverse selection explanation for the bid-ask spread. Copeland & Galai (1983) is an empirical study of the bid-ask spread in options markets.
- Economics professors sometimes make use of self-selection for student exams. One of my colleagues put the following instructions on an MBA exam, after stating that either Question 5 or 6 must be answered.

  "The value of Question 5 is less than that of Question 6. Question 5, however, is straightforward and the average student may expect to answer it correctly. Question 6 is more tricky: only those who have understood and absorbed the content of the course well will be able to answer it correctly. . . For a candidate to earn a final course grade of A or higher, it will be *necessary* for him to answer Question 6 successfully."

  Making the question even more self-referential, he asked the students for an explanation of its purpose.

  Another one of my colleagues took the approach of asking whether anyone in the class would be willing to skip the final exam and settle for an A−. Those students who were willing received an A−. The others got A's. But nobody had to take the exam. (This method did upset a few people.)

  More formally, Guasch & Weiss (1980) have looked at adverse selection and the willingness of workers with different abilities to take tests.
- Nalebuff & Scharfstein (1987) have written on testing, generalizing Mirrlees (1974), who showed how a forcing contract in which output is costlessly observed might attain efficiency by punishing only for very low output. In Nalebuff & Scharfstein, testing is costly and agents are risk averse. They develop an equilibrium in which the employer tests workers with small probability, using high-quality tests and heavy punishments to attain almost the first-best. Under a condition which implies that large expenditures on each test can eliminate false accusations, they show that the principal will test workers with small probability, but use expensive, accurate tests when he does test a worker, and impose a heavy punishment for lying.

Problems

9.1: Insurance with Equations and Diagrams.

The text analyzes Insurance Game III using diagrams. Here, let us use equations too. Let $U(t) = \log(t)$.

(9.1a) Give the numeric values (x, y) for the full-information separating contracts C_3 and C_4 from figure 9.6. What are the coordinates of C_3 and C_4?

(9.1b) Why is it not necessary to use the $U(t) = \log(t)$ function to find the values?

(9.1c) At the separating contract under incomplete information, C_5, $x = 2.01$. What is y? Justify the value 2.01 for x. What are the coordinates of C_5?

(9.1d) What is a contract C_6 that might be profitable and that would lure both types away from C_3 and C_5?

9.2: Testing and Commitment.

Fraction β of workers are talented, with output $a_t = 5$, and fraction $(1 - \beta)$ are untalented, with output $a_u = 0$. Both types have a reservation wage of 1 and are risk neutral. At a cost of 2 to itself and 1 to the job applicant, employer Apex can test a job applicant and discover his true ability with probability θ, which takes a value of something over 0.5. There is just one period of work. Let $\beta = 0.001$. Suppose that Apex can commit itself to a wage schedule before the workers take the test, and that Apex must test all applicants and pay all the workers it hires the same wage, to avoid grumbling among the workers and corruption in the personnel division.

(9.2a) What is the lowest wage, w_t, that will induce talented workers to apply? What is the lowest wage, w_u, that will induce untalented workers to apply? Which is greater?

(9.2b) What is the minimum accuracy value θ that will induce Apex to use the test? What are the firm's expected profits per worker who applies?

(9.2c) Now suppose that Apex can pay w_p to workers who pass the test and w_f to workers who flunk. What are w_p and w_f? What is the minimum accuracy value θ that will induce Apex to use the test? What are the firm's expected profits per worker who applies?

(9.2d) What happens if Apex cannot commit to paying the advertised wage, and can decide each applicant's wage individually?

(9.2e) If Apex cannot commit to testing every applicant, why is there no equilibrium in which either untalented workers do not apply or the firm tests every applicant?

9.3: Finding the Mixed-Strategy Equilibrium in a Testing Game.

Half of all high school graduates are talented, producing output $a = x$, and half are untalented, producing output $a = 0$. Both types have a reservation wage of 1 and are risk neutral. At a cost of 2 to himself and 1 to the job applicant, an employer can test a graduate and discover his true ability. Employers compete with each other in offering wages but they cooperate in revealing test results, so that an employer knows if an applicant has already been tested and failed. There is just one period of work. The employer cannot commit to testing every applicant.

(9.3a) Why is there no equilibrium in which either untalented workers do not apply or the employer tests every applicant?

(9.3b) In equilibrium, the employer tests workers with probability γ and pays those who pass the test w, the talented workers all present themselves for testing, and the untalented workers present themselves with probability α. Find an expression for the equilibrium value of α in terms of w. Explain why α is independent of x.

(9.3c) If $x = 8$, what are the equilibrium values of α, γ, and w?

9.4: Two-Time Losers.[2]

Some people are strictly principled and will commit no robberies, even if there is no penalty. Others are incorrigible criminals and will commit two robberies, regardless of the penalty. Society wishes to inflict a certain penalty on criminals as retribution. Retribution requires an expected penalty of 15 per crime (15 if detection is sure, 150 if it has probability 0.1, etc.). Innocent people are sometimes falsely convicted, as shown in table 9.2.

Table 9.2 Two-Time Losers

		Convictions		
		0	1	2
Robberies	0	0.81	0.18	0.01
	2	0.49	0.42	0.09

Two systems are proposed: (i) a penalty of X for each conviction, and (ii) a penalty of 0 for the first conviction, and some amount P for the second conviction.

(9.4a) What must X and P be to achieve the desired amount of retribution?

(9.4b) Which system inflicts the lesser cost on innocent people? How much is the cost in each case?

9.5: Insurance and State-Space Diagrams.

Two types of risk-averse people, clean-living and dissolute, would like to buy health insurance. Clean-living people become sick with probability 0.3, and dissolute people with probability 0.9. In state-space diagrams with the person's wealth if he is healthy on the vertical axis and if he is sick on the horizontal, every person's initial endowment is (5, 10), because his initial wealth is 10 and the cost of medical treatment is 5.

(9.5a) What is the expected wealth of each type of person?

(9.5b) Draw a state-space diagram with the indifference curves for a risk-neutral insurance company that insures each type of person separately. Draw in the post-insurance allocations C_1 for the dissolute and C_2 for the clean-living under the assumption that a person's type is contractible.

[2] Compare this with problem 8.2. How are they different? See Rubinstein (1979).

(9.5c) Draw a new state-space diagram with the initial endowment and the indifference curves for the two types of people that go through that point.

(9.5d) Explain why, under asymmetric information, no pooling contract C_3 can be part of a Nash equilibrium.

(9.5e) If the insurance company is a monopoly, can a pooling contract be part of a Nash equilibrium?

10 Signalling

10.1 The Informed Player Moves First: Signalling

Signalling is a way for an agent to communicate his type under adverse selection. The signalling contract specifies a wage that depends on an observable characteristic—the signal—which the agent chooses for himself after Nature chooses his type. Figures 7.1d and 7.1e showed the extensive forms of two kinds of models with signals. If the agent chooses his signal before the contract is offered, he is signalling to the principal. If he chooses the signal afterwards, the principal is screening him. Not only will it become apparent that this difference in the order of moves is important, it will also be seen that signalling costs must differ between agent types for signalling to be useful, and the outcome is often inefficient.

We begin with signalling models in which workers choose education levels to signal their abilities. Section 10.1 lays out the fundamental properties of a signalling model, and section 10.2 shows how the details of the model affect the equilibrium. Section 10.3 steps back from the technical detail to more practical considerations in applying the model to education. Section 10.4 turns the game into a screening model. Section 10.5 switches to diagrams and applies signalling to new stock issues to show how two signals need to be used when the agent has two unobservable characteristics.

Spence (1973) introduced the idea of signalling in the context of education. We will construct a series of models which formalize the notion that education has no direct effect on a person's ability to be productive in the real world but is useful for demonstrating his ability to employers. Let half of the workers have the type "high ability" and half "low ability," where ability is a number denoting the dollar value of his output. Output is assumed to be a noncontractible variable and there is no uncertainty. If output is contractible, it should be in the contract, as we have seen in chapter 7. Lack of uncertainty is a simplifying assumption, imposed so that the contracts are functions only of the signals rather than a combination of the signal and the output.

Employers do not observe the worker's ability, but they do know the distribution of abilities, and they observe the worker's education. To simplify, we will specify that the players are one worker and two employers. The employers compete profits down to zero and the worker receives the gains from trade. The worker's strategy is his education level and his choice of employer. The employers' strategies are the contracts they offer giving wages as functions of education level. The key to the model is that the signal, education, is less costly for workers with higher ability.

In the first four variants of the game, workers choose their education levels before employers decide how pay should vary with education.

Education I

Players

A worker and two employers.

Order of Play

(0) Nature chooses the worker's ability $a \in \{2, 5.5\}$, the *Low* and *High* ability each having probability 0.5. The variable a is observed by the worker, but not by the employers.
(1) The worker chooses education level $s \in \{0, 1\}$.
(2) The employers each offer a wage contract $w(s)$.
(3) The worker accepts a contract, or rejects both of them.
(4) Output equals a.

Payoffs

The worker's payoff is his wage minus his cost of education, and the employer's is his profit.

$$\pi_{worker} = \begin{cases} w - 8s/a & \text{if the worker accepts contract } w. \\ 0 & \text{if the worker rejects both contracts.} \end{cases}$$

$$\pi_{employer} = \begin{cases} a - w & \text{for the employer whose contract is accepted.} \\ 0 & \text{for the other employer.} \end{cases}$$

The payoffs assume that education is more costly for a worker if his ability takes a lower value, which is what permits separation to occur. As in any hidden knowledge game, we must think about both pooling and separating equilibria. Education I has both. In the pooling equilibrium, which we will call PE 1.1, both types of workers pick zero education and the employers pay the zero-profit wage of 3.75 (= [2 + 5.5]/2) regardless of the education level.

Pooling Equilibrium 1.1 (PE 1.1)

$$\begin{cases} s(Low) = s(High) = 0 \\ w(0) = w(1) = 3.75 \\ Prob(a = Low|s = 1) = 0.5 \end{cases}$$

PE 1.1 needs to be specified as a perfect Bayesian equilibrium rather than simply a Nash equilibrium because of the importance of the interpretation that the uninformed player puts on out-of-equilibrium behavior. The equilibrium needs to specify the employer's beliefs when he observes $s = 1$, since that is never observed in equilibrium. In PE 1.1, the beliefs are passive conjectures (see section 6.2): employers believe that a worker who chooses $s = 1$ is *Low* with the prior probability, 0.5. Given this belief, both types of workers realize that education is useless, and the model reaches the unsurprising outcome that workers do not bother to acquire unproductive education.

Under other beliefs, the pooling equilibrium breaks down. Under the belief $Prob(a = Low|s = 1) = 0$, for example, employers believe that any worker who acquired education is a *High*, so pooling is not Nash because the *High* workers are tempted to deviate and acquire education. This leads to the separating equilibrium for which signalling is best known, in which the high-ability worker acquires education to prove to employers that he really has high ability.

Separating Equilibrium 1.2 (SE 1.2)
$$\begin{cases} s(Low) = 0, \; s(High) = 1 \\ w(0) = 2, \; w(1) = 5.5 \end{cases}$$

Following the method used in chapters 7 and 8, we will show that SE 1.2 is a perfect Bayesian equilibrium by using the standard constraints which an equilibrium must satisfy. A pair of separating contracts must maximize the utility of the *Highs* and the *Lows* subject to two constraints: (a) the participation constraints that the firms can offer the contracts without making losses; and (b) the self-selection constraints that the *Lows* are not attracted to the *High* contract, and the *Highs* are not attracted to the *Low* contract. The participation constraints for the employers require that

$$w(0) \leq a_L = 2 \text{ and } w(1) \leq a_H = 5.5. \tag{1}$$

Competition between the employers makes the expressions in (1) hold as equalities. The self-selection constraint of the *Lows* is

$$U_L(s = 0) \geq U_L(s = 1), \tag{2}$$

which in Education I is

$$w(0) - 0 \geq w(1) - \frac{8}{2}. \tag{3}$$

Since in SE 1.2 the separating wage of the *Lows* is 2 and the separating wage of the *Highs* is 5.5 from (1), the self-selection constraint (3) is satisfied.

The self-selection constraint of the *Highs* is

$$U_H(s = 1) \geq U_H(s = 0), \tag{4}$$

which in Education I is

$$w(1) - \frac{8}{5.5} \geq w(0) - 0. \tag{5}$$

Constraint (5) is satisfied by SE 1.2.

There is another conceivable pooling equilibrium for Education I, in which $s(Low) = s(High) = 1$, but this turns out not to be an equilibrium, because the *Lows* would deviate to zero education. Even if such a deviation caused the employer to believe they were low-ability with probability 1 and reduce their wage to 2, the low-ability workers would still prefer to deviate, because

$$U_L(s = 0) = 2 \geq U_L(s = 1) = 3.75 - \frac{8(1)}{2}. \tag{6}$$

Thus, a pooling equilibrium with $s = 1$ would violate incentive compatibility for the *Low* workers.

Notice that we do not need to worry about a nonpooling constraint for this game, unlike in the case of the games of chapter 9. One might think that because employers compete for workers, competition between them might result in their offering a pooling contract that the high-ability workers would prefer to the separating contract. The reason this does not matter is that the employers do not compete by offering contracts, but by reacting to workers who have acquired education. That is why this is signalling and not screening: the employers cannot offer contracts in advance that change the workers' incentives to acquire education.

We can test the equilibrium by looking at the best responses. Given the worker's strategy and the other employer's strategy, an employer must pay the worker his full output or lose him to the other employer. Given the employers' contracts, the *Low* has a choice between the payoff 2 ($= 2 - 0$) for ignorance and 1.5 ($= 5.5 - 8/2$) for education, so he picks ignorance. The *High* has a choice between the payoff 2 ($= 2 - 0$) for ignorance and 4.05 ($= 5.5 - 8/5.5$, rounded) for education, so he picks education.

Unlike the pooling equilibrium, the separating equilibrium does not need to specify beliefs. Either of the two education levels might be observed in equilibrium, so Bayes' Rule always tells the employers how to interpret what they see. If they see that an agent has acquired education, they deduce that his ability is *High*, and if they see that he has not, they deduce that it is *Low*. A worker is free to deviate from the education level appropriate to his type, but the employers' beliefs will continue to be based on equilibrium behavior. If a *High* worker deviates by choosing $s = 0$ and tells the employers he is a *High* who would rather pool than separate, the employers disbelieve him and offer him the *Low* wage of 2 that is appropriate to $s = 0$, not the pooling wage of 3.75 or the *High* wage of 5.5.

Separation is possible because education is more costly for workers if their ability is lower. If education cost the same for both types of worker, education would not work as a signal, because the low-ability workers would imitate the high-ability workers. This requirement of different signalling costs is known as the **single-crossing property**, since when the costs are depicted graphically, as in section 10.4, the indifference curves of the two types intersect a single time.

A strong case can be made that the beliefs required for the pooling equilibria are not sensible. Harking back to the equilibrium refinements of section 6.2, recall that one suggestion (from Cho & Kreps [1987]) is to inquire into whether one type of player could not possibly benefit from deviating, no matter how the uninformed player changed his beliefs as a result. Here, the *Low* worker could never benefit from deviating from PE 1.1. Under the passive conjectures specified, the *Low* has a payoff of 3.75 in equilibrium versus -0.25 ($=3.75$-8/2) if he deviates and becomes educated. Under the belief that most encourages deviation—that a worker who deviates is *High* with probability one—the *Low* would get a wage of 5.5 if he deviated, but his payoff from deviating would only be 1.5 (= 5.5 − 8/2), which is less than 2. The more reasonable belief seems to be that a worker who acquires education is a *High*, which does not support the pooling equilibrium.

The nature of the separating equilibrium lends support to the claim that education *per se* is useless or even pernicious, because it imposes social costs but does not increase total output. While we may be reassured by the fact that Professor Spence himself thought it worthwhile to become Dean of Harvard College, the implications are disturbing and suggest that we should think seriously about how well the model applies to the real world. We will do that in section 10.3. For now, note that in the model, unlike most real-world situations, information about the agent's talent has no social value, because all agents would be hired and employed at the same task even under full information. Also, if side payments are not possible, SE 1.2 is second-best efficient in the sense that a social planner could not make both types of workers better off. Separation helps the high-ability workers even though it hurts the low-ability workers.

10.2 Variants of the Signalling Model of Education

Although Education I is a curious and important model, it does not exhaust the implications of signalling which can be discovered in simple models. This section will start with Education II, which will show an alternative to the arbitrary assumption of beliefs in the perfect Bayesian equilibrium concept. Education III will be the same as Education I except for its parameter value, and it will have two pooling equilibria rather than one separating equilibrium and one pooling equilibrium. Education IV will allow a continuum of education levels, and will unify Education I and Education III by showing how all of their equilibria and more can be obtained in a model with a less restricted strategy space.

Education II: Modelling Trembles so Nothing is Out of Equilibrium

The pooling equilibrium of Education I required the modeller to specify the employers' out-of-equilibrium beliefs. An equivalent model constructs the game tree to support the beliefs instead of introducing them via the equilibrium concept. This approach was briefly mentioned in connection with the game of PhD Admissions in section 6.2. The advantage is that the assumptions on beliefs are put in the rules of the game along with the other assumptions. So let us replace Nature's move in Education I and modify the payoffs as follows.

Education II

(0) Nature chooses worker ability $a \in \{2, 5.5\}$, each ability having probability 0.5. (a is observed by the worker, but not by the employer.) With probability 0.001, Nature endows a worker with free education.

...

Payoffs

$$\pi_{worker} = \begin{cases} w - 8s/a & \text{if the worker accepts contract } w \text{ (ordinarily)} \\ w & \text{if the worker accepts contract } w \text{ (with free education)} \\ 0 & \text{if the worker does not accept a contract} \end{cases}$$

With probability 0.001 the worker receives free education regardless of his ability. If the employer sees a worker with education, he knows that the worker might be one of this rare type, in which case the probability that the worker is *Low* is 0.5. Both $s = 0$ and $s = 1$ can be observed in any equilibrium and Education II has almost the same two equilibria as Education I, without the need to specify beliefs. The separating equilibrium did not depend on beliefs, and remains an equilibrium. What was Pooling Equilibrium 1.1 becomes "almost" a pooling equilibrium—almost all workers behave the same, but the small number with free education behave differently. The two types of greatest interest—the *High* and the *Low*—are not separated, but the ordinary workers are separated from the workers whose education is free. Even that small amount of separation allows the employers to use Bayes' Rule and eliminates the need for exogenous beliefs.

Education III: No Separating Equilibrium, Two Pooling Equilibria

Let us next modify Education I by changing the possible worker abilities from {2, 5.5} to {2, 12}. The separating equilibrium vanishes, but a new pooling equilibrium emerges. In equilibria PE 3.1 and PE 3.2, both pooling contracts pay the same zero-profit wage of 7 (= [2 + 12]/2), and both types of agents acquire the same amount of education, but the amount depends on the equilibrium.

Pooling Equilibrium 3.1
(PE 3.1)

$$\begin{cases} s(Low) = s(High) = 0 \\ w(0) = w(1) = 7 \\ Prob(a = Low|s = 1) = 0.5 \\ \text{(passive conjectures)} \end{cases}$$

Pooling Equilibrium 3.2
(PE 3.2)

$$\begin{cases} s(Low) = s(High) = 1 \\ w(0) = 2,\ w(1) = 7 \\ Prob(a = Low|s = 0) = 1 \end{cases}$$

PE 3.1 is-similar to the pooling equilibrium in Education I and II, but PE 3.2 is inefficient. Both types of workers receive the same wage, but they incur the education costs anyway. Each type is frightened to do without education because the employer would pay him not as if his ability were average, but as if he were known to be *Low*.

Examination of PE 3.2 shows why a separating equilibrium no longer exists. Any separating equilibrium would require $w(0) = 2$ and $w(1) = 7$, but this is the contract that leads to PE 3.2. The self-selection and zero-profit constraints cannot be satisfied simultaneously, because the *Low* type is willing to acquire $s = 1$ to obtain the high wage.

It is not surprising that information problems create inefficiencies in the sense that first-best efficiency is lost. Indeed, the surprise is that in some games with asymmetric information, such as Broadway Game I in section 7.4, the first-best can still be achieved by tricks such as boiling-in-oil contracts. More often, we discover that the outcome is second-best efficient: given the informational constraints, a social planner could not alter the equilibrium without hurting some type of player. PE 3.2 is not even second-best efficient, because PE 3.1 and PE 3.2 result in the exact same wages and allocation of workers to tasks. The inefficiency is purely a problem of unfortunate expectations, like the inefficiency from choosing the dominated equilibrium in Ranked Coordination.

PE 3.2 also illustrates a fine point of the definition of pooling, because although the two types of workers adopt the same strategies, the equilibrium contract offers different wages for different education. The implied threat to pay a low wage to an uneducated worker never needs to be carried out, so the equilibrium is still called a pooling equilibrium. Notice that perfectness does not rule out threats based on beliefs. The model imposes these beliefs on the employer, and he would carry out his threats, because he believes they are best responses. The employer receives a higher payoff under some beliefs than under others, but he is not free to choose his beliefs.

Following the approach of Education II, we could eliminate PE 3.2 by adding an exogenous probability of 0.001 that either type is completely unable to buy education. Then no behavior is never observed in equilibrium and we end up with PE 3.1 because the only rational belief is that if $s = 0$ is observed, the worker has equal probability of being *High* or being *Low*. To eliminate PE 3.1 requires less reasonable beliefs; for example, a probability of 0.001 that a *Low* gets free education together with a probability of 0 that a *High* does.

These first three games illustrate the basics of signalling: (a) separating and pooling equilibria both may exist, (b) out-of-equilibrium beliefs matter, and (c) sometimes one perfect Bayesian equilibrium can Pareto dominate others. These results are robust, but Education IV will illustrate some dangers of using simplified games with binary strategy spaces instead of continuous and unbounded strategies. So far education has been limited to $s = 0$ or $s = 1$; Education IV allows it to take greater or intermediate values.

Education IV: Continuous Signals and Continua of Equilibria

Let us now return to Education I, with the one change that education s can take any level on the continuum between 0 and infinity.

The game now has continua of pooling and separating equilibria which differ according to the value of education chosen. In the pooling equilibria, the equilibrium education level is s^*, where each s^* in the interval $[0, \bar{s}]$ supports a different equilibrium. The out-of-equilibrium belief most likely to support a pooling equilibrium is $Prob(a = Low|s \neq s^*) = 1$, so let us use this to find the value of $\bar{s}$, the greatest amount of education that can be generated by a pooling equilibrium. The equilibrium is PE 4.1, where $s^* \in [0, \bar{s}]$.

Pooling Equilibrium 4.1
(PE 4.1)

$$\begin{cases} s(Low) = s(High) = s^* \\ w(s^*) = 3.75 \\ w(s \neq s^*) = 2 \\ Prob(a = Low|s \neq s^*) = 1 \end{cases}$$

The critical value $\bar{s}$ can be discovered from the incentive-compatibility constraint of the *Low* type, which is binding if $s^* = \bar{s}$. The most tempting deviation is to zero education, so that is the deviation that appears in the constraint.

$$U_L(s = 0) = 2 \leq U_L(s = \bar{s}) = 3.75 - \frac{8\bar{s}}{2}. \tag{7}$$

Equation (7) yields $\bar{s} = 7/16$. Any value of s^* less than 7/16 will also support a pooling equilibrium. Note that the incentive-compatibility constraint of the *High* type is not binding. If a *High* deviates to $s = 0$, he, too, will be thought to be a *Low*, so

$$U_H(s = 0) = 2 \leq U_H\left(s = \frac{7}{16}\right) = 3.75 - \frac{8\bar{s}}{5.5} \approx 3.1. \tag{8}$$

In the separating equilibria, the education levels chosen in equilibrium are 0 for the *Lows* and s^* for the *Highs*, where each s^* in the interval $[\bar{s}, \bar{\bar{s}}]$ supports a different equilibrium. A difference from the case of separating equilibria in games with binary strategy spaces is that now there are possible out-of-equilibrium actions even in a separating equilibrium. The two types of workers will separate to two education levels, but that leaves an infinite number of out-of-equilibrium education levels. As before, let us use the most extreme belief for the employers' beliefs after observing an out-of-equilibrium education level: that $Prob(a = Low|s \neq s^*) = 1$. The equilibrium is SE 4.2, where $s^* \in [\bar{s}, s]$.

Separating Equilibrium 4.2
(SE 4.2)

$$\begin{cases} s(Low) = 0,\ s(High) = s^* \\ w(s^*) = 5.5 \\ w(s \neq s^*) = 2 \\ Prob(a = Low|s \notin \{0, s^*\}) = 1 \end{cases}$$

The critical value $\bar{s}$ can be discovered from the incentive-compatibility constraint of the *Low*, which is binding if $s^* = \bar{s}$.

$$U_L(s = 0) = 2 \geq U_L(s = \bar{s}) = 5.5 - \frac{8\bar{s}}{2}. \tag{9}$$

Equation (9) yields $\bar{s} = 7/8$. Any value of s^* greater than 7/8 will also deter the *Low* workers from acquiring education. If the education needed for the wage of 5.5 is too great, the *High* workers will give up on education too. Their incentive compatibility constraint requires that

$$U_H(s = 0) = 2 \leq U_H(s = \bar{\bar{s}}) = 5.5 - \frac{8\bar{\bar{s}}}{5.5}. \tag{10}$$

Equation (9) yields $\bar{\bar{s}} = 77/32$. s^* can take any value lower than 77/32 and the *Highs* will be willing to acquire education.

The big difference from Education I is that Education IV has Pareto-ranked equilibria. Pooling can occur not just at zero education but at positive levels, as in Education III, and the pooling equilibria with positive education levels are all Pareto inferior. Also, the separating equilibria can be Pareto ranked, since separation with $s^* = \bar{s}$ dominates separation with $s^* = \bar{\bar{s}}$. Using a binary strategy space instead of a continuum conceals this problem.

Education IV also shows how restricting the strategy space can alter the kinds of equilibria that are possible. Education III had no separating equilibrium because at the maximum possible signal, $s = 1$, the *Lows* were still willing to imitate the *Highs*. Education IV would not have any separating equilibria either if the strategy space were restricted to allow only education levels less than 7/8. Using a bounded strategy space eliminates possibly realistic equilibria.

This is not to say that models with binary strategy sets are always misleading. Education I is a fine model for showing how signalling can be used to separate agents of different types; it become misleading only when used to reach a conclusion such as "If a separating equilibrium exists, it is unique." As with any assumption, one must be careful not to narrow the model so much as to render vacuous the question it is designed to answer.

10.3 General Comments on Signalling in Education

Signalling and Similar Phenomena

The distinguishing feature of signalling is that the agent's action, although not directly related to output, is useful because it is related to ability. For the signal to work, it must be less costly for an agent with higher ability. Separation can occur in Education I because when the principal pays a greater wage to the educated workers, only the *Highs*, whose utility costs of education are lower, are willing to acquire it. That is why a signal works where a simple message would not: actions speak louder than words.

Signalling is outwardly similar to other solutions to adverse selection. The high-ability agent finds it cheaper than the low-ability one to build a reputation, but the reputation-building actions are based directly on his high ability. In a typical reputation model he shows ability by producing high output period after period. Also, the nature of reputation is to require several periods of play, which signalling does not.

Another form of communication is possible when some observable variable not under the control of the worker is correlated with ability. Age, for example, is correlated with reliability, so an employer pays older workers more, but the correlation does not arise because it is easier for reliable workers to acquire the attribute of age. Because age is not an action chosen by the worker, we would not need game theory to model it.

Problems in Applying Signalling to Education

On the empirical level, the first question to ask of a signalling model of education is, "What is education?" For operational purposes this means, "In what units is education measured?" Two possible answers are "years of education" and "grade-point average." If the sacrifice of a year of earnings is greater for a low-ability worker, years of education can serve as a signal. If less intelligent students must work harder to get straight A's, then grade-point-average can also be a signal.

Layard & Psacharopoulos (1974) give three rationales for rejecting signalling as an important motive for education. First, dropouts get as high a rate of return on education as those who complete degrees, so the signal is not the diploma, although it might be the years of education. Second, wage differentials between different education levels rise with age, although one would expect the signal to be less important after the employer has acquired more observations on the worker's output. Third, testing is not widely used for hiring, despite its low cost relative to education. Tests are available, but unused: students commonly take tests like the American SAT whose results they could credibly communicate to employers, and their scores correlate highly with subsequent grade-point average. One would also expect an employer to prefer to pay an 18-year-old low wages for four years to determine his ability, rather than waiting to see what grades he gets as a history major.

Productive Signalling

Even if education is largely signalling, we might not want to close the schools. Signalling might be wasteful in a pooling equilibrium like PE 3.2, but in a separating equilibrium it can be second-best efficient for at least three reasons. First, it allows the employer to match workers with jobs suited to their talents. If the only jobs available were "professor" and "typist," then, in a pooling equilibrium, both *High* and *Low* workers would be employed, but they would be randomly allocated to the two jobs. Given the principle of comparative advantage, typing might improve, but I think, pridefully, that research would suffer.

Second, signalling keeps talented workers from moving to jobs where their productivity is lower but their talent is known. Without signalling, a talented worker might leave a corporation and start his own company, where he would be less productive but better paid. The naive observer would see that corporations hire only one type of worker (*Low*), and imagine there was no welfare loss.

Third, if ability is endogenous—moral hazard rather than adverse selection—signalling encourages workers to acquire ability. One of my teachers said that you always understand your next-to-last econometrics class. Suppose that solidly learning econometrics increases the student's ability, but a grade of A is not enough to show that he solidly learned the material. To signal his newly acquired ability, the student must also take "Time Series," which he cannot pass without a solid understanding of econometrics. "Time Series" might be useless in itself, but if it did not exist, the student would not be able to show he had learned basic econometrics.

10.4 The Informed Player Moves Second: Screening

In screening games, the informed player moves second, which means that he moves in response to contracts offered by the uninformed player. Having the uninformed player make the offers is important because his offer conveys no information about himself, unlike what happens in a signalling model.

Education V: Screening with a Discrete Signal

Players

A worker and two employers.

Order of Play

(0) Nature chooses worker ability $a \in \{2, 5.5\}$, each ability having probability 0.5. Employers do not observe ability, but the worker does.
(1) Each employer offers a wage contract $w(s)$.
(2) The worker chooses education level $s \in \{0, 1\}$.
(3) The worker accepts a contract, or rejects both of them.
(4) Output equals a.

Payoffs

$$\pi_{worker} = \begin{cases} w - 8s/a & \text{if the worker accepts contract } w. \\ 0 & \text{if the worker rejects both contracts.} \end{cases}$$

$$\pi_{employer} = \begin{cases} a - w & \text{for the employer whose contract is accepted.} \\ 0 & \text{for the other employer.} \end{cases}$$

Education V has no pooling equilibrium, because if one employer tried to offer the zero profit pooling contract, $w(0) = 3.75$, the other employer would offer $w(1) = 5.5$ and draw away all the *Highs*. The unique equilibrium is

Separating Equilibrium 5.1 (SE 5.1)
$$\begin{cases} s(Low) = 0, \; s(High) = 1 \\ w(0) = 2, \; w(1) = 5.5 \end{cases}$$

Beliefs do not need to be specified in a screening model. The uninformed player moves first, so his beliefs after seeing the move of the informed player are irrelevant. The informed player is fully informed, so his beliefs are not affected by what he observes. This is much like simple adverse selection, in which the uninformed player moves first, offering a set of contracts, after which the informed player chooses one of them. The modeller does not need to refine perfectness in a screening model, although he might be tempted to abandon it altogether in favor of the reactive or Wilson equilibrium concepts described in section 8.5. The similarity between adverse selection and screening is strong enough that Education V would not have been out of place in chapter 9, but it is presented here because the context is so similar to the signalling models of education.

Education VI allows a continuum of education levels, in a game which is otherwise the same as Education V.

Education VI: Screening with a Continuous Signal

Players

A worker and two employers.

Order of Play

(0) Nature chooses worker ability $a \in \{2, 5.5\}$, each ability having probability 0.5. Employers do not observe ability, but the worker does.
(1) Each employer offers a wage contract $w(s)$.
(2) The worker choose education level $s \in [0, 1]$.
(3) The worker chooses a contract, or rejects both of them.
(4) Output equals a.

Payoffs

$$\pi_{worker} = \begin{cases} w - 8s/a & \text{if the worker accepts contract } w. \\ 0 & \text{if the worker rejects both contracts.} \end{cases}$$

$$\pi_{employer} = \begin{cases} a - w & \text{for the employer whose contract is accepted.} \\ 0 & \text{for the other employer.} \end{cases}$$

Pooling equilibria generally do not exist in screening games with continuous signals, and sometimes separating equilibria in pure strategies do not exist either—recall Insurance Game III from section 9.4. Education VI, however, does have a separating Nash equilibrium, with a unique equilibrium path.

Separating Equilibrium 6.1
(SE 6.1)

$$\begin{cases} s(Low) = 0, \ s(High) = 0.875 \\ w = \begin{cases} 2 & \text{if } s < 0.875 \\ 5.5 & \text{if } s \geq 0.875 \end{cases} \end{cases}$$

In any separating contract, the *Lows* must be paid a wage of 2 for an education of 0, because this is the most attractive contract that breaks even. The separating contract for the *Highs* must maximize their utility subject to the constraints discussed in Education I. When the signal is continuous, the constraints are especially useful to the modeller for calculating the equilibrium. The participation constraints for the employers require that

$$w(0) \leq a_L = 2 \text{ and } w(s^*) \leq a_H = 5.5, \tag{11}$$

where s^* is the separating value of education that we are trying to find. Competition turns the inequalities in (11) into equalities. The self-selection constraint for the low-ability workers is

$$U_L\,(s = 0) \geq U_L(s = s^*), \tag{12}$$

which in Education VI is

$$w(0) - 0 \geq w(s^*) - \frac{8s^*}{2}. \tag{13}$$

Since the separating wage is 2 for the *Low's* and 5.5 for the *High's*, constraint (13) is satisfied as an equality if $s^* = 0.875$, which is the crucial education level in SE 6.1.

$$U_H(s = 0) = w(0) \leq U_H(s = s^*) = w(s^*) - \frac{8s^*}{5.5}. \tag{14}$$

If $s^* = 0.875$, inequality (14) is true, and it would also be true for higher values of s^*. Unlike the case of the continuous-strategy signalling game, Education IV, however, the equilibrium contract in Education VI is unique, because the employers compete to offer the most attractive contract that satisfies the participation and incentive compatibility constraints. The most attractive is the separating contract that Pareto dominates the other separating contracts by requiring the relatively low separating signal of $s^* = 0.875$.

Similarly, competition in offering attractive contracts rules out pooling contracts. The nonpooling constraint, required by competition between employers, is

$$U_H(s = s^*) \geq U_H(pooling), \tag{15}$$

which, for Education VI, is, using the most attractive possible pooling contract,

$$w(s^*) - \frac{8s^*}{5.5} \geq 3.75. \tag{16}$$

Since the payoff of *Highs* in the separating contract is 4.23 ($= 5.5 - 8 \cdot 0.875/5.5$, rounded), the nonpooling constraint is satisfied.

No Pooling Equilibrium in Education VI

Education VI lacks a pooling equilibrium, which would require the outcome $\{s = 0, w(0) = 3.75\}$, shown as C_1 in figure 10.1. If one employer offered a pooling contract requiring more than zero education (such as the inefficient PE 3.2), the other employer could make the more attractive offer of the same wage for zero education. The wage is 3.75 to ensure zero profits. The rest of the wage function—the wages for positive education levels—can take a variety of shapes, so long as the wage does not rise so fast with education that the *Highs* are tempted to become educated.

But no equilibrium has these characteristics. In a Nash equilibrium, no employer can offer a pooling contract, because the other employer could always profit by offering a separating contract paying more to the educated. One such separating contract is C_2 in figure 10.1, which pays 5 to workers with an education of $s = 0.5$ and yields a payoff of 4.89 ($= 5 - [8 \cdot 0.5]/5.5$, rounded) to the *Highs* and 3 ($= 5 - 8 \cdot 0.5/2$) to the *Lows*. Only *Highs* prefer C_2 to the pooling contract C_1, which yields payoffs of 3.75 to both *High* and *Low*, and if only *Highs* accept C_2, it yields positive profits to the employer.

Nonexistence of a pooling equilibrium in screening models without continuous strategy spaces is a general result. The linearity of the curves in Education VI is special, but in any screening model the *Low's* would have greater costs of education, which is equivalent to steeper indifference curves. This is the **single-crossing property** alluded to in Education I. Any pooling equilibrium must, like

Figure 10.1 Education VI: No Pooling Nash Equilibrium

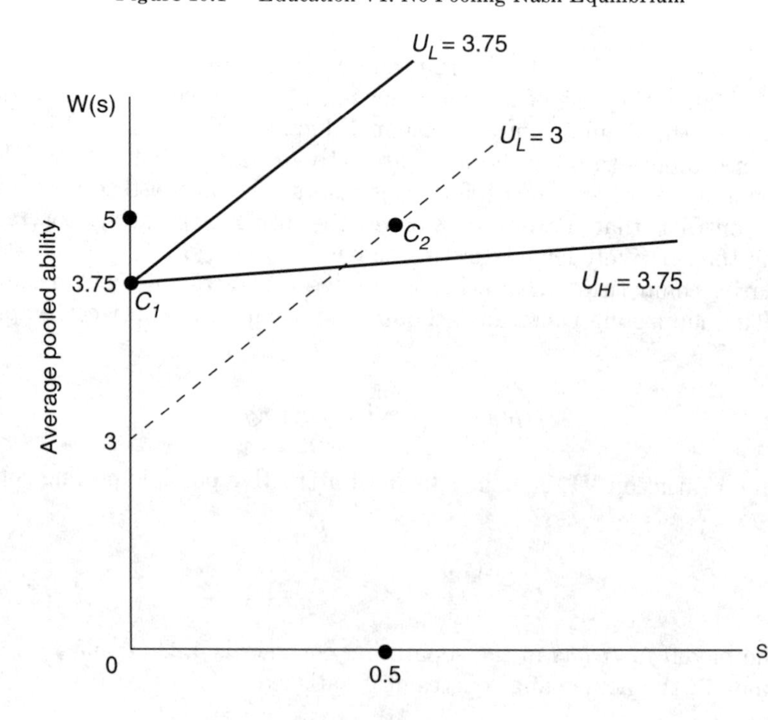

C_1, lie on the vertical axis where education is zero and the wage equals the average ability. A separating contract like C_2 can always be found to the northeast of the pooling contract, between the indifference curves of the two types, and it will yield positive profits by attracting only the *Highs*.

Education VII: No Nash Equilibrium

In Education VI we showed that screening models have no pooling equilibria. In Education VII the parameters are changed a little to eliminate even the separating equilibrium. Let the proportion of *Highs* be 0.9 instead of 0.5, so the zero-profit pooling wage is 5.15 (= 0.9[5.5] + 0.1[2]) instead of 3.75. Consider the separating contracts C_3 and C_4, shown in figure 10.2, calculated in the same way as SE 5.1. The pair (C_3, C_4) is the most attractive pair of contracts that separates *High's* from *Low's* by satisfying constraint (7). *Low* workers accept contract C_3, obtain $s = 0$, and receive a wage of 2, their ability. *Highs* accept contract C_4, obtain $s = 0.875$, and receive a wage of 5.5, their ability. Education is not attractive to *Low's* because the *Low* payoff from pretending to be *High* is 2 ($= 5.5 - 8 \cdot 0.875/2$), no better than the *Low* payoff of 2 from C_3 ($= 2 - 8 \cdot 0/2$).

The wage of the pooling contract C_5 is 5.15, so that even the *Highs* strictly prefer C_5 to (C_3, C_4). But our reasoning that no pooling equilibrium exists is still valid; some contract C_6 would attract all the *Highs* from C_5. No Nash equilibrium in pure strategies exists, either separating or pooling.

Figure 10.2 Education VII: No Nash Equilibrium

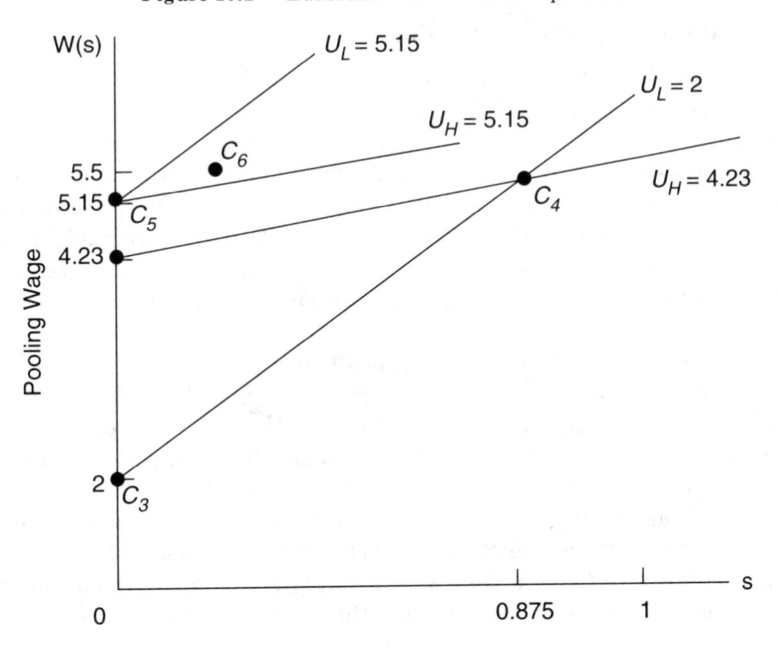

Screening and Adverse Selection

We also encountered a nonexistence problem in the adverse selection games of section 9.4, where, just as in screening, the informed player takes no action until after the uninformed player has offered a set of contracts. Screening models behave much the same way as simple adverse selection models, in contrast to signalling models.

Compare Education VI, the first screening model of education, with the Rothschild & Stiglitz (1976) Insurance Game III of section 9.4. Both have contracts which give something of value in exchange for costs incurred to communicate the informed players' private information. In Education VI the contracts specified a higher wage if the worker acquired more education. In Insurance Game III the contracts specified a lower premium if the customer accepted more risk by accepting a higher coinsurance rate. In both models, the two types of informed player separate, picking two different levels of the costly variable. The outcomes are only second-best efficient because in Insurance Game III the *Safe* type is not fully insured and in Education VI the *High* type incurs the cost of education. Insurance Game III specifies two distinct contracts, as opposed to the $w(s)$ function of Education V, but since the worker only picks one of two education levels in equilibrium, the difference is more apparent than real. The fact that Education VI is a game of certainty and Insurance Game III is not is also unimportant.

What then is the difference between the two games? The main difference is that in Education VI, it is possible to conceive of education apart from the wage contract, whereas in Insurance Game III the signal is communicated by the choice of the insurance contract. The inefficiency is more striking in the screening game, where the cost of communication is distinct from the act of accepting a particular contract.

Wilson Equilibrium and Reactive Equilibrium

As in Insurance Game III, it is possible to go beyond Nash equilibrium to find an equilibrium of some other kind for Education VII. Is it reasonable to say that a pooling equilibrium could always be broken by a contract which draws away the *High's*? After the *Highs* departed, the old pooling contract, which would still soak up all the *Lows*, would be unprofitable. In figure 10.2, if C_5 is withdrawn after C_6 is offered, the *Lows* prefer C_6 to the zero they obtain from unemployment, and C_6 becomes a pooling contract. This is irrelevant to the question of whether C_5 is a Nash equilibrium, but it might lead one to doubt the wisdom of the equilibrium concept.

Under the concept of the **Wilson equilibrium** from section 9.5, the pooling equilibrium is legitimate, because an employer thinking about introducing the new equilibrium-breaking contract would realize that the new contract would be unprofitable once the old contract was withdrawn. **Reactive equilibrium** can also be applied, and generates a separating equilibrium. Under its reasoning, the separating equilibrium cannot be broken by a pooling contract, because the pooling contract would in turn be broken by a second separating contract. The Wilson C_5 and the reactive (C_3, C_4) are the two clear candidates for equilibrium in figure 10.2. Alternately, we could restructure the model so that the worker moves first—the assumption in sections 10.1 and 10.2. While avoiding the existence problem, that introduces the need to think about out-of-equilibrium beliefs, and it really models a different situation, in which workers cannot change their education in response to employers' contracts.

A Summary of the Education Models

Because of signalling's complexity, most of this chapter has been devoted to elaboration of the education model. We began with Education I, which showed how with two types and two signal levels the perfect Bayesian equilibrium could be either separating or pooling. Education II took the same model and replaced the specification of out-of-equilibrium beliefs with an additional move by Nature, while Education III changed the parameters in Education I to increase the difference between the types and to show how signalling could continue with pooling. Education IV changed Education I by allowing a continuum of education levels, which resulted in a continuum of inefficient equilibria, each with a different signal level. After a purely verbal discussion of how to apply signalling models, we looked at screening, in which the employer moves first. Education V was a screening reprise of Education I, while Education VI broadened the model to allow a continuous signal, which eliminates pooling equilibria. Education VII modified the parameters of Education VI to show that sometimes no pure-strategy Nash equilibrium exists at all.

Throughout it was implicitly assumed that all the players were risk neutral. Risk neutrality is unimportant, because there is no uncertainty in the model and the agents bear no risk. If the workers were risk averse and they differed in their degrees of risk aversion, the contracts could try to use the difference to support a separating equilibrium because willingness to accept risk might act as a signal. If the principal were risk averse he might offer a wage less than the average productivity in the pooling equilibrium, but he is under no risk at all in the separating equilibrium, because it is fully revealing. The models are also games of certainty, and this too is unimportant. If output were uncertain, agents would just make use of the expected payoffs rather than the raw payoffs and very little would change.

We could extend the education models further—allowing more than two levels of ability would be a high priority—but instead, let us turn to the financial markets and look graphically at a model with two continuous characteristics of type and two continuous signals.

10.5 Two Signals: Game of Underpricing New Stock Issues

One signal might not be enough when there is not one but two characteristics of an agent that he wishes to communicate to the principal. This has been generally analyzed in Engers (1987), and multiple signal models have been especially popular in financial economics, with models of warranty issue by Matthews & Moore (1981) and of the role of investment bankers in new stock issues by Hughes (1986). We will use a model of initial public offerings of stock as the example in this section.

Empirically, it has been found that companies consistently issue stock at a price so low that it rises sharply in the days after the issue, an abnormal return estimated to average 11.4 percent (Copeland & Weston [1988], p. 377). The game of Underpricing New Stock Issues tries to explain this using the percentage of the stock retained by the original owner and the amount of underpricing as two signals. The two characteristics being signalled are the mean of the value of the new stock, which is of obvious concern to the potential buyers, and the variance, the importance of which will be explained later.

Underpricing New Stock Issues

(Grinblatt & Hwang [1989])

Players

The entrepreneur and many investors.

Order of Play

(See Figure 2.3a for a time line.)

(0) Nature chooses the expected value (μ) and variance (σ^2) of a share of the firm using some distribution F.
(1) The entrepreneur retains fraction α of the stock and offers to sell the rest at a price per share of P_0.
(2) The investors decide whether to accept or reject the offer.
(3) The market price becomes P_1, the investors' estimate of μ.
(4) Nature chooses the value V of a share using some distribution G such that μ is the mean of V and σ^2 is the variance. With probability θ, V is revealed to the investors and becomes the market price.
(5) The entrepreneur sells his remaining shares at the market price.

Payoffs

$$\pi_{entrepreneur} = U([1-\alpha]P_0 + \alpha[\theta V + (1-\theta)P_1]), \text{ where } U' > 0 \text{ and } U'' < 0.$$

$$\pi_{investors} = (1-\alpha)(V - P_0) + \alpha(1-\theta)(V - P_1).$$

The entrepreneur's payoff is the utility of the value of the shares he issues at P_0 plus the value of those he sells later at the price P_1 or V. The investors' payoff is the true value of the shares they buy minus the price they pay.

Underpricing New Stock Issues subsumes the simpler model of Leland & Pyle (1977), in which σ^2 is common knowledge and in which if the entrepreneur chooses to retain a large fraction of the shares, the investors deduce that the stock value is high. The one signal in that model is fully revealing because holding a larger fraction exposes the undiversified entrepreneur to a larger amount of risk, which he is unwilling to accept unless the stock value is greater than investors would guess without the signal.

If the variance of the project is high, that also increases the risk to the undiversified entrepreneur, which is important even though the investors are risk neutral and do not care directly about the value of σ^2. Since the risk is greater when variance is high, the signal α is more effective; retaining a smaller amount allows the entrepreneur to sell the remainder at the same price as a larger amount for a lower-variance firm. Even though the investors are diversified and do not care directly about firm-specific risk, they are interested in the variance because it tells them something about the effectiveness of entrepreneur-retained shares as a signal of share value. Figure 10.3 shows the signalling schedules for two variance levels.

In the game of Underpricing New Stock Issues, σ^2 is not known to the investors, so the signal is no longer fully revealing. An α equal to 0.1 could mean either that the firm has a low value with low variance, or a high value with high vari-

Figure 10.3 How the Signal Changes with the Variance

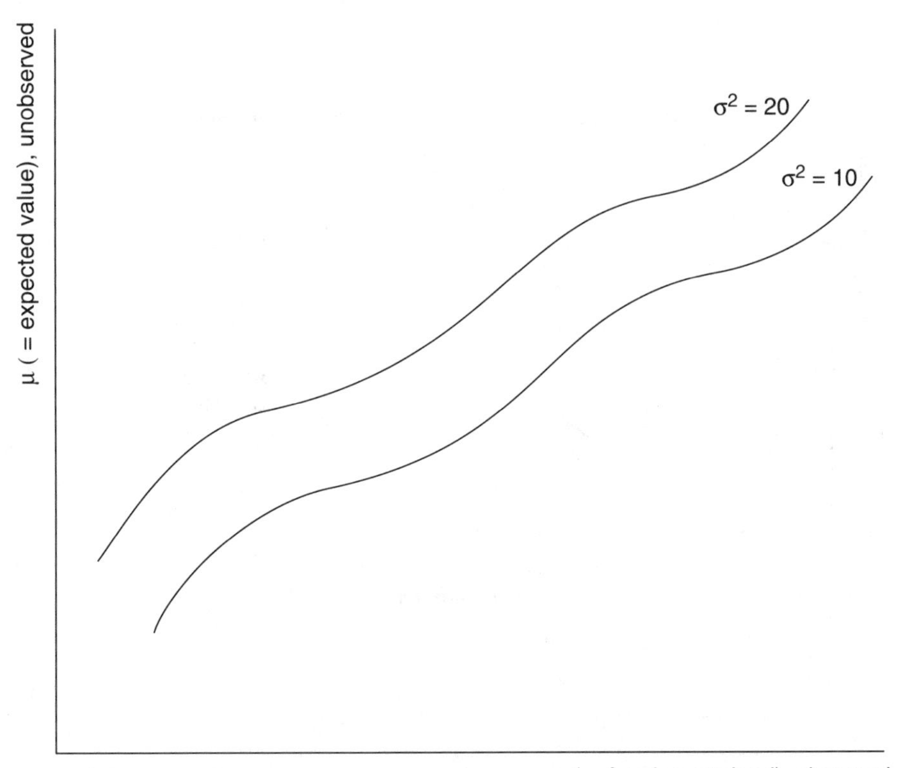

ance. But the entrepreneur can use a second signal, the price at which the stock is issued, and by observing α and P_0, the investors can deduce μ and σ^2.

Using particular numbers for concreteness, consider the entrepreneur's alternatives if he decides to retain $\alpha = 0.1$. He could offer the stock without a discount, at $P_0 = 90$, in which case the price would not rise afterwards, or he could offer a discount, setting $P_0 = 80$, after which it would rise to $P_1 = \mu = 120$. In discounting, he would not do so well on the fraction $1 - \alpha$ that he sold initially, but when in move (5) he sold the fraction he retained, he would do better on average. If $\mu = 90$, the entrepreneur would not underprice, because he would only get 80 in the initial sale, and with probability θ the true value would be revealed by Nature and the retained shares could only be sold at an expected price of 90. If the variance of the stock were lower, the entrepreneur would pick a smaller discount and be willing to hold a larger fraction. Figure 10.4 shows the different combinations of initial price and fraction retained that might be used.

This model explains why new stock is issued at a low price. The entrepreneur knows that the price will rise, but only if he issues it at a low initial price to show that the variance is high. The price discount shows that signalling by holding a large fraction of stock is unusually costly, but he is nonetheless willing to signal. The discount is costly because he is selling stock at less than its true value, and retaining stock is costly because he bears extra risk, but both are necessary to signal that the stock is valuable.

Figure 10.4 Different Ways of Signalling a Given μ

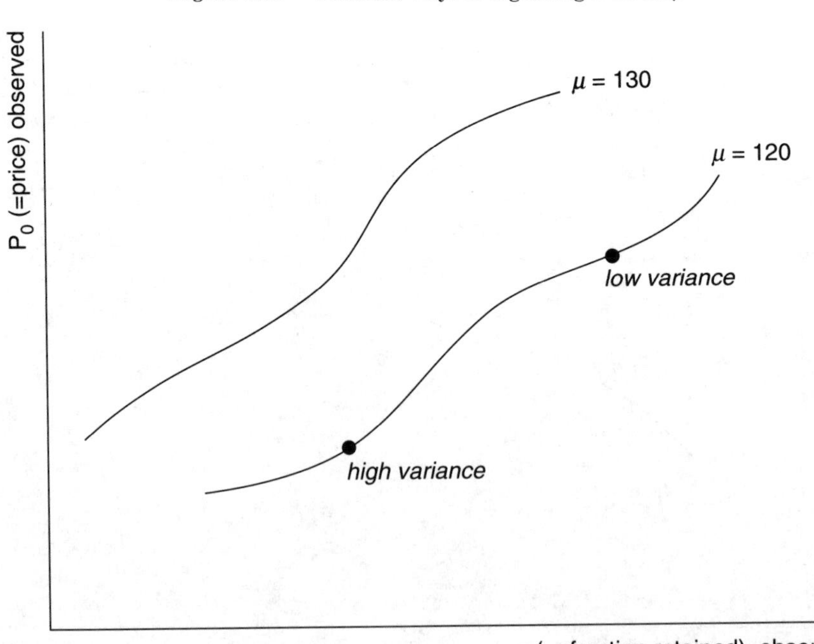

Notes

N10.1 The Informed Player Moves First: Signalling

- The term "signalling" was introduced by Spence (1973). The games in this book take advantage of hindsight to build simpler and more rational models of education than in his original article, which used a rather strange equilibrium concept: a strategy profile from which no worker has incentive to deviate and under which the employer's profits are 0. Under that concept, the firm's incentives to deviate are irrelevant.

  The distinction between signalling and screening has been attributed to Stiglitz & Weiss (1989). The literature has shown wide variation in the use of both terms, and "signal" is such a useful word that it is often used in models that have no signalling of the kind discussed in this chapter.
- One convention sometimes used in signalling models is to call the signalling player (the agent), the **sender** and the player signalled to (the principal), the **receiver**.
- The applications of signalling are too many to properly list. A few examples are the use of prices in C. Wilson (1980) and Stiglitz (1987), the payment of dividends in Ross (1977), bargaining (section 11.5), and greenmail (section 15.2). Banks (1990) has written a short book surveying signalling models in political science. Empirical papers include Layard & Psacharopoulos (1974) on education and Staten & Umbeck (1986) on occupational diseases.
- Legal bargaining is one area of application for signalling. See Grossman & Katz (1983). Reinganum (1988) has a nice example of the value of precommitment in legal signalling. In her model, a prosecutor who wishes to punish the guilty and release the innocent wishes, if parameters are such that most defendants are guilty, to commit to a pooling strategy in which his plea bargaining offer is the same whatever the probability that a particular defendant would be found guilty.
- The peacock's tail may be a signal. Zahavi (1975) suggests that a large tail may benefit the peacock because, by hampering him, it demonstrates to potential mates that he is fit enough to survive even with a handicap.
- **Advertising**
  Advertising is a natural application for signalling. The literature includes Nelson (1974), written before signalling was well-known, Kihlstrom & Riordan (1984) and Milgrom & Roberts (1986). I will briefly describe a model based on Nelson's. Firms are of one of two types, low-quality or high-quality. Consumers do not know that a firm exists until they receive an advertisement from it, and they do not know its quality until they buy its product. They are unwilling to pay more than zero for low quality, but any product is costly to produce. This is not a reputation model, because it is finite in length and quality is exogenous.

  If the cost of an advertisement is greater than the profit from one sale, but less than the profit from repeat sales, then high rates of advertising are associated with high product quality. A firm with low quality would not advertise, but a firm with high quality would.

  The model can work even if consumers do not understand the market and do not make rational deductions from the firm's incentives, so it does not have to be a signalling model. If consumers react passively and sample the product of any firm from whom they receive an advertisement, it is still true that the high-quality firm advertises more, because the customers it attracts become repeat customers. If consumers do understand the firms' incentives, signalling reinforces the result. Consumers know that firms which advertise must have high quality, so they are willing to try them. This understanding is important, because if consumers knew that 90 percent of firms were low-quality but did not understand that only high-quality firms advertise, they would not respond to the advertisements which they received. This should bring to mind section 6.2's game of PhD Admissions.
- If there are just two workers in the population, the model is different depending on whether:
  (1) Each is *High* ability with objective probability 0.5, so possibly both are *High* ability;
  or
  (2) One of them is *High* and the other is *Low*, so only the subjective probability is 0.5.

  The outcomes are different because in case (2) if one worker credibly signals he is *High* ability, the employer knows the other one must be *Low* ability.

Problems

10.1: Is Lower Ability Better?

Change Education I so that the two possible worker abilities are $a \in \{1, 4\}$.

(10.1a) What are the equilibria of this game? What are the payoffs of the workers (and the payoffs averaged across workers) in each equilibrium?

(10.1b) Apply the Intuitive Criterion (see N6.2). Are the equilibria the same?

(10.1c) What happens to the equilibrium worker payoffs if the high-ability is 5 instead of 4?

(10.1d) Apply the Intuitive Criterion to the new game. Are the equilibria the same?

(10.1e) Could it be that a rise in the maximum ability reduces the average worker's payoff? Can it hurt all the workers?

10.2: Productive Education and Nonexistence of Equilibrium.

Change Education I so that the two equally likely abilities are $a_L = 2$ and $a_H = 5$ and education is productive: the payoff of the employer whose contract is accepted is $\pi_{employer} = a + 2s - w$. The worker's utility function remains $U = w - 8s/a$.

(10.2a) Under full information, what are the wages for educated and uneducated workers of each type, and who acquires education?

(10.2b) Show that with incomplete information the equilibrium is unique (except for beliefs and wages out of equilibrium) but unreasonable.

10.3: Price and Quality.

Consumers have prior beliefs that Apex produces low-quality goods with probability 0.4 and high-quality with probability 0.6. A unit of output costs 1 to produce in either case, and it is worth 10 to the consumer if it is high-quality and 0 if low-quality. The consumer, who is risk neutral, decides whether to buy in each of two periods, but he does not know the quality until he buys. There is no discounting.

(10.3a) What is Apex' price and profit if it must choose one price, p^*, for both periods?

(10.3b) What is Apex' price and profit if it can choose two prices, p_1 and p_2, for the two periods, but it cannot commit ahead to p_2?

(10.3c) What is the answer to part (b) if the discount rate is $r = 0.1$?

(10.3d) Returning to $r = 0$, what if Apex can commit to p_2?

(10.3e) How do the answers to (a) and (b) change if the probability of low quality is 0.95 instead of 0.4? (There is a twist to this question.)

10.4: Signalling with a Continuous Signal.

Suppose that with equal probability a worker's ability is $a_L = 1$ or $a_H = 5$, and that the worker chooses any amount of education $y \in [0, \infty)$. Let $U_{worker} = w - 8y/a$ and $\pi_{employer} = a - w$.

(10.4a) There is a continuum of pooling equilibria, with different levels of y^*, the amount of education necessary to obtain the high wage. What education levels, y^*, and wages, $w(y)$, are paid in the pooling equilibria, and what is the set of out-of-equilibrium beliefs that supports them? What are the incentive compatibility constraints?

(10.4b) There is a continuum of separating equilibria, with different levels of y^*. What are the education levels and wages in the separating equilibria? Why are out-of-equilibrium beliefs needed, and what beliefs support the suggested equilibria? What are the self-selection constraints for these equilibria?

(10.4c) If you were forced to predict the one equilibrium which will be the one played out, which would it be?

10.5: Advertising.

Brydox introduces a new shampoo which is actually very good, but is believed by consumers to be good with only a probability of 0.5. A consumer would pay 10 for high quality and 0 for low quality, and the shampoo costs 6 per unit to produce. The firm may spend as much as it likes on stupid TV commercials showing happy people washing their hair, but the potential market consists of 100 cold-blooded economists who are not taken in by psychological tricks. The market can be divided into two periods.

(10.5a) If advertising is banned, will Brydox go out of business?

(10.5b) If there are two periods of consumer purchase, and consumers discover the quality of the shampoo if they purchase in the first period, show that Brydox might spend substantial amounts on stupid commercials.

(10.5c) What is the minimum and maximum that Brydox will spend on advertising, if it spends a positive amount?

Part III

Applications

11 Bargaining

The Basic Bargaining Problem: Splitting a Pie

Part III of this book is designed to stretch your muscles by providing a large number of applications of the techniques from Parts I and II. The next five chapters lack the unity of the earlier chapters in two ways: they may be read in any order, and they hit only the highlights of each topic. Moreover, while chapter 11 (bargaining) and chapter 12 (auctions) cover reasonably well-defined areas of research, chapters 13, 14, and 15 are selections of models from the sprawling literature of industrial organization. What the chapters have in common is that they use new theory to answer old questions.

Bargaining theory attacks a kind of price determination ill-described by standard economic theory. In markets with many participants on one side or the other, standard theory does a good job of explaining prices. In competitive markets we find the intersection of the supply and demand curves, while in markets monopolized on one side we find the monopoly or monopsony output. The problem is when there are few players on each side. Early on in one's study of economics, one learns that under bilateral monopoly (one buyer and one seller), standard economic theory is inapplicable because the traders must bargain. In the chapters on asymmetric information we would have come across this repeatedly except for our assumption that either the principal or the agent faced competition.

Sections 11.1 and 11.2 introduce the archetypal bargaining problem, Splitting a Pie, ever more complicated versions of which make up the rest of the chapter. Section 11.2, where we take the original rules of the game and apply the Nash bargaining solution, is our one dip into cooperative game theory in this book. Section 11.3 looks at bargaining as a finitely repeated process of offers and counteroffers, and section 11.4 views it as an infinitely repeated process. Section 11.5 returns to a finite number of repetitions (two, in fact), but with incomplete information.

Splitting a Pie

Players

Smith and Jones.

Order of Play

The players choose shares θ_s and θ_j of the pie simultaneously.

Payoffs

If $\theta_s + \theta_j \leq 1$, each player gets the fraction he chose: $\begin{cases} \pi_s = \theta_s. \\ \pi_j = \theta_j. \end{cases}$

If $\theta_s + \theta_j > 1$, then $\pi_s = \pi_j = 0$.

Splitting a Pie resembles the game of Chicken except that it has a continuum of Nash equilibria: any strategy profile (θ_s, θ_j) such that $\theta_s + \theta_j = 1$ is Nash. The Nash concept is at its worst here, because the assumption that the equilibrium being played is common knowledge is very strong when there is a continuum of equilibria. The idea of the focal point (section 1.5) might help to choose a single Nash equilibrium. The strategy space of Chicken is discrete and it has no symmetric pure-strategy equilibrium, but the strategy space of Splitting a Pie is continuous, which permits a symmetric pure-strategy equilibrium to exist. That equilibrium is the even split, (0.5, 0.5), which is a focal point.

If the players moved in sequence, the game would have a tremendous first-mover advantage. If Jones moved first, the unique Nash outcome would be (0, 1), although only weakly, because Smith would be indifferent as to his action. (This is the same open-set problem that was discussed in section 4.3.) Smith accepts the offer if he chooses θ_s to make $\theta_s + \theta_j = 1$, but if we added only an epsilon worth of ill-will to the model, he would pick $\theta_s > 0$ and reject the offer.

In many applications this version of the game of Splitting a Pie is unacceptably simple, because if the two players find their fractions add to more than 1 they have a chance to change their minds. In labor negotiations, for example, if manager Jones makes an offer which union Smith rejects, they do not immediately forfeit the gains from combining capital and labor. They lose a week's production and make new offers. The recent trend in research has been to model such a sequence of offers, but before we do that let us see how cooperative game theory deals with the original game.

11.2 The Nash Bargaining Solution

When game theory was young and games were static, a favorite approach was to decide upon some characteristics an equilibrium should have based on notions of fairness or efficiency, then mathematicize the characteristics, and maybe add a few other axioms to make the equilibrium turn out neatly. Nash (1950a) did this for the bargaining problem in what is perhaps the best known application of cooperative game theory. Nash's objective was to pick axioms that would characterize the agreement the two players would anticipate making with each other. He used a game only a little more complicated than Splitting a Pie. In the Nash model, the two players can have different utilities if they do not come to an agreement, and the utility functions can be nonlinear in terms of shares of the pie. Figures 11.1a and 11.1b compare the two games.

Figure 11.1 (a) Splitting a Pie; (b) Nash Bargaining Game

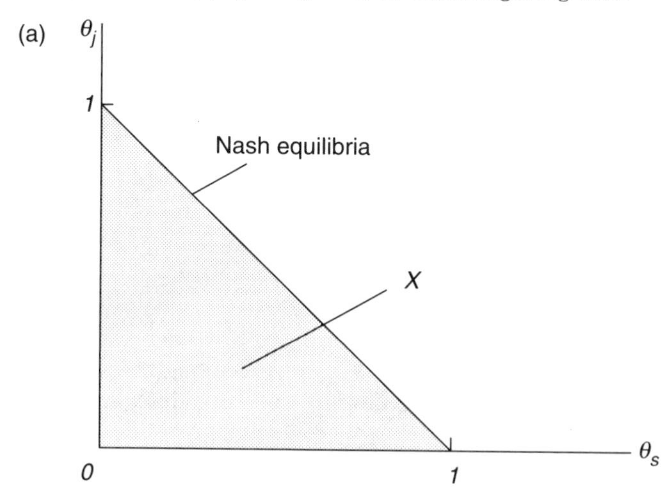

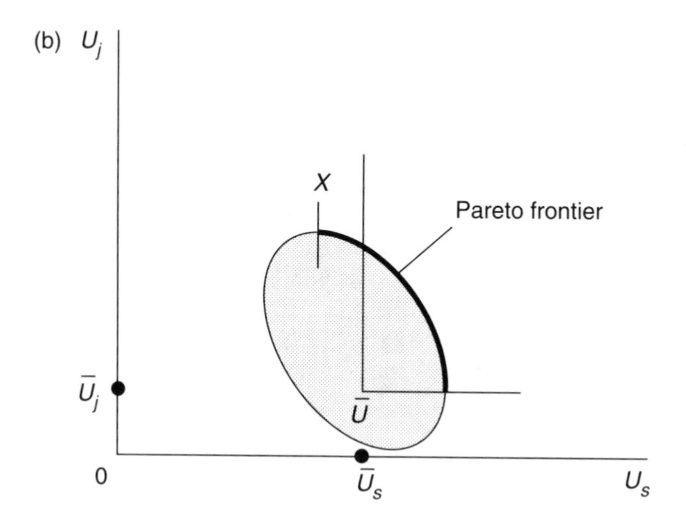

In Figure 11.1, the shaded region denoted by X is the set of feasible payoffs, which we will assume to be convex. The disagreement point is $\bar{U} = (\bar{U}_s, \bar{U}_j)$. The Nash bargaining solution, $U^* = (U_s^*, U_j^*)$, is a function of $\bar{U}$ and X. The axioms that generate the concept are as follows:

(1) Invariance.

For any strictly increasing linear function F,

$$U^*[F(\bar{U}), F(X)] = F[U^*(\bar{U}, X)]. \tag{1}$$

This says that the solution is independent of the units in which utility is measured.

(2) Efficiency.

The solution is Pareto optimal, so that not both players can be made better off. In mathematical terms,

$$(U_s, U_j) > U^* \Rightarrow (U_s, U_j) \notin X. \tag{2}$$

(3) Independence of Irrelevant Alternatives.

If we drop some possible utility profiles from X, leaving the smaller set Y, then, if U^* was not one of the dropped points, U^* does not change.

$$U^*(\bar{U}, X) \in Y \subseteq X \Rightarrow U^*(\bar{U}, Y) = U^*(\bar{U}, X). \tag{3}$$

(4) Anonymity (or Symmetry).

Switching the labels on players Smith and Jones does not affect the solution.

The axiom of Independence of Irrelevant Alternatives is the most debated of the four, but if I were to complain, it would be about the axiomatic approach, which depends heavily on the intuition behind the axioms. Everyday intuition says that the outcome should be efficient and symmetric, so that other outcomes can be ruled out *a priori*. But most of the games in the earlier chapters of this book turn out to have reasonable but inefficient outcomes, and games like Chicken have reasonable asymmetric outcomes.

Whatever their drawbacks, these axioms fully characterize the Nash solution. It can be proven that if U^* satisfies the four axioms above, then it is the unique strategy profile such that

$$U^* = \underset{U \in X, U \geq \bar{U}}{Argmax} \, (U_s - \bar{U}_s)(U_j - \bar{U}_j). \tag{4}$$

Splitting a Pie is a simple enough game so that not all the axioms are needed to generate a solution. If we put the game in this context, however, problem (4) becomes

$$\underset{\theta_s, \theta_j | \theta_s + \theta_j \leq 1}{Maximize} (\theta_s - 0)(\theta_j - 0), \tag{5}$$

which generates the first order conditions

$$\theta_s - \lambda = 0, \text{ and } \theta_j - \lambda = 0, \tag{6}$$

where λ is the Lagrange multiplier on the constraint. From (6) and the constraint, we obtain $\theta_s = \theta_j = 1/2$, the even split that we found as a focal point of the noncooperative game.

Although Nash's objective was simply to characterize the anticipations of the players, I perceive a heavier note of morality in cooperative game theory than in noncooperative game theory. Cooperative outcomes are neat, fair, beautiful, and efficient. In the next few sections we will look at noncooperative bargaining models that, while plausible, lack every one of those features. Cooperative game theory may be useful for ethical decisions, but its attractive features are often inappropriate for economic situations, and the spirit of the axiomatic approach is very different from the utility maximization of economic theory.

It should be kept in mind, however, that the ethical component of cooperative game theory can also be realistic, because people are often ethical, or pretend to be. People very often follow the rules they believe represent virtuous behavior, even at some monetary cost. In bargaining experiments in which one player is given the ability to make a take-it-or-leave it offer, it is very commonly found that he offers a 50-50 split. Presumably this is because he either wishes to be fair or he fears a spiteful response from the other player to a smaller offer. If the subjects are made to feel that they had "earned" the right to be the offering party, they behave much more like the players in noncooperative game theory (Hoffman & Spitzer [1985]). Frank (1988) and Thaler (1992) describe numerous occasions where simple games fail to describe real-world or experimental results. People's payoffs include more than their monetary rewards, and sometimes knowing the cultural disutility of actions is more important than knowing the dollar rewards. This is one reason why it is helpful to a modeller to keep his games simple: when he actually applies them to the real world, the model must not be so unwieldy that he cannot combine it with his knowledge of the particular setting.

11.3 Alternating Offers over Finite Time

In the games of the next two sections, the actions are the same as in Splitting a Pie, but with many periods of offers and counteroffers. This means that strategies are no longer just actions, but rather rules for choosing actions based on the actions chosen in earlier periods.

Alternating Offers

Players

Smith and Jones.

Order of Play

(1) Smith makes an offer θ_1.
(1*) Jones accepts or rejects.
(2) Jones makes an offer θ_2.
(2*) Smith accepts or rejects.
...
(T) Smith offers θ_T.
(T*) Jones accepts or rejects.

Payoffs

The discount factor is $\delta \leq 1$.
If Smith's offer is accepted by Jones in round m,

$$\pi_s = \delta^m \theta_m,$$
$$\pi_j = \delta^m (1 - \theta_m).$$

(If Jones' offer is accepted, reverse the subscripts.)
If no offer is ever accepted, both payoffs equal zero.

When a game has many rounds we need to decide whether discounting is appropriate (see section 4.5). Recall that if the discount rate is r then the discount factor is $\delta = 1/(1 + r)$, so, without discounting, $r = 0$ and $\delta = 1$. Whether discounting is appropriate to the situation being modelled depends on whether delay should matter to the payoffs because the bargaining occurs over real time (section 2.2) or the game might suddenly end (section 5.2). The game of Alternating Offers can be interpreted in either of two ways, depending on whether or not it occurs over real time. If the players made all the offers and counteroffers between dawn and dusk of a single day, discounting would be inconsequential because, essentially, no time had passed. If each offer consumed a week of time, on the other hand, the delay before the pie was finally consumed would be important to the players and their payoffs should be discounted.

Consider first the game without discounting. There is a unique subgame perfect outcome—Smith gets the entire pie—which is supported by a number of different equilibria. In each equilibrium, Smith offers $\theta_s = 1$ in each period, but each equilibrium is different in terms of when Jones accepts the offer. All of them are weak equilibria because Jones is indifferent between accepting and rejecting, and they differ only in the timing of Jones' final acceptance.

Smith owes his success to his ability to make the last offer. When Smith claims the entire pie in the last period, Jones gains nothing by refusing to accept. What we have here is not really a first-mover advantage, but a last-mover advantage in offering, a difference not apparent in the one-period model.

In the game with discounting, the total value of the pie is 1 in the first period, δ in the second, and so forth. In period T, if it is reached, Smith would offer 0 to Jones, keeping 1 for himself, and Jones would accept under our assumption on indifferent players. In period $T - 1$, Jones could offer Smith δ, keeping $(1 - \delta)$ for himself, and Smith would accept, although he could receive a greater share by refusing, because that greater share would arrive later and be discounted.

By the same token, in period $T - 2$, Smith would offer Jones $\delta(1 - \delta)$, keeping $1 - \delta(1 - \delta)$ for himself, and Jones would accept, since with a positive share Jones also prefers the game to end soon. In period $T - 3$, Jones would offer Smith $\delta[1 - \delta(1 - \delta)]$, keeping $1 - \delta[1 - \delta(1 - \delta)]$ for himself, and Smith would accept, again to prevent delay. Table 11.1 shows the progression of Smith's shares when $\delta = 0.9$.

Table 11.1 Alternating Offers over Finite Time

Round	Smith's share	Jones' share	Total value	Offerer
$T - 3$	0.819	0.181	0.9^{T-4}	Jones
$T - 2$	0.91	0.09	0.9^{T-3}	Smith
$T - 1$	0.9	0.1	0.9^{T-2}	Jones
T	1	0	0.9^{T-1}	Smith

As we work back from the end, Smith always does a little better when he makes the offer than when Jones does, but if we consider just the class of periods in which Smith makes the offer, Smith's share falls. If we were to continue to work back for a large number of periods, Smith's offer in a period in which he makes the offer would approach $1/(1 + \delta)$, which equals about 0.53 if $\delta = 0.9$. The reasoning behind that precise expression is given in the next section. In equilibrium, the very first offer would be accepted, since it is chosen precisely so that the other player can do no better by waiting.

11.4 Alternating Offers over Infinite Time

The Folk Theorem of section 5.2 says that when discounting is low and a game is repeated an infinite number of times, there are many equilibrium outcomes. That does not apply to the bargaining game, however, because it is not a repeated game. It ends when one player accepts an offer, and only the accepted offer is relevant to the payoffs, and not the earlier proposals. In particular, there are no out-of-equilibrium punishments such as enforce the Folk Theorem's outcomes.

Let players Smith and Jones have discount factors of δ_s and δ_j which are not necessarily equal, but are strictly positive and no greater than one. In the unique subgame perfect outcome for the infinite-period bargaining game, Smith's share is

$$\theta_s = \frac{1 - \delta_j}{1 - \delta_s \delta_j}, \tag{7}$$

which, if $\delta_s = \delta_j = \delta$, is equivalent to

$$\theta_s = \frac{1}{1 + \delta}. \tag{8}$$

If the discount rate is high, Smith gets most of the pie: a 1,000 percent discount rate ($r = 10$) makes $\delta = 0.091$ and $\theta_s = 0.92$ (rounded), which makes sense, since under such extreme discounting the second period hardly matters and we are almost back to the simple game of section 11.1. At the other extreme, if r is small, the pie is split almost evenly: if $r = 0.01$, then $\delta \approx 0.99$ and $\theta_s \approx 0.503$.

It is crucial that the discount rate be strictly greater than 0, even if only by a little. Otherwise, the game has the same continuum of perfect equilibria as in section 11.1. Since nothing changes over time, there is no incentive to come to an early agreement. When discount rates are equal, the intuition behind the result is that since a player's cost of delay is proportional to his share of the pie, if Smith were to offer a grossly unequal split, such as (0.7, 0.3), Jones, with less to lose by delay, would reject the offer. Only if the split is close to even would Jones accept, as we will now prove.

Proposition 11.1 (Rubinstein [1982])

In the discounted infinite game, the unique perfect equilibrium outcome is $\theta_s = (1 - \delta_j)/(1 - \delta_s\delta_j)$, *where Smith is the first mover.*

Proof

We found that in the T-period game Smith gets a larger share in a period in which he makes the offer. Denote by M the maximum nondiscounted share, taken over all the perfect equilibria that might exist, that Smith can obtain in a period in which he makes the offer. Consider the game starting at t. Smith is sure to get no more than M, as noted in table 11.2. (Jones would thus get 1 − M, but that is not relevant to the proof.)

Table 11.2 Alternating Offers over Infinite Time

Round	Smith's share	Jones' share	Offerer
$T - 2$	$1 - \delta_j(1 - \delta_s M)$		Smith
$T - 1$		$1 - \delta_s M$	Jones
T	M		Smith

The trick is to find a way besides M to represent the maximum Smith can obtain. Consider the offer made by Jones at $t - 1$. Smith will accept any offer which gives him more than the discounted value of M received one period later, so Jones can make an offer of $\delta_s M$ to Smith, retaining $1 - \delta_s M$ for himself. At $t - 2$, Smith knows that Jones will turn down any offer less than the discounted value of the minimum Jones can look forward to receiving at $t - 1$. Smith, therefore, cannot offer any less than $\delta_j(1 - \delta_s M)$, or retain for himself any more than $1 - \delta_j(1 - \delta_s M)$ at $t - 2$.

Now we have two expressions for "the maximum which Smith can receive," which we can set equal to each other:

$$M = 1 - \delta_j(1 - \delta_s M). \tag{9}$$

Solving equation (9) for M, we obtain

$$M = \frac{1 - \delta_j}{1 - \delta_s \delta_j}. \tag{10}$$

We can repeat the argument using m, the minimum of Smith's share. If Smith can expect at least m at t, Jones cannot receive more than $1 - \delta_s m$ at $t - 1$. At $t - 2$ Smith knows that if he offers Jones the discounted value of that amount, Jones will accept, so Smith can guarantee himself $1 - \delta_j(1 - \delta_s m)$, which is the same as the expression we found for M. The smallest perfect equilibrium share that Smith can receive is the same as the largest, so the equilibrium outcome must be unique.

No Discounting, But a Fixed Bargaining Cost

There are two ways to model bargaining costs per period: as proportional to the remaining value of the pie (the way used above), or as fixed costs in each period, which is analyzed next (again following Rubinstein [1982]). To understand the difference, think of labor negotiations during a construction project. If a strike slows down completion, there are two kinds of losses. One is the loss from delay in renting or selling the new building, a loss proportional to its value. The other is the loss from late-completion penalties in the contract, which often take the form of a fixed penalty each week. The two kinds of costs have very different effects on the bargaining process.

To represent this second kind of cost, assume that there is no discounting, but whenever Smith or Jones makes an offer, he incurs the cost c_s or c_j. In every subgame perfect equilibrium, Smith makes an offer and Jones accepts, but there are three possible cases.

(1) Delay Costs are Equal

$$c_s = c_j = c.$$

The Nash indeterminacy of section 11.1 remains almost as bad; any fraction such that each player gets at least c is supported by some perfect equilibrium.

(2) Delay Hurts Jones More

$$c_s < c_j.$$

Smith gets the entire pie. Jones has more to lose than Smith by delaying, and delay does not change the situation except by diminishing the wealth of the players. The game is stationary, because it looks the same to both players no matter how many periods have already elapsed. If in any period t Jones offered Smith x, in period $(t - 1)$ Smith could offer Jones $(1 - x - c_j)$, keeping $(x + c_j)$ for himself. In period $(t - 2)$, Jones would offer Smith $(x + c_j - c_s)$, keeping $(1 - x - c_j + c_s)$ for himself, and in periods $(t - 4)$ and $(t - 6)$ Jones would offer $(1 - x - 2c_j + 2c_s)$ and $(1 - x - 3c_j + 3c_s)$. As we work backwards, Smith's advantage rises to $\gamma(c_j - c_s)$ for an arbitrarily large integer γ. Looking ahead from the start of the game, Jones is willing to give up and accept zero.

(3) Delay Hurts Smith More

$$c_s > c_j.$$

Smith gets a share worth c_j and Jones gets $(1 - c_j)$. The cost c_j is a lower bound on the share of Smith, the first mover, because if Smith knows Jones will offer (0, 1) in the second period, Smith can offer $(c_j, 1 - c_j)$ in the first period and Jones will accept.

11.5 Incomplete Information

Instant agreement has characterized even the multiperiod games of complete information discussed so far. Under incomplete information, knowledge can change over the course of the game and bargaining can last more than one period in equilibrium, a result that might be called inefficient but is certainly realistic. Models with complete information have difficulty explaining such things as strikes or wars, but if over time an uninformed player can learn the type of the informed player by observing what offers are made or rejected, such unfortunate outcomes can arise. The literature on bargaining under incomplete information is vast. For this section, I have chosen to use a model based on the first part of Fudenberg & Tirole (1983), but it is only a particular example of how one could construct such a model, and not a good indicator of what results are to be expected from bargaining.

Bargaining with Incomplete Information

Players

A seller, and a buyer called Buyer_{100} or Buyer_{150} depending on his type.

Order of Play

(0) Nature picks the buyer's type, his valuation of the object being sold, which is $b = 100$ with probability γ and $b = 150$ with probability $(1 - \gamma)$.
(1) The seller offers price p_1.
(2) The buyer accepts or rejects p_1 (acceptance ends the game).
(3) The seller offers a second price p_2.
(4) The buyer accepts or rejects p_2.

Payoffs

$$\pi_{seller} = \begin{cases} p_1 & \text{if } p_1 \text{ is accepted} \\ \delta p_2 & \text{if } p_2 \text{ is accepted} \\ 0 & \text{if no offer is accepted} \end{cases}$$

$$\pi_{buyer} = \begin{cases} b - p_1 & \text{if } p_1 \text{ is accepted} \\ \delta(b - p_2) & \text{if } p_2 \text{ is accepted} \\ 0 & \text{if no offer is accepted} \end{cases}$$

If p_2 is accepted, the buyer's payoff is $\delta(b - p_2)$, rather than $\delta b - p_2$, because the present value of cash paid in the second period is less than that of cash paid in the first period. Consumption in the second period provides less pleasure, but payment provides less pain. Let us set $\delta = 0.9$ for the numerical computations.

As described, the game tells of an encounter between one buyer and one seller. Another interpretation is that the game is between a continuum of buyers and one seller, in which case the analysis works out the same way, but the verbal translation is different. As noted in section 3.2, models with a continuum of buyers are sometimes easier to understand, because the two types of buyers can be interpeted as different fractions of the population, and mixed strategies can be interpreted as fractions of the total population using different pure strategies.

If this were a game of complete information, then in equilibrium the seller, who has the last-mover advantage, would choose $p_1 = 100$ or $p_1 = 150$ and the buyer would accept. If the buyer did not accept, the seller would offer the same price again, and the buyer would accept that second offer. With incomplete information, however, the probability that the buyer is a $Buyer_{100}$ determines whether the equilibrium is pooling at a low price, or separating at a high price with some offers rejected.

The incomplete information game is a screening model like those we saw in chapter 10, in the sense that the uninformed player moves first. Screening models with a continuum of types do not have pooling equilibria, but the assumption that there are only two types will permit one in this model. If the buyer made the offers, instead of the seller, this would be a signalling game, in which out-of-equilibrium beliefs would be more important. The game is not a pure screening model, however, because it lasts two periods and the uninformed player does have a chance to make his second move after the informed player has moved and possibly, thereby, has revealed information. This gives it the character of mixed screening and signalling.

A Case of Many Low-Valuation Buyers: $\gamma = 0.5$

We will start by assuming that $\gamma = 0.5$, so the probability is fairly high that the buyer has the valuation 100.

Equilibrium (Pooling)

In the first period, $p_1 = 100$, Buyer$_{100}$ accepts $p_1 \leq 100$, and Buyer$_{150}$ accepts $p_1 \leq 105$. In the second period, $p_2 = 100$, Buyer$_{100}$ accepts $p_2 \leq 100$, and Buyer$_{150}$ accepts $p_2 \leq 150$. The seller's beliefs out of equilibrium are that if a buyer rejects $p_1 = 100$, he is Buyer$_{100}$ with probability γ (passive conjectures). The outcome is that $p_1 = 100$ and the buyer accepts.

Let us test that this is a perfect Bayesian equilibrium. As always, work back from the end. Both types of buyers have a dominant strategy for the last move: accept any offer below b. Given the parameters, the seller should not raise p_2 above 100, because with probability $\gamma = 0.5$ he would lose a profit of 100 and his potential revenue is no greater than 150.

Buyer$_{150}$, although looking ahead to $p_2 = 100$, is willing to pay more than 100 in period one because of discounting. His payoff is the same from accepting $p_1 = 105$ as from accepting $p_2 = 100$, because his nominal surplus of 50 from accepting the lower price is discounted to a utility value of 45. Buyer$_{100}$, however, is never willing to pay more than 100, and discounting is irrelevant because he has no surplus to look forward to anyway.

The seller knows that even if $p_1 = 105$ he will still sell to Buyer$_{150}$, but if he tries that and finds no buyer in the first period, he has delayed receipt of his payment, which is discounted. Since $100 > 97.5$ $(= [1 - \gamma] \cdot [105] + \gamma \cdot \delta \cdot [100])$, the seller prefers the safe present price of 100 to the alternative of a gamble between a present 105 and a future 100.

Out-of-equilibrium beliefs are specified for the equilibrium, but they do not really matter. Whatever inference the seller may draw if the buyer refuses $p_1 = 100$, the inference never induces the buyer to change his actions. It might be that the seller believes a refusal indicates that the buyer's value to be 150, so $p_2 = 150$, but that does not change the buyer's incentive to accept $p_1 = 100$.

A Case with Few Low-Valuation Buyers: $\gamma = 0.05$

If the proportion of low-valuation buyers is as small as $\gamma = 0.05$, the equilibrium is separating and in mixed strategies.

Equilibrium (Separating, in Mixed Strategies)

In the first period, $p_1 = 150$, Buyer$_{100}$ accepts $p_1 \leq 100$, and Buyer$_{150}$ accepts p_1 with probability $m(p_1)$, where

$$\begin{cases} m = 1 & \text{if } p_1 \leq 105 \\ m = \alpha & \text{if } 105 < p_1 \leq 150 \text{ (where } 0 \leq \alpha \leq 0.89) \\ m = 0 & \text{if } p_1 > 150 \end{cases}$$

In the second period, $p_2 = 150$ if the seller believes that he faces a Buyer$_{100}$ with probability less than ⅓, and otherwise $p_2 = 100$. Buyer$_{100}$ accepts $p_2 \leq 100$, and Buyer$_{150}$ accepts $p_2 \leq 150$. The outcome is that $p_1 = 150$, which is sometimes accepted by Buyer$_{150}$; $p_2 = 150$, which is accepted by Buyer$_{150}$; and Buyer$_{100}$ never accepts an offer.

The observed outcome is simple—the price always stays at 150, and some buyers accept in each period while other buyers never accept—but the equilibrium strategies are quite complicated. As we will see, the equilibrium is not even fully determined, because the mixing probability α can take any of a continuum of values.

The strategies in the second-period game are simple enough. In the second period, the buyer accepts if the price is less than his valuation and the seller trades off a safe 100 against a gamble between 0 and 150. He is indifferent between them if

$$100 = 0 \cdot Prob(\text{Buyer}_{100}) + 150 \cdot [1 - Prob(Buyer_{100})], \tag{11}$$

which yields a critical value of $Prob(\text{Buyer}_{100}) = ⅓$. If neither type of buyer accepted first-period offers, the second-period belief would be $Prob(Buyer_{100}) = \gamma$, which we assumed to be 0.05, so the second-period price would be 150.

The first-period strategies are more complicated. The first-period strategy of Buyer$_{150}$ is not the pure strategy of accepting the offer $p_1 = 150$, because if he always accepted in the first period the seller would lower the price in the second period, knowing that a buyer who rejected the first-period offer must be a Buyer$_{100}$. Anticipating the price drop, Buyer$_{150}$ would refuse $p_1 = 150$, which contradicts the reason for the drop.

In equilibrium, it must be that after a refusal in the first period the seller puts a high enough probability on Buyer$_{150}$ that he decides to keep the price high in the second period. For the seller to want to keep $p_2 = 150$, the probability that a buyer who refused the first-period offer is a Buyer$_{150}$ must be at least ⅔, from equation (11). If both p_1 and p_2 are equal to 150, the buyer will be indifferent as to when he accepts, so he is willing to follow a mixed strategy. We can calculate the mixing probability $m(150)$ by finding the value that makes the seller willing to keep the price at 150 in the second period. The probability that a buyer is a Buyer$_{150}$ is equal to $1 - \gamma$ in the first period, but a Buyer$_{150}$ only rejects the first-period offer with probability $1 - m(150)$. Therefore, in the second period, using Bayes' rule,

$$Prob(\text{Buyer}_{100}) = \frac{\gamma}{\gamma + [1 - m(150)][1 - \gamma]}. \tag{12}$$

Plugging in $\gamma = 0.05$ and $Prob(\text{Buyer}_{100} = 1/3$, equation (12) can be solved to yield $m(150) = 0.89$ (rounded). The calculation has ensured that if the Buyer_{150} accepts the first offer with probability 0.89, the probability that a second-period buyer has valuation 100 is 1/3. The value $\alpha = 0.89$ is the maximum equilibrium probability that a Buyer_{150} refuses to buy in the first period, but a smaller value for α would support an equilibrium *a fortiori* since the probability that a refuser had valuation 150 would be even greater than 2/3.

There is a continuum of equilibria, which differ in terms of their values for α. Two values are focal points, 0 and 0.89. The value of 0 is a pure strategy, and has the attraction of simplicity. The value of 0.89 is Pareto efficient, because it is the highest equilibrium probability that the buyer accepts immediately, which avoids the lost utility from delay. The qualitative difference between the two equilibria is that in the pure-strategy equilibrium no buyers accept the offer in the first period.

Out-of-Equilibrium Behavior

The description just given describes only part of the separating equilibrium. A full description would specify each player's actions at each node of the game tree given the past history of the game, including out-of-equilibrium paths that start with deviations such as $p_1 = 140$. As we will see in describing the path starting with the seller deviation $p_1 = 140$, there is an overwhelming range of possible out-of-equilibrium behavior.

Consider what happens if the seller offers a price of 140 in the first period. For the same reasons as we described for $p_1 = 150$, the equilibrium cannot be in pure strategies. The equilibrium strategies in the out-of-equilibrium subgame are for the Buyer_{150} to mix between accepting and rejecting, and for the seller to mix between $p_2 = 100$ and $p_2 = 150$. Here, unlike the case of the equilibrium path, the seller must also mix, because otherwise the buyer would strongly prefer to accept 140 rather than wait for 150. The seller is willing to mix only if he believes that there is exactly a one-third probability that the buyer is a Buyer_{100}, so the buyer's strategy is $m(150) = 0.89$, as before. Denote the seller's mixing probability of $p_2 = 100$ by μ. It must take a value that makes the buyer indifferent between accepting and rejecting, so

$$150 - p_1 = .9\mu\,(150 - 100) + (1 - \mu) \cdot 0, \tag{13}$$

which solves to $\mu = 3\frac{1}{3} - \frac{p_1}{45}$, or $\mu = 0.22$ for $p_1 = 140$.

The consequences of this kind of deviation are unimportant on the equilibrium path. Out-of-equilibrium behavior is important when player Smith deviating might induce player Jones to change his behavior in a way that makes Smith's deviation profitable. This was the case in the entry deterrence games of chapter 4 that were used to illustrate perfectness: if the entrant deviated from a strategy profile in which he was supposed to stay out, the incumbent would change his fighting behavior to collusion, which would make the entrant's entry profitable.

Here, however, the seller deviating by charging a price of $p_1 = 140$ does not call a bluff in any way; it is simply a blunder. The complications in the equilibrium description do not arise from the buyer's immediate response, but in the seller's second-period response to his own deviation. Descriptions of equilibria are customarily incomplete because they do not specify a full strategy for each player, including his way of responding to his own deviations from equilibrium, but the off-equilibrium paths that are not described have no importance to the analysis.

These calculations have been intricate and hard, though they provide good examples of the kind of care that must be taken with out-of-equilibrium behavior. But the most important lesson of this model is that bargaining can lead to inefficiency. In the separating equilibrium, some of the Buyer_{150}'s delay their transactions until the second period, which is inefficient since the payoffs are discounted. Moreover, the Buyer_{100}'s never buy at all, and the potential gains from trade are lost.

A more technical conclusion is that the price the buyer pays depends heavily on the seller's equilibrium beliefs. If the seller thinks that the buyer has a high valuation with probability 0.5, the price is 100, but if he thinks the probability is 0.05, the price rises to 150. This implies that a buyer is unfortunate if he is part of a group which is believed to have high valuations more often; even if his own valuation is low, what we might call his bargaining power is low when he is part of a high-valuing group. Ayres (1991) found that when he hired testers to pose as customers at car dealerships, their success depended on their race and gender even though they were given identical predetermined bargaining strategies to follow. Since the testers did as badly even when faced with salesmen of their own race and gender, it seems likely that they were hurt by being members of groups that usually can be induced to pay higher prices.

Notes

N11.2 The Nash Bargaining Solution

- See Binmore, Rubinstein, & Wolinsky (1986) for a comparison of the cooperative and noncooperative approaches to bargaining. For overviews of cooperative game theory see Luce & Raiffa (1957) and Shubik (1982).
- While the Nash bargaining solution can be generalized to n players (see Harsanyi [1977], p. 196), the possibility of interaction between coalitions of players introduces new complexities. Solutions such as the Shapley value (Shapley [1953b] try to account for these complexities.

  The **Shapley value** satisfies the properties of Invariance, Anonymity, Efficiency, and Linearity in the variables from which it is calculated. Let S_i denote a **coalition** containing player i; that is, a group of players including i that makes a sharing agreement. Let $v(S_i)$ denote the sum of the values of the utilities of the players in coalition S_i, and $v(S_i - \{i\})$ the sum of the utilities in the coalition created by removing i from S_i. Finally, let $c(s)$ be the number of coalitions of size s containing player i. The Shapley value for player i is then

$$\phi_i = \frac{1}{n} \sum_{s=1}^{n} \frac{1}{c(s)} \sum_{S_i} [v(S_i) - v(S_i - \{i\})]. \tag{14}$$

  where the S_i are of size s. The motivation for the Shapley value is that player i receives the average of his marginal contributions to different coalitions that might form. Gul (1989) has provided a noncooperative interpretation.

N11.4 Alternating Offers over Infinite Time

- The proof of Proposition 11.1 is not from the original Rubinstein (1982), but is adapted from Shaked & Sutton (1984). The maximum rather than the supremum can be used because of the assumption that indifferent players always accept offers.
- In extending the game of Alternating Offers to three players, there is no obviously best way of specifying how players make and accept offers. Haller (1986) shows that for at least one specification, the outcome is not similar to the Rubinstein (1982) outcome, but rather is a return to the indeterminacy of the game without discounting.

N11.5. Incomplete Information

- Bargaining under asymmetric information has inspired a large literature. In early articles, Fudenberg & Tirole (1983) uses a two-period model with two types of buyers and two types of sellers. Sobel & Takahashi (1983) builds a model with either T or infinite periods, a continuum of types of buyers, and one type of seller. Cramton (1984) uses an infinite number of periods, a continuum of types of buyers, and a continuum of types of sellers. Rubinstein (1985a) uses an infinite number of periods, two types of buyers, and one type of seller, but the types of buyers differ not in their valuations, but in their discount rates. Rubinstein (1985b) puts emphasis on the choice of out-of-equilibrium conjectures. W. Samuelson (1984) looks at the case where one bargainer knows the size of the pie better than the other bargainer. Perry (1986) uses a model with fixed bargaining costs and asymmetric information in which each bargainer makes an offer in turn, rather than one offering and the other accepting or rejecting. For an update and overview, see the excellent survey of Kennan & R. Wilson (1993).
- The asymmetric information model in section 11.5 has **one-sided** asymmetry in the information: only the buyer's type is private information. Fudenberg & Tirole (1983) and others have also built models with **two-sided** asymmetry, in which buyers' and sellers' types are both private information. In such models a multiplicity of perfect Bayesian equilibria can be supported for a given set of parameter values. Out-of-equilibrium beliefs then become quite important, and provided much of the motivation for the exotic refinements mentioned in section 6.2.
- There is no separating equilibrium if, instead of discounting, the asymmetric information model has fixed-size per-period bargaining costs, unless the bargaining cost is higher for the high-valuation buyer than for the low-valuation. If, for example, there is no discounting, but a cost of c is incurred each period that bargaining continues, no separating equilibrium is possible. That is the typical signalling result. In a separating equilibrium the buyer tries to signal a low valuation by holding out, which fails unless it really is less costly for a low-valuation buyer to hold out. See Perry (1986) for a model with fixed bargaining costs which ends after one round of bargaining.

Problems

11.1: A Fixed Cost of Bargaining and Grudges.

Smith and Jones are trying to split 100 dollars. In bargaining round 1, Smith makes an offer at cost 0, proposing to keep S_1 for himself and Jones either accepts (ending the game) or rejects. In round 2, Jones makes an offer at cost 10 of S_2 for Smith and Smith either accepts or rejects. In round 3, Smith makes an offer of S_3 at cost c, and Jones either accepts or rejects. If no offer is ever accepted, the 100 dollars goes to a third player, Dobbs.
(11.1a) If $c = 0$, what is the equilibrium outcome?
(11.1b) If $c = 80$, what is the equilibrium outcome?
(11.1c) If $c = 10$, what is the equilibrium outcome?

(11.1d) What happens if $c = 0$, but Jones is very emotional and would spit in Smith's face and throw the 100 dollars to Dobbs if Smith proposes $S = 100$? Assume that Smith knows Jones' personality perfectly.

11.2: Selling Cars.

A car dealer must pay $10,000 to the manufacturer for each car he adds to his inventory. He faces three buyers. From the point of view of the dealer, Smith's valuation is uniformly distributed between $11,000 and $21,000, Jones' is between $9,000 and $11,000, and Brown's is between $4,000 and $12,000. The dealer's policy is to make a single take-it-or-leave-it offer to each customer, and he is smart enough to avoid making differerent offers to customers who could resell to each other. Use the notation that the maximum valuation is $\overline{V}$ and the range of valuations is R.

(11.2a) What will the offers be?

(11.2b) Who is most likely to buy a car? How does this compare with the outcome with perfect price discrimination under full information? How does it compare with the outcome when the dealer charges $10,000 to each customer?

(11.2c) What happens to the equilibrium prices if, with probability 0.25, each buyer has a valuation of $0, but the probability distribution remains otherwise the same?

11.3: The Nash Bargaining Solution.

Smith and Jones, shipwrecked on a desert island, are trying to split 100 pounds of cornmeal and 100 pints of molasses, their only supplies. Smith's utility function is $U_s = C + 0.5M$ and Jones' is $U_j = 3.5C + 3.5M$. If they cannot agree, they fight to the death, with $U = 0$ for the loser. Jones wins with probability 0.8.

(11.3a) What is the threat point?

(11.3b) With a 50-50 split of the supplies, what are the utilities if the two players do not recontract? Is this efficient?

(11.3c) Draw the threat point and the Pareto frontier in utility space (put U_s on the horizontal axis).

(11.3d) According to the Nash bargaining solution, what are the utilities? How are the goods split?

(11.3e) Suppose Smith discovers a cookbook full of recipes for a variety of molasses candies and corn muffins, and his utility function becomes $U_s = 10C + 5M$. Show that the split of goods in part (d) remains the same despite his improved utility function.

11.4: Incomplete Information.

(11.4a) What is the equilibrium in the game of Bargaining with Incomplete Information if the probability of a low-valuation buyer is $\gamma = 0.1$, instead of 0.05 or 0.5?

(11.4b) What level of γ marks the boundary between separating and pooling equilibria?

11.5: A Fixed Cost of Bargaining and Incomplete Information.

Smith and Jones are trying to split 100 dollars. In bargaining round 1, Smith makes an offer at cost c, proposing to keep S_1 for himself. Jones either accepts (ending the game) or rejects. In round 2, Jones makes an offer of S_2 for Smith, at cost 10, and Smith either accepts or rejects. In round 3, Smith makes an offer of S_3 at cost c, and Jones either accepts or rejects. If no offer is ever accepted, the 100 dollars goes to a third player, Parker.

(11.5a) If $c = 0$, what is the equilibrium outcome?

(11.5b) If $c = 80$, what is the equilibrium outcome?

(11.5c) If Jones' priors are that $c = 0$ and $c = 80$ are equally likely, but only Smith knows the true value, what is the equilibrium outcome? (Hint: the equilibrium uses mixed strategies.)

11.6: A Fixed Bargaining Cost, Again.

Apex and Brydox are entering into a joint venture that will yield 500 million dollars, but they must negotiate the split first. In bargaining round 1, Apex makes an offer at cost 0, proposing to keep A_1 for itself. Brydox either accepts (ending the game) or rejects. In Round 2, Brydox makes an offer at cost 10 million of A_2 for Apex, and Apex either accepts or rejects. In Round 3, Apex makes an offer of A_3 at cost c, and Brydox either accepts or rejects. If no offer is ever accepted, the joint venture is cancelled.

(11.6a) If $c = 0$, what is the equilibrium? What is the equilibrium outcome?

(11.6b) If $c = 10$, what is the equilibrium? What is the equilibrium outcome?

(11.6c) If $c = 300$, what is the equilibrium? What is the equilibrium outcome?

12 Auctions

12.1 Auction Classification and Private-Value Strategies

Because auctions are stylized markets with well-defined rules, modelling them with game theory is particularly appropriate. Moreover, several of the motivations behind auctions are similar to the motivations behind the asymmetric information contracts of part II of this book. Besides the mundane reasons such as speed of sale that make auctions important, auctions are useful for a variety of informational purposes. Often the buyers know more than the seller about the value of what is being sold, and the seller, not wanting to suggest a price first, uses an auction as a way to extract information. Art auctions are a good example, because the value of a painting depends on the buyer's tastes, which are known only to himself.

Auctions are also useful for agency reasons, because they hinder dishonest dealing between the seller's agent and the buyer. If the mayor were free to offer a price for building the new city hall and accept the first contractor who showed up, the lucky contractor would probably be the one who made the biggest political contribution. If the contract is put up for auction, cheating the public is more costly, and the difficulty of rigging the bids may outweigh the political gain.

We will spend most of this chapter on the effectiveness of different kinds of auction rules in extracting surplus from buyers, which requires considering the strategies with which they respond to the rules. Section 12.2 classifies auctions based on the relationships between different buyers' valuations of what is being auctioned, and explains the possible auction rules and the bidding strategies optimal for each rule. Section 12.3 compares the outcomes under the various rules. Section 12.4 discusses optimal strategies under common-value information, which can lead bidders into the "winner's curse" if they are not careful. Section 12.5 is about information asymmetry in common-value auctions.

Private, Common, and Correlated Values

Auctions differ enough for an intricate classification to be useful. One way to classify auctions is based on differences in the values buyers put on what is being auctioned. We will call the dollar value of the utility that player i receives from an object its **value** to him, V_i, and we will call his *estimate* of its value his **valuation**, $\hat{V}_i$.

In a **private-value** auction, each player knows his value with certainty, although he may still have to estimate the values of the other players. An example is the sale of antique chairs to people who will not resell them. The values need not be independent. If it were common knowledge that the values were either all high or all low, for example, that could affect the choice of auction rules by the seller and the estimates made by buyers of each others' values. A player's value equals his valuation in a private-value auction.

If an auction is to be private-value, it cannot be followed by costless resale of the object. If there were resale, a bidder's valuation would depend on the price at which he could resell, which would depend on the other players' valuations.

What is special about a private-value auction is that a player cannot extract any information about his own value from the valuations of the other players. Knowing all the other bids in advance would not change his valuation, although it might well change his strategy. The outcomes would be similar even if he had to estimate his own value, so long as the behavior of other players did not help him to estimate it, so this kind of auction could just as well be called a "private-valuation auction."

In a **common-value** auction, the players have identical values, but each player forms his own valuation by estimating on the basis of his private information. An example is bidding for US Treasury bills. A player's valuation would change if he could sneak a look at the other players' valuations, because they are all trying to estimate the same true value.

The **correlated-value** auction is a general category which includes the common-value auction as an extreme case. In this auction, the valuations of the different players are correlated, but their values may differ. Practically every auction we see is correlated-value, but, as always in modelling, we must trade off descriptive accuracy against simplicity, and private-value versus common-value is an appropriate simplification.

Auction Rules and Private-Value Strategies

Auctions have as many different sets of rules as poker games do. Cassady's 1967 book describes a myriad of rules, but here I will just list the main varieties and describe the equilibrium private-value strategies. In teaching this material, I ask each student to pick a valuation between 80 and 100, after which we conduct the various kinds of auctions. I advise the reader to try this. Pick two valuations and try out sample strategy profiles for the different auctions as they are described. Even though the values are private, it will immediately become clear that the best-response bids still depend on the strategies the bidder thinks the other players have adopted.

The types of auctions to be described are:

(1) English (first-price open-cry).
(2) First-price sealed-bid.

(3) Second-price sealed-bid (Vickrey).
(4) Dutch (descending).

(1) English (first-price open-cry)

Rules. Each bidder is free to revise his bid upwards. When no bidder wishes to revise his bid further, the highest bidder wins the object and pays his bid.

Strategies. A player's strategy is his series of bids as a function of (1) his value, (2) his prior estimate of the other players' valuations, and (3) the past bids of all the players. His bid can therefore be updated as his information set changes.

Payoffs. The winner's payoff is his value minus his highest bid.

A player's dominant strategy in a private-value English auction is to keep bidding some small amount ε more than the previous high bid until he reaches his valuation, and then to stop. This is optimal because he always wants to buy the object if the price is less than its value to him, but he wants to pay the lowest price possible. All bidding ends when the price reaches the valuation of the player with the second-highest valuation. The optimal strategy is independent of risk neutrality if players know their own values with certainty rather than having to estimate them, although risk-averse players who must estimate their values should be more conservative in bidding.

In correlated-value open-cry auctions, the bidding procedure is important. The most common procedures are (1) for the auctioneer to raise prices at a constant rate, (2) for him to raise prices at whatever rate he thinks appropriate, and (3) for the bidders to raise prices as specified in the rules above. A fourth procedure is often the easiest to model: the **open-exit** auction, in which the price rises continuously and players must publicly announce that they are dropping out (and cannot reenter) when the price becomes unacceptably high. In an open-exit auction the players have more evidence available about each others' valuations than when they can drop out secretly.

(2) First-price sealed-bid

Rules. Each bidder submits one bid, in ignorance of the other bids. The highest bidder pays his bid and wins the object.

Strategies. A player's strategy is his bid as a function of his value and his prior beliefs about other players' valuations.

Payoffs. The winner's payoff is his value minus his bid.

Suppose Smith's value is 100. If he bid 100 and won when the second bid was 80, he would wish that he had bid less. If it is common knowledge that the second-highest value is 80, Smith's bid should be $80 + \varepsilon$. If he is not sure about the second-highest value, the problem is difficult and no general solution has been discovered. The tradeoff is between bidding high—thus winning more often—and bidding low—thus benefiting more if the bid wins. The optimal strategy, whatever it may be, depends on risk neutrality and beliefs about the other bidders, so the equilibrium is less robust than the equilibria of English and second-price auctions.

Nash equilibria can be found for more specific first-price auctions. Suppose that there are N risk-neutral bidders, and that Nature assigns them values independently using a uniform density from 0 to some amount $\bar{v}$. Denote player i's value by v_i, and let us consider the strategy for player 1. If some other player has a higher value, then in a symmetric equilibrium, player 1 is going to lose the auction anyway, so we can ignore that possibility in finding his optimal bid. Player 1's equilibrium strategy is to bid ε above his expectation of the second-highest value, conditional on his bid being the highest (i.e., assuming that no other bidder has a value over v_1).

If we assume that v_1 is the highest value, the probability that player 2's value, which is uniformly distributed between 0 and v_1, equals v is $1/v_1$, and the probability that v_2 is less than or equal to v is v/v_1. The probability that v_2 equals v and is the second-highest value is

$$Prob(v_2 = v) \cdot Prob(v_3 \leq v) \cdot Prob(v_4 \leq v) \cdots Prob(v_N \leq v), \tag{1}$$

which equals

$$\left(\frac{1}{v_1}\right)\left(\frac{v}{v_1}\right)^{N-2}. \tag{2}$$

Since there are $N - 1$ players besides player 1, the probability that one of them has the value v, and that v is the second-highest is $N - 1$ times expression (2). The expectation of v is the integral of v over the range 0 to v_1,

$$\begin{aligned} E(v) &= \int_0^{v_1} v(N-1)\left(\frac{1}{v_1}\right)\left(\frac{v}{v_1}\right)^{N-2} dv \\ &= \frac{N-1}{v_1^{N-1}} \int_0^{v_1} v^{N-1} dv \\ &= \frac{(N-1)v_1}{N}. \end{aligned} \tag{3}$$

Thus we find that player 1 ought to bid a fraction $(N - 1)/N$ of his own value, plus ε.

The previous example is an elegant result, but it is not a general rule. Suppose Smith knows that Brown's value is 0 or 100 with equal probability, and Smith's value of 400 is known by both players. Brown bids either 0 or 100 in equilibrium, and Smith always bids $(100 + \varepsilon)$, because his value is so high that winning is more important than paying a low price.

If Smith's value were 102 instead of 400, the equilibrium would be much different. Smith would use a mixed strategy, and while Brown would still offer 0 if his value were 0, if his value were 100 he would use a mixed strategy too. No pure strategy can be part of a Nash equilibrium, because if Smith always bid a value $x < 100$, Brown would always bid $x + \varepsilon$, in which case Smith would deviate to $x + 2\varepsilon$, and if Smith bid $x \geq 100$ he would be paying 100 more than necessary half the time.

(3) Second-price sealed-bid (Vickrey)

Rules. Each bidder submits one bid, in ignorance of the other bids. The bids are opened, and the highest bidder pays the amount of the second-highest bid and wins the object.

Strategies. A player's strategy is his bid as a function of his value and his prior belief about other players' valuations.

Payoffs. The winner's payoff is his value minus the second-highest bid that was made.

Second-price auctions are similar to English auctions. They are rarely used in reality, but are useful for modelling. Bidding one's valuation is the dominant strategy: a player who bids less is more likely to lose the auction, but pays the same price if he does win. The structure of the payoffs is reminiscent of the Groves mechanism of section 9.6, because in both games a player's strategy affects some major event (who wins the auction or whether the project is undertaken), but his strategy affects his own payoff only via that event. In the auction's equilibrium, each player bids his value and the winner ends up paying the second-highest value. If players know their own values, the outcome does not depend on risk neutrality.

(4) Dutch (descending)

Rules. The seller announces a bid, which he continuously lowers until some buyer stops him and takes the object at that price.

Strategies. A player's strategy is when to stop the bidding as a function of his valuation and his prior beliefs as to the other players' valuations.

Payoffs. The winner's payoff is his value minus his bid.

The Dutch auction is **strategically equivalent** to the first-price sealed-bid auction, which means that there is a one-to-one mapping between the strategy sets and the equilibria of the two games. The reason for the strategic equivalence is that no relevant information is disclosed in the course of the auction, only at the end, when it is too late to change anybody's behavior. In the first-price auction a player's bid is irrelevant unless it is the highest, and in the Dutch auction a player's stopping price is also irrelevant unless it is the highest. The equilibrium price is calculated in the same way for both auctions.

Dutch auctions are actually used. One example is the Ontario tobacco auction, which uses a clock four feet in diameter marked with quarter-cent gradations. Each of six or so buyers has a stop button. The clock hand drops a quarter cent at a time, and the stop buttons are registered so that ties cannot occur (tobacco buyers need reflexes like race-car drivers). The farmers who are selling their tobacco watch from an adjoining room and can later reject the bids if they feel they are too low (a form of reserve price). 2,500,000 lb. a day can be sold using the clock (Cassady [1967] p. 200).

Dutch auctions are common in less obvious forms. Filene's is one of the biggest stores in Boston, and Filene's Basement is its most famous department. In the basement are a variety of marked-down items formerly in the regular store, each with a price and date attached. The price customers pay at the register is the price on the tag minus a discount which depends on how long ago the item was dated. As time passes and the item remains unsold, the discount rises from 10 to 50 to 70 percent. The idea of predictable time discounting has also recently been used by bookstores ("Waldenbooks to Cut Some Book Prices in Stages in Test of New Selling Tactic," *Wall Street Journal*, March 29, 1988, p. 34).

12.2 Comparing Auction Rules

Equivalence Theorems

When one mentions auction theory to an economic theorist, the first thing that springs to his mind is the idea that, in some sense, different kinds of auctions are really the same. Milgrom & Weber (1982) give a good summary of how and why this is true. Regardless of the information structure, the Dutch and first-price sealed-bid auctions are the same in the sense that the strategies and the payoffs associated with the strategies are the same. That equivalence does not depend on risk neutrality, but let us assume that all players are risk neutral for the next few paragraphs.

In private, independent-value auctions, the second-price sealed-bid and English auctions are the same in the sense that the bidder who values the object most highly wins and pays the valuation of the bidder who values it second-highest, but the strategies are different in the two auctions. In all four kinds of private independent-value auctions discussed, the seller's expected price is the same. This fact is the biggest result in auction theory: the **revenue-equivalence theorem** (Vickrey [1961]).

The revenue-equivalence theorem does not imply that in every realization of the game all four auction rules yield the same price, only that the expected price is the same. The difference arises because in the Dutch and first-price sealed-bid auctions, the winning bidder has estimated the value of the second-highest bidder, and that estimate, while correct on average, is above or below the true value in particular realizations. The variance of the price is higher in those auctions because of the additional estimation, which means that a risk-averse seller should use the English or second-price auction.

Whether the auction is private-value or not, the Dutch and first-price sealed-bid auctions are strategically equivalent. If the auction is correlated-value and there are three or more bidders, the open-exit English auction leads to greater revenue than the second-price sealed-bid auction, and both yield greater revenue than the first-price sealed-bid auction (Milgrom & Weber [1982]). If there are just two bidders, however, the open-exit English auction is no better than the second-price sealed-bid auction, because the open-exit feature—knowing when non-bidding players drop out—is irrelevant.

A question of less practical interest is whether an auction form is Pareto-optimal; that is, does the auctioned object end up in the hands of whoever values it most? In a common-value auction this is not an interesting question, because all bidders value the object equally. In a private-value auction, all the different auction forms—first-price, second-price, Dutch, and English—are Pareto optimal. They are also optimal in a correlated-value auction if all the players draw their information from the same distribution and if the equilibrium is in symmetric strategies.

Auctions with risk-averse bidders are difficult to analyze. One finding is that in a private-value auction the first-price sealed-bid auction yields a greater expected revenue than the English or second-price auctions. That is because by increasing his bid from the level optimal for a risk-neutral bidder, the risk-averse bidder insures himself. If he wins, his surplus is slightly less because of the higher price, but he is more likely to win and avoid a surplus of zero. Thus, the buyers' risk aversion helps the seller.

Hindering Buyer Collusion

As I mentioned at the start of this chapter, one motivation for auctions is to discourage collusion between players. Some auctions are more vulnerable to this than others. Robinson (1985) has pointed out that whether the auction is private-value or common-value, the first-price sealed-bid auction is superior to the second-price sealed-bid or English auctions for the purpose of deterring collusion among bidders.

Consider a buyer's cartel in which buyer Smith has a private value of 20, the other buyers' values are each 18, and they agree that everybody will bid 5 except Smith, who will bid 6. (We will not consider the rationality of this choice of bids, which might be based on avoiding legal penalties.) In an English auction this is self-enforcing, because if somebody cheats and bids 7, Smith is willing to go all the way up to 20 and the cheater will end up with no gain from his deviation. Enforcement is also easy in a second-price sealed-bid auction, because the cartel agreement can be that Smith bids 20 and everyone else bids 6.

In a first-price sealed-bid auction, however, it is hard to prevent buyers from cheating on their agreement in a one-shot game. Smith does not want to bid 20, because he would have to pay 20, but if he bids anything less than the other players' value of 18 he risks their overbidding him. The buyer will end up paying a price of 18, rather than the 6 he would receive in an English auction with collusion. The seller therefore will use the first-price sealed-bid auction if he fears collusion.

12.3 Common-Value Auctions and the Winner's Curse

In section 12.2 we distinguished private-value auctions from common-value auctions, in which the values of the players are identical but their valuations may differ. All four sets of rules discussed in section 12.2 can be used for common-value auctions, but the optimal strategies are different. In common-value auctions, each player can extract useful information about the object's value to himself from the bids of the other players. Surprisingly enough, a buyer can use the information from other buyers' bids even in a sealed-bid auction, as will be explained below.

When I teach this material I bring a jar of pennies to class and ask the students to bid for it in an English auction. All but two of the students get to look at the jar before the bidding starts, and everybody is told that the jar contains more than 5 and less than 100 pennies. Before the bidding starts, I ask each student to write down his best guess of the number of pennies. The two students who do not get to see the jar are like "technical analysts," who try to forecast stock prices using charts showing the past movements of the stock while remaining ignorant of the stock's "fundamentals."

A common-value auction in which all the bidders knew the value would not be very interesting, but more commonly, as in the penny jar example, the bidders must estimate the common value. The obvious strategy, especially following our discussion of private-value auctions, is for a player to bid up to his unbiased estimate of the number of pennies in the jar. But this strategy makes the winner's payoff negative, because the winner is the bidder who has made the largest positive error in his valuation. The bidders who underestimated the number of pennies, on the other hand, lose the auction, but their payoff is limited to a downside value of zero, which they would receive even if the true value were common knowledge. Only the winner suffers from overbidding: he has stumbled into the **winner's curse**. When other players are better informed, it is even worse for an uninformed player to win. Anyone, for example, who wins an auction against 50 experts should worry about why they all bid less.

To avoid the winner's curse, players should scale down their estimates in forming their bids. The mental process is a little like deciding how much to bid in a private-value, first-price sealed-bid auction, in which bidder Smith estimates the second-highest value conditional upon himself having the highest value and winning. In the common-value auction, Smith estimates his own value, not the second-highest, conditional upon himself winning the auction. He knows that if he wins using his unbiased estimate, he probably bid too high, so after winning with such a bid he would like to retract it. Ideally, he would submit a bid of [X *if I lose, but* $(X - Y)$ *if I win*], where X is his valuation conditional upon losing and $(X - Y)$ is his lower valuation conditional upon winning. If he still won with a bid of $(X - Y)$ he would be happy; if he lost, he would be relieved. But Smith can achieve the same effect by simply submitting the bid $(X - Y)$ in the first place, since the size of losing bids is irrelevant.

Another explanation of the winner's curse can be devised from the Milgrom definition of "bad news" (Milgrom [1981b], Appendix B). Suppose that the government is auctioning off the mineral rights to a plot of land with common value V and that bidder i has valuation $\hat{V}_i$. Suppose also that the bidders are identical in everything but their valuations, which are based on the various information sets Nature has assigned them, and that the equilibrium is symmetric, so the equilibrium bid function $b(\hat{V}_i)$ is the same for each player. If Bidder 1 wins with a bid $b(\hat{V}_1)$ that is based on his prior valuation $\hat{V}_1$, his posterior valuation $\tilde{V}_1$ is

$$\tilde{V}_1 = E(V|\hat{V}_1, b(\hat{V}_2) < b(\hat{V}_1), \ldots, b(\hat{V}_n) < b(\hat{V}_1)). \tag{4}$$

The news that $b(\hat{V}_2) < \infty$ would be neither good nor bad, since it conveys no information, but the information that $b(\hat{V}_2) < b(\hat{V}_1)$ is bad news, since it rules out values of b more likely to be produced by large values of $\hat{V}_2$. In fact, the lower the value of $b(\hat{V}_1)$, the worse is the news of having won. Hence,

$$\tilde{V}_1 < E(V|\hat{V}_1) = \hat{V}_1, \tag{5}$$

and if Bidder 1 had bid $b(\hat{V}_1) = \hat{V}_1$ he would immediately regret having won. If his winning bid were enough below $\hat{V}_1$, however, he would be pleased to win.

Deciding how much to scale down the bid is a hard problem because the amount depends on how much all the other players scale down. In a second-price auction a player calculates the value of $\tilde{V}_1$ using equation (4), but that equation hides considerable complexity under the disguise of the term $b(\hat{V}_2)$, which is itself calculated as a function of $b(\hat{V}_1)$ using an equation like (4).

Oil Tracts and the Winner's Curse

The best known example of the winner's curse is from bidding for offshore oil tracts. Offshore drilling can be unprofitable even if oil is discovered, because something must be paid the government for the mineral rights. Capen, Clapp, & Campbell (1971) suggest that bidders' ignorance of the winner's curse caused overbidding in US government auctions of the 1960s. If the oil companies had bid close to what their engineers estimated the tracts were worth, rather than scaling down their bids, the winning companies would have lost on their investments. The hundredfold difference in the sizes of some of the bids in the sealed-bid auctions shown in table 12.1 lends some plausibility to the view that this is what happened.

Later studies such as Mead et al. (1984) that actually looked at profitability conclude that the rates of return from offshore drilling were not abnormally low, so perhaps the oil companies did scale down their bids rationally. The spread in bids is surprisingly wide, but that does not mean that the bidders did not properly scale down their estimates. Although expected profits are zero under optimal bidding, realized profits could be either positive or negative. With some probability, one bidder makes a large over-estimate which results in too high a bid even after rationally adjusting for the winner's curse. The knowledge of how to bid optimally does not eliminate bad luck; it only mitigates its effects.

Table 12.1 Bids by Serious Competitors in Oil Auctions

Offshore Louisiana 1967 Tract SS 207	**Santa Barbara Channel 1968** Tract 375	**Offshore Texas 1968** Tract 506	**Alaska North Slope 1969** Tract 253
32.5	43.5	43.5	10.5
17.7	32.1	15.5	5.2
11.1	18.1	11.6	2.1
7.1	10.2	8.5	1.4
5.6	6.3	8.1	0.5
4.1		5.6	0.4
3.3		4.7	
		2.8	
		2.6	
		0.7	
		0.7	
		0.4	

[*Source:* Capen, Clapp & Campbell (1971)]
Note: All bids are in millions of dollars.

Another consideration is the irrationality of the other bidders. If bidder Apex has figured out the winner's curse, but bidders Brydox and Central have not, what should Apex do? Its rivals will overbid, which affects Apex' best response. Apex should scale down its bid even further, because the winner's curse is intensified against over-optimistic rivals. If Apex wins against a rival who commonly overbids, Apex has very likely overestimated the value.

Risk aversion affects bidding in a surprisingly similar way. If all the players are equally risk averse, the bids would be lower, because the asset is a gamble whose value is lower for the risk averse. If Smith is more risk averse than Brown, then Smith should be more cautious for two reasons: the direct reason that the gamble is worth less to Smith, and the indirect reason that when Smith wins against a rival like Brown who regularly bids more, Smith probably overestimated the value. Parallel reasoning holds if the players are risk neutral, but the private value of the object differs among them.

Asymmetric equilibria can even arise when the players are identical. Second-price two-person common-value auctions usually have many asymmetric equilibria besides the symmetric equilibrium we have been discussing (see Milgrom [1981c] and Bikhchandani [1988]). Suppose that Smith and Brown have identical payoff functions, but Smith thinks Brown is going to bid aggressively. The winner's curse is intensified for Smith, who would probably have overestimated if he won against an aggressive bidder like Brown, so Smith bids more cautiously. But if Smith bids cautiously, Brown is safe in bidding aggressively, and there is an asymmetric equilibrium. For this reason, acquiring a reputation for aggressiveness is valuable.

Oddly enough, if there are three or more players the sealed-bid, second-price, common-value auction has a unique equilibrium, which is also symmetric. The open-exit auction is different: it has asymmetric equilibria, because after one bidder drops out, the two remaining bidders know that they are alone together in a subgame which is a two-player auction. Regardless of the number of players, first-price sealed-bid auctions do not have this kind of asymmetric equilibrium. Threats in a first-price auction are costly because the high bidder pays his bid even if his rival decides to bid less in response. Thus, a bidder's aggressiveness is not made safer by intimidation of another bidder.

The winner's curse crops up in situations seemingly far removed from auctions. An employer must beware of hiring a worker passed over by other employers. Someone renting an apartment must hope that he is not the first visitor who arrived when the neighboring trumpeter was asleep. A firm considering a new project must worry that the project has been considered and rejected by competitors. The winner's curse can even be applied to political theory, where certain issues keep popping up. Opinions are like estimates, and one interpretation of different valuations is that everyone gets the same data, but they analyze it differently.

On a more mundane level, in 1987 there were four major candidates—Bush, Kemp, Dole, and Other—running for the Republican nomination for President of the United States. Consider an entrepreneur auctioning off four certificates, each paying one dollar if its particular candidate wins the nomination. If every bidder is rational, the entrepreneur should receive a maximum of one dollar in total revenue from these four auctions, and less if bidders are risk averse. But if the auction were held in a bar full of partisans, how much do you think he would actually receive?

12.4 Information in Common-Value Auctions

The Seller's Information

Milgrom & Weber (1982) have found that honesty is the best policy as far as the seller is concerned. If it is common knowledge that he has private information, he should release it before the auction. The reason is not that the bidders are risk averse (though perhaps this strengthens the result), but the "No news is bad news" result of section 8.1. If the seller refuses to disclose something, buyers know that the information must be unfavorable, and an unravelling argument tells us that the quality must be the very worst possible.

Quite apart from unravelling, another reason to disclose information is to mitigate the winner's curse, even if the information just reduces uncertainty over the value without changing its expectation. In trying to avoid the winner's curse, bidders lower their bids, so anything which makes it less of a danger raises their bids.

Asymmetric Information Among the Buyers

Suppose that Smith and Brown are two of many bidders in a common-value auction. If Smith knows he has uniformly worse information than Brown (that is, if his information partition is finer than Brown's), he should stay out of the auction: his expected payoff is negative if Brown expects zero profits.

If Smith's information is not uniformly worse, he can still benefit by entering the auction. Having independent information, in fact, is more valuable than having good information. Consider a common-value first-price sealed-bid auction with four bidders. Bidders Smith and Black have the same good information, Brown has that same information plus an extra signal, and Jones usually has only a poor estimate, but one different from any other bidder's. Smith and Black should drop out of the auction—they can never beat Brown without overpaying. But Jones will sometimes win, and his expected surplus is positive. If, for example, real-estate tracts are being sold, and Jones is quite ignorant of land values, he can still do well if, on rare occasions, he has inside information concerning the location of a new freeway, even though ordinarily he should refrain from bidding. If Smith and Black both use the same appraisal formula, they will compete each other's profits away, and if Brown uses the formula plus extra private information, he drives their profits negative by taking some of the best deals away from them and leaving the worst ones.

In general, a bidder should bid less if there are more bidders or if his information is absolutely worse (that is, if his information partition is coarser). He should also bid less if parts of his information partition are coarser than those of his rivals, even if his information is not uniformly worse. These considerations are most important in sealed-bid auctions, because in an open-cry auction, information is revealed by the bids while other bidders still have time to act on it.

Notes

N12.1 Auction Classification and Private-Value Strategies

- McAfee & McMillan (1987) and Milgrom (1987) are excellent surveys of the literature and theory of auctions. Both articles take some pains to relate the material to models of asymmetric information. Milgrom & Weber (1982) is a classic article that covers many aspects of auctions.
- **The Dollar Auction.** Auctions look like tournaments in that the winner is the player who chooses the largest amount for some costly variable, but in auctions the losers generally do not incur costs proportional to their bids. Shubik (1971), however, has suggested an auction for a dollar bill in which both the first- and the second-highest bidders pay the second price. If both players begin with infinite wealth, the game illustrates why equilibrium might not exist if strategy sets are unbounded. Once one bidder has started bidding against another, both of them do best by continuing to bid so as to win the dollar as well as pay the bid. This auction may seem absurd, but it has some similarity to patent races (see section 14.1) and arms races.

N12.2 Comparing Auction Rules

- Cassady (1967) is an excellent source of institutional detail on auctions. The appendix to his book includes advertisements and sets of auction rules, and he cites numerous newspaper articles. See also *New York Times*, 26 July 1985, p. C23 and 31 July 1985, pp. A1, C20; *Wall Street Journal*, 24 August 1984, pp. 1, 16; and "The Crackdown on Colluding Roadbuilders," *Fortune*, 3 October 1983, p. 79.
- One might think that a second-price open-cry auction would come to the same results as a first-price open-cry auction, because if the price advances by ε at each bid, the first and second bids

are practically the same. But the second-price auction can be manipulated. If somebody initially bids $10 for something worth $80, another bidder could safely bid $1,000. No one else would bid more, and he would pay only the second price: $10.

- In one variant of the English auction, the auctioneer announces each new price and a bidder can hold up a card to indicate he is willing to bid that price. This set of rules is more practical to administer in large crowds and it also allows the seller to act strategically during the course of the auction. If, for example, the two highest valuations are 100 and 120, this kind of auction could yield a price of 110, while the usual rules would only allow a price of $100 + \varepsilon$.
- Vickrey (1961) notes that a Dutch auction could be set up as a second-price auction. When the first bidder presses his button, he primes a buzzer that goes off when a second bidder presses a button.
- Auctions are especially suitable for empirical study because they are so stylized and generate masses of data. Tenorio (1993) is a nice example of empirical work using data from real auctions—in his case, the Zambian foreign exchange market.
- Second-price auctions have actually been used in a computer operating system. An operating system must assign a computer's resources to different tasks, and researchers at Xerox Corporation designed the Spawn system, under which users allocate "money" in a second-price sealed bid auction for computer resources. See "Improving a Computer Network's Efficiency," *The New York Times*, 29 March 1989, p. 35.
- After the last bid of an open-cry art auction in France, the representative of the Louvre has the right to raise his hand and shout "préemption de l'état," after which he takes the painting at the highest price bid (*The Economist*, May 23, 1987, p. 98). How does that affect the equilibrium strategies? What would happen if the Louvre could resell?
- **Share Auctions.** In a share auction each buyer submits a bid for both a quantity and a price. The bidder with the highest price receives the quantity for which he bid at that price. If any of the product being auctioned remains, the bidder with the second-highest price takes the quantity he bid for, and so forth. The rules of a share auction can allow each buyer to submit several bids, often called a **schedule** of bids. The details of share auctions vary, and they can be either first-price or second-price. Models of share auctions are very complicated; see R. Wilson (1979).
- **Reserve prices** are prices below which the seller refuses to sell. They can increase the seller's revenue, and their effect is to make the auction more like a regular fixed-price market. For discussion, see Milgrom & Weber (1982). They are also useful when buyers collude, a situation of bilateral monopoly (see "At Many Auctions, Illegal Bidding Thrives as a Longtime Practice Among Dealers," *Wall Street Journal*, February 19, 1988, p. 21).

  In some real-world English auctions, the auctioneer does not announce the reserve price in advance, and he starts the bidding below it. This can be explained as a way by allowing bidders to show each other that their valuations are greater than the starting price, even though it may turn out that they are all lower than the reserve price.
- Concerning auctions with risk-averse players, see Maskin & Riley (1984).

N12.4 Information in Common-Value Auctions

- Even if valuations are correlated, the optimal bidding strategies can still be the same as in private-value auctions if the values are independent. If everyone overestimates their values by 10 percent, a player can still extract no information about his value by seeing other players' valuations.
- "Getting carried away" may be a rational feature of a common-value auction. If a bidder has a high private value and then learns from the course of the bidding that the common value is larger than he thought, he may well end up paying more than he had planned, although he would not regret it afterwards. Other explanations for why bidders seem to pay too much are the winner's curse and the fact that in every auction all but one or two of the bidders think that the winning bid is greater than the value of the object.
- Milgrom & Weber (1982) use the concept of **affiliated** variables in classifying auctions. Roughly speaking, random variables X and Y are affiliated if a larger value of X means that a larger value of Y is more likely, or, at the least, no less likely. Independent random variables are affiliated.

Problems

12.1: Rent-Seeking.

Two risk-neutral neighbors in 16th century England, Smith and Jones, have gone to court and are considering bribing a judge. Each of them makes a gift, and the one whose gift is the largest is awarded property worth £2,000. If both bribe the same amount, the chances are 50 percent for each of them to win the lawsuit. Gifts must be either £0, £900, or £2,000.

(12.1a) What is the unique pure-strategy equilibrium for this game?

(12.1b) Suppose that it is also possible to give a £1500 gift. Why does there no longer exist a pure-strategy equilibrium?

(12.1c) What is the symmetric mixed-strategy equilibrium for the expanded game? What is the judge's expected payoff?

(12.1d) In the expanded game, if the losing litigant gets back his gift, what are the two equilibria? Would the judge prefer this rule?

12.2: The Founding of Hong Kong.[1]

The Tai-Pan and Mr. Brock are bidding in an English auction for a parcel of land on a knoll in Hong Kong. They must bid integer values, and the Tai-Pan bids first. Tying bids cannot be made, and bids cannot be withdrawn once they are made. The direct value of the land is 1 to Brock and 2 to the Tai-Pan, but the Tai-Pan has said publicly that he wants it, so if Brock gets it, he receives 5 in "face" and the Tai-Pan loses 10. Moreover, Brock hates the Tai-Pan and receives 1 in utility for each 1 that the Tai-Pan pays out to get the land.

(12.2a) Fill in the entries in table 12.2.

Table 12.2 The Tai-Pan Game

Winning bid	1	2	3	4	5	6	7	8	9	10	11	12
If Brock wins: π_{Brock} $\pi_{Tai\text{-}Pan}$												
If Brock loses: π_{Brock} $\pi_{Tai\text{-}Pan}$												

(12.2b) In equilibrium, who wins, and at what bid?

(12.2c) What happens if the Tai-Pan can precommit to a strategy?

(12.2d) What happens if the Tai-Pan cannot precommit, but he also hates Brock, and gets 1 in utility for each 1 that Brock pays out to get the land?

[1] See James Clavell, *Tai-Pan*. In the novel, the Tai-Pan's son, realizing the danger to his house, manipulates the official in charge of the auction so as to cancel it and prevent the bidding war.

12.3: Government and Monopoly.

Incumbent Apex and potential entrant Brydox are bidding for government favors in the widget market. Apex wants to defeat a bill that would require it to share its widget patent rights with Brydox. Brydox wants the bill to pass. Whoever offers the chairman of the House Telecommunications Committee more campaign contributions wins, and the loser pays nothing. The market demand curve is $P = 25 - Q$, and marginal cost is constant at 1.

(12.3a) Who will bid higher if duopolists follow Bertrand behavior? How much will the winner bid?

(12.3b) Who will bid higher if duopolists follow Cournot behavior? How much will the winner bid?

(12.3c) What happens under Cournot behavior if Apex can commit to giving away its patent freely to everyone in the world if the entry bill passes? How much will Apex bid?

12.4: An Auction with Stupid Bidders.

Smith's value for an object has a private component equal to 1 and a component common with Jones and Brown. Jones' and Brown's private components both equal zero. Each player estimates the common component Z independently, and player i's estimate is either x_i above the true value or x_i below, with equal probability. Jones and Brown are naive and always bid their valuations. The auction is English.

(12.4a) If $x_{Smith} = 0$, what is Smith's dominant strategy if his estimate of Z equals 20?

(12.4b) *If* $x_i = 8$ for all players and Smith estimates $Z = 20$, what are the probabilities that he puts on different values of Z?

(12.4c) If $x_i = 8$ but Smith knows that $Z = 12$ with certainty, what are the probabilities he puts on the different combinations of bids by Jones and Brown?

(12.4d) Why is 9 a better upper limit on bids for Smith than 21, if his estimate of Z is 20, and $x_i = 8$ for all three players?

13 Pricing

13.1 Quantities as Strategies: Cournot Equilibrium Revisited

Chapter 13 is about how firms with market power set prices. Section 13.1 generalizes the Cournot Game of section 3.5, in which two firms choose the quantities they sell, while section 13.2 sets out the Bertrand model of firms choosing prices. Both the Bertrand and Cournot models are then expanded to allow for differentiated products. Section 13.3 goes back to the origins of product differentiation, and develops two Hotelling location models. Section 13.4 shows how to do comparative statics in games, using the differentiated Bertrand model as an example and supermodularity and the implicit function theorem as tools. Section 13.5 shows that even if a firm is a monopolist, if it sells a durable good, it suffers competition from its future self.

Cournot Behavior with General Cost and Demand Functions

In the next few sections, sellers compete against each other while moving simultaneously. We will start by generalizing the Cournot Game of section 3.5 from linear demand and zero costs to a wider class of functions. The two players are firms Apex and Brydox, and their strategies are their choices of the quantities q_a and q_b. The payoffs are based on the total cost functions, $c(q_a)$ and $c(q_b)$, and the demand function, $p(q)$, where $q = q_a + q_b$. This specification says that only the sum of the outputs affects the price. The implication is that the firms produce an identical product, because whether it is Apex or Brydox that produces an extra unit, the effect on the price is the same.

Let us take the point of view of Apex. In the Cournot-Nash analysis, Apex chooses its output of q_a for a given level of q_b as if its choice did not affect q_b. From its point of view, q_a is a function of q_b, but q_b is exogenous. Apex sees the effect of its output on price as

$$\frac{\partial p}{\partial q_a} = \frac{dp}{dq}\frac{\partial q}{\partial q_a} = \frac{dp}{dq}. \tag{1}$$

Apex's payoff function is

$$\pi_a = p(q)q_a - c(q_a). \tag{2}$$

To find Apex' reaction function, we differentiate with respect to its strategy to obtain

$$\frac{d\pi_a}{dq_a} = p + \frac{dp}{dq}q_a - \frac{dc}{dq_a} = 0, \tag{3}$$

which implies

$$q_a = \frac{\frac{dc}{dq_a} - p}{\frac{dp}{dq}}, \tag{4}$$

or, simplifying the notation,

$$q_a = \frac{c' - p}{p'}. \tag{5}$$

If particular functional forms for $p(q)$ and $c(q_a)$ are available, equation (5) can be solved to find q_a as a function of q_b. More generally, to find the change in Apex' best response for an exogenous change in Brydox' output, differentiate (5) with respect to q_b, remembering that q_b exerts not only a direct effect, but possibly an indirect effect on q_a.

$$\frac{dq_a}{dq_b} = \frac{(p - c')\left(p'' + p''\frac{dq_a}{dq_b}\right)}{p'^2} + \frac{c''\frac{dq_a}{dq_b} - p' - p'\frac{dq_a}{dq_b}}{p'}. \tag{6}$$

Equation (6) can be solved for dq_a/dq_b to obtain the slope of the reaction function,

$$\frac{dq_a}{dq_b} = \frac{(p - c')p'' - p'^2}{2p'^2 - c''p' - (p - c')p''}. \tag{7}$$

If both costs and demand are linear, as in section 3.5, then $c'' = 0$ and $p'' = 0$, so equation (7) becomes

$$\frac{dq_a}{dq_b} = -\frac{p'^2}{2p'^2} = -\frac{1}{2}.$$

The general model faces two problems that did not arise in the linear model: nonuniqueness and nonexistence. If demand is concave and costs are convex, which implies that $p'' < 0$ and $c'' > 0$, then all is well as far as existence goes.

Figure 13.1 Multiple Cournot-Nash Equilibria

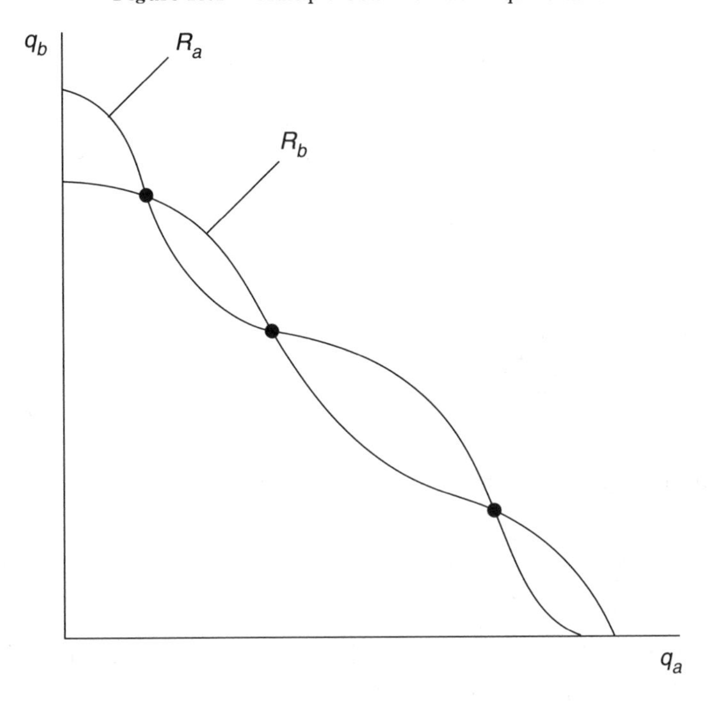

Since price is greater than marginal cost ($p > c'$), equation (7) tells us that the reaction functions are downward sloping, because $2p'^2 - c''p' - (p - c')p''$ is positive and both $(p - c')p''$ and $-p'^2$ are negative. If the reaction curves are downward sloping, they cross and an equilibrium exists, as was shown in figure 3.1 for the linear case represented by equation (8). We usually do assume that costs are at least weakly convex, since that is the result of diminishing or constant returns, but there is no reason to believe that demand is either concave or convex. If the demand curves are not linear, the contorted reaction functions of equation (7) might give rise to multiple Cournot equilibria, as in figure 13.1.

If demand is convex or costs are concave, so $p'' > 0$ or $c'' < 0$, the reaction functions can be upward sloping, in which case they might never cross and no equilibrium would exist. The problem can also be seen from Apex' payoff function, equation (2). If $p(q)$ is convex, the payoff function might not be concave, in which case standard maximization techniques break down. The problems of the general Cournot model teach a lesson to modellers: sometimes simple assumptions such as linearity generate atypical results.

Many Oligopolists

Let us return to the simpler game in which production costs are zero and demand is linear. For concreteness, we will use the particular inverse demand function

$$p(q) = 120 - q. \tag{9}$$

Using (9), the payoff function (2) becomes

$$\pi_a = 120q_a - q_a^2 - q_b q_a. \tag{10}$$

In section 3.5, firms picked outputs of 40 apiece given demand function (9). This generated a price of 40. With n firms instead of two, the demand function is

$$p\left(\sum_{i=1}^{n} q_i\right) = 120 - \sum_{i=1}^{n} q_i, \tag{11}$$

and firm j's payoff function is

$$\pi_j = 120q_j - q_j^2 - q_j \sum_{i \neq j} q_i. \tag{12}$$

Differentiating j's payoff function with respect to q_j yields

$$\frac{d\pi_j}{dq_j} = 120 - 2q_j - \sum_{i \neq j} q_i = 0. \tag{13}$$

The first step in finding the equilibrium is to guess that it is symmetric, so that $q_j = q_i$, $(i = 1, \ldots, n)$. This is an educated guess, since every player faces a first-order condition like (13). By symmetry, equation (13) becomes $120 - (n + 1)q_j = 0$, so that

$$q_j = \frac{120}{n + 1}. \tag{14}$$

Consider several different values for n. If $n = 1$, then $q_j = 60$, the monopoly optimum; and if $n = 2$ then $q_j = 40$, the Cournot output found in section 3.5. If $n = 5$, $q_j = 20$; and as n rises, individual output shrinks to 0. Moreover, the total output of $nq_j = 120n/(n + 1)$ gradually approaches 120, the competitive output, and the market price falls to 0, the marginal cost of production. As the number of firms increases, profits fall.

Conjectural Variation

Conjectural variation, an equilibrium concept different in flavor from any that has yet appeared in this book, is a way to quantify the degree of cooperation between oligopolists. Let us continue to specify the strategies as quantities. In a Nash equilibrium, no player wants to deviate, and his beliefs about how the other players would behave are confirmed whatever nodes are reached. Under conjectural variation, a player believes, for reasons outside the model, that if he deviated, the other players would deviate in specified ways. This should seem quite an unnatural idea to anyone who has read this far in the book, since it violates the basic assumptions of Bayesian games and it is rather hazy about what is happening in this simultaneous-move game. The idea may be clearer in an example. Returning to the two-player model, we can use equation (3) to write Apex' self-perceived first order condition as

$$\frac{d\pi_a}{dq_a} = p + \left(\frac{dp}{dq}\right)\left(\frac{dq}{dq_a}\right)q_a - \frac{dc}{dq_a} = 0. \tag{15}$$

The difference between the first-order conditions (3) and (15) is that (15) contains

$$\frac{dq}{dq_a} = 1 + \frac{dq_b}{dq_a}. \tag{16}$$

Equation (16) says that the expected effect on industry output of an increase in q_a by one unit has two components: a direct increase of one unit, and an indirect increase from Brydox increasing his output in response. The first-order condition (15) must be qualified by "self-perceived" because Apex might be mistaken in his beliefs about Brydox' response. The belief implicit in Nash equilibrium, that Apex' deviation is not followed by a response from Brydox, is the only belief that supports an equilibrium in which one player or the other is not mistaken. But if consistency of beliefs is not required, other beliefs are possible that lead to different behavior.

Firm i's **conjectural variation** *is the rate dq_{-i}/dq_i at which he conjectures that the output of other firms would change if i's own output changed.*

$CV = 0$
In a Cournot-Nash equilibrium, Apex believes that if he deviated by producing more, Brydox would not deviate, so the conjectural variation equals 0.

$CV = -1$
If Apex believes that an increase in his output is matched by a decrease in Brydox' output, so the total industry output is left unchanged, the conjectural variation is -1. If both firms use this conjectural variation, the industry output is the competitive level; firms ignore the effect of their output in depressing the price. Of course, if both firms use a negative value, their beliefs are inconsistent.

$CV = 1$
If Apex believes that Brydox would exactly match his output changes, the conjectural variation is 1. With two firms with identical cost curves, industry output is at the cartel level, though an n-player game would need $CV = n - 1$ to achieve that level.

In Stackelberg equilibrium (section 3.5), the conjectural variation of the Stackelberg follower is between 0 and 1, and takes the value given by a reaction function like equation (7).

In the world oil market, fringe producers like Britain face the OPEC cartel. If Britain's conjectural variation equals -1, Britain believes that producing more would make OPEC cut back an equal amount; if 0, that OPEC would ignore Britain; if 0.5, that OPEC would follow with a smaller increase; if 1, that OPEC would match every increase; and if 10, that OPEC would respond by flooding the market. Setting up equations with the appropriate values for the conjectural variations of all the players, we could solve for the equilibrium output. The idea is useful for organizing different models of duopoly and it is simple enough to be empirically estimated. Even without knowing the correct theory, an estimate could be made of how much OPEC actually does respond to Britain.

13.2 Prices as Strategies: Bertrand Equilibrium

The Bertrand (1883) duopoly model seems to be only slightly different from the Cournot model, but it reaches radically different conclusions. The Bertrand solution is nothing other than a Nash equilibrium in prices rather than quantities. We will use the same two-player, zero-cost, linear-demand world as before, but now the strategy spaces will be the prices, not the quantities. We will also use the same demand function, equation (9), which implies that if p is the lowest price, $q = 120 - p$. In the Cournot model, firms chose quantities but allowed the market price to vary freely; in the Bertrand model, they choose prices and sell as much as they can. The strategies for Apex and Brydox are p_a and p_b. The payoff function for Apex (and analogously for Brydox) is

$$\pi_a = \begin{cases} p_a(120 - p_a) & \text{if } p_i < p_b \\ \dfrac{p_a(120 - p_a)}{2} & \text{if } p_a = p_b \\ 0 & \text{if } p_a > p_b \end{cases}$$

The Bertrand game has a unique Nash equilibrium: $p_a = p_b = 0$. No other pair of prices could be an equilibrium, because one firm could capture the entire market by slightly undercutting the other's price. The only pair of prices where undercutting is not a temptation is (0, 0). Duopoly profits are not just less than monopoly profits, they are zero.

Like the surprising outcome of Prisoner's Dilemma, the Bertrand equilibrium is less surprising once one thinks about the model's limitations. What it shows is that duopoly profits do not arise just because there are two firms. Profits arise from something else, such as multiple periods, incomplete information, or differentiated products.

Both the Bertrand and Cournot models are in common use. The Bertrand model can be awkward mathematically because of the discontinuous jump from a market share of 0 to 100 percent after a slight price cut. The Cournot model is useful as a simple model that avoids this problem and which predicts that the price will fall gradually as more firms enter the market. There are also ways to modify the Bertrand model to obtain intermediate prices and gradual effects of entry, and we will proceed to look at some of these modifications.

Capacity Constraints: The Edgeworth Paradox

Let us start by altering the Bertrand model by constraining each firm to sell no more than $K = 70$ units. The industry capacity of 140 exceeds the competitive output, but do profits continue to be zero?

When capacities are limited we require additional assumptions because of the new possibility that a firm with a lower price might attract more customers than it can supply. We need to specify a **rationing rule** telling us which customers are served at the low price and which must buy from the high-price firm. The rationing rule is unimportant to the payoff of the low-price firm, but crucial to the high-price firm. One possible rule is

Intensity rationing. *The customers that value the product most buy from the firm with the lower price.*

The inverse demand function from equation (9) is $p = 120 - q$, and under intensity rationing, the K customers with the strongest demand buy from the low-price firm. Suppose that Brydox is the low-price firm, charging a price of 30 so that 90 consumers wish to buy from it. The residual demand facing Apex is then

$$q_a = 120 - p_a - K. \tag{17}$$

The demand curve is shown in figure 13.2a.

Under intensity rationing, the payoff functions are, given that $K = 70$,

$$\pi_a = \begin{cases} p_a \cdot Min\{120 - p_a, 70\} & \text{if } p_a < p_a \quad (a) \\ \dfrac{p_a(120 - p_a)}{2} & \text{if } p_a = p_b \quad (b) \\ 0 & \text{if } p_a > p_b,\ p_b \geq 50 \ (c) \\ p_a(120 - p_a - 70) & \text{if } p_a > p_b,\ p_b < 50 \ (d) \end{cases} \tag{18}$$

The appropriate rationing rule depends on what is being modelled. Intensity rationing is appropriate if buyers with more intense demand make greater efforts to obtain low prices. If the intense buyers are wealthy people who are unwilling to wait in line, the least intense buyers might end up at the low-price firm which is the case of **inverse-intensity rationing**. An intermediate rule is proportional rationing, under which every type of consumer is equally likely to be able to buy at the low price.

Proportional rationing. *Each customer has the same probability of being able to buy from the low-price firm.*

Figure 13.2 Rationing Rules (a) intensity rationing if $K = 70$; (b) proportional rationing

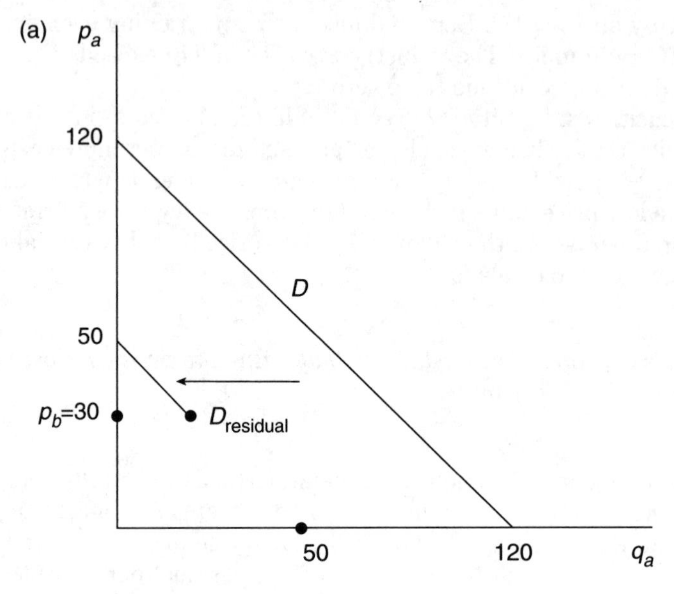

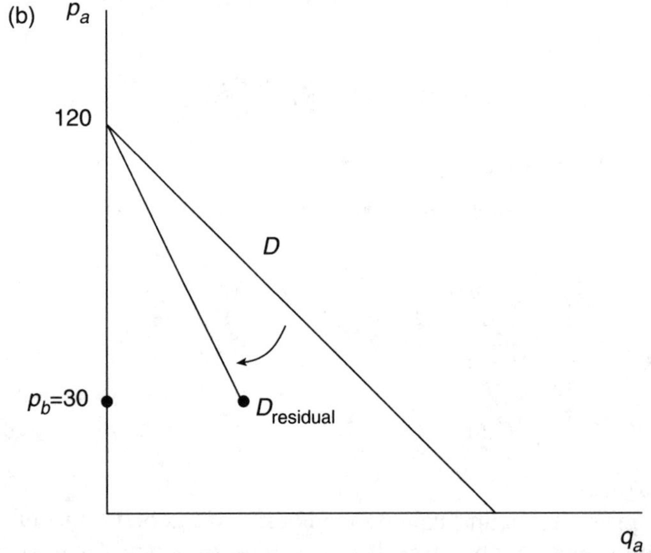

Under proportional rationing, if $K = 70$ and 90 customers wanted to buy from Brydox, 2/9 ($= (q(p_b) - K)/q(p_b)$) of each type of customer will be forced to buy from Apex (for example, 2/9 of the type who are willing to pay 120). The residual demand curve facing Apex, shown in figure 13.2b and equation (19), intercepts the price axis at 120, but slopes down at a rate three times as fast as market demand because there are only 2/9 as many remaining customers of each type.

$$q_a = (120 - p_a)\left(\frac{120 - p_b - K}{120 - p_b}\right) \tag{19}$$

The capacity constraint has a very important effect: (0, 0) is no longer a Nash equilibrium in prices. Consider Apex' best response when Brydox charges a price of zero. If Apex raises his price above zero, he retains most of his customers (because Brydox is already producing at capacity), but his profits rise from zero to some positive number, regardless of the rationing rule. In any equilibrium, both players must charge prices within some small amount ε of each other, or the one with the lower price would deviate by raising his price. But if the prices are equal, then both players have unused capacity, and each has an incentive to undercut the other. No pure-strategy equilibrium exists under either rationing rule. This is known as the **Edgeworth paradox**, after Edgeworth (1897).

A mixed-strategy equilibrium does exist, calculated using intensity rationing by Levitan & Shubik (1972) and analyzed in Dasgupta & Maskin (1986b). Expected profits are positive, because the firms charge positive prices. Under proportional rationing, as under intensity rationing, profits are positive in equilibrium, but the high-price firm does better with proportional rationing. The high-price firm would do best with **inverse-intensity rationing**, under which the customers with the least intense demand are served at the low-price firm, leaving the ones willing to pay more at the mercy of the high-price firm.

Even if capacity were made endogenous, the outcome would be inefficient, either because firms would charge prices higher than marginal cost (if their capacity were low), or they would invest in excess capacity (even though they price at marginal cost).

Product Differentiation

The Bertrand model without capacity constraints generates zero profits because only slight price discounts are needed to bid away customers. The assumption behind this is that the two firms sell identical goods, so if Apex' price is slightly higher than Brydox', all the customers go to Brydox. If customers have brand loyalty or poor price information, the equilibrium is different and the demand curves facing Apex and Brydox might be

$$q_a = 24 - 2p_a + p_b \tag{20}$$

and

$$q_b = 24 - 2p_b + p_a. \tag{21}$$

The greater the difference in the coefficients on prices in demand curves like these, the less substitutable are the products. As with standard demand curves like (9), we have made implicit assumptions about the extreme points of (20) and (21). These equations only apply if the quantities demanded turn out to be nonnegative, and we might also want to restrict them to prices below some ceiling, since otherwise the demand facing one firm becomes infinite as the other's price rises to infinity. With those restrictions, the payoffs are

$$\pi_a = p_a\,(24 - 2p_a + p_b) \tag{22}$$

and

$$\pi_b = p_b\,(24 - 2p_b + p_a). \tag{23}$$

Maximizing Apex' payoff, we obtain the first-order condition

$$\frac{d\pi_a}{dp_a} = 24 - 4p_a + p_b = 0, \tag{24}$$

and the reaction function

$$p_a = 6 + p_b/4. \tag{25}$$

Since Brydox has a parallel first-order condition, the equilibrium occurs where $p_a = p_b = 8$. The quantity each firm produces is 16, which is below the 24 each would produce at prices of zero. Figure 13.3 shows that the reaction functions intersect. Apex' demand curve has the elasticity

$$\left(\frac{\partial q_a}{\partial p_a}\right) \cdot \left(\frac{p_a}{q_a}\right) = -2\left(\frac{p_a}{q_a}\right), \tag{26}$$

which is finite even when $p_a = p_b$, unlike the case of the standard Bertrand model.

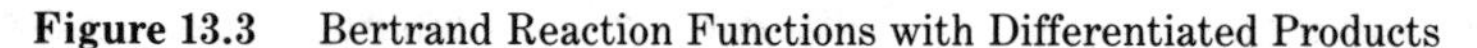

Figure 13.3 Bertrand Reaction Functions with Differentiated Products

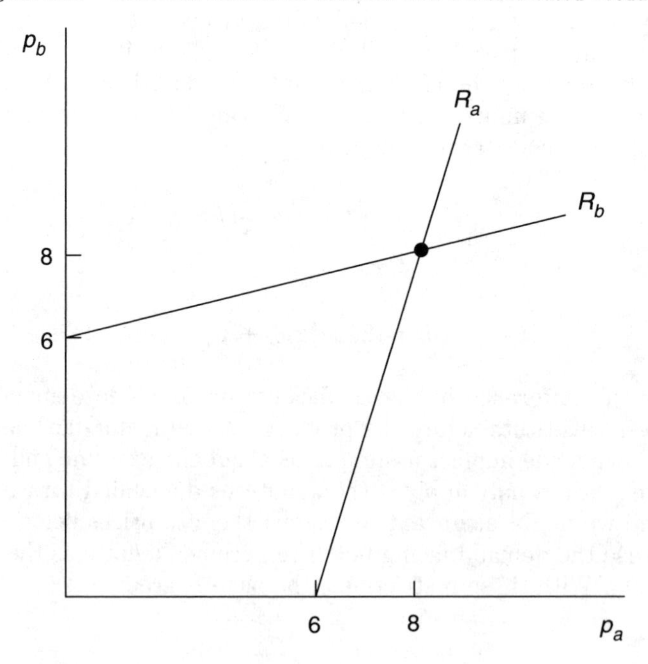

Cournot Equilibrium with Differentiated Products

We can also work out the Cournot equilibrium for the demand functions (20) and (21), but product differentiation does not affect it by much. Start by expressing the price in terms of quantities alone, obtaining

$$p_a = 12 - \frac{1}{2}q_a + \frac{1}{2}p_b \tag{27}$$

and

$$p_b = 12 - \frac{1}{2}q_b + \frac{1}{2}p_a. \tag{28}$$

After substituting from (28) into (27) and solving for p_a, we obtain

$$p_a = 24 - \frac{2q_a}{3} - \frac{q_b}{3}. \tag{29}$$

The first-order condition for Apex' maximization problem is

$$\frac{d\pi_a}{dq_a} = 24 - \frac{4q_a}{3} - \frac{q_b}{3} = 0, \tag{30}$$

which gives rise to the reaction function

$$q_a = 18 - \frac{q_b}{4}. \tag{31}$$

We can guess that $q_a = q_b$. It follows from (3) that $q_a = 14.4$ and the market price is 9.6. On checking, you would find this to indeed be a Nash equilibrium.

13.3 Location Models

In section 13.2 we analyzed the Bertrand model with differentiated products using demand functions whose arguments were the prices of both firms. Such a model is suspect because it is not based on primitive assumptions. In particular, the demand functions might not be generated by maximizing any possible utility function. A demand curve with a constant elasticity less than one, for example, is impossible because as the price goes to zero, the amount spent on the commodity goes to infinity. Also, the demand functions (20) and (21) were restricted to prices below a certain level, and it would be good to be able to justify that restriction.

Location models construct demand functions like (20) and (21) from primitive assumptions. In location models, a differentiated product's characteristics are points in a space. If cars differ only in terms of their mileage, the space is a one-

Figure 13.4 Location Models

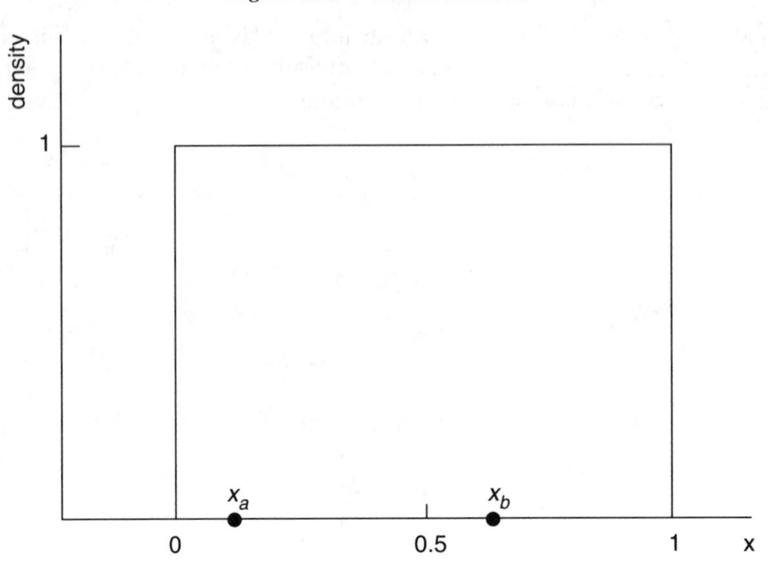

dimensional line. If acceleration is also important, the space is a two-dimensional plane. An easy way to think about this approach is to consider the location where a product is sold. The product "gasoline sold at the corner of Wilshire and Westwood," is different from "gasoline sold at the corner of Wilshire and Fourth." Depending on where consumers live, they have different preferences over the two, but, if prices diverge enough, they will be willing to switch from one gas station to the other.

Location models form a literature in themselves. We will look at the first two models analyzed in the classic article of Hotelling (1929), a model of price choice and a model of location choice. Figure 13.4 shows what is common to both. Two firms are located at points x_a and x_b along a line running from zero to one, with a constant density of consumers throughout. In the Hotelling Pricing Game, firms choose prices for given locations. In the Hotelling Location Game, prices are fixed and the firms choose the locations.

The Hotelling Pricing Game

(Hotelling [1929])

Players

Sellers Apex and Brydox, located at x_a and x_b, where $x_a < x_b$, and a continuum of buyers indexed by location $x \in [0, 1]$.

Order of Play

(1) The sellers simultaneously choose prices p_a and p_b.
(2) Each buyer chooses a seller.

Payoffs

Demand is uniformly distributed on the interval [0, 1] with a density equal to one (think of each consumer as buying one unit). Production costs are zero. Each consumer always buys, so his problem is to minimize the sum of the price plus the linear transport cost, which is θ per unit distance travelled.

$$\pi_{buyer\ at\ x} = -Min\{\theta \mid x_a - x \mid + p_a, \theta \mid x_b - x \mid + p_b\}. \tag{32}$$

$$\pi_a = \begin{cases} 0 & \text{if } p_a - p_b > \theta(x_b - x_a) \text{ (Brydox captures entire market)} & \text{(33a)} \\ p_a & \text{if } p_b - p_a > \theta(x_b - x_a) \text{ (Apex captures entire market)} & \text{(33b)} \\ p_a\left(\frac{1}{2\theta}[(p_b - p_a) + \theta(x_a + x_b)]\right) & \text{otherwise (market is divided)} & \text{(33c)} \end{cases}$$

Brydox has analogous payoffs.

The payoffs result from buyer behavior. A buyer's utility depends on the price he pays and the distance he travels. Price aside, Apex is most attractive to the customer at $x = 0$ ("Customer 0") and least attractive to the customer at $x = 1$ ("Customer 1"). Customer 0 will buy from Apex so long as

$$\theta x_a + p_a < \theta x_b + p_b, \tag{34}$$

which implies that

$$p_a - p_b < \theta(x_b - x_a), \tag{35}$$

which yields payoff (33a). Customer 1 will buy from Brydox if

$$\theta(1 - x_a) + p_a > \theta(1 - x_b) + p_b, \tag{36}$$

which implies that

$$p_b - p_a < \theta(x_b - x_a), \tag{37}$$

which yields payoff (33b).

Very likely, inequalities (35) and (36) are both satisfied, in which case Customer 0 goes to Apex and Customer 1 goes to Brydox. This is the case represented by payoff (33c), and the next task is to find the location of Customer x^*, defined as the customer who is at the boundary between the two markets, indifferent between Apex and Brydox. First, notice that if Apex attracts Customer x_b, he also attracts all $x > x_b$, because beyond x_b the customers' distances from both sellers increase at the same rate. So we know that if there is an indifferent consumer, he is between x_a and x_b. Knowing this, (32) tells us that

$$\theta(x^* - x_a) + p_a = \theta(x_b - x^*) + p_b, \tag{38}$$

so that

$$p_b - p_a = \theta(2x^* - x_a - x_b), \tag{39}$$

and

$$x^* = \frac{1}{2\theta}[(p_b - p_a) + \theta(x_a + x_b)]. \tag{40}$$

Since Apex keeps all the customers between 0 and x^*, equation (40) is the demand function facing Apex so long as he does not set his price so far above Brydox' that he loses even Customer 0. The demand facing Brydox equals $(1 - x^*)$. Note that if $p_b = p_a$, then from (40), $x^* = (x_a + x_b)/2$, independent of θ, which is just what we would expect. Demand is linear in the prices of both firms, and looks similar to demand curves (20) and (21), which were used in section 13.2 for the Bertrand game with differentiated products.

Now that we have found the demand functions, the Nash equilibrium can be calculated in the same way as in section 13.2, by setting up the profit functions for each firm, differentiating with respect to the price of each, and solving the two first-order conditions for the two prices. Subject to the proviso that the firms are willing to pick prices to satisfy inequalities (35) and (37), the resulting Nash equilibrium is

$$p_a = \frac{(2 + x_a + x_b)\theta}{3}, \; p_b = \frac{(4 - x_a - x_b)\theta}{3}. \tag{41}$$

From (41) one can see that Apex charges a higher price if a large x_a gives it more safe customers or if a large x_b makes the number of contestable customers greater. The simplest case is when $x_a = 0$ and $x_b = 1$, which is when (41) tells us that both firms charge a price equal to θ. Profits are positive and increasing in the transportation cost, unless $\theta = 0$, in which case we have returned to the basic Bertrand model.

We cannot rest satisfied with the neat equilibrium of equation (41), because the proviso that firms are willing to charge prices to satisfy (35) and (37) is often violated. Hotelling did not notice this, and fell into a trap which is very easy to fall into using game theory. Economists are accustomed to models in which the calculus approach gives an answer that is both the local optimum and the global optimum. In games like this one, however, the local optimum is not global, because of the discontinuity in the objective function. Vickrey (1964) and D'Aspremont, Gabszewicz, & Thisse (1979) have shown that if x_a and x_b are close together, no pure-strategy equilibrium exists, for reasons similar to why none exists in the Bertrand model with capacity constraints. If both firms charge nonrandom prices, neither would deviate to a slightly different price, but one might deviate to a much lower price that would capture every single customer. But if both firms charged that low price, each would deviate by raising his price slightly. It turns out that if Apex and Brydox are located symmetrically around the center of the interval, then, if $x_a \geq 0.25$ and $x_b \leq 0.75$, no pure-strategy equilibrium exists, although Dasgupta & Maskin (1986b) show that a mixed-strategy equilibrium does exist.

Let us now turn to the choice of location. We will simplify the model by pushing consumers into the background and imposing a single exogenous price on all firms.

The Hotelling Location Game

(Hotelling [1929])

Players

n Sellers.

Order of Play

The sellers simultaneously choose locations $x_i \in [0, 1]$.

Payoffs

Consumers are distributed along the interval [0, 1] with a uniform density equal to one. The price equals one, and production costs are zero. The sellers are ordered by their location so $x_1 \leq x_2 \leq \ldots \leq x_n$, $x_0 \equiv 0$ and $x_{n+1} \equiv 1$. Seller i attracts half the customers from the gaps on each side of him, so that his payoff is

$$\pi_1 = x_1 + \frac{x_2 - x_1}{2}, \tag{42}$$

$$\pi_n = \frac{x_n - x_{n-1}}{2} + 1 - x_n, \tag{43}$$

or, for $i = 2, \ldots n-1$,

$$\pi_i = \frac{x_i - x_{i-1}}{2} + \frac{x_{i+1} - x_i}{2}. \tag{44}$$

With **one seller**, the location does not matter in this model, since the customers are captive. If price were a choice variable and demand were elastic, we would expect the monopolist to locate at $x = 0.5$.

With **two sellers**, both firms locate at $x = 0.5$, regardless of whether or not demand is elastic. This is a stable Nash equilibrium, as can be seen by inspecting figure 13.4 and imagining best responses to each other's location. The best response is always to locate ε closer to the center of the interval than one's rival. When both firms do this, they end up splitting the market since both of them end up exactly at the center.

With **three sellers** the model does not have a Nash equilibrium in pure strategies. Consider any strategy profile in which each player locates at a separate point. Such a strategy profile is not an equilibrium, because the two players nearest the ends would edge in to squeeze the middle player's market share. But if a strategy profile has any two players at the same point, the third player would be able to acquire a share of at least $(0.5 - \varepsilon)$ by moving next to them; and if the third player's share is that large, one of the doubled-up players would deviate by jumping to his other side and capturing his entire market share. The only equilibrium is in mixed strategies. Shaked (1982) has computed the symmetric mixing probability density $m(x)$ to be

$$m(x) = \begin{cases} 2 & \text{if } \frac{1}{4} \leq x \leq \frac{3}{4} \\ 0 & \text{otherwise} \end{cases} \tag{45}$$

Strangely enough, three is a special number. With **more than three sellers**, an equilibrium in pure strategies does exist (Eaton & Lipsey [1975]). Dasgupta & Maskin (1986b), as amended by Simon (1987), have also shown that an equilibrium, possibly in mixed strategies, exists for any number of players n in a space of any dimension m.

Since prices are inflexible, the competitive market does not achieve efficiency. A benevolent social planner or a monopolist who could charge higher prices if he located his outlets closer to more consumers would choose different locations than competing firms. In particular, when two competing firms both locate in the center of the line, consumers are no better off than if there were just one firm. The average distance of a consumer from a seller would be minimized by setting $x_1 = 0.25$ and $x_2 = 0.75$, the locations that would be chosen either by the social planner or the monopolist.

13.4 Comparative Statics and Supermodular Games

Comparative statics is the analysis of what happens to endogenous variables in a model when the exogenous variables change. This is a central part of economics. When wages rise, for example, we wish to know how the price of steel will change in response. Game theory presents special problems for comparative statics, since when a parameter changes, not only does Smith's equilibrium strategy change in response, but Jones' strategy changes as a result of Smith's change as well. A small change in the parameter might produce a large change in the equilibrium because of feedback between the different players' strategies.

Let us use a differentiated Bertrand game as an example. Suppose there are N firms, and for firm n the demand curve is

$$Q_n = Max\{\alpha - \beta_n p_n + \sum_{m \neq n} \gamma_m p_m, 0\}, \tag{46}$$

with $\alpha \in (0, \infty)$, $\beta_n \in (0, \infty)$, and $\gamma_n \in (0, \infty)$ for all n. Assume that the effect of p_n on firm n's sales is larger than the effect of the other firms' prices, so that

$$\beta_n > \sum_{m \neq n} \gamma_m. \tag{47}$$

Let firm n have constant marginal cost κc_n, where $\kappa \in \{1, 2\}$ and $c_n \in (0, \infty)$, and let us assume that firm n's costs are low enough that it does operate in equilibrium. The shift variable κ represents the effect of the political regime on costs. The payoff function for firm n is

$$\pi_n = (p_n - \kappa c_n)(\alpha - \beta_n p_n + \sum_{m \neq n} \gamma_m p_m). \tag{48}$$

Firms choose prices simultaneously.

Does this game have an equilibrium? Does it have several equilibria? What happens to the equilibrium price if a parameter such as c_n or κ changes? These are difficult questions because if c_n increases, the immediate effect is to change firm n's price, but the other firms will react to the price change, which in turn will affect n's price. Moreover, this is not a symmetric game—the costs and demand curves differ from firm to firm, which could make algebraic solution of the Nash equilibrium quite messy. It is not even clear whether the equilibrium is unique.

Two approaches to comparative statics can be used here: the implicit function theorem, and supermodularity. We will look at each in turn.

The Implicit Function Theorem

The implicit-function theorem says that if $f(x, y) = 0$,

$$\frac{\partial x}{\partial y} = -\left(\frac{\partial f}{\partial y} \Big/ \frac{\partial f}{\partial x}\right). \tag{49}$$

This is especially useful if x is a choice variable and y a parameter, because then the first-order condition takes the form $f(x, y) = 0$, and the second-order condition determines the sign of $\partial f/\partial x$. One only has to make certain that the solution is an interior solution, so the first- and second-order conditions are valid.

In the differentiated Bertrand game, equilibrium prices will lie inside the interval $(c_n, \bar{p})$ for some large number $\bar{p}$, because a price of c_n would yield zero profits, rather than the positive profits of a slightly higher price, and $\bar{p}$ can be chosen to yield zero quantity demanded and hence zero profits. The equilibrium or equilibria are, therefore, interior solutions, in which case in this well-behaved problem they satisfy the first-order condition,

$$\frac{\partial \pi_n}{\partial p_n} = \alpha - 2\beta_n p_n + \sum_{m \neq n} \gamma_m p_m + \kappa c_n \beta_n = 0, \tag{50}$$

and the second-order condition,

$$\frac{\partial^2 \pi_n}{\partial p_n^2} = -2\, \beta_n < 0. \tag{51}$$

We can apply the implicit function theorem by letting $\partial \pi_n(p_n, c_n)/\partial p_n = 0$ from equation (50) be our $f(x, y) = 0$ and using equation (49). Then

$$\begin{aligned} \frac{\partial p_n}{\partial c_n} &= \left(\frac{\frac{\partial^2 \pi_n}{\partial p_n \partial c_n}}{\frac{\partial^2 \pi_n}{\partial p_n^2}} \right) \\ &= -\left(\frac{\kappa \beta_n}{-2\beta_n} \right) \\ &= \frac{\kappa}{2}. \end{aligned} \tag{52}$$

Thus, an increase in firm n's individual cost parameter increases its price at a rate of $\kappa/2$. Keep in mind, however, that the implicit-function theorem only tells us about infinitesimal changes, not finite changes. If c_n increases enough, the nature of the equilibrium changes drastically, because firm n goes out of business.

We cannot go on to discover the effect of changing κ on p_n, because κ is a discrete variable, and the implicit-function theorem applies only to continuous variables. The implicit-function theorem is nonetheless very useful when it does apply. This is a simple example, but the approach can be used even when the functions involved are very complicated. In complicated cases, knowing that the second-order condition holds allows the modeller to avoid having to determine the sign of the denominator if all that interests him is the sign of the relationship between the two variables.

Supermodularity

The second approach uses the idea of the supermodular game, an idea related to that of strategic complements. Suppose that there are N players in a game, subscripted by m and n, and that player n has a strategy consisting of k_n elements, subscripted by i and j, so his strategy is the vector $x_n = (x_{n1}, \ldots, x_{nk_n})$. Let his strategy set be S_n and his payoff function $\pi_n(x_n, x_{-n}; \tau)$, where τ represents a fixed parameter. We say that the game is a **smooth supermodular game** if the following four conditions are satisfied:

(A1′) The strategy set is an interval in $\boldsymbol{R}^{kn}$:

$$S_n = [\underline{x}_n, \overline{x}_n]. \tag{53}$$

(A2′) π_n is twice continuously differentiable on S_n.

(A3′) (supermodularity) Increasing one component of player n's strategy does not decrease the net marginal benefit of any other component: for all n, and all i and j such that $1 \leq i < j \leq k_n$,

$$\frac{\partial^2 \pi_n}{\partial x_{ni} \partial x_{nj}} \geq 0. \tag{54}$$

(A4′) (increasing differences in one's own and other strategies) Increasing one component of n's strategy does not decrease the net marginal benefit of increasing any component of player m's strategy: for all $n \neq m$, and all i and j such that $1 \leq i \leq k_n$ and $1 \leq j \leq k_m$,

$$\frac{\partial^2 \pi_n}{\partial x_{ni} \partial x_{mj}} \geq 0. \tag{55}$$

In addition, we will be able to talk about the comparative statics of smooth supermodular games if a fifth condition is satisfied, the increasing differences condition, (A5′).

(A5′) (increasing differences in one's own strategies and parameters) Increasing parameter c does not decrease the net marginal benefit to player n of any component of his own strategy: for all n, and all i such that $1 \leq i \leq k_n$,

$$\frac{\partial^2 \pi_n}{\partial x_{ni} \partial \tau} \geq 0. \tag{56}$$

The heart of supermodularity is in assumptions (A3′) and (A4′). Assumption (A3′) says that the components of player n's strategies are all **complementary inputs**; when one component increases, it is worth increasing the other components too. This means that even if a strategy is a complicated one, one can still arrive at qualitative results about the strategy, because all the components of the optimal strategy will move in the same direction together. Assumption (A4′) says that the strategies of players m and n are **strategic complements**; when player m increases a component of his strategy, player n will want to do so also. When the strategies of the players reinforce each other in this way, the feedback between them is less tangled than if they undermined each other.

I have put primes on the assumptions because they are the special cases, for smooth games, of the general definition of supermodular games (see list in Appendix B). Smooth games use differentiable functions, but the supermodularity theorems apply more generally. One condition that is relevant here is condition (A5):

(A5) π_n has increasing differences in x_n and τ, for fixed x_{-n}; for all $x_n \geq x_n'$, the difference $\pi_n(x_n, x_{-n}, \tau) - \pi_n(x_n', x_{-n}, \tau)$ is nondecreasing with respect to τ.

Is the differentiated Bertrand game supermodular? The strategy set can be restricted to $[c_n, \bar{p}]$ for player n, so (A1′) is satisfied. π_n is twice continuously differentiable on the interval $[c_n, \bar{p}]$, so (A2′) is satisfied. A player's strategy has just one component, p_n, so (A3′) is immediately satisfied. The following inequality is true:

$$\frac{\partial^2 \pi_n}{\partial p_n \partial p_m} = \gamma_m > 0, \tag{57}$$

so (A4′) is satisfied. And it is also true that

$$\frac{\partial^2 \pi_n}{\partial p_n \partial c_n} = \kappa \beta_n > 0, \tag{58}$$

so (A5′) is satisfied for c_n.

From equation (50), $\partial \pi_n / \partial p_n$ is increasing in κ, so $\pi_n(p_n, p_{-n}, \kappa) - \pi_n(p_n', p_{-n}, \kappa$ is nondecreasing in κ for $p_n > p_n'$, and (A5′) is satisfied for κ.

Thus, all the assumptions are satisfied. This being the case, a number of theorems can be applied. Two of them are Theorems 13.1 and 13.2.

Theorem 13.1

If the game is supermodular, there exists a largest and a smallest Nash equilibrium in pure strategies.

Theorem 13.2

If the game is supermodular and assumption (A5) or (A5′) is satisfied, the largest and the smallest equilibrium are nondecreasing functions of the parameter τ.

Applying Theorems 13.1 and 13.2 yields the following results for the differentiated Bertrand game:

(1) There exists a largest and a smallest Nash equilibrium in pure strategies (Theorem 13.1).
(2) The largest and smallest equilibrium prices for firm n are nondecreasing functions of the cost parameters c_n and κ (Theorem 13.2).

Note that supermodularity has yielded comparative statics on κ, unlike the implicit-function theorem. It yields weaker comparative statics on c_n, however, because it just finds the effect of c_n on p_n^* to be nondecreasing, rather than telling us its value.

Theorem 13.2 is also useful in proving that the equilibrium here is, in fact, unique—the largest and the smallest equilibrium are one and the same. Since

$$\frac{\partial^2 \pi_n}{\partial p_n \partial p_m} = \gamma_m, \tag{59}$$

it will be true that

$$-\left(\frac{\partial^2 \pi_n}{\partial p_n^2}\right) > \sum_{m \neq n} \frac{\partial^2 \pi_n}{\partial p_n \partial p_m}. \tag{60}$$

Condition (60) is what is commonly called a **dominant-diagonal condition.** It says that direct effects on profits are more important than all the indirect effects, so if one expresses the second derivatives in matrix form, the main diagonal would have the largest elements. For a three-firm case that matrix would be

$$\begin{bmatrix} \frac{\partial^2 \pi_1}{\partial p_1^2} & \frac{\partial^2 \pi_1}{\partial p_1 \partial p_2} & \frac{\partial^2 \pi_1}{\partial p_1 \partial p_3} \\ \frac{\partial^2 \pi_2}{\partial p_2 \partial p_1} & \frac{\partial^2 \pi_2}{\partial p_2^2} & \frac{\partial^2 \pi_2}{\partial p_2 \partial p_3} \\ \frac{\partial^2 \pi_3}{\partial p_3 \partial p_1} & \frac{\partial^2 \pi_3}{\partial p_3 \partial p_2} & \frac{\partial^2 \pi_3}{\partial p_3^2} \end{bmatrix} \tag{61}$$

Suppose there were two equilibrium price profiles, p and $\hat{p}$. Theorem 13.1 says that the largest and smallest equilibria can be ranked, so for every strategy in the strategy profile, it would be true that $\hat{p} \geq p$. But because the first-order condition applies at both equilibria, we know that

$$\frac{\partial \pi_n(p)}{\partial p_n} - \frac{\partial \pi_n(\hat{p})}{\partial p_n} = 0. \tag{62}$$

One can rewrite equation (62) differently. Starting at equilibrium p and moving to $\hat{p}$, the first derivative would change as all the components of p changed. If we use t to index the slow changes in the components, we can write these changes as

$$\int_0^1 \left\{ \left((\hat{p}_n - p_n) \cdot \frac{\partial^2 \pi_n[t\hat{p} + (1-t)p]}{\partial p_n^2} \right) + \left(\sum_{m \neq n} (\hat{p}_m - p_m) \cdot \frac{\partial^2 \pi_n[t\hat{p} + (1-t)p]}{\partial p_n \partial p_m} \right) \right\} dt. \tag{63}$$

Expression (63) equals expression (62). But from equation (60), expression (63) must be negative, and equation (62) equals zero. This is a contradiction, so there cannot really be two different equilibria. The biggest and smallest equilibria are one and the same, and the equilibrium is unique.

13.5 Durable Monopoly

Introductory economics courses are often vague on the issue of the time period over which transactions take place. When a diagram shows the supply and demand for widgets, the x-axis is labelled "widgets," not "widgets per week" or "widgets per year." Also, the diagram splits off one time period from future time periods, using the implicit assumption that supply and demand in one period is unaffected by events of future periods. One problem with this on the demand side is that the purchase of a good which lasts for more than one use is an investment; although the price is paid now, the utility from the good continues into the future. If Smith buys a house, he is buying not just the right to live in the house tomorrow, but the right to live in it for many years to come, or even to live in it for a few years and then to sell the remaining years to someone else. The continuing utility he receives from this durable good is called its **service flow**. Even though he may not intend to rent out the house, it is an investment decision for him because it trades off present expenditure for future utility. Since even a shirt produces a service flow over more than an instant of time, the durability of goods presents difficult definitional problems for national income accounts. Houses are counted as part of national investment (and an estimate of their service flow as part of services consumption), automobiles as durable goods consumption, and shirts as nondurable goods consumption, but all are to some extent durable investments.

In microeconomic theory, "durable monopoly" refers not to monopolies that last a long time, but to monopolies that sell durable goods. These present a curious problem. When a monopolist sells something like a refrigerator to a consumer, that consumer drops out of the market until the refrigerator wears out. The demand curve is, therefore, changing over time as a result of the monopolist's choice of price, which means that the modeller should not make his decisions in one period and ignore future periods. Demand is not **time separable**, because a rise in price at time t_1 affects the quantity demanded at time t_2.

The durable monopolist has a special problem because in a sense he does have a competitor—himself in the later periods. If he were to set a high price in the first period, thereby removing high-demand buyers from the market, he would be tempted to set a lower price in the next period to take advantage of the remaining consumers. But if it were known that he would lower the price, the high-demand buyers would not buy at a high price in the first period. The threat of the future low price forces the monopolist to keep his current price low.

To formalize this situation, let the seller have a monopoly on a durable good which lasts two periods. He must set a price for each period, and the buyer must decide what quantity to buy in each period. Because this one buyer is meant to represent the entire market demand, the moves are ordered so that he has no market power, as in the principal-agent models in section 7.3. Alternatively, the buyer can be viewed as representing a continuum of consumers (see Coase [1972] and Bulow [1982]). In this interpretation, instead of "the buyer" buying q_1 in the first period, q_1 of the buyers each buy one unit in the first period.

Durable Monopoly

Players

A buyer and a seller.

Order of Play

(1) The seller picks the first-period price, p_1.
(2) The buyer buys quantity q_1 and consumes the service flow q_1.
(3) The seller picks the second-period price, p_2.
(4) The buyer buys an additional quantity q_2 and consumes the service flow $(q_1 + q_2)$.

Payoffs

Production cost is zero and there is no discounting. The seller's payoff is his revenue, and the buyer's payoff is the sum across periods of his benefits from consumption minus his expenditure. His benefits arise from his being willing to pay as much as

$$B(q_t) = 60 - \frac{q_t}{2} \tag{64}$$

for the marginal unit service flow consumed in period t, as shown in figure 13.5. The payoffs are therefore

$$\pi_{seller} = q_1 p_1 + q_2 p_2 \tag{65}$$

and

$$\pi_{buyer} = [consumer\ surplus_1] + [consumer\ surplus_2]$$
$$= [total\ benefit_1 - expenditure_1] + [total\ benefit_2 - expenditure_2] \tag{66}$$
$$= \left[\frac{(60 - B(q_1))q_1}{2} + B(q_1)q_1 - p_1q_1\right]$$
$$+ \left[\frac{60 - B(q_1 + q_2)}{2}(q_1 + q_2) + B(q_1 + q_2)(q_1 + q_2) - p_2q_2\right].$$

Thinking about durable monopoly is hard because we are used to one-period models in which the demand curve, which relates the price to the quantity demanded, is identical to the marginal-benefit curve, which relates the marginal-benefit to the quantity consumed. Here, the two curves are different. The marginal benefit curve is the same in each period, since it is part of the rules of the game, relating consumption to utility. The demand curve will change over time and depends on the equilibrium strategies, depending as it does on the number of periods left in which to consume the good's services, expected future prices, and the quantity already owned. Marginal benefit is a given for the buyer; quantity demanded is his strategy.

The buyer's total benefit in period 1 is the dollar value of his utility from his purchase of q_1, which equals the amount he would have been willing to pay to rent q_1. This is composed of the two areas shown in figure 13.5a, the upper tri-

Figure 13.5 Buyer's Marginal Benefit per Period in The Game of Durable Monopoly

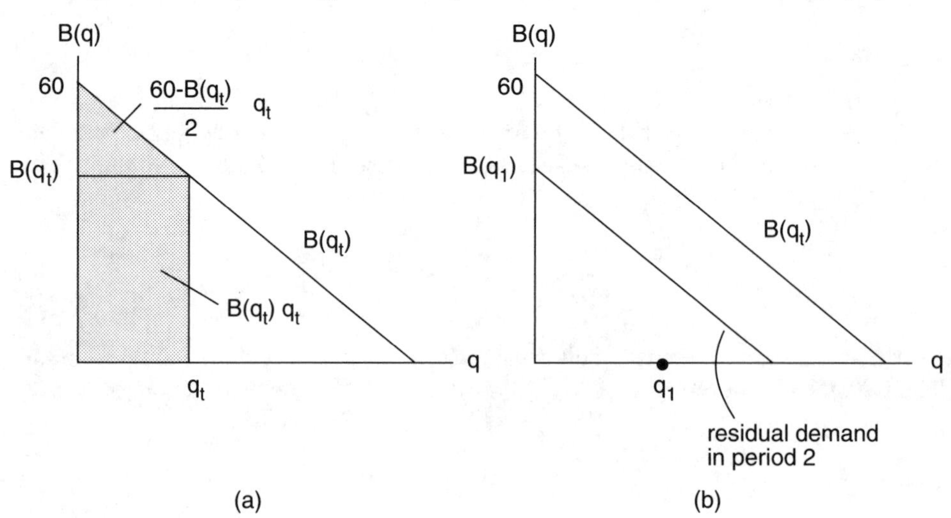

angle of area (½) $(q_1 + q_2)(60 - B(q_1 + q_2))$ and the lower rectangle of area $(q_1 + q_2)B(q_1 + q_2)$. From this must be subtracted his expenditure in period 1, p_1q_1, to obtain what we might call his consumer surplus in the first period. Note that p_1q_1 will not be the lower rectangle, unless by some strange accident, and the "consumer surplus" might easily be negative, since the expenditure in period 1 will also yield utility in period 2 because the good is durable.

To find the equilibrium price path one cannot simply differentiate the seller's utility with respect to p_1 and p_2, because that would violate the sequential rationality of the seller and the rational response of the buyer. Instead, one must look for a subgame perfect equilibrium, which means starting in the second period and discovering how much the buyer would purchase given his first-period purchase of q_1, and what second-period price the seller would charge given the buyer's second-period demand function.

In the first period, the marginal unit consumed was the q_1-th. In the second period, it will be the $(q_1 + q_2)$-th. The residual demand curve after the first period's purchases is shown in figure 13.5b. It is a demand curve very much like the demand curve resulting from intensity rationing in the capacity-constrained Bertrand game of section 13.2, as shown in figure 13.2a. The most intense portion of the buyer's demand, up to q_1 units, has already been satisfied, and what is left begins with a marginal benefit of $B(q_1)$, and falls at the same slope as the original marginal benefit curve. The equation for the residual demand is therefore, using equation (66),

$$p_2 = B(q_1) - \frac{q_2}{2} = 60 - \frac{q_1}{2} - \frac{q_2}{2}. \tag{67}$$

Solving for the monopoly quantity, q_2^*, the seller maximizes q_2p_2, solving the problem

$$\underset{q_2}{Maximize}\ q_2 \left(60 - \frac{q_1 + q_2}{2}\right), \tag{68}$$

which generates the first-order condition

$$60 - q_2 - \frac{q_1}{2} = 0, \tag{69}$$

so that

$$q_2^* = 60 - \frac{q_1}{2}. \tag{70}$$

From equations (64) and (70), it can be seen that $p_2^* = 30 - q_1/4$.

We must now find q_1^*. In period one, the buyer looks ahead to the possibility of buying in period two at a lower price. Buying in the first period has two benefits: consumption of the service flow in the first period and consumption of the service flow in the second period. The price he would pay for a unit in period one cannot exceed the marginal benefit from the first-period service flow in period one plus the foreseen value of p_2, which, from (70) is $30 - q_1/4$. If the seller chooses to sell q_1 in the first period, therefore, he can do so at the price

$$\begin{aligned} p_1(q_1) &= B(q_1) + p_2 \\ &= \left(60 - \frac{q_1}{2}\right) + \left(30 - \frac{q_1}{4}\right), \\ &= 90 - \frac{3}{4}q_1. \end{aligned} \tag{71}$$

Knowing that in the second period he will choose q_2 according to (70), the seller combines (70) with (71) to give the maximand in the problem of choosing q_1 to maximize profit over the two periods, which is

$$\begin{aligned} (p_1q_1 + p_2q_2) &= \left(90 - \frac{3}{4}q_1\right)q_1 + \left(30 - \frac{q_1}{4}\right)\left(60 - \frac{q_1}{2}\right) \\ &= 1800 + 60q_1 - \frac{5}{8}q_1^2, \end{aligned} \tag{72}$$

which has the first-order condition

$$60 - \frac{5}{4}q_1 = 0, \tag{73}$$

so that

$$q_1^* = 48 \tag{74}$$

and, making use of (71), $p_1^* = 54$.

It follows from (70) that $q_2^* = 36$ and $p_2 = 18$. The seller's profits over the two periods are $\pi_s = 3{,}240$ $(= 54(48) + 18(36))$.

The purpose of these calculations is to compare the situation with three other market structures: a competitive market, a monopolist who rents instead of selling, and a monopolist who commits to selling only in the first period.

A *competitive market* bids down the price to the marginal cost of zero. Then, $p_1 = 0$ and $q_1 = 120$ from (66), and profits equal zero.

If the monopolist *rents* instead of selling, then equation (66) is like an ordinary demand equation, because the monopolist is effectively selling the good's services separately each period. He could rent a quantity of 60 each period at a rental fee of 30 and his profits would sum to $\pi_s = 3{,}600$. That is higher than 3,240, so profits are higher from renting than from selling outright. The problem with selling outright is that the first-period price cannot be very high or the buyer knows that the seller will be tempted to lower the price once the buyer has bought in the first period. Renting avoids this problem.

If the monopolist can *commit to not producing in the second period*, he will do just as well as the monopolist who rents, since he can sell a quantity of 60 at a price of 60, the sum of the rents for the two periods. An example is the artist who breaks the plates for his engravings after a production run of some announced size. We must also assume that the artist can convince the market that he has broken the plates. People joke that the best way an artist can increase the value of his work is by dying, and that, too, fits the model.

If the modeller ignored sequential rationality and simply looked for the Nash equilibrium that maximized the payoff of the seller by his choice of p_1 and p_2, he would come to the commitment result. An example of such an equilibrium is ($p_1 = 60$, $p_2 = 200$, *Buyer purchases according to* $q_1 = 120 - p_1$, *and* $q_2 = 0$). This is Nash because neither player has incentive to deviate given the other's strategy, but it fails to be subgame perfect, because the seller should realize that if he deviates and chooses a lower price once the second period is reached, the buyer will respond by deviating from $q_2 = 0$ and will buy more units.

With more than two periods, the difficulties of the durable-goods monopolist become even more striking. In an infinite-period model without discounting, if the marginal cost of production is zero, the equilibrium price for outright sale instead of renting is constant—at zero! Think about this in the context of a model with many buyers. Early consumers foresee that the monopolist has an incentive to cut the price after they buy, in order to sell to the remaining consumers who value the product less. In fact, the monopolist would continue to cut the price and sell more and more units to consumers with weaker and weaker demand until the price fell to marginal cost. Without discounting, even the high-valuation consumers refuse to buy at a high price, because they know they could wait until the price falls to zero. And this is not a trick of infinity: a large number of periods generates a price close to zero.

We can also use the durable monopoly model to think about the durability of the product. If the seller can develop a product so flimsy that it only lasts one period, that is equivalent to renting. A consumer is willing to pay the same price to own a one-hoss shay that he knows will break down in one year as he would pay to rent it for a year. Low durability leads to the same output and profits as renting, which explains why a firm with market power might produce goods that wear out quickly. The explanation is not that the monopolist can use his market power to inflict lower quality on consumers—after all, the price he receives is lower too—but that the lower durability makes it so credible to high-valuation buyers that the seller expects their business in the future and will not lower his price.

Notes

N13.1 Quantities as Strategies: the Cournot Equilibrium Revisited

- Articles on the existence of a pure-strategy equilibrium in the Cournot model include Novshek (1985) and Roberts & Sonnenschein (1976).
- **Merger in a Cournot Model.** A problem with the Cournot model is that a firm's best policy is often to split up into separate firms. Apex gets half the industry profits in a duopoly game. If Apex split into firms $Apex_1$ and $Apex_2$, it would get two-thirds of the profit in the Cournot triopoly game, even though industry profit falls.

This point was made by Salant, Switzer, & Reynolds (1983) and is the subject of problem 13.2. It is interesting that nobody noted this earlier, given the intense interest in Cournot models. The insight comes from approaching the problem by asking whether a player could improve his lot if his strategy space were expanded in reasonable ways.

- An ingenious look at how the number of firms in a market affects the price is Bresnahan & Reiss (1991), which looks empirically at a number of very small markets with one, two, three or more competing firms. They find a big decline in the price from one to two firms, a smaller decline from two to three, and not much change thereafter.

  Exemplifying theory, as discussed in the Introduction to this book, lends itself to explaining particular cases, but it is much less useful for making generalizations across industries. Empirical work associated with exemplifying theory tends to consist of historical anecdote rather than the linear regressions to which economics has become accustomed. Generalization and econometrics are still often useful in industrial organization, however, as Bresnahan & Reiss (1991) shows. The most ambitious attempt to connect general data with the modern theory of industrial organization is Sutton's 1991 book, *Sunk Costs and Market Structure*, which is an extraordinarily well-balanced mix of theory, history, and numerical data.
- The idea of conjectural variation is attributed to Bowley (1924) and is discussed in Jacquemin (1985) and Varian (1992, p. 302).
- Do not confuse a conjectural variation of -1 with perfect competition, even though both may lead to the efficient output. In perfect competition, individuals do not believe that they affect the rest of the market, but if $CV = -1$, a firm believes that other firms will cut back when it produces more. Perfect competition is more like a game with players so small relative to the market that even though $CV = 0$, as in Nash equilibrium, each player correctly believes that his actions have a trivial effect on the market price.

N13.2 Prices as Strategies: the Bertrand Equilibrium

- Intensity rationing has also been called **efficient rationing**. Sometimes, however, as in section 13.2, this rationing rule is inefficient. Some low-intensity customers left facing the high price decide not to buy the product even though their benefit is greater than its marginal cost. The reason intensity rationing has been thought to be efficient is that it is efficient if the rationed-out customers are unable to buy at any price.
- OPEC has tried both price and quantity controls ("OPEC, Seeking Flexibility, May Choose Not to Set Oil Prices, but to Fix Output," *Wall Street Journal*, October 8, 1987, pp. 2, 29; "Saudi King Fahd is Urged by Aides To Link Oil Prices to Spot Markets," *Wall Street Journal*, October 7, 1987, p. 2). Weitzman (1974) is the classic reference on price vs. quantity control by regulators, although he does not use the context of oligopoly. The decision rests partly on enforceability, and OPEC has also hired accounting firms to monitor prices ("Dutch Accountants Take On a Formidable Task: Ferreting Out 'Cheaters' in the Ranks of OPEC," *Wall Street Journal*, February 26, 1985, p. 39).
- Kreps & Scheinkman (1985) show how capacity choice and Bertrand pricing can lead to a Cournot outcome. Two firms face downward-sloping market demand. In the first stage of the game, they simultaneously choose capacities, and in the second stage they simultaneously choose prices (possibly by mixed strategies). If a firm cannot satisfy the demand facing it in the second stage (because of the capacity limit), it uses intensity rationing. (The results depend on this.) The unique subgame perfect equilibrium is for each firm to choose the Cournot capacity and price.
- Haltiwanger & Waldman (unpub) have suggested a dichotomy applicable to many different games between players who are **responders**, choosing their actions flexibly, and those who are **nonresponders**, who are inflexible. A player might be a nonresponder because he is irrational, because he moves first, or simply because his strategy set is small. The categories are used in a second dichotomy, between games exhibiting **synergism**, in which responders choose to do whatever the majority do (upward sloping reaction curves), and games exhibiting **congestion**, in which responders want to join the minority (downward sloping reaction curves). Under synergism, the equilibrium is more like what it would be if all the players were nonresponders; under congestion, the responders have more influence. Haltiwanger & Waldman apply the dichotomies to network externalities, efficiency wages, and reputation.

  If the reaction functions of two firms are upward sloping, it has been said that the actions are **strategic complements**, and if they are downward sloping, **strategic substitutes** (Bulow, Geanakoplos, & Klemperer [1985]). Gal-Or (1985) notes that if reaction curves slope downwards (as in Cournot) there is a first-mover advantage, while if they slope upwards (as in differentiated Bertrand) there is a second-mover advantage.

- Section 13.3 shows how to generate demand curves (20) and (21) using a location model, but they can also be generated directly by a quadratic utility function. Dixit (1979) states that with respect to three goods 0, 1, and 2, the utility function

$$U = q_0 + \alpha_1 q_1 + \alpha_2 q_2 - \frac{1}{2}\left(\beta_1 q_1^2 + 2\gamma q_1 q_2 + \beta_2 q_2^2\right) \tag{75}$$

(where the constants α_1, α_2, β_1, and β_2 are positive and $\gamma^2 \leq \beta_1\beta_2$) generates the inverse demand functions

$$p_1 = \alpha_1 - \beta_1 q_1 - \gamma q_2 \tag{76}$$

and

$$p_2 = \alpha_2 - \beta_2 q_2 - \gamma q_1. \tag{77}$$

N13.3 Location Models

- For a booklength treatment of location models, see Greenhut & Ohta (1975).
- Vickrey notes the possible absence of a pure-strategy equilibrium in Hotelling's model in pp. 323–24 of his 1964 book *Microstatics*, but does not go on to consider the mixed-strategy equilibrium that can be found in D'Aspremont et al. (1979).
- Location models and switching-cost models are attempts to go beyond the notion of a market price. Antitrust cases are good sources for descriptions of the complexities of pricing in particular markets. See, for example, Sultan's 1974 book on electrical equipment in the 1950's, or antitrust opinions such as *US v. Addyston Pipe & Steel Co. et al.*, 85 Fed 271.
- It is important in location models whether the positions of the players on the line are moveable. See, for example, Lane (1980).
- The location games in this chapter model use a one-dimensional space with end points, i.e., a line segment. Another kind of one-dimensional space is a circle (not to be confused with a disk). The difference is that no point on a circle is distinctive, so no consumer preference can be called extreme. It is, if you like, Peoria versus Berkeley. The circle might be used for modelling convenience or because it fits a situation: e.g., airline flights spread over the 24 hours of the day. With two players, the Hotelling location game on a circle has a continuum of pure-strategy equilibria that are of one of two types: both players locating at the same spot, versus players separated from each other by 180°. The three-player model also has a continuum of pure-strategy equilibria, each player separated from another by 120°, in contrast to the nonexistence of a pure-strategy equilibrium when the game is played on a line segment.
- Characteristics such as the color of cars could be modelled as location, but only on a player-by-player basis, because they have no natural ordering. While Smith's ranking of (red = 1, yellow = 2, blue = 10) could be depicted on a line, if Brown's ranking is (red = 1, blue = 5, yellow = 6) we cannot use the same line for him. In the text, the characteristic was something like physical location, about which people may have different preferences but agree on what positions are close to what other positions.

N13.5 Durable Monopoly

- The proposition that price falls to marginal cost in a durable monopoly with no discounting and infinite time is called the "Coase Conjecture," after Coase (1972). It is really a proposition and not a conjecture, but alliteration was too strong to resist.
- Gaskins (1974) has written a well-known article on the problem of the durable monopolist who foresees that he will be creating his own future competition in the future because his product can recycled, using the context of the aluminum market.

- Leasing by a durable monopoly was the main issue in the antitrust case *US v. United Shoe Machinery Corporation*, 110 F. Supp. 295 (1953), but not because it increased monopoly profits. The complaint was rather that long-term leasing impeded entry by new sellers of shoe machinery, a curious idea when the proposed alternative was outright sale. More likely, leasing was used as a form of financing for the machinery customers; by leasing, they did not need to borrow as they would have to do if it was a matter of financing a purchase. See Wiley *et al.* (1990).
- Another way out of the durable monopolist's problem is to give best-price guarantees to customers, promising to refund part of the purchase price if any future customer gets a lower price. Perversely, this hurts consumers, because it stops the seller from being tempted to lower his price. The "most-favored-nation" contract, which is the analogous contract in markets with several sellers, is analyzed by Holt & Scheffman (1987), for example, who demonstrate how it can maintain high prices, and Png & D. Hirshleifer (1987), who show how it can be used to price discriminate between different types of buyers.
- The durable monopoly model should remind you of bargaining under incomplete information. Both situations can be modelled using two periods, and in both situations the problem for the seller is that he is tempted to offer a low price in the second period after having offered a high price in the first period. In the durable monopoly model this would happen if the high-valuation sellers bought in the first period and thus were absent from consideration by the second period. In the bargaining model this would happen if the seller rejected the first-period offer and could conclude that he must have a low-valuation and act accordingly in the second period. With a rational buyer, neither of these things can happen, and the models' complications arise from the attempt of the seller to get around the problem.

  In the durable-monopoly problem, this would make the high-valuation buyers wait until the second period to buy, and in the bargaining problem, it would make them wait until the second period. For further discussion, see the survey by Kennan & R. Wilson (1993).

Problems

13.1: Differentiated Bertrand with Advertising.

Two firms that produce substitutes are competing with demand curves

$$q_1 = 10 - \alpha p_1 + \beta p_2 \tag{78}$$

and

$$q_2 = 10 - \alpha p_2 + \beta p_1. \tag{79}$$

Marginal cost is constant at $c = 3$. A player's strategy is his price. Assume that $\alpha > \beta/2$.

(13.1a) What is the reaction function for Firm 1? Draw the reaction curves for both firms.

(13.1b) What is the equilibrium? What is the equilibrium quantity for Firm 1?

(13.1c) Show how Firm 2's reaction function changes when β increases. What happens to the reaction curves in the diagram?

(13.1d) Suppose that an advertising campaign could increase the value of β by one, and that this would increase the profits of each firm by more than the cost of the campaign. What does this mean? If either firm could pay for this campaign, what game would result between them?

13.2: Cournot Mergers.[1]

There are three identical firms in an industry with demand given by $P = 1 - Q$, where $Q = q_1 + q_2 + q_3$. The marginal cost is zero.

(13.2a) Compute the Cournot equilibrium price and quantities.

(13.2b) How do you know that there are no asymmetric Cournot equilibria, in which one firm produces a different amount than the others?

(13.2c) Show that if two of the firms merge, their shareholders are worse off.

13.3: Differentiated Bertrand.

Two firms that produce substitutes have the demand curves

$$q_1 = 1 - \alpha p_1 + \beta(p_2 - p_1) \tag{80}$$

and

$$q_2 = 1 - \alpha p_2 + \beta(p_1 - p_2), \tag{81}$$

where $\alpha - \beta$. Marginal cost is constant at c, where $c < 1/\alpha$. A player's strategy is his price.

(13.3a) What are the equations for the reaction curves $p_1(p_2)$ and $p_2(p_1)$? Draw them.

(13.3b) What is the pure-strategy equilibrium for this game?

(13.3c) What happens to prices if α, β, or c increase?

(13.3d) What happens to each firm's price if α increases, but only Firm 2 realizes it (and Firm 2 knows that Firm 1 is uninformed)? Would Firm 2 reveal the change to Firm 1?

Problem 13.4: Asymmetric Cournot Duopoly.

Apex has variable costs of q_a^2 and a fixed cost of 1000, while Brydox has variable costs of $2q_b^2$ and no fixed cost. Demand is $p = 115 - q_a - q_b$.

(13.4a) What is the equation for Apex' Cournot reaction function?

(13.4b) What is the equation for Brydox' Cournot reaction function?

(13.4c) What are the outputs and profits in the Cournot equilibrium?

[1] See Salant, Switzer, & Reynolds (1983).

14 Entry

14.1 Innovation and Patent Races

Introduction

How firms come to enter particular industries is a major topic in industrial organization. Of the many potential products that might be produced, firms decide that only a relatively small number would be profitable. Even if some firms are producing a particular product, many potential producers have decided not to for one reason or another. Information and strategic behavior are especially important in borderline industries in which only one or two firms are active in production, so the subject of entry is a good one for illustrating techniques in game theory.

This chapter begins with a discussion of innovation with the complications of imitation by other firms and patent protection by the government. Section 14.2 moves on to the question of how an innovator might use its informational advantage to protect its monopoly status by using limit pricing as a form of signal jamming. Section 14.3 analyzes a more traditional form of entry deterrence, predatory pricing, using a Gang of Four model of a repeated game under incomplete information. Section 14.4 returns to a simpler model of predatory pricing, but shows how the ability of the incumbent to credibly engage in a price war can actually backfire by inducing entry for buyout.

Market Power as a Precursor of Innovation

Market power is not always inimical to social welfare. Although restrictive monopoly output is inefficient, the profits it generates encourage innovation, an important source of both additional market power and economic growth. The importance of innovation, however, is diminished because of imitation, which can so severely diminish innovation's rewards as to entirely prevent it. An innovator generally incurs some research cost, but a discovery instantly imitated can yield zero net revenues. Table 14.1 shows how the payoffs look if the firm that innovates incurs a cost of 1 but imitation is costless and results in Bertrand competition. Innovation is a dominated strategy.

Table 14.1 Imitation with Bertrand Pricing

		Brydox	
		Innovate	*Imitate*
Apex:	*Innovate*	−1, −1	−1,0
	Imitate	0, −1	**0,0**

Payoffs to: (Apex, Brydox)

Under different assumptions, innovation occurs even with costless imitation. The key is what happens in the product market. Suppose there are two firms in the industry and they behave according to some rule such as Cournot or perfect collusion that can be represented ordinally by the payoffs in table 14.2 (a version of the game of Chicken). Although the firm that innovates pays the entire cost and keeps only half the benefit, imitation is no longer dominant. Apex imitates if Brydox innovates, but not if Brydox imitates. If Apex could move first, it would bind itself not to innovate, perhaps by disbanding its research laboratory.

Table 14.2 Imitation with Profits in the Product Market

		Brydox	
		Innovate	*Imitate*
Apex:	*Innovate*	1,1	**1,2**
	Imitate	**2,1**	0,0

Payoffs to: (Apex, Brydox)

Without a first-mover advantage, the game has two pure-strategy Nash equilibria, (*Innovate, Imitate*) and (*Imitate, Innovate*), and a symmetric Nash equilibrium in mixed strategies in which each firm innovates with probability 0.5. The mixed-strategy equilibrium is inefficient, since sometimes both firms innovate and sometimes neither.

History might provide a focal point or explain why one player moves first. Japan was for many years incapable of doing much basic scientific research, and does relatively little even today. The United States, therefore, had to innovate rather than imitate in the past, and today continues to do more basic research.

Much of the literature on innovation compares the relative merits of monopoly and competition. One reason a monopoly might innovate more is because it can capture more of the benefits, capturing the entire benefit if perfect price discrimination is possible (otherwise, some of the benefit goes to consumers). In addition, the monopoly avoids a second inefficiency: entrants innovating solely to steal the old innovator's rents without much increasing consumer surplus. The welfare aspects of innovation theory—and, indeed, all aspects—are intricate, and the interested reader is referred to the surveys by Kamien & Schwartz (1982) and Reinganum (1989).

Patent Races

One way that governments respond to imitation is by issuing patents: exclusive rights to make, use, or sell an innovation. If a firm patents its discovery, other firms cannot imitate, or even make the discovery independently. Research effort therefore has a discontinuous payoff: if the researcher is the first to make a discovery, he receives the patent; if he is second, nothing. This makes patents examples of the tournaments discussed in section 8.5, although if no player exerts any effort, none of them will get the reward, unlike in tournaments. Patents are also special because they lose their value if consumers find a substitute and stop buying the patented product. Moreover, the effort in tournaments is usually exerted over a fixed time period, whereas research usually has an endogenous time period, ending when the discovery is made. Because of this endogeneity, we call the competition a **patent race**.

We will consider two models of patents. On the technical side, the first model shows how to derive a continuous mixed-strategies probability distribution, instead of just the single number derived in section 3.3. On the substantive side, it shows how patent races lead to inefficiency.

Patent Race for a New Market

Players

Three identical firms, Apex, Brydox, and Central.

Order of Play

Each firm simultaneously chooses research spending $x_i \geq 0$, $(i = a, b, c)$.

Payoffs

Firms are risk neutral and the discount rate is zero. Innovation occurs at time $T(x_i)$ where $T' < 0$. The value of the patent is V, and if several players innovate simultaneously they share its value.

$$\pi_i = \begin{cases} V - x_i & \text{if } T(x_i) < T(x_j)\ (\forall j \neq i) & \text{(Firm } i \text{ gets patent)} \\ \dfrac{V}{1+m} - x_i & \text{if } T(x_i) = T(x_j) & \text{(Firm } i \text{ shares patent with } m = 1 \text{ or } 2 \text{ other firms)} \\ -x_i & \text{if } T(x_i) > T(x_j) \text{ for some } j & \text{(Firm } i \text{ does not get patent)} \end{cases}$$

The game does not have any pure-strategy Nash equilibria, because the payoff functions are discontinuous. A slight difference in research by one player can make a big difference in the payoffs, as shown in figure 14.1 for fixed values of x_b and x_c. The research levels shown in figure 14.1 are not equilibrium values. If

Figure 14.1 Payoffs in The Game of Patent Race for a New Market

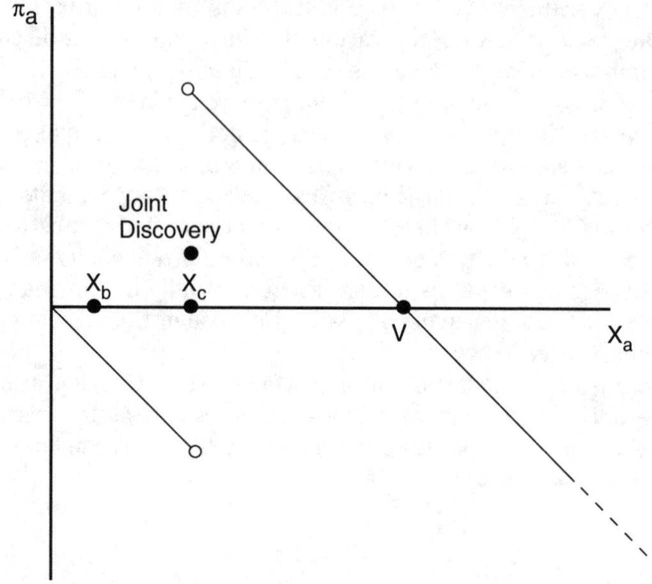

Apex chose any research level x_a less than V, Brydox would respond with $x_a + \varepsilon$ and win the patent. If Apex chose $x_a = V$, Brydox and Central would respond with $x_b = 0$ and $x_c = 0$, which would make Apex want to switch to $x_a = \varepsilon$.

There does exist a symmetric mixed-strategy equilibrium. We will derive $M_i(x)$, the cumulative density function for the equilibrium mixed strategy, rather than the density function itself. The probability with which firm i chooses a research level less than or equal to x will be $M_i(x)$. In a mixed-strategy equilibrium a player is indifferent between any of the pure strategies among which he is mixing. Since we know that the pure strategies $x_a = 0$ and $x_a = V$ yield zero payoffs, so must the expected payoff for every strategy mixed between, if Apex mixes over the support $[0, V]$. The expected payoff from the pure strategy x_a is the expected value of winning minus the cost of research. Letting x stand for nonrandom and X for random variables, this is

$$V \cdot Pr(x_a \geq X_b, x_a \geq X_c) - x_a = 0, \tag{1}$$

which can be rewritten as

$$V \cdot Pr(X_b \leq x_a)\, Pr(X_c \leq x_a) - x_a = 0, \tag{2}$$

or

$$V \cdot M_b(x_a) M_c(x_a) - x_a = 0. \tag{3}$$

We can rearrange equation (3) to obtain

$$M_b(x_a)M_c(x_a) = \frac{x_a}{V}. \tag{4}$$

If all three firms choose the same mixing distribution M, then

$$M(x) = \left(\frac{x}{V}\right)^{1/2} \text{ for } 0 \leq x \leq V. \tag{5}$$

What is noteworthy about a patent race is not the nonexistence of a pure-strategy equilibrium but the overexpenditure on research. All three players have expected payoffs of zero, because the patent value V is completely dissipated in the race. As in Brecht's *The Threepenny Opera*, "When all race after happiness/ Happiness comes in last." To be sure, the innovation is made earlier than it would have been by a monopolist, but hurrying the innovation is not worth the cost, from society's point of view, a result that would persist even if the discount rate were positive. The patent race is an example of **rent-seeking** (see Posner [1975] and Tullock [1967]), in which players dissipate the value of monopoly rents in the struggle to acquire them. Indeed, Rogerson (1982) uses a game very similar to Patent Race for a New Market to analyze competition for a government monopoly franchise.

The second patent race we will analyze is asymmetric because one player is an incumbent and the other an entrant. The aims are to discover which firm spends more and to explain why firms acquire valuable patents they do not use. A typical story of a sleeping innovation (though not in this case patented) is the story of synthetic caviar. In 1976, Romanoff Caviar Co. said that they had developed synthetic caviar as a "defensive marketing weapon" that they would not introduce in the US unless the Soviet Union introduced a synthetic caviar they claimed to have developed. The new product would sell for one quarter of the old price and *Business Week* said that the reason Romanoff did not introduce it was to avoid cannibalizing its own market (*Business Week*, June 28, 1976, p. 51). The game-theoretic aspects of this situation put the claims of all its players in doubt, but this dubious validity makes it all the more typical of sleeping patent stories.

However difficult to validate, there do exist good theoretical models of sleeping patents, the best known of which is Gilbert & Newbery (1982), in which the incumbent firm does research and acquires a sleeping patent, while the entrant does no research. We will look at a slightly more complicated model which does not reach such an extreme result.

Patent Race for an Old Market

Players

An incumbent and an entrant.

Order of Play

(1) The firms simultaneously choose research spending x_i and x_e, which result in research achievements $f(x_i)$ and $f(x_e)$, where $f' > 0$ and $f'' < 0$.
(2) Nature chooses which player wins the patent using a function g that maps research achievement to a probability between zero and one.

$$Prob(incumbent\ wins\ patent) = g[f(x_i) - f(x_e)], \tag{6}$$

where $g' > 0$, $g(0) = 0.5$, and $0 \leq g \leq 1$.
(3) The winner of the patent decides whether to spend Z to implement it.

Payoffs

The old patent yields revenue y and the new patent yields v. The payoffs are shown in table 14.3.

Table 14.3 Patent Race for an Old Market: Payoffs

Outcome	$\pi_{incumbent}$	$\pi_{entrant}$
The entrant wins and implements	$-x_i$	$v - x_e - Z$
The incumbent wins and implements	$v - x_i - Z$	$-x_e$
Neither player implements	$y - x_i$	$-x_e$

Equation (6) specifies the function $g[f(x_i) - f(x_e)]$ to capture the three ideas of (a) diminishing returns to inputs, (b) rivalry, and (c) winning a patent race as a probability. The $f(x)$ functions model dimishing returns because f increases at a decreasing rate in the input x. Using the difference between the f functions for each firm makes it relative effort that matters. The $g(\cdot)$ function turns this measure of relative effective input into a probability between zero and one.

The entrant will do no research unless he plans to implement, so we will disregard his strongly dominated strategy, ($x_e > 0$, *no implementation*). The incumbent wins with probability g and the entrant with probability $1 - g$, so from table 14.3 the expected payoff functions are

$$\begin{aligned}\pi_{incumbent} = &(1 - g[f(x_i) - f(x_e)])\,(-x_i)\\ &+ g[f(x_i) - f(x_e)]Max\{v - x_i - Z,\, y - x_i\}\end{aligned} \tag{7}$$

and

$$\pi_{entrant} = (1 - g[f(x_i) - f(x_e)])(v - x_e - Z) + g[f(x_i) - f(x_e)](-x_e). \tag{8}$$

On differentiating, and simplifying the notation, we obtain the first-order conditions

$$\frac{d\pi_i}{dx_i} = -(1 - g[f_i - f_e]) - g'f_i'(-x_i) \\ + g'f_i'Max\{v - x_i - Z, y - x_i\} - g[f_i - f_e] = 0 \tag{9}$$

and

$$\frac{d\pi_e}{dx_e} = -(1 - g[f_i - f_e]) + g'f_e'(v - x_e - Z) \\ - g[f_i - f_e] + g'f_e'x_e = 0. \tag{10}$$

Equating (9) and (10), which both equal zero, we obtain

$$-(1 - g) + g'f_i'x_i + g'f_i'Max\{v - x_i - Z, y - x_i\} - g \\ = -(1 - g) + g'f_e'(v - x_e - Z) - g + g'f_e'x_e, \tag{11}$$

which simplifies to

$$f_i'[x_i + Max\{v - x_i - Z, y - x_i\}] = f_e'[v - x_e - Z + x_e], \tag{12}$$

or

$$\frac{f_i'}{f_e'} = \frac{v - Z}{Max\{v - Z, y\}}. \tag{13}$$

We can use equation (13) to show that different parameters generate two qualitatively different outcomes.

Outcome 1.

The entrant and incumbent spend equal amounts, and each implements if successful. This happens if there is a big gain from patent implementation, that is, if

$$v - Z \geq y, \tag{14}$$

so that equation (13) becomes

$$\frac{f_i'}{f_e'} = \frac{v - Z}{v - Z} = 1, \tag{15}$$

which implies that $x_i = x_e$.

Outcome 2.

The incumbent spends more and does not implement if he is successful (he acquires a sleeping patent). This happens if the gain from implementation is small, that is, if

$$v - Z < y, \tag{16}$$

so that equation (13) becomes

$$\frac{f_i'}{f_e'} = \frac{v - Z}{y} < 1, \tag{17}$$

which implies that $f_i' < f_e'$. Since we assumed that $f'' < 0$, f' is decreasing in x, and it follows that $x_i > x_e$.

This model shows that the presence of another player can stimulate the incumbent to do research he otherwise would not, and that he may or may not implement the discovery. The incumbent has at least as much incentive for research as the entrant, because a large part of a successful entrant's payoff comes at the incumbent's expense. The benefit to the incumbent is the maximum of the benefit from implementing and the benefit from stopping the entrant, but the entrant's benefit can only come from implementing. Contrary to the popular belief that sleeping patents are bad, here they can help society by eliminating wasteful implementation.

14.2 Signal Jamming

One form of innovation is to enter a new market, producing something that everybody knows can be produced but which might not be profitable because demand might be too weak. Patents do not protect this kind of innovation, and imitation is very easy, but the firm might be able to protect itself by strategic behavior. In traditional industrial organization, the three most discussed kinds of strategic entry deterrence are government restrictions on rivals, limit pricing, and predatory pricing. Government restrictions, which include patents as well as regulatory barriers, were discussed in the previous section. Limit pricing is charging a low price before entry occurs and predatory pricing is charging a high price after it occurs. Chapter 4 showed why the obvious explanation for predatory pricing is unsatisfactory because it requires noncredible threats. Limit pricing is even more puzzling, since there is no direct connection between what the incumbent charges before entry and what he charges after entry. Modern industrial organization has suggested various more plausible reasons why these tactics might work to deter entry. In this section, we will consider a form of limit pricing motivated by the incumbent's desire to protect its informational advantage.

This book has examined a number of models in which an informed player tries to convey information to an uninformed player by some means or other—by entering into an incentive contract, or by signalling. Sometimes, however, the informed party has the opposite problem: his natural behavior would convey his private information but he wants to keep it secret. This happens, for example, if one firm is informed about its poor ability to compete successfully, and it wants to conceal this information from a rival. The informed player may then engage in costly actions, just as in signalling, but now the costly action will be **signal jamming** (a term coined in Fudenberg & Tirole [1986c]): preventing information from appearing rather than generating information.

The model used to illustrate signal jamming here is a model of limit pricing which some readers may have seen as Rasmusen (unpublished). This might occur for a variety of reasons, including to signal that the incumbent has such low costs that rivals would regret entering (see problem 6.2 and Milgrom & Roberts [1982a]). Here the explanation will be signal jamming: by keeping profits low, the incumbent keeps it unclear to the rival whether the market is big enough to accommodate two firms profitably. In the model, the incumbent can control MR, the net monopoly revenue after subtracting variable costs. The duopoly revenue might or might not be big enough to cover the fixed cost C that each operating firm must incur.

Limit Pricing as Signal Jamming

Players

The incumbent and the rival.

Order of Play

(0) Nature chooses the market to be *Small* with probability θ and *Large* with probability $(1 - \theta)$, observed only by the incumbent.

(1) The incumbent chooses his first-period net revenue MR' to equal MR_0 or MR_1 if the market is small, MR_1 or MR_2 if it is large, where M is the monopoly premium and $R_0 < R_1 < R_2$. Both players observe the value of R'.

(2) The rival decides whether to be *In* or *Out* of the market.

(3) Second-period net duopoly revenue R'' equals R_1 if the market is small and R_2 if it is large. If the rival chose *In*, each firm earns R''. If the rival chose *Out*, the incumbent earns MR'' and the entrant earns 0.

Payoffs

If the rival does not enter, the payoffs are $\pi_{incumbent} = (MR' - C) + (MR'' - C)$ and $\pi_{rival} = 0$.
If the rival does enter, the payoffs are $\pi_{incumbent} = (MR' - C) + (R'' - C)$ and $\pi_{rival} = R'' - C$.
Assume that $R_2 - C > 0$, and $R_1 - C < 0$.

There are four equilibria, each appropriate to a different parameter region in figure 14.2. If the premium from being a monopoly is small enough, a nonstrategic equilibrium exists in which the incumbent simply maximizes profits in each period

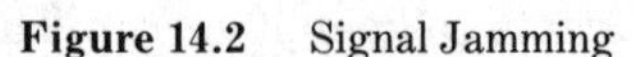

Figure 14.2 Signal Jamming

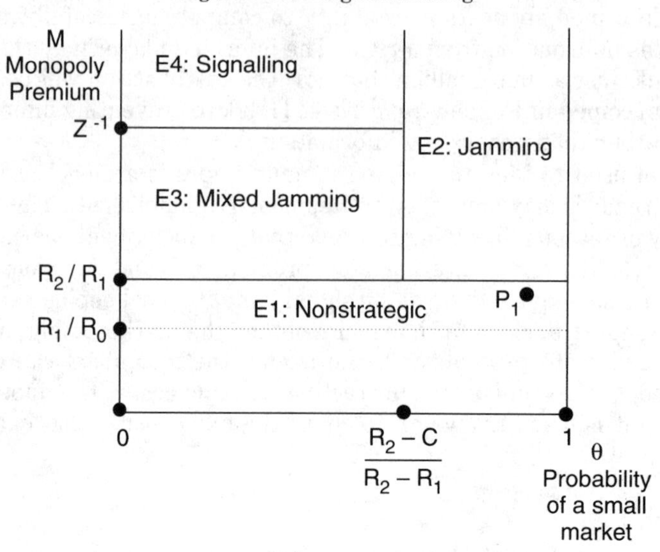

separately. This equilibrium is: (E1: NONSTRATEGIC. $R_2|Large$, $R_1|Small$, $Out|R_0$, $Out|R_1$, $In|R_2$). The incumbent's equilibrium payoff in a large market is $\pi_I(R_2|Large) = (MR_2 - C) + (R_2 - C)$, compared with the deviation payoff of $\pi_I(R_1|Large) = (MR_1 - C) + (MR_2 - C)$. The incumbent has no incentive to deviate if $\pi_I(R_2|Large) - \pi_I(R_1|Large) = MR_2 + R_2 - MR_1 - MR_2 \geq 0$, which is equivalent to

$$M \leq \frac{R_2}{R_1}, \tag{18}$$

as shown in figure 14.2. The rival will not deviate, because the incumbent's choice fully reveals the size of the market.

Signal jamming occurs if monopoly profits are somewhat higher and if the rival would refrain from entering the market unless he decides it is more profitable than his prior beliefs would indicate. The equilibrium is (E2: PURE SIGNAL JAMMING. $R_1|Large$, $R_1|Small$, $Out|R_0$, $Out|R_1$, $In|R_2$). The rival's strategy is the same as in E1, so the incumbent's optimal behavior remains the same, and he chooses R_1 if the opposite of condition (18) is true. As for the rival, if he stays out, his second-period payoff is zero, and if he enters, its expected value is $\theta(R_1 - C) + (1 - \theta)(R_2 - C)$. Hence, as shown in figure 14.2, he will follow the equilibrium behavior of staying out if

$$\theta \geq \frac{R_2 - C}{R_2 - R_1}. \tag{19}$$

A mixed form of signal jamming occurs if the probability of a small market is unlikely, so if the signal of first-period revenues were jammed completely, the rival would enter anyway. This equilibrium is (E3: MIXED SIGNAL JAMMING.

($R_1|Small$, $R_1|Large$ *with probability* α, $R_2|Large$ *with probability* $(1 - \alpha)$, $Out|R_0$, $In|R_1$ *with probability* β, $Out|R_1$ *with probability* $(1 - \beta)$, $In|R_2$). If the incumbent played $R_2|Large$ and $R_1|Small$, the rival would interpret R_1 as indicating a small market—an interpretation which would give the incumbent incentive to play $R_1|Large$. But if the incumbent always plays R_1, the rival would enter even after observing R_1, knowing that there was a high probability that the market was really large. Hence, the equilibrium must be in mixed strategies, which is equilibrium E3, or the incumbent must convince the rival to stay out by playing R_0, which is equilibrium E4.

For the rival to mix, he must be indifferent between the second-period payoffs of $\pi_E(In|R_1) = (\theta/[\theta + (1 - \theta)\alpha])(R_1 - C) + ([(1 - \theta)\alpha]/[\theta + (1 - \theta)\alpha])(R_2 - C)$ and $\pi_E(Out|R_1) = 0$. Equating these two payoffs and solving for α gives $\alpha = [\theta/(1 - \theta)][(C - R_1)/(R_2 - C)]$, which is always non-negative, but avoids equalling one only if condition (19) is false.

For the incumbent to mix when the market is large, he must be indifferent between $\pi_I(R_2|Large) = (MR_2 - C) + (R_2 - C)$ and $\pi_I(R_1|Large) = (MR_1 - C) + \beta(R_2 - C) + (1 - \beta)(MR_2 - C)$. Equating these two payoffs and solving for β gives $\beta = MR_1 - R_2|(M - 1)R_2$, which is strictly less than 1, and which is nonnegative if condition (19) is false.

If the market is small, the incumbent's alternative payoffs are the equilibrium payoff of $\pi_I(R_1|Small) = (MR_1 - C) + \beta(R_1 - C) + (1 - \beta)(MR_1 - C)$ and the deviation payoff of $\pi_I(R_0|Small) = (MR_0 - C) + (MR_1 - C)$. The difference is

$$\pi_I(R_1|Small) - \pi_I(R_0|Small) = \\ [MR_1 + \beta R_1 + (1 - \beta)MR_1] - [MR_0 + MR_1]. \tag{20}$$

Expression (20) is nonnegative under either of two conditions. The first is if R_0 is small enough; that is, if

$$R_0 \leq R_1\left(1 - \frac{R_1}{R_2}\right). \tag{21}$$

The second is if M is no greater than some amount Z^{-1} defined so that

$$M \leq \left(\frac{R_1}{R_2} - 1 + \frac{R_0}{R_1}\right)^{-1} = Z^{-1}. \tag{22}$$

If condition (21) is false, then $Z^{-1} > R_2/R_1$, because $Z < R_1/R_2$ and $Z > 0$. Thus, we can draw region E3 as it is shown in figure 14.2.

It follows that if condition (22) is replaced by its converse, the unique equilibrium is for the incumbent to choose $R_0|Small$, and the equilibrium is (E4: SIGNALLING. $R_0|Small$, $R_2|Large$, $Out|R_0$, $In|R_1$, $In|R_2$). Passive conjectures will support this pooling signalling equilibrium, as will the out-of-equilibrium belief that if the rival observes R_1, he believes the market is large with probability $[(1 - \theta)\alpha]/[\theta + (1 - \theta)\alpha]$, as in equilibrium E3.

The signalling equilibrium is also an equilibrium for other parameter regions outside of E4, though less reasonable beliefs are required. Let the out-of-equilibrium belief be $Prob(Large|R_1) = 1$. The equilibrium payoff is $\pi_I(R_0|Small) = (MR_0 - C) + (MR_1 - C)$ and the deviation payoff is $\pi_I(R_1|Small) = (MR_1 - C) + (R_1 - C)$. The signalling equilibrium remains an equilibrium so long as $M \geq R_1/R_0$.

The signalling equilibrium is an interesting one, because it turns the asymmetric information problem full circle. The informed player wants to conceal his private information by costly signal jamming if the information is *Large*, so when the information is *Small*, he must signal at some cost that he is not signal jamming. If E4 is the equilibrium, the incumbent is hurt by the possibility of signal jamming; he would much prefer a simpler world in which it was illegal or nobody considered the possibility. This is often the case: strategic behavior can help a player in some circumstances, but given that the other players know that he might be behaving strategically, everyone would prefer a world in which everyone is honest and nonstrategic.

14.3 Predatory Pricing: The Kreps-Wilson Model

The second of the traditional forms of strategic entry deterrence is **predatory pricing**, in which the firm seeking to acquire the market charges a low price in direct competition with its rival, rather than before entry. We have looked at predation already in chapters 4, 5 and 6 in the Entry Deterrence games. The major problem with entry deterrence under complete information is the Chainstore Paradox. The heart of the paradox is the sequential rationality problem faced by an incumbent who wishes to threaten a prospective entrant with low post-entry prices. The incumbent can respond to entry in two ways. He can collude with the entrant and share the profits, or he can fight by lowering his price so that both firms make losses. We have already seen that the incumbent would not fight in a perfect equilibrium if the game has complete information. Foreseeing the incumbent's accommodation, the potential entrant ignores the threats.

In Kreps & Wilson (1982a), an application of the Gang of Four model of chapter 6, incomplete information allows the threat of predatory pricing to successfully deter entry. A monopolist with outlets in N towns faces an entrant who can enter each town. In our adaption of the model, we will start by assuming that the order in which the towns can be entered is common knowledge, and that if the entrant passes up his chance to enter a town, he cannot enter it later. The incomplete information takes the form of a small probability that the monopolist is "strong" and has nothing but *Fight* in his action set: he is an uncontrolled manager who gratifies his passions in squelching entry instead of maximizing profits.

Predatory Pricing

(Kreps & Wilson [1982a])

Players

The entrant and the monopolist.

Order of Play

(0) Nature chooses the monopolist to be *Strong* with low probability θ and *Weak* with high probability $(1 - \theta)$. Only the monopolist observes Nature's move.
(1) The entrant chooses *Enter* or *Stay Out* for the first town.
(2) The monopolist chooses *Collude* or *Fight* if he is weak, *Fight* if he is strong.
(3) Steps (1) and (2) are repeated for towns 2 through N.

Payoffs

The discount rate is zero. Table 14.4 gives the payoffs per period, which are the same as in table 4.1.

Table 14.4 Predatory Pricing

		Weak Incumbent	
		Collude	*Fight*
Entrant:	*Enter*	**40,50**	−10,0
	Stay out	0,100	**0,100**

Payoffs to: (Entrant, Incumbent)

In describing the equilibrium, we will for concreteness denote towns by names such as i_{-30} and i_{-5}, where the numbers are to be taken to be purely ordinal. The entrant has an opportunity to enter town i_{-30} before i_{-5}, but there are not necessarily 25 towns between the last town where the incumbent always chooses *Fight* and the first town where he always chooses *Collude*. The actual gap depends on θ but not N.

Part of the Equilibrium for the Game of Predatory Pricing

Entrant: Enter first at town i_{-10}. If entry has occurred before i_{-10} and been answered with *Collude*, enter every town after the first one entered.

Strong monopolist: Always fight entry.

Weak monopolist: Fight any entry up to i_{-30}. Fight the first entry after i_{-30} with a probability $m(i)$ that diminishes until it reaches zero at i_{-5}. If *Collude* is ever chosen instead, always collude thereafter. If *Fight* was chosen in response to the first attempt at entry, increase the mixing probability $m(i)$ in subsequent towns.

This description, which is illustrated by figure 14.3, only covers the equilibrium path and small deviations. Note that out-of-equilibrium beliefs do not have to be specified (unlike in the original model of Kreps and Wilson), since whenever a monopolist colludes, in or out of equilibrium, Bayes' Rule says that the entrant must believe him to be *Weak*.

The entrant will certainly stay out until i_{-30}. If no town is entered until i_{-5} and the monopolist is *Weak*, then entry at i_{-5} is undoubtedly profitable. But entry is attempted at i_{-10}, because since $m(i)$ is diminishing in i, the weak monopolist probably would not fight even there.

Out of equilibrium, if an entrant were to enter at i_{-90}, the weak monopolist would be willing to fight, to maintain i_{-10} as the next town to be entered. If he did not, then the entrant, realizing that he could not possibly be facing a strong monopolist, would enter every subsequent town from i_{-89} to i_{-1}. If no town were entered until i_{-5}, the weak monopolist would be unwilling to fight in that town, because too few towns would be left to protect. If a town between i_{-30} and i_{-5} has been entered and fought, the monopolist raises the mixing probability that he fights in the next town entered because he has a more valuable reputation to defend. By fighting in the first town he has increased the belief that he is strong and increased the gap until the next town is entered.

What if the entrant deviated and entered town i_{-20}? The equilibrium calls for a mixed-strategy response beginning with i_{-30}, so the weak monopolist must be indifferent between fighting and not fighting. If he fights, he loses current revenue but the entrant's posterior belief that he is strong rises, rising more if the fight occurs late in the game. The entrant knows that in equilibrium the weak monopolist would fight with a probability of, say, 0.9 in town i_{-20}, so fighting there would not much increase the belief that he was strong, but if he fought in town i_{-13}, where the mixing probability has fallen to 0.2, the belief would rise much more. On the other hand, the gain from a given reputation diminishes as fewer towns remain to be protected, so the mixing probability falls over time.

The description of the equilibrium strategies is incomplete because describing what happens after unsuccessful entry becomes rather intricate. Even in the simultaneous-move games of chapter 3, we saw that games with mixed-strategy

Figure 14.3 Equilibrium in The Game of Predatory Pricing

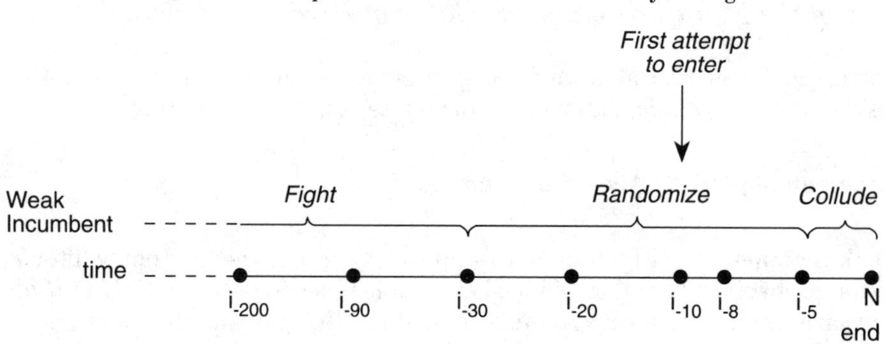

equilibria have many different possible realizations. In repeated games like Predatory Pricing, the number of possible realizations makes an exact description very complicated indeed. If, for example, the entrant entered town i_{-20} and the monopolist chose *Fight*, the entrant's belief that he was strong would rise, pushing the next town entered to i_{-8} instead of i_{-10}. A complete description of the strategies would say what would happen for every possible history of the game, which is impractical.

Moreover, because of mixing, even the equilibrium path becomes nonunique after i_{-10}, when the first town is entered. When the entrant enters at i_{-10}, the weak monopolist chooses randomly whether to fight, so the entrant's belief that the monopolist is strong increases if he is fought. As a result, the next entry might be not at i_{-9}, but i_{-7}.

As a final note, let us return to the initial assumption that if the entrant decided not to enter town i, he could not change his mind later. We have seen that no towns will be entered until the last one is approached, because the incumbent wants to protect his reputation for strength. But if the entrant can change his mind, the last town is never approached. The entrant knows he would take losses in the first $(N - 30)$ towns, and it is not worth his while to reduce the number to 30 so that the monopolist would start choosing *Collude*. Paradoxically, allowing the entrant many chances to enter helps not him but the incumbent.

14.4 Entry for Buyout

The previous section suggested that predatory pricing might actually be a credible threat if information were slightly incomplete, because the incumbent might be willing to suffer losses fighting the first entrant in order to deter future entry. This is not the end of the story, however, because even if entry costs exceed operating revenues, entry might still be profitable if the entrant is bought out by the incumbent.

To see this most simply, let us start by thinking how entry might be deterred under complete information. The incumbent needs some way to precommit himself to unprofitable post-entry pricing. Spence (1977) and Dixit (1980) suggest that the incumbent could enlarge his initial capacity so that the post-entry price would naturally drop to below average cost. The post-entry price would still be above average variable cost, so, having already sunk the capacity cost, the incumbent fights entry without further expense. The entrant's capacity cost is not yet sunk, so he refrains from entry.

In the model with the extensive form of figure 14.4, the incumbent has the additional option of buying out the entrant. An incumbent who fights entry bears two costs: the loss from selling at a price below average total cost, and the opportunity cost of not earning monopoly profits. He can make the first a sunk cost, but not the second. The entrant, foreseeing that the incumbent will buy him out, enters despite knowing that the duopoly price will be less than average total cost. The incumbent faces a second perfectness problem, for while he may try to deter entry by threatening not to buy out the entrant, the threat is not credible.

Figure 14.4 Entry for Buyout

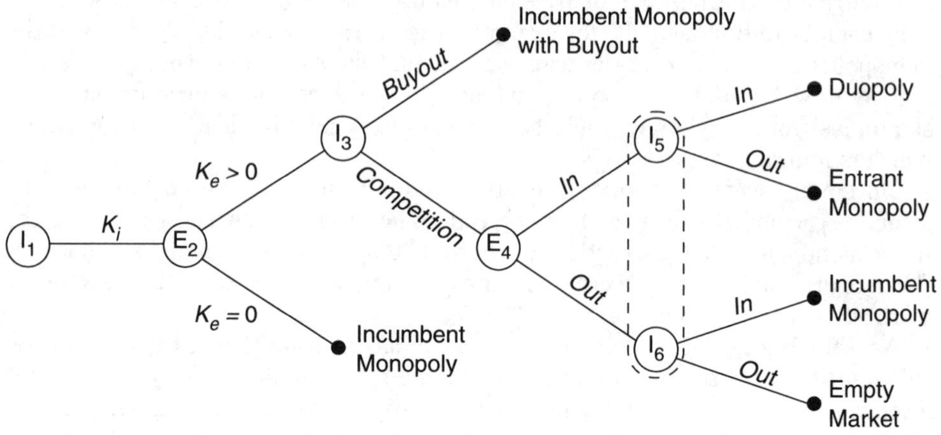

Entry for Buyout

(Rasmusen [1988a])

Players

The incumbent and the entrant.

Order of Play

(1) The incumbent selects capacity K_i.
(2) The entrant decides whether to enter or stay out, choosing a capacity $K_e \geqslant 0$.
(3) If the entrant picks a positive capacity, the incumbent decides whether to buy him out at price B.
(4) If the entrant has been bought out, the incumbent selects output $q_i \leqslant K_i + K_e$.
(5) If the entrant has not been bought out, each player decides whether to stay in the market or exit.
(6) If a player has remained in the market, he selects the output $q_i \leq K_i$ or $q_e \leq K_e$.

Payoffs

Each unit of capacity costs a, the constant marginal cost is c, a firm that stays in the market incurs fixed cost F, and there is no discounting. There is only one period of production.

If no entry occurs, $\pi_i = [p(q_i) - c]q_i - aK_i - F$ and $\pi_e = 0$.
If entry occurs and is bought out, $\pi_i = [p(q_i) - c]q_i - aK_i - B - F$ and $\pi_e = B - aK_e$.
Otherwise,

$$\pi_{incumbent} = \begin{cases} [p(q_i, q_e) - c]q_i - aK_i - F & \text{if the incumbent stays.} \\ -aK_i & \text{if the incumbent exits.} \end{cases}$$

$$\pi_{entrant} = \begin{cases} [p(q_i, q_e) - c]q_e - aK_e - F & \text{if the entrant stays.} \\ -aK_e & \text{if the entrant exits.} \end{cases}$$

Two things have yet to be specified: the buyout price B and the price function $p(q_i, q_e)$. To specify them requires particular solution concepts for bargaining and duopoly, which chapters 11 and 13 have shown are not uncontroversial. Here, they are subsidiary to the main point and can be chosen according to the taste of the modeller. We have "blackboxed" the pricing and bargaining subgames so as not to deflect attention to subsidiary parts of the model. The numerical example below will name specific functions for those subgames, but other numerical examples could use different functions to illustrate the same points.

Numerical Example

Assume that the market demand curve is

$$p = 100 - q_i - q_e. \tag{23}$$

Let the cost per unit of capacity be $a = 10$, the marginal cost of output $c = 10$, and the fixed cost $F = 601$. Assume that output follows Cournot behavior and that the bargaining solution splits the surplus equally, in accordance with the Nash bargaining solution and Rubinstein (1982).

If the incumbent faced no threat of entry, he would behave as a simple monopolist, choosing a capacity equal to the output which solves

$$\underset{q_i}{\text{Maximize}}\ (100 - q_i)q_i - 10q_i - 10q_i. \tag{24}$$

Problem (24) has the first order condition

$$80 - 2q_i = 0, \tag{25}$$

so the monopoly capacity and output would both equal 40, yielding a net operating revenue of 1,399 ($= [p - c]q_i - F$), well above the capacity cost of 400.

We will not go into details, but under the parameters assumed the incumbent chooses the same output and capacity of 40 even if entry is possible but buyout is not. If the potential entrant were to enter, he could do no better than to choose

$K_e = 30$, which costs 300. With capacities $K_i = 40$ and $K_e = 30$, Cournot behavior leads the two firms to solve

$$\underset{q_i}{Maximize}\ (100 - q_i - q_e)q_i - 10q_i \text{ such that } q_i \leq 40 \tag{26}$$

and

$$\underset{q_e}{Maximize}\ (100 - q_i - q_e)q_e - 10q_e \text{ such that } q_e \leq 30, \tag{27}$$

which have the first-order conditions

$$90 - 2q_i - q_e = 0 \tag{28}$$

and

$$90 - q_i - 2q_e = 0. \tag{29}$$

The Cournot outputs both equal 30, yielding a price of 40 and net revenues of $R_i^d = R_e^d = 299$ ($= [p - c]q_i - F$). The entrant's profit net of capacity cost would be -1 ($= R_e^d - 30a$), which is less than the zero from not entering.

What if both entry and buyout are possible, but the incumbent still chooses $K_i = 40$? If the entrant chooses $K_e = 30$ again, $R_e^d = R_i^d = 299$, just as above. If he buys out the entrant, the incumbent, having increased his capacity to 70, produces a monopoly output of 45. Half of the surplus from buyout is

$$\begin{aligned} B &= \frac{1}{2}\left[\underset{q_i}{Maximum}\ \{[p(q_i) - c]q_i | q_i \leq 70\} - F - (R_e^d + R_i^d)\right] \\ &= \frac{1}{2}[(55 - 10)45 - 601 - (299 + 299)] = 413. \end{aligned} \tag{30}$$

The entrant is bought out for his Cournot revenue of 299 plus the 413 which is his share of the buyout surplus, a total buyout price of 712. Since 712 exceeds the entrant's capacity cost of 300, buyout induces entry which would otherwise have been deterred. Nor can the incumbent deter entry by picking a different capacity. Choosing any K_i greater than 30 leads to the same Cournot output of 60 and the same buyout price of 712. Choosing K_i less than 30 allows the entrant to make a profit even without being bought out.

Realizing that entry cannot be deterred, the incumbent would choose a smaller initial capacity. A Cournot player whose capacity is less than 30 would produce a quantity equal to his capacity. Since buyout will occur, if a firm starts with a capacity less than 30 and adds one unit, the marginal cost of capacity is 10 and the marginal benefit is the increase (for the entrant) or decrease (for the incumbent) in the buyout price. If it is the entrant who adds a unit of capacity, R_e^d rises by at least $(40 - 10)$, the lowest possible Cournot price minus the marginal cost of output. Moreover, R_i^d falls because the entrant's extra output lowers the market price, so under our bargaining solution the buyout price rises by more than 15 (= 40 − 10/2) and the entrant should add extra capacity up to $K_e = 30$. A parallel argument shows why the incumbent should build a capacity of at least 30. Increasing the capacities any further leaves the buyout price unchanged, because the duopoly net revenues are unaffected, so both firms choose exactly 30.

The industry capacity equals 60 when buyout is allowed, but after the buyout only 45 is used. Industry profits in the absence of possible entry would have been 999 (= 1,399 − 400), but with buyout they are 824 (= 1,424 − 600), so buyout has decreased industry profits by 175. Consumer surplus has risen from 800 ($= 0.5[100 - p(q|K = 40)][q|K = 40]$) to 1,012.5 ($= 0.5[100 - p(q|K = 60)][q|K = 60]$), a gain of 212.5, so buyout raises total welfare in this example. The increase in output outweighs the inefficiency of the entrant's investment in capacity, an outcome that depends on the particular parameters we chose.

This model is a tangle of paradoxes. The central paradox is that the ability of the incumbent to destroy industry profits after entry ends up hurting rather than helping him, because it increases the buyout price. This has a flavor similar to the "judo economics" of Gelman & Salop (1983): the incumbent's very size and influence weighs against him. In the numerical example, allowing the incumbent to buy out the entrant raised total welfare, even though it solidified monopoly power and resulted in wasteful excess capacity. Under other parameters, the effect of excess capacity dominates, and allowing buyout would lower welfare—but only because it encourages entry, of which we usually approve. Adding more potential entrants would also have perverse effects. If the incumbent's excess capacity can deter one entrant, it can deter any number. We have seen that a single entrant might enter anyway, for the sake of the buyout price. But if there are many potential entrants, it is easier to deter entry. Buying out a single entrant would not do the incumbent much good, so he would only be willing to pay a small buyout price, and the small price would discourage any entrant from being the first. The game becomes complicated, but clearly the multiplicity of potential entrants makes entry more difficult for any of them.

Notes

N14.1 Innovation and Patent Races

- The idea of the patent race is described by Barzel (1968), although his model showed the same effect of overhasty innovation even without patents.
- The Brecht quotation is from Act III, scene 7 of *The Threepenny Opera*, translated by John Willett (Bertolt Brecht, *Collected Works*, London: Eyre Methuen, 1987).
- Reinganum (1985) has shown that an important element of patent races is whether increased research hastens the arrival of the patent or just affects whether it is acquired. If more research hastens the innovation, the incumbent might spend less than the entrant because the incumbent is enjoying a stream of profits from his present position that the new innovation destroys.
- **Uncertainty in Innovation.** The game of Patent Race for an Old Market is only one way to model innovation under uncertainty. A more common way to model uncertainty, used by Loury (1979) and Dasgupta & Stiglitz (1980), uses continuous time with discrete discoveries and specifies that discoveries arrive as a Poisson process with parameter $\lambda(X)$, where X is research expenditure, $\lambda' > 0$, and $\lambda'' < 0$. Then

$$\begin{aligned} &Prob(invention\ at\ t) &&= \lambda e^{-\lambda(X)t}; \\ &Prob(invention\ before\ t) &&= 1 - e^{-\lambda(X)t}. \end{aligned} \tag{31}$$

A little algebra gives us the current value of the firm, V_0, as a function of the innovation rate, the interest rate, the post-innovation value, V_1, and the current revenue flow, R_0. The return on the firm equals the current cash flow plus the probability of a capital gain.

$$rV_0 = R_0 - X + \lambda(V_1 - V_0), \tag{32}$$

which implies

$$V_0 = \frac{\lambda V_1 + R_0 - X}{\lambda + r}. \tag{33}$$

Expression (33) is frequently useful in modelling.

- A common theme in entry models is what has been called the **fat-cat effect** by Fudenberg & Tirole (1986a, p. 23). Consider a two-stage game, in the first stage of which an incumbent firm chooses its advertising level, and in the second stage plays a Bertrand subgame with an entrant. If the advertising in the first stage gives the incumbent a base of captive customers who have inelastic demand, he will choose a higher price than the entrant. The incumbent has become a "fat cat." The effect is present in many models. In the Hotelling Pricing Game of section 13.3, a firm so situated that it has a large "safe" market would choose a higher price. In the game of Customer Switching Costs of section 5.5, a firm that has old customers locked in would choose a higher price than a fresh entrant in the last period of a finitely repeated game.

N14.3 Predatory Pricing: The Kreps-Wilson Model

- Books on the new theoretical organization include Fudenberg & Tirole (1986a), Jacquemin (1985), Krouse (1990) the collection edited by Stiglitz & Mathewson (1986), Schmalensee & Willig (1989), and Tirole (1988). On the theory of regulation, see Spulber (1989) and Laffont & Tirole (1993).
- Kreps & Wilson (1982a) do not simply assume that one type of monopolist always chooses *Fight*. They make the more elaborate but primitive assumption that his payoff function makes fighting a dominant strategy. Table 14.5 shows a set of payoffs for the strong monopolist which generate this result.

Table 14.5 Predatory Pricing with a Dominant Strategy

		Strong Incumbent	
		Collude	*Fight*
Entrant:	*Enter*	20,10	−10,40
	Stay out	0,100	**0,100**

Payoffs to: (Entrant, Incumbent)

- Under the Kreps-Wilson assumption, the strong monopolist would actually choose to collude in the early periods of the game in some perfect Bayesian equilibria. Such an equilibrium could be supported by out-of-equilibrium beliefs which the authors point out are absurd: if the monopolist fights in the early periods, the entrant believes he must be a weak monopolist.

15 The New Industrial Organization

15.1 Credit and the Age of the Firm: The Diamond Model

Introduction

This chapter contains a variety of unrelated applications of game theory to industrial organization. Industrial organization has traditionally been much concerned with the kinds of questions addressed in chapters 13 and 14—how firms chose prices and output, how easy it is to enter a market, and how innovation occurs. These are still active topics of research—notice that they occupy two chapters—but the most exciting development of recent years has been to use economic theory to explain more special features of different markets, especially institutional features. This is a return to the topics of the older and despised days before World War II when many of the scholars called economists described the workings of different industries without trying to explain them. The difference is that where an economist in 1920 might have contented himself with the claim that banks give cheaper credit to older firms and that the government allows excess profits to low-cost public utilities, perhaps with a hint that institutions are often inefficient, the economist of 1990 looks at the same phenomenon and tries to explain why maximizing individuals might reach those outcomes.

The first model we will look at is Diamond's model of credit terms, which seeks to explain why older firms get cheaper credit using a game similar to the Gang of Four models of chapters 6 and 14. Section 15.2 will analyze takeovers, looking at different models that attempt to explain why managers resist takeovers. Section 15.3 looks deeper into the formation of stock prices, and presents a basic model of market microstructure—how it is that stock traders decide to set their prices, knowing that new information may be flowing into the market. Section 15.4 concludes the book with a model that can illustrate either government procurement or public utility regulation, and shows why the government may decide to allow the utility returns above the market rate.

Credit and the Age of the Firm: the Diamond Model

Telser (1966) has suggested that predatory pricing would be a credible threat if the incumbent had access to cheaper credit than the entrant, and so could hold out for more periods of losses before going bankrupt. While one might wonder whether this is effective protection against entry—what if the entrant is a large old firm from another industry?—we shall focus on how better-established firms might get cheaper credit.

D. Diamond (1989) aims to explain why old firms are less likely than young firms to default on debt. His model has both adverse selection, because firms differ in type, and moral hazard, because they take hidden actions. The three types of firms, R, S, and RS, are "born" at time zero and borrow to finance projects at the start of each of T periods. We must imagine that there are overlapping generations of firms, so that at any point in time a variety of ages are coexisting, but the model looks at the lifecycle of only one generation. All the players are risk neutral. Type RS firms can choose independently risky projects with negative expected values or safe projects with low but positive expected values. Although the risky projects are worse in expectation, if they are successful the return is much higher than from safe projects. Type R firms can choose only risky projects, and type S firms only safe projects. At the end of each period the projects bring in their profits and loans are repaid, after which new loans and projects are chosen for the next period. Lenders cannot tell which project is chosen or what a firm's current profits are, but they can seize the firm's assets if a loan is not repaid, which always happens if the risky project was chosen and turned out unsuccessfully.

This game should remind the reader of two other models of credit that have been described in this book: The Repossession Game of section 8.3 and the Stiglitz-Weiss model of section 9.7. Both of those were one-shot games in which the bank worried about not being repaid; in The Repossession Game because the borrower did not exert enough effort, and in the Stiglitz-Weiss model because he was of an undesirable type that could not repay. The Diamond model is a mixture of adverse selection and moral hazard: the borrowers differ in type, but some borrowers have a choice of action.

The equilibrium path has three parts. The RS firms start by choosing risky projects. Their downside risk is limited by bankruptcy, but if the project is successful the firm keeps large residual profits after repaying the loan. Over time, the number of firms with access to the risky project (the RS's and R's) diminishes through bankruptcy, while the number of S's remains unchanged. Lenders can therefore maintain zero profits while lowering their interest rates. When the interest rate falls, the value of a stream of safe investment profits minus interest payments rises relative to the expected value of the few periods of risky returns minus interest payments before bankruptcy. After the interest rate has fallen sufficiently, the second phase of the game begins when the RS firms switch to safe projects at a period we will call t_1. Only the tiny and diminishing group of type R firms continue to choose risky projects. Since the lenders know that the RS firms switch, the interest rate can fall sharply at t_1. A firm that is older is less likely to be a type R, so it is charged a lower interest rate. Figure 15.1 shows the path of the interest rate over time.

Figure 15.1 Interest Rates over Time

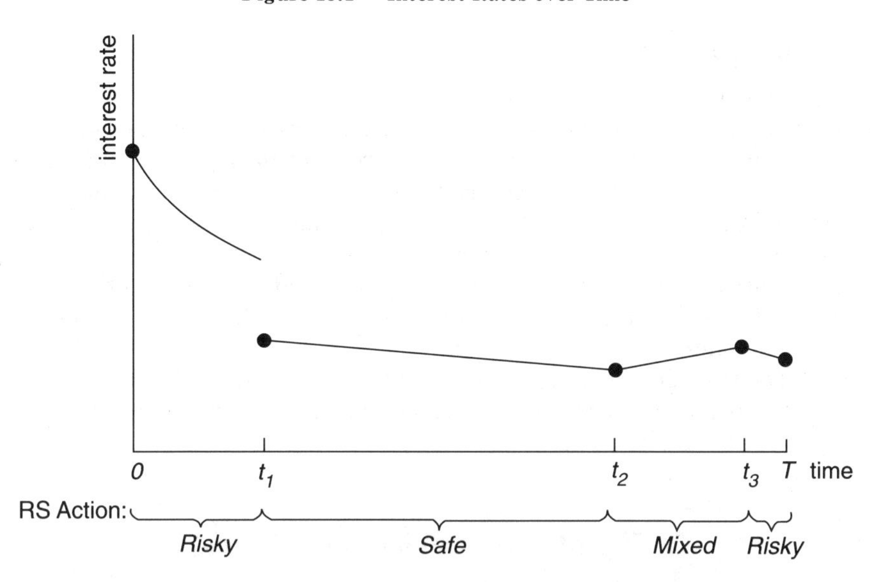

Towards period T the value of future profits from safe projects declines and even with a low interest rate the RS's are again tempted to choose risky projects. They do not all switch at once, however, unlike what happens in period t_1. Before period t_1, if a few RS's had decided to switch to safe projects, the lenders would have been willing to lower the interest rate, which would have made switching even more attractive. If a few firms switch to risky projects at some period t_2, on the other hand, the interest rate rises and switching to risky projects becomes more attractive—a result reminiscent of the Lemons model in chapter 9. Between t_2 and t_3, the RS's follow a mixed strategy, an increasing number of them choosing risky projects as time passes. The increasing proportion of risky projects causes the interest rate to rise. At t_3, the interest rate is high enough and the end of the game is close enough that the RS's revert to the pure strategy of choosing risky projects. The interest rate declines during this last phase as the number of RS's diminishes because of failed risky projects.

One might ask, in the spirit of modelling by example, why the model contains three types of firms rather than just two. Types S and RS are clearly needed, but why type R? The little extra detail in the game description allows simplification of the equilibrium, because with three types bankruptcy is never out-of-equilibrium behavior, since the failing firm might be a type R. Bayes' Rule can, therefore, always be applied, eliminating the problem of ruling out peculiar beliefs and absurd perfect Bayesian equilibria.

This is a Gang of Four model but differs from previous examples in an important respect: the Diamond model is not stationary, and as time progresses, some firms of types R and RS go bankrupt, which changes the lenders' payoff functions. Thus, it is not, strictly speaking, a repeated game.

15.2 Takeovers and Greenmail

The Free Rider Problem

Game theory is well suited to modelling takeovers because the takeover process depends crucially on information and includes a number of sharply delineated actions and events. Suppose that under its current mismanagement, a firm has a value per share of v, but no shareholder has enough shares to justify the expense of a proxy fight to throw out the current managers, although doing so would raise the value to $(v + x)$. An outside bidder makes a tender offer, for any shares offered, conditional upon obtaining a majority. Any bid p between v and $(v + x)$ can make both the bidder and the shareholders better off. But do the shareholders accept such an offer?

We will see that they do not. Quite simply, the only reason the bidder makes a tender offer is that the value would rise higher than his bid, so no shareholder should accept any bid he makes.

The Free Rider Problem in Takeovers

(Grossman & Hart [1980])

Players

A bidder and a continuum of shareholders, holding a total amount m of shares.

Order of Play

(1) The bidder offers p per share for the m shares.
(2) Each shareholder decides whether to accept the bid (denote by θ the fraction that accept).
(3) If $\theta \geq 0.5$, the bid price is paid out, and the value of the firm rises from v to $(v + x)$ per share.

Payoffs

If $\theta < 0.5$, the takeover fails, the bidder's payoff is zero, and the shareholder's payoff is v per share. Otherwise,

$$\pi_{bidder} = \theta m(v + x - p), \text{ if } \theta \geq 0.5.$$

$$\pi_{shareholder} = \begin{cases} p & \text{if the shareholder accepts} \\ v + x & \text{if the shareholder rejects} \end{cases}$$

In any iterated dominant-strategy equilibrium, the bidder's payoff equals zero. Bids above $(v + x)$ are dominated strategies, since the bidder could not possibly profit from them. But if the bid is any lower, an individual shareholder should hold out for the new value $(v + x)$ rather than accepting p. To be sure, when they all do that, the offer fails and they end up with v, but no individual wants to accept if he thinks the offer will succeed. The only equilibria are the many strategy profiles that lead to a failed takeover, or a bid of $p = (v + x)$ accepted by a majority, which succeeds but yields a payoff of zero to the bidder. If organizing an offer has even the slightest cost, the bidder would not do it.

The free rider problem is clearest where there is a continuum of shareholders, so that the decision of any individual does not affect the success of the tender offer. If there were, instead, nine players with one share each, then, in one asymmetric equilibrium, five of them tender at a price just slightly above the old market price and four hold out. Each of the five tenderers knows that if he were to hold out, the offer would fail and his payoff would be zero. This is an example of the discontinuity problem of section 8.7.

In practice, the free rider problem is not quite so severe even with a continuum of shareholders. If the bidder can quietly buy a sizeable number of shares without driving up the price (something severely restricted in the United States by the Williams Act), then his capital gains on those shares can make a takeover profitable even if he makes nothing from the shares bought in the public offer. Dilution tactics such as freeze-out mergers (see Macey & McChesney [1985]) also help the bidder. In a freeze-out, the bidder buys 51 percent of the shares and merges the new acquisition with another firm he owns, at a price below its full value. If dilution is strong enough, the shareholders are willing to sell at a price less than $(v + x)$.

Still another takeover tactic is the two-tier tender offer, a nice application of the Prisoner's Dilemma. Suppose that the underlying value of the firm is 30, which is the initial stock price. A monopolistic bidder offers a price of 10 for 51 percent of the stock and 5 for the other 49 percent, conditional upon 51 percent tendering. It is then a dominant strategy to tender, even though all the shareholders would be better off refusing to sell.

Greenmail

Greenmail occurs when managers buy out some shareholders at an inflated stock price to stop them from taking over. Opponents of greenmail explain this using the Corrupt Managers model. Suppose that a little dilution is possible, or that the bidder owns some shares to start with, so that he can take over the firm but would lose most of the gains to the other shareholders. The managers are willing to pay the bidder a large amount of greenmail to keep their jobs, and both manager and bidder prefer greenmail to an actual takeover, despite the fact that the other shareholders are considerably worse off. The most common objection to this model is that it fails to explain why the corporate charter does not prohibit greenmail.

Managers often use what we might call the Noble Managers model to justify greenmail. In this model, current management knows the true value of the firm, which is greater than both the current stock price and the takeover bid. They pay greenmail to protect the shareholders from selling their mistakenly undervalued shares. This implies either that the shareholders are irrational or that the stock price rises after greenmail because shareholders know that the greenmail signal (giving up the benefits of a takeover) is more costly for a firm which really is not worth more than the takeover bid.

Shleifer & Vishny (1986) have constructed a more sophisticated model in which greenmail is in the interest of the shareholders. The idea is that greenmail encourages potential bidders to investigate the firm, eventually leading to a takeover at a higher price than the initial offer. Greenmail is costly, but for that very reason it is an effective signal that the managers think a better offer could come along later. (Like the model of Noble Managers, this assumes that the manager acts in the interests of the shareholders.) I will present a numerical example in the spirit of Shleifer & Vishny rather than following them exactly, since their exposition is not directed towards the behavior of the stock price.

The story behind the model is that a manager has been approached by a bidder, and he must decide whether to pay him greenmail in the hopes that other bidders—"white knights"—will appear. The manager has better information than the market as a whole about the probability of other bidders appearing, and some other bidders can only appear after they undertake costly investigation, which they will not do if they think the takeover price will be bid up by competition with the first bidder. The manager pays greenmail to encourage new bidders by getting rid of their competition.

Greenmail to Attract White Knights
(Shleifer & Vishny [1986])

Players

The manager, the market, and bidder Brydox. (Bidders Raider and Apex do not make decisions.)

Order of Play

Figure 15.2 shows the game tree. After each time t, the market picks a share price p_t.

(0) Unobserved by any player, Nature picks the state to be (A), (B), (C), or (D), with probabilities 0.1, 0.3, 0.1, and 0.5, unobserved by any player.

(1) Unless the state is (D), the Raider appears and offers a price of 15. The manager's information partition becomes $\{(A), (B, C), (D)\}$; everyone else's becomes $\{(A, B, C), (D)\}$.

(2) The manager decides whether to pay greenmail and extinguish the Raider's offer at a cost of 5 per share.

(3) If the state is (A), Apex appears and offers a price of 25 if greenmail was paid, and 30 otherwise.

(4) If the state is (B), Brydox decides whether to buy information at a cost of 8 per share. If he does, he can make an offer of 20 if the Raider has been paid greenmail, or 27 if he must compete with the Raider.

(5) Shareholders accept the best offer outstanding, which is the final value of a share. If no offer is outstanding, the final value is 5 if greenmail was paid, 10 otherwise.

Payoffs

The manager maximizes the final value.
The market minimizes the squared difference between p_t and the final value.
If he buys information, Brydox receives 23 (= 31 − 8) minus the value of his offer; otherwise he receives zero.

The payoffs specify that the manager should maximize the final value of the firm, rather than a weighted average of the prices p_0 through p_5. This assumption is reasonable because the only shareholders to benefit from a high value of p_t are those who sell their stock at t. The manager cannot say: "The stock is overvalued: Sell!", because the market would learn the overvaluation too, and refuse to buy.

The prices 15, 20, 27, and 30 are assumed to be the results of blackboxed bargaining games between the manager and the bidders. Assuming that the value of the firm to Brydox is 31 ensures that he will not buy information if he foresees that he would have to compete with the Raider. Since Brydox has a dominant strategy—buy information if the Raider has been paid greenmail, otherwise do not—our focus will be on the market price and the decision of whether to pay greenmail. This model is also not designed to answer the question of why the

Figure 15.2 The Game Tree for Greenmail to Attract White Knights

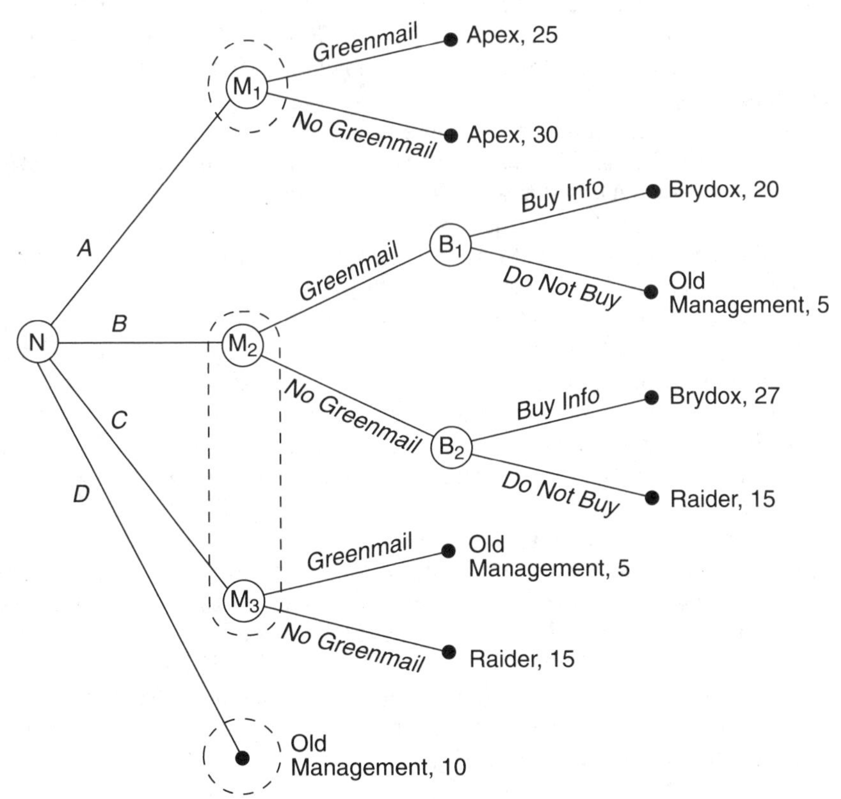

Raider appears. His behavior is exogenous. As the model stands, his expected profit is positive since he is sometimes paid greenmail, but if he actually had to buy the firm he would regret it in states B and C, since the final value of the firm would be 10.

We will see that in equilibrium the manager pays greenmail in states (B) and (C), but not in (A) or (D). Table 15.1 shows the equilibrium path of the market price.

Table 15.1 Game of Greenmail to Attract White Knights: The Equilibrium Prices

State	Probability	p_0	p_1	p_2	p_3	p_4	p_5	Final Management
(A)	0.1	14.5	19	30	30	30	30	Apex
(B)	0.3	14.5	19	16.25	16.25	20	20	Brydox
(C)	0.1	14.5	19	16.25	16.25	5	5	Old management
(D)	0.5	14.5	10	10	10	10	10	Old management

The market's optimal strategy amounts to estimating the final value. Before the market receives any information, its prior beliefs estimate the final value to be 14.5 (= 0.1[30] + 0.3[20] + 0.1[5] + 0.5[10]). If state (D) is ruled out by the arrival of the Raider, the price rises to 19 (= 0.2[30] + 0.6[20] + 0.2[5]). If the Raider does not appear, it becomes common knowledge that the state is (D), and the price falls to 10.

If the state is (A), the manager knows it and refuses to pay greenmail in expectation of Apex' offer of 30. Observing the lack of greenmail, the market deduces that the state is (A), and the price immediately rises to 30.

If the state is (B) or (C) the manager does pay greenmail and the market, ruling out (A), uses Bayes' Rule to assign probabilities of 0.75 to (B) and 0.25 to (C). The price falls from 19 to 16.25 (= 0.75[20] + 0.25[5]).

It is clear that the manager should not pay greenmail in state (A) or in state (D), when the manager knows that Brydox is not around to investigate. What if the manager deviates in the information set (B, C) and refuses to pay greenmail? The market would initially believe that the state was (A), so the price would rise to $p_2 = 30$. But the price would fall again after Apex failed to make an offer and the market realized that the manager had deviated. Brydox would refuse to enter at time 3, and the Raider's offer of 15 would be accepted. The payoff of 15 would be less than the expected payoff of 16.25 from paying greenmail.

The model does not say that greenmail is always good for the shareholders, only that it can be good *ex ante*. If the true state turns out to be (C), then greenmail was a mistake, *ex post*, but since state (B) is more likely, the manager is correct to pay greenmail in information set (B, C). What is noteworthy is that greenmail is optimal even though it drives down the stock price from 19 to 16.25. Greenmail communicates the bad news that Apex is not around, but makes the best of that misfortune by attracting Brydox.

15.3 Market Microstructure

The prices of securities such as stocks depend on what investors believe are the values of the assets that underlie them. The values are highly uncertain, and new information about them is constantly being generated. The market microstructure literature is concerned with how new information enters the market. In the paradigmatic situation, an informed trader has private information about the value which he hopes to use to make profitable trades, but other traders know that someone might have private information. This is a situation of adverse selection, because the informed trader has better information on the value of the stock, and no uninformed trader wants to trade with an informed trader. An institution that many markets have developed is that of the marketmaker or specialist, a trader in a particular stock who is always willing to buy or sell to keep the market going. Other traders feel safer in trading with the marketmaker than with a potentially informed trader, but this just transfers the adverse selection problem to the marketmaker.

The two models in this section will look at how a marketmaker deals with the problem of informed trading. Both are descendants of the verbal model in Bagehot (1971).[1] In The Bagehot Model, which is most closely related to Glosten & Milgrom (1985), there may or may not be one or more informed traders, but the informed traders as a group have a trade of fixed size if they are present. The marketmaker must decide how big a bid-ask spread to charge. In the Kyle Model (Kyle [1985]) there is a single informed trader, and it is he who decides how much to trade. On observing the imbalance of orders, the marketmaker decides what price to offer.

The Bagehot model is perhaps a better explanation of why marketmakers might charge a bid/ask spread even under competitive conditions and with zero transactions costs. Its assumption is that the marketmaker cannot change the price depending on volume, but must instead offer a price, and then accept whatever order comes along—a buy order, or a sell order.

The Bagehot Model

Players

The informed trader and two competing marketmakers.

Order of Play

(0) Nature chooses the asset value v to be either $p_0 - \delta$ or $p_0 + \delta$ with equal probability. The marketmakers never observe the asset value, nor do they observe whether anyone else observes it, but the informed trader observes v with probability θ.

[1] "Bagehot," pronounced "badget", is a pseudonym for Jack Treynor.

(1) The marketmakers choose their spreads s, offering prices $p_{bid} = p_0 - s/2$ at which they will buy the security and $p_{ask} = p_0 + s/2$ for which they will sell it.[2]
(2) The informed trader decides whether to buy one unit, to sell one unit, or to do nothing.
(3) Noise traders buy n units and sell n units.

Payoffs

Everyone is risk neutral. The informed trader's payoff is $v - p_{ask}$ if he buys, $p_{bid} - v$ if he sells, and zero if he does nothing. The marketmaker who offers the highest p_{bid} trades with all the customers who wish to sell, and the marketmaker who offers the lowest p_{ask} trades with all the customers who wish to buy. If the marketmakers set equal prices, they split the market evenly. A marketmaker who sells x units gets a payoff of $x(p_{ask} - v)$, and a marketmaker who buys x units gets a payoff of $x(v - p_{bid})$.

This is a very simple game. Competition between the marketmakers will make their prices identical and their profits zero. The informed trader should buy if $v > p_{ask}$ and sell if $v < p_{bid}$. He has no incentive to trade if $v \in [p_{bid}, p_{ask}]$.

A marketmaker will always lose money trading with the informed trader, but if $s > 0$, so $p_{ask} > p_0$ and $p_{bid} < p_0$, he will earn positive expected profits trading with the noise traders. The expected profit from the $2n$ noise traders will equal ns, since the marketmaker is perfectly hedged from risk in dealing with them. He earns a profit equal to the spread s from each buy/sell pair of noise traders that he matches. For profits to be zero, it must therefore be true that

$$0.5(\theta)\left[\left(p_0 + \frac{s}{2}\right) - (p_0 + \delta)\right] + 0.5(\theta)\left[(p_0 - \delta) - \left(p_0 - \frac{s}{2}\right)\right] + ns = 0. \quad (1)$$

Equation (1) implies that $\theta(s - 2\delta) + ns = 0$, so

$$s^* = \frac{2\delta\theta}{n} + \theta. \quad (2)$$

Equation (2) has a number of implications. First, the spread s^* is strictly positive. Even though marketmakers compete and have zero transactions costs, they charge a different price when buying and when selling. They make money dealing with the noise traders but lose money with the informed trader, if he is present. The comparative statics reflect this. The spread s^* rises in δ, the variance of the true value, because divergent true values increase losses from trading with the informed trader. The spread s^* falls in n, which reflects the number of noise traders relative to informed traders, because when there are more noise traders, the profits from trading with them are greater. The spread s^* rises in θ, the

[2] Note that for simplicity the game constrains the marketmaker to choose p_{bid} and p_{ask} symmetrically around p_0.

probability that the informed trader really has inside information, which is also intuitive but requires a little calculus to demonstrate starting from equation (2):

$$\frac{\partial s^*}{\partial \theta} = \frac{2\delta}{n + \theta} - \frac{2\delta\theta}{(n + \theta)^2} = \left[\frac{1}{(n + \theta)^2}\right](2\delta n + 2\delta\theta - 2\delta\theta) > 0. \quad (3)$$

The second model of market microstructure, important because it is commonly used as a foundation for more complicated models, is the Kyle model, which focuses on the decision of the informed trader, not the marketmaker. The Kyle model is set up so that the marketmaker observes the trade volume before he chooses the price.

The Kyle Model

(Kyle [1985])

Players

The informed trader and two competing marketmakers.

Order of Play

(0) Nature chooses the asset value v from a normal distribution with mean p_0 and variance σ_v^2, observed by the informed trader but not by the marketmakers.

(1) The informed trader offers a trade of size $x(v)$, which is a purchase if positive and a sale if negative, unobserved by the marketmaker.

(2) Nature chooses a trade of size u by noise traders, unobserved by the marketmaker, where u is distributed normally with zero mean and variance σ_u^2.

(3) The marketmakers observe the total market trade offer $y = x + u$, and choose prices $p(y)$.

(4) Trades are executed. If y is positive (the market wants to purchase, in net), whichever marketmaker offers the lowest price executes the trades; if y is negative (the market wants to sell, in net), whichever marketmaker offers the highest price executes the trades. v is then revealed to everyone.

Payoffs

All players are risk neutral. The informed trader's payoff is $(v - p)x$. The marketmaker's payoff is zero if he does not trade and $(p - v)y$ if he does.

An equilibrium for this game is the strategy profile

$$x(v) = (v - p_0)\left(\frac{\sigma_u}{\sigma_v}\right) \quad (4)$$

and

$$p(y) = p_0 + \left(\frac{\sigma_v}{2\sigma_u}\right)y. \tag{5}$$

This is reasonable. It says that the informed trader will increase the size of his trade as v gets bigger relative to p_0 (and he will sell, not buy, if $v - p_0 < 0$), and the marketmaker will increase the price he charges for selling if y is bigger, meaning that more people want to sell, which is an indicator that the informed trader might be trading heavily. The variance of the asset value (σ_v^2) and of the noise trading (σ_u^2) enter as one would expect, and they matter only in their relation to each other. If σ_v^2/σ_u^2 is large, then the asset value fluctuates more than the amount of noise trading, and it is difficult for the informed trader to conceal his trades under the noise. The informed trader will trade less, and a given amount of trading will cause a greater response from the marketmaker. One might say that the market is less "liquid": a trade of given size will have a greater impact on the price.

I will not (and cannot) prove uniqueness of the equilibrium, since it is very hard to check all possible profiles of nonlinear strategies, but I will show that {(4), (5)} is Nash and is the unique linear equilibrium. To start, hypothesize that the informed trader uses a linear strategy, so

$$x(v) = \alpha + \beta v \tag{6}$$

for some constants α and β. Competition between the marketmakers means that their expected profits will be zero, which requires that the price they offer be the expected value of v. Thus, their equilibrium strategy $p(y)$ will be an unbiased estimate of v given their data y, where they know that y is normally distributed and that

$$\begin{aligned} y &= x + u \\ &= \alpha + \beta v + u. \end{aligned} \tag{7}$$

This means that their best estimate of v given the data y is, following the usual regression rule (which readers unfamiliar with statistics must accept on faith),

$$\begin{aligned} E(v|y) &= E(v) + \left(\frac{cov(v, y)}{var(y)}\right)y \\ &= p_0 + \left(\frac{\beta\sigma_v^2}{\beta^2\sigma_v^2 + \sigma_u^2}\right)y \qquad (8) \\ &= p_0 + \lambda y, \end{aligned}$$

where λ is a new shorthand variable to save writing out in what follows the term in parentheses in the next-to-last line.

Under our assumption that x is a linear function of v, $p(y)$ will be a linear function of y. Given that $p(y) = p_0 + \lambda y$, what must next be shown is that x will indeed be a linear function of v. Start by writing the informed trader's expected payoff, which is

$$\begin{aligned} E\pi_i &= E([v - p(y)]x) \\ &= E([v - p_0 - \lambda(x + u)]x) \\ &= [v - p_0 - \lambda(x + 0)]x, \end{aligned} \tag{9}$$

since $E(u) = 0$. Maximizing the expected payoff with respect to x yields the first order condition

$$v - p_0 - 2\lambda x = 0, \tag{10}$$

which, on rearranging, becomes

$$x = -\frac{p_0}{2\lambda} + \left(\frac{1}{2\lambda}\right)v. \tag{11}$$

Equation (11) establishes that $x(v)$ is linear, given that $p(y)$ is linear. All that is left is to find the value of λ. Note by comparing (11) and (6) that $\beta = 1/2\lambda$. Substituting this β into the expression for λ given in (8) results in

$$\lambda = \frac{\beta\sigma_v^2}{\beta^2\sigma_v^2 + \sigma_u^2} = \frac{\frac{\sigma_v^2}{2\lambda}}{\frac{\sigma_v^2}{(4\lambda^2)} + \sigma_u^2}, \tag{12}$$

which, upon solving for λ yields $\lambda = \sigma_v/2\sigma_u$. Since $\beta = 1/2\lambda$, it follows that $\beta = \sigma_u/\sigma_v$. These values of λ and β together with equation (11) give the strategies asserted at the start in equations (4) and (5).

15.4 Rate of Return Regulation and Government Procurement

The central idea in both government procurement and the regulation of natural monopolies is that the government is trying to induce a private firm to efficiently provide a good to the public while covering the cost of production. If information is symmetric, this is an easy problem; the government simply pays the firm the cost of producing the good efficiently, whether the good be a new tank or electric power. Usually, however, the firm has better information than the government. The variety of ways the firm might have better information and the government might extract it has given rise to a large literature in which moral hazard with hidden actions, moral hazard with hidden knowledge, adverse selection, and signalling all appear. Suppose, for example, that the government wishes the firm to provide cable television service to a city. The firm knows more about its costs before agreeing to accept the franchise (adverse selection), discovers more after accepting it and beginning operations (moral hazard with hidden knowledge), and exerts greater or smaller effort to keep costs low (moral hazard with hidden actions). The government's problem is to acquire cable service at the lowest cost. It wants to be generous enough to induce the firm to accept the franchise in the first place but no more generous than necessary. It cannot simply agree to cover the firm's costs, because the firm would always claim high costs and exert low effort. Instead, the government might auction off the right to provide the service, might allow the firm a maximum price (**price cap**), or might agree to compensate the firm to varying degrees for different levels of cost (**rate-of-return regulation**).

The problems of regulatory franchises and government procurement are the same in many ways. If the government wants to purchase an aircraft carrier, it also has the problem of how much to offer the firm. Roughly speaking, the equivalent of a price cap is a flat price, and the equivalent of rate-of-return regulation is a cost-plus contract, although the details differ in interesting ways between regulation and procurement. (A price cap allows downwards flexibility in prices, and rate-of-return regulation allows an expected but not guaranteed profit, for example.)

Although the literature on mechanism design can be traced back to Mirrlees (1971), its true blossoming has occurred since Baron & Myerson's 1982 article, "Regulating a Monopolist with Unknown Costs." Spulber (1989) and Laffont & Tirole (1993) provide 690-page and 702-page treatments of the confusing array of possible models and policies in their books on government regulation. Here we will content ourselves with one model, a version of the model Laffont and Tirole use to introduce their book on pages 55 to 62. This is a two-type model in which the cost parameter and effort of a firm is its private information, but its realized cost is public and nonstochastic—an adverse selection model, essentially. The government will reimburse the firm's costs, but also fixes a subsidy (which, if negative, becomes a tax) depending on the level of the firm's costs. The questions the model hopes to answer are whether effort will be too high or too low and whether the subsidy is positive and rises with costs.

Government Procurement

Players

The government and the firm.

Order of Play

(0) Nature assigns the firm a cost parameter β. The low cost $\beta = L$ has probability θ and the high cost $\beta = H$ has probability $(1 - \theta)$.

(1) The government offers a contract $s(c)$ agreeing to cover the firm's costs of producing a space station and specifying the subsidy for each cost level that the firm might report.

(2) The firm accepts or rejects the contract.

(3) If the firm accepts, it chooses an effort level e.

(4) The firm finishes the space station at a cost of $c = \beta - e$. The government reimburses the cost and pays the appropriate subsidy.

Payoffs

Both firm and government are risk neutral, and both receive payoffs of zero if the firm rejects the contract. If the firm accepts, its payoff is

$$\pi_{firm} = s - d(e), \tag{13}$$

where $d(e)$, the disutility of effort, is increasing and convex, so $d' > 0$ and $d'' > 0$, and, for technical convenience, is increasingly convex, so $d''' > 0$.[3] The government's payoff is

$$\pi_{government} = B - (1 + \lambda)c - \lambda s - d(e), \tag{14}$$

where B is the benefit from the space station and λ is the deadweight loss from the taxation needed for government spending.[4]

Assume for the moment that B is sufficiently large that the government definitely wishes to build the station. (How large will become apparent later.) Note that cost, not output, is the focus of this model. The output is one space station regardless of agency problems, but the government wants to minimize the cost of producing the station.

This model differs from most principal-agent models in this book in that the government, unlike previous principals, is altruistic towards the agent. If the government were selfish, its payoff would be $B - (1 + \lambda)c - (1 + \lambda)s$. Instead, it maximizes social welfare, which includes the welfare of both the citizenry and the firm. The welfare of the citizenry is $B - (1 + \lambda)c - (1 + \lambda)s$ and that of the firm is $s - d(e)$. Summing these yields equation (14).

Note also that the *Low* type of firm is good, not bad, unlike in previous models, because here the type refers to cost, not to ability or effort.

Government Procurement I

In the first version of the game, β is observed by the government, which can assign different contracts to the two types of firms. The government pays subsidies of s_1 to a low-cost firm of type $\beta = L$ ("Firm L") for the low cost $\underline{c}$, s_2 to a firm of type $\beta = H$ ("Firm H") for the high cost $\bar{c}$, and a boiling-in-oil subsidy of $s = -\infty$ to a firm that does not choose the cost level assigned to it.

The participation constraints will be binding for both types of firms, and to make a firm's payoffs zero the the government will provide subsidies that exactly

3 The assumption that $d''' > 0$ allows the use of first-order conditions by making the maximand in (25) concave. See p. 58 of Laffont & Tirole (1993).

4 This loss is estimated to be around $0.3 for each $1 of tax revenue raised, at the margin for the United States (Hausman & Poterba [1987]).

cover the firm's disutility of effort. Since there is no uncertainty, we can invert the cost equation and write it as $e = \beta - c$. The subsidies will be $s_1 = d(L - \underline{c})$ and $s_2 = d(H - \bar{c})$. Substituting these into the government's payoff function yields

$$\pi_{government} = B - (1 + \lambda)\underline{c} - \lambda d(L - \underline{c}) - d(L - \underline{c}) \tag{15}$$

for firm L. Since $d'' > 0$, the government's payoff function is concave, and the standard optimization technique can be used. The first-order condition for the optimal level of $\bar{c}$ is

$$\frac{\partial \pi_{government}}{\partial \underline{c}} = -(1 + \lambda) + \lambda d'(L - \underline{c}) + d'(L - \underline{c}) = 0, \tag{16}$$

so

$$d'(L - \underline{c}) = 1. \tag{17}$$

Equation (17) says that at the efficient effort level, the marginal disutility of effort equals the marginal reduction in cost because of effort. Exactly the same is true for firm H, so $L - \underline{c} = H - \bar{c}$ and it follows that $s_1 = s_2$. The cost targets assigned to each firm are $\underline{c} = L - e^*$ and $\bar{c} = H - e^*$. The two firms exert the same efficient effort level and are paid the same positive subsidy as a return to their disutility of effort. Let us call this effort level e^* and the subsidy level s^*. The assumption that B is sufficiently large can now be made more specific: it is that $B - (1 + \lambda)(H - e^* - d(e^*)) \geq 0$.

Government Procurement II

In the second variant of the game, β is not observed by the government, which must therefore provide incentives for the firms to volunteer their types if it wants firm L to produce at lower costs than firm H.

The government could use a pooling contract, simply providing a contract with a subsidy of $d(e^*)$ for a cost of $H - e^*$, just enough to compensate firm H for its effort, and an infinitely negative subsidy for any other cost. Both types would accept this, but firm L could exert effort less than e^* and still receive the subsidy. This shows that if the government would build the space station under full information knowing that the firm has high costs, it would also build it under incomplete information, when the firm might or might not have high costs, but this, it turns out, is not the optimal contract. Instead, the government will pay a premium greater than the additional effort to a firm that reveals that its costs are actually low, since even though this will give the firm an above-market profit, it will save on the amount of cost that the government must reimburse.

Let us find the optimal contract with values $(\underline{c}, s_1)$ and $(\bar{c}, s_2)$ and heavy punishments for other cost levels. Although we have not yet established that the optimal contract is separating, we will go through the analysis for two separate contracts, and if a pooling contract is optimal, we will find that $\underline{c} = \bar{c}$. The contract must satisfy participation constraints and incentive compatibility constraints for each type of firm. The participation constraint for firm L is

$$s_1 - d(L - \underline{c}) \geq 0 \tag{18}$$

and for firm H it is

$$s_2 - d(H - \bar{c}) \geq 0. \tag{19}$$

The incentive-compatibility constraint for firm L is

$$s_1 - d(L - \underline{c}) \geq s_2 - d(L - \bar{c}) \tag{20}$$

and for firm H it is

$$s_2 - d(H - \bar{c}) \geq s_1 - d(H - \underline{c}). \tag{21}$$

Since firm L can imitate firm H if it wishes, if constraint (19) is satisfied, so is (18). Constraint (19) will be binding (and therefore satisfied as an equality), because the government will reduce the subsidy as much as possible in order to avoid the deadweight loss of taxation that exists because $\lambda > 0$. The incentive compatibility constraint for firm L must also be binding, because if the pair ($\underline{c}$, s_1) is strictly more attractive for firm L, the government could reduce the subsidy s_1. Constraint (20) is therefore satisfied as an equality. (The same argument does not hold for firm H, because if s_2 were reduced, the participation constraint would be violated.) Knowing that these two constraints are binding, we can write

$$s_2 = d(H - \bar{c}) \tag{22}$$

and, making use of both (20) and (22),

$$s_1 = d(L - \underline{c}) + d(H - \bar{c}) - d(L - \bar{c}). \tag{23}$$

From (14), the government's maximization problem under incomplete information is

$$\begin{aligned} \underset{\underline{c},\bar{c},s_1,s_2}{Maximize}\ \theta[B - (1 + \lambda)\underline{c} - \lambda s_1 - d(L - \underline{c})] \\ + [1 - \theta][B - (1 + \lambda)\bar{c} - \lambda s_2 - d(H - \bar{c})]. \end{aligned} \tag{24}$$

Substituting for s_1 and s_2 from (22) and (23) simplifies the problem to

$$\begin{aligned} \underset{\underline{c},\bar{c}}{Maximize}\ \theta[B - (1 + \lambda)\underline{c} - \lambda(d(L - \underline{c}) + d(H - \bar{c}) - d(L - \bar{c})) \\ - d(L - \underline{c})] + [1 - \theta][B - (1 + \lambda)\bar{c} \\ - \lambda d(H - \bar{c}) - d(H - \bar{c})]. \end{aligned} \tag{25}$$

The first-order condition with respect to $\underline{c}$ is

$$\theta[-(1+\lambda) + \lambda d'(L-\underline{c}) + d'(L-\underline{c})] = 0, \tag{26}$$

which simplifies to

$$d'(L-\underline{c}) = 1. \tag{27}$$

Thus, firm L chooses the efficient effort level e^* in equilibrium, and $\underline{c}$ takes the same value as it did in Government Procurement I. From the definition of $s^* = d(e^*)$ in that game, equation (23) can be rewritten as

$$s_1 = s^* + d(H-\bar{c}) - d(L-\bar{c}). \tag{28}$$

Because $d(H-\bar{c}) > d(L-\bar{c})$, equation (28) shows that $s_1 > s^*$. Incomplete information increases the subsidy to the low-cost firm, which earns more than its reservation utility in the game with incomplete information. Since the high-cost firm will earn exactly its reservation utility, this means that the government is on average providing its supplier with an above-market rate of return—not because of corruption or political influence, but because that is the way to induce low-cost suppliers to reveal that their costs are low. This should be kept in mind as an alternative explanation to the product quality model of chapter 5 and the efficiency wage model of chapter 8 for why above-average rates of return persist.

Turning now to the contract alternative to be chosen by the high-cost firm, the first-order condition for maximizing the government payoff with respect to $\bar{c}$ is

$$\begin{aligned}\theta[-\lambda(-d'(H-\bar{c}) + d'(L-\bar{c}))] + [1-\theta][-(1+\lambda)\\ + \lambda d'(H-\bar{c}) + d'(H-\bar{c})] = 0.\end{aligned} \tag{29}$$

This can be rewritten as

$$d'(H-\bar{c}) = 1 - \left(\frac{\lambda}{1+\lambda}\right)\left(\frac{\theta}{1-\theta}\right)[d'(H-\bar{c}) + d'(L-\bar{c})]. \tag{30}$$

Since the right-hand-side of equation (30) is less than one, firm H has a lower level of d' than firm L, and must be exerting effort less than e^*, since $d'' > 0$. Perhaps this explains the expression, "good enough for government work." Also, since the participation constraint (22) is satisfied as an equality, it must also be true that $s_2 < s^*$. The high-cost firm's subsidy is lower than under full information, although since its effort is also lower, its payoff stays the same.

We must also verify that the incentive compatibility constraint for firm H is satisfied as a weak inequality; the high-cost firm is not near being tempted to pick the low-cost firm's contract. This is a bit subtle. Setting the left-hand side of the incentive compatibility constraint (21) equal to zero because the participation constraint is binding for firm H, substituting in for s_1 from equation (23), and rearranging, gives

$$d(H-\underline{c}) - d(L-\underline{c}) > d(H-\bar{c}) - d(L-\bar{c}). \tag{31}$$

This is true as a strict inequality, because $d'' > 0$ and the arguments of d on the left-hand-side of equation (31) take larger values than on the right-hand side.

The game of Government Procurement illustrates that there is a tradeoff between the government's two objectives of inducing the correct amount of effort and minimizing the subsidy to the firm. Even under complete information, the government cannot provide a subsidy of zero, or no firm will build the space station. Under incomplete information, not only must the subsidies be positive but the low-cost firm earns **informational rents**; the government offers a contract that pays the low-cost firm more than under complete information to prevent it from mimicking the high-cost firm by choosing an inefficiently low effort. The high-cost firm, however, does choose an inefficiently low effort, because if it were assigned greater effort it would have to be paid a greater subsidy, which would tempt the low-cost firm to imitate it. In equilibrium, the government has compromised by having some probability of an inefficiently high subsidy *ex post*, and some probability of inefficiently low effort.

A little reflection will provide a host of ways to alter this model. What if the firm only discovers its costs after accepting the contract? What if two firms bid against each other for the contract? What if the firm can bribe the government? What if the firm and the government bargain over the gains from the project, instead of the government being able to make a take-it-or-leave-it contract offer? What if the game is repeated, so the government can use the information it acquires in the second period? If it is repeated, can the government commit to long-term contracts? Can it commit not to renegotiate? See Spulber (1989) and Laffont & Tirole (1993) if these questions interest you.

Appendix A: Answers to Odd-Numbered Problems

This appendix contains answers to the odd-numbered problems in the book. The answers to the even-numbered problems are available from Blackwell Publishers or via Internet. Use telnet or ftp to reach my account at Indiana University. The machine name is rasmusen.bus.indiana.edu, the IP number is 129.79.122.177, the account is 'guest' and the password is 'guest'. This is a Unix account, so remember to use lowercase letters and use the command 'ls' to list files. It can be reached by telnet using the command "telnet rasmusen.bus.indiana.edu" or by telephoning 812-855-4211 to reach Indiana University's 2400 baud modem and typing 'connect rasmusen'. (The number for the 9600 baud modem is 812-855-9681.) The answers are written in ASCII using LaTeX commands. This means that you can download them to your computer easily, and load them into your own word processor, but if you do not know LaTeX you will have to do some work interpreting the answers. I may also put DVI files for the answers in that account. These are image files, which cannot be read as text, but which might be printable on a PostScript printer. I encourage readers to submit additional homework problems as well as errors and frustrations. They can either be put in the guest file, or sent to me by Internet e-mail at Erasmuse@Indiana.edu.

Other books which contain exercises with answers include Bierman & Fernandez (1993), Binmore (1992), Fudenberg & Tirole (1991a), J. Hirshleifer & Riley (1992), and Moulin (1986b). I must ask pardon of any authors from whom I have borrowed without attribution in the problems below; these are the descendants of problems that I wrote for teaching without careful attention to my sources.

Chapter 1

1.1: 2-by-2 Games. Find examples of 2-by-2 games with the following properties:

(1.1a) No Nash equilibrium (you can ignore mixed strategies).
Answer. See Simple Cycling (Table A.1).

(1.1b) No weakly Pareto-dominant strategy profile.
Answer. See Simple Cycling (Table A.1).

Table A.1 Simple Cycling

		Jones		
		Left		*Right*
	Up	1,0	→	0,1
Smith:		↑		↓
	Down	0,1	←	1,0

Payoffs to: (Smith, Jones)

(1.1c) At least two Nash equilibria, including one equilibrium that Pareto dominates all other strategy profiles.
Answer. In Ranked Coordination (Table 1.7), (*Large*, *Large*) has uniformly higher payoffs than (*Small*, *Small*).

(1.1d) At least three Nash equilibria.
Answer. In The Swiss Cheese Game, a 2-by-2 game in which every payoff profile is (0,0), every strategy profile is a Nash equilibrium.

1.3: Timmy and Scarface. Players Timmy and Scarface are caught in a game like the Prisoner's Dilemma except that Scarface already has a criminal record, so he will always get a prison term at least 5 years greater than Timmy will get, regardless of who finks and who denies. Construct an outcome matrix (with Scarface as Row) and find the Nash equilibrium for this game. (Note: There are at least two games that reasonably fit this story.)

Answer. The story is too vague to tell us exactly which game Scarface and Timmy are playing, so I will give two possibilities. Table A.2 is constructed by subtracting 5 from each of Scarface's payoffs in the original Prisoner's Dilemma in Table 1.1, except for subtracting 15 from his (*Confess*, *Deny*) payoff. In equilibrium, Scarface denies and Timmy confesses.

Table A.2 Scarface I

		Timmy		
		Deny		*Confess*
	Deny	−6, −1	→	**−15,0**
Scarface:		↑		↑
	Confess	−15, −10	←	−13, −8

Payoffs to: (Scarface, Timmy)

Table A.2 is a little far-fetched, because it implies that when Scarface confesses, Timmy's denial increases Scarface's punishment, as well as Timmy's. This is possible, if the judge wants to punish Timmy more (for denying), but must always punish Scarface more than Timmy. But Table A.3 shows another game to fit the story, one which preserves the Prisoner's Dilemma property that a prisoner is treated more leniently for providing useful evidence. Here, (*Confess*, *Confess*) is the Nash equilibrium, even though *Confess* is not a dominant strategy for Scarface (he would *Deny* if he thought Timmy would go along with him).

Table A.3 Scarface II

		Timmy		
		Deny		*Confess*
Scarface:	*Deny*	−6, −1	→	−30,0
		↑		↓
	Confess	−13, −8	→	**−20, −5**

Payoffs to: (Scarface, Timmy)

1.5: Discoordination. Suppose that a man and a woman each choose whether to go to a prize fight or to a ballet. The man would rather go to the prize fight, and the woman to the ballet. What is more important, however, is that the man wants to show up at the same event as the woman and that she wants to avoid him.

(1.5a) Construct a game matrix to illustrate this game, choosing numbers to fit the preferences described verbally.
Answer. See Battle of the Sexes with Unrequited Love (Table A.4).

Table A.4 Battle of the Sexes with Unrequited Love[1]

		Woman		
		Prize Fight		*Ballet*
Man	*Prize Fight*	20, −2	→	−10,2
		↑		↓
	Ballet	−20,1	←	10, −1

Payoffs to: (Man, Woman)

(1.5b) If the woman moves first, what will happen?
Answer. (*Ballet*, *Ballet*). The woman knows the only alternative is (*Prize Fight*, *Prize Fight*), which has a lower payoff.

(1.5c) Does the game have a first-mover advantage?
Answer. No—it has a first-mover disadvantage.

(1.5d) Show that there is no Nash equilibrium if the players move simultaneously.
Answer. (*Prize Fight*, *Ballet*) and (*Ballet*, *Prize Fight*) are not Nash because the man would deviate; (*Prize Fight*, *Prize Fight*) and (*Ballet*, *Ballet*) are not, because the woman would.[2]

1 Or, for the 1990's, The Sexual Harassment Game.
2 There does exist a Nash equilibrium in mixed strategies, which will be discussed in Chapter 3.

Chapter 2

2.1: The Monty Hall Problem. You are a contestant on the TV show, "Let's Make a Deal." You face three curtains, labelled A, B and C. Behind two of them are toasters, and behind the third is a Mazda Miata car. You choose A, and the TV showmaster says, pulling curtain B aside to reveal a toaster, "You're lucky you didn't choose B, but before I show you what is behind the other two curtains, would you like to change from curtain A to curtain C?" Should you switch? What is the exact probability that curtain C hides the Miata?

Answer. You should switch to curtain C, because

$$
\begin{aligned}
Prob(Miata\ behind\ C|Host\ chose\ B) &= \frac{Prob(Host\ chose\ B|Miata\ behind\ C)Prob(Miata\ behind\ C)}{Prob(Host\ chose\ B)} \\
&= \frac{(1)\left(\frac{1}{3}\right)}{(1)\left(\frac{1}{3}\right) + \left(\frac{1}{2}\right)\left(\frac{1}{3}\right)}. \\
&= \frac{2}{3}.
\end{aligned}
$$

The key is to remember that this is a game. The host's action has revealed more than that the Miata is not behind B; it has also revealed that the host did not want to choose curtain C. If the Miata were behind B or C, he would pull aside the curtain it was not behind. Otherwise, he would pull aside a curtain randomly. His choice tells you nothing new about the probability that the Miata is behind curtain A, which remains ⅓, so the probability of it being behind C must rise to ⅔ (to make the total probability equal one).

2.3: Cancer Tests. Imagine that you are being tested for cancer, using a test that is 98% accurate. If you indeed have cancer, the test shows positive (indicating cancer) 98% of the time. If you do not have cancer, it shows negative 98% of the time. You have heard that 1 in 20 people in the population actually have cancer. Now your doctor tells you that you tested positive, but you shouldn't worry because his last 19 patients all died. How worried should you be? What is the probability you have cancer?

Answer. Doctors, of course, are not mathematicians. Using Bayes' Rule:

$$
\begin{aligned}
Prob(Cancer|Positive) &= \frac{Prob(Positive|Cancer)Prob(Cancer)}{Prob(Positive)} \\
&= \frac{0.98(0.05)}{0.98(0.05) + 0.02(0.95)} \qquad (1) \\
&\approx 0.72.
\end{aligned}
$$

With a 72 percent chance of cancer, you should be very worried. But at least it is not 98 percent.

Here is another way to see the answer. Suppose 10,000 tests are done. Of these, an average of 500 people have cancer. Of these, 98% test positive on average—490 people. Of the 9,500 cancer-free people, 2% test positive on average—190 people. Thus there are 680 positive tests, of which 490 are true positives. The probability of having cancer if you test positive is 490/680, about 72%.

2.5: Joint Ventures. Software Inc. and Hardware Inc. have formed a joint venture. Each can exert either high or low effort, which is equivalent to costs of 20 and 0. Hardware moves first, but Software cannot observe his effort. Revenues are split equally at the end, and the two firms are risk neutral. If both firms exert low effort, total revenues are 100. If the parts are defective, the total revenue is 100; otherwise, if both exert high effort, revenue is 200, but if only one player does, revenue is 100 with probability 0.9 and 200 with probability 0.1. Before they start, both players believe that the probability of defective parts is 0.7. Hardware discovers the truth about the parts by observation before he chooses effort, but Software does not.

(2.5a) Draw the extensive form and put dashed lines around the information sets of Software at any of the nodes at which he moves.

Answer. See Figure A.1.

Figure A.1 The Extensive Form for the Joint Ventures Game

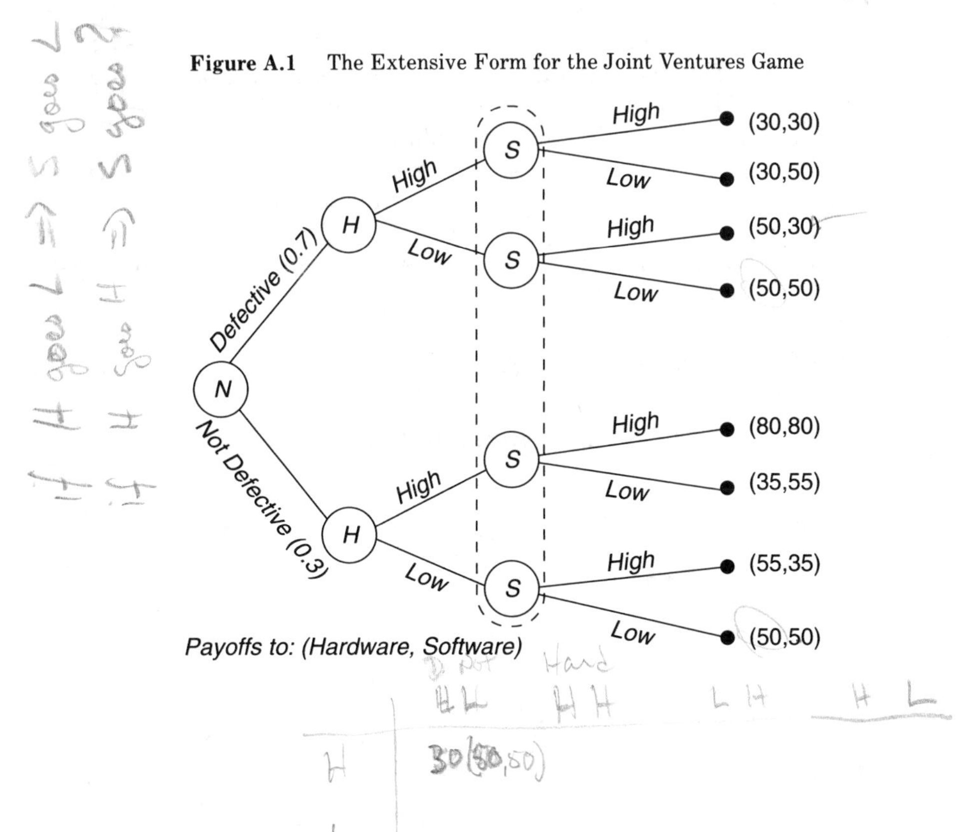

(2.5b) What is the Nash equilibrium?
Answer. (*Low* if defective parts, *Low* if not defective parts, *Low*).

(2.5c) What is Software's belief, in equilibrium, as to the probability that Hardware chooses low effort?
Answer. One. In equilibrium, Hardware always chooses *Low*.

(2.5d) If Software sees that revenue is 100, what probability does he assign to defective parts if he himself exerted high effort and he believes that Hardware chose low effort?
Answer. 0.72 (= (1) (0.7)/[(1)(0.7) + (0.9)(0.3)]).

Chapter 3

3.1: Presidential Primaries. Smith and Jones are fighting it out for the Democratic nomination for President of the United States. The more months they keep fighting, the more money they spend, because a candidate must spend one million dollars a month in order to stay in the race. If one of them drops out, the other one wins the nomination, which is worth 11 million dollars. The discount rate is r per month. To simplify the problem, you may assume that this battle could go on forever if neither of them drops out. Let θ denote the probability that an individual player will drop out each month in the mixed-strategy equilibrium.

(3.1a) In the mixed-strategy equilibrium, what is the probability θ each month that Smith will drop out? What happens if r changes from 0.1 to 0.15?
Answer. The value of exiting is zero. The value of staying in is $V = \theta(10) + (1 - \theta)(-1 + V/(1 + r))$. Thus, $V - (1 - \theta)V/(1 + r) = 10\theta - 1 + \theta$, and $V = (11\theta - 1)(1 + r)/(r + \theta)$. Thus, $\theta = 1/11$ in equilibrium.

The discount rate does not affect the equilibrium outcome, so a change in r produces no observable effect.

(3.1b) What are the two pure-strategy equilibria?
Answer. (Smith drops out, Jones stays in no matter what) and (Jones drops out, Smith stays in no matter what).

(3.1c) If the game lasts only one period, and the Republican wins the general election (for Democrat payoffs of zero) if both Democrats refuse to exit, what is the probability γ with which each candidate exits in a symmetric equilibrium?
Answer. The payoff matrix is shown in Table A.5.

Table A.5 Fighting Democrats

		Jones		
		Exit (γ)		*Stay* ($1-\gamma$)
Smith	*Exit* (γ)	0,0	→	0,10
		↓		↑
	Stay ($1-\gamma$)	10,0	←	−1,−1

The value of exiting is $V(exit) = 0$. The value of staying in is $V(Stay) = 10\gamma + (-1)(1 - \gamma) = 11\gamma - 1$. Hence, each player stays in with probability $\gamma = 1/11$, the same as in the war of attrition of part (a).

3.3: Uniqueness in Matching Pennies. In the game of Matching Pennies, Smith and Jones each show a penny with either heads or tails up. If they choose the same side of the penny, Smith gets both pennies; otherwise, Jones gets them.
(3.3a) Draw the outcome matrix for Matching Pennies.

Table A.6 Matching Pennies

		Jones		
		Heads (θ)		*Tails* ($1-\theta$)
Smith:	*Heads* (γ)	1, −1	→	−1, 1
		↑		↓
	Tails ($1-\gamma$)	−1, 1	←	1, −1

Payoffs to: (Smith, Jones)

(3.3b) Show that there is no Nash equilibrium in pure strategies.
Answer. (*Heads*, *Heads*) is not Nash, because Jones would deviate to *Tails*. (*Heads*, *Tails*) is not Nash, because Smith would deviate to *Tails*. (*Tails*, *Tails*) is not Nash, because Jones would deviate to *Heads*. (*Tails*, *Heads*) is not Nash, because Smith would deviate to *Heads*.

(3.3c) Find the mixed-strategy equilibrium, denoting Smith's probability of *Heads* by γ and Jones' by θ.
Answer. Equate the pure-strategy payoffs. Then for Smith, $\pi(Heads) = \pi(Tails)$, and

$$\theta(1) + (1 - \theta)(-1) = \theta(-1) + (1 - \theta)(1), \quad (2)$$

which tells us that $2\theta - 1 = -2\theta + 1$, and $\theta = 0.5$. For Jones, $\pi(Heads) = \pi(Tails)$, so

$$\gamma(-1) + (1 - \gamma)(1) = \gamma(1) + (1 - \gamma)(-1), \quad (3)$$

which tells us that $1 - 2\gamma = 2\gamma - 1$ and $\gamma = 0.5$.

(3.3d) Prove that there is only one mixed-strategy equilibrium.
Answer. Suppose $\theta > 0.5$. Then Smith will choose *Heads* as a pure strategy. Suppose $\theta < 0.5$. Then Smith will choose *Tails* as a pure

strategy. Similarly, if $\gamma > 0.5$, Jones will choose *Tails* as a pure strategy, and if $\gamma < 0.5$, Jones will choose *Heads* as a pure strategy. This leaves (0.5, 0.5) as the only possible mixed-strategy equilibrium.

Compare this with the multiple equilibria in problem 3.5. In that problem, there are three player, not two. Should that make a difference?

3.5: A Voting Paradox. Adam, Charles, and Vladimir are the only three voters in Podunk. Only Adam owns property. There is a proposition on the ballot to tax property-holders 120 dollars and distribute the proceeds equally among all citizens who do not own property. Each citizen dislikes having to go to the polling place and vote (despite the short lines), and would pay 20 dollars to avoid voting. They all must decide whether to vote before going to work. The proposition fails if the vote is tied. Assume that in equilibrium Adam votes with probability θ and Charles and Vladimir each vote with the same probability γ, but they decide to vote independently of each other.

(3.5a) What is the probability that the proposition will pass, as a function of θ and γ?

Answer. The probability that Adam loses can be decomposed into three probabilities—that all three vote, that Adam does not vote but that one of the other two does, and that Adam does not vote but both of the others do. These sum to $\theta\gamma^2 + (1 - \theta)2\gamma(1 - \gamma) + (1 - \theta)\gamma^2$, which is, rearranged, $\gamma(2\gamma\theta - 2\theta + 2 - \gamma)$.

(3.5b) What are the two possible equilibrium probabilities γ_1 and γ_2 with which Charles might vote? Why, intuitively, are there two symmetric equilibria?

Answer. The equilibrium is in mixed strategies, so each player must have equal payoffs from his pure strategies. Let us start with Adam's payoffs. If he votes, he loses 20 immediately, and 120 more if both Charles and Vladimir have voted:

$$\pi_a(Vote) = -20 + \gamma^2(-120). \tag{4}$$

If Adam does not vote, he loses 120 if either Charles or Vladimir vote, or if both vote:

$$\pi_a(Not\ Vote) = (2\gamma(1 - \gamma) + \gamma^2)(-120). \tag{5}$$

Equating $\pi_a(Vote)$ and $\pi_a(Not\ Vote)$ gives

$$0 = 20 - 240\gamma + 240\gamma^2. \tag{6}$$

The quadratic formula solves for γ:

$$\gamma = \frac{12 \pm \sqrt{144 - 4 \cdot 1 \cdot 12}}{24}. \tag{7}$$

This equations has two solutions, $\gamma_1 = 0.09$ (rounded) and $\gamma_2 = 0.91$ (rounded).

Why are there two solutions? If Charles and Vladimir are sure not to vote, Adam will not vote, because if he does not vote he will win, 0 to 0. If Charles and Vladimir are sure to vote, Adam will not vote, because if he does not vote he will lose, 2 to 0, but if he does vote, he will lose anyway, 2 to 1. Adam only wants to vote if Charles and Vladimir vote with moderate probabilities. Thus, for him to be indifferent between voting and not voting, it suffices either for γ to be low or to be high—it just cannot be moderate.

(3.5c) What is the probability θ that Adam will vote in each of the two symmetric equilibria?

Answer. Now use the payoffs for Charles, which depend on whether Adam and Vladimir vote:

$$\pi_c(Vote) = -20 + 60[\gamma + (1 - \gamma)(1 - \theta)] \tag{8}$$

$$\pi_c(Not\ Vote) = 60\gamma(1 - \theta). \tag{9}$$

Equating these and using $\gamma^* = 0.09$ gives $\theta = 0.70$ (rounded). Equating these and using $\gamma^* = 0.91$ gives $\theta = 0.30$ (rounded).

(3.5d) What is the probability that the proposition will pass?

Answer. The probability that Adam will lose his property is, using the equation in part (a) and the values already discovered, either 0.06 (rounded) ($= (0.7)(0.09)^2 + (0.3)(2(0.09)(0.91) + (0.09)^2)$) or 0.94 (rounded) ($= (0.3)(0.91)^2 + (0.7)(2(0.91)(0.09) + (0.91)^2)$).

Chapter 4

4.1: Repeated Entry Deterrence. Consider two repetitions without discounting of the game of Entry Deterrence I from section 4.2. Assume that there is one entrant, who sequentially decides whether to enter two markets that have the same incumbent.

(4.1a) Draw the extensive form of this game.

Answer. See figure A.2. If the entrant does not enter, the incumbent's response to entry in that period is unimportant.

Figure A.2 Repeated Entry Deterrence

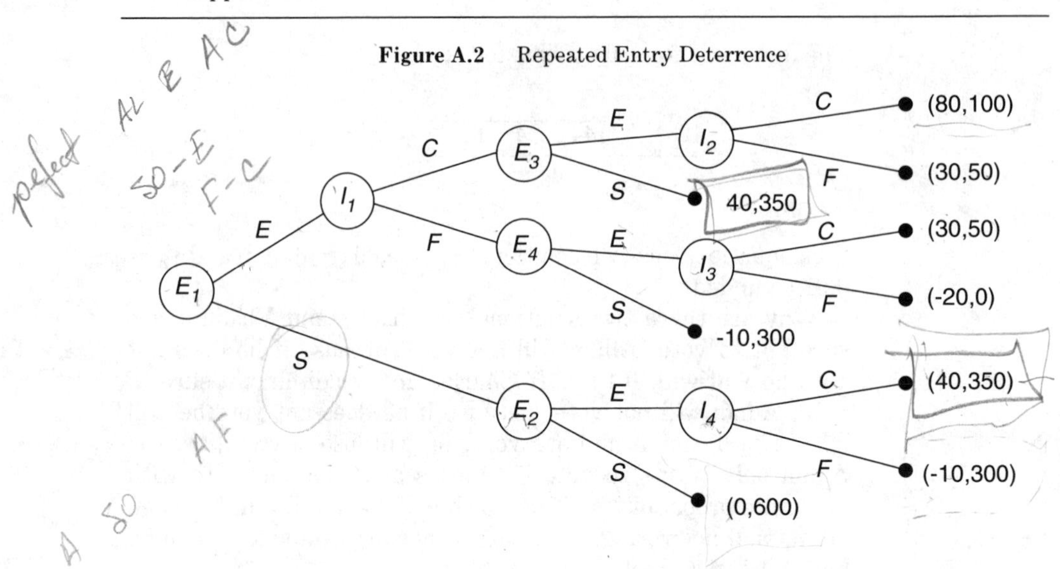

(4.1b) What are the 16 elements of the strategy set of the entrant?

Answer. The entrant makes a binary decision at four nodes, so his strategy must have four components, strictly speaking, and the number of possible arrangements is (2)(2)(2)(2) = 16. Table A.7 shows the strategy space, with *E* for *Enter* and *S* for *Stay out.* Usually modellers are not so careful. Table A.7 includes action

Table A.7 The Entrant's Strategy Set

Strategy	E_1	E_2	E_3	E_4
1	*E*	*E*	*E*	*E*
2	*E*	*E*	*S*	*E*
3	*E*	*E*	*E*	*S*
4	*E*	*E*	*S*	*S*
5	*E*	*S*	*S*	*S*
6	*E*	*S*	*E*	*E*
7	*E*	*S*	*S*	*E*
8	*E*	*S*	*E*	*S*
9	*S*	*E*	*E*	*E*
10	*S*	*S*	*E*	*E*
11	*S*	*S*	*S*	*E*
12	*S*	*S*	*S*	*S*
13	*S*	*E*	*S*	*S*
14	*S*	*E*	*S*	*E*
15	*S*	*E*	*E*	*S*
16	*S*	*S*	*E*	*S*

rules for the Entrant to follow at nodes that cannot be reached unless the Entrant trembles, somehow deviating from its own strategy. If the Entrant chooses Strategy 16, for example, nodes E_3 and E_4 cannot possibly be reached, even if the Incumbent deviates, so one might think that the parts of the strategy dealing with those nodes are unimportant. Table A.8 removes the unimportant parts of the strategy, and Table A.9 condenses the strategy set down to its six significantly distinct strategies.

Table A.8 The Entrant's Strategy Set, Abridged Version I

Strategy	E_1	E_2	E_3	E_4
1	E	–	E	E
2	E	–	S	E
3	E	–	E	S
4	E	–	S	S
5	E	–	S	S
6	E	–	E	E
7	E	–	S	E
8	E	–	E	S
9	S	E	–	–
10	S	S	–	–
11	S	S	–	–
12	S	S	–	–
13	S	E	–	–
14	S	E	–	–
15	S	E	–	–
16	S	S	–	–

Table A.9 The Entrant's Strategy Set, Abridged Version II

Strategy	E_1	E_2	E_3	E_4
1	E	–	E	E
3	E	–	E	S
4	E	–	S	S
7	E	–	S	E
9	S	E	–	–
10	S	S	–	–

(4.1c) What is the subgame perfect equilibrium?
Answer. The entrant always enters and the incumbent always colludes.
(4.1d) What is one of the non-perfect Nash equilibria?
Answer. The entrant stays out in the first period, and enters in the second period. The incumbent fights any entry that might occur in the first period, and colludes in the second period.

4.3: Pliny and the Freedmen's Trial. (Pliny, 1963, pp. 221–4, Riker, 1986, pp. 78–88). Afranius Dexter died mysteriously, perhaps dead by his own hand, perhaps killed by his freedmen (servants a step above slaves), or perhaps killed by his freedmen on his own orders. The freedmen went on trial before the Roman Senate. Assume that 45 percent of the senators favor acquittal, 35 percent favor banishment, and 20 percent favor execution, and that the preference rankings in the three groups are $A > B > E$, $B > A > E$, and $E > B > A$. Also assume that each group has a leader and votes as a bloc.
(4.3a) Modern legal procedure requires the court to decide guilt first and then assign a penalty if the accused is found guilty. Draw a tree to represent the sequence of events (this will not be a game tree, since it will represent the actions of groups of players, not of individuals). What is the outcome in a perfect equilibrium?
Answer. Guilty would win in the first round by a vote of 55 to 45, and banishment would win in the second by 80 to 20. See figure A.3.

Figure A.3 Modern Legal Procedure

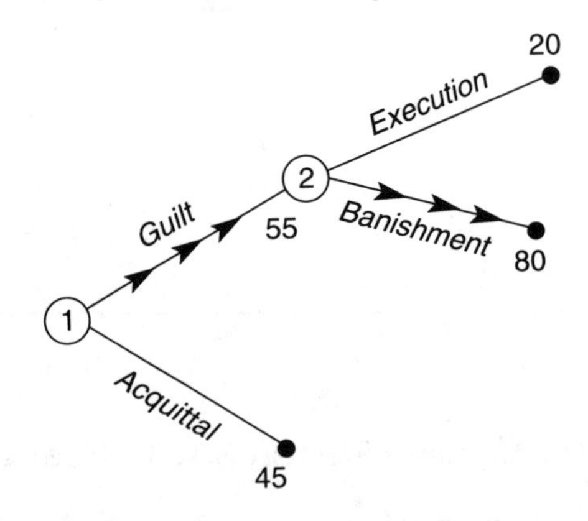

(4.3b) Suppose that the acquittal bloc can precommit to how they will vote in the second round if guilt wins in the first round. What will they do, and what will happen? What would the execution bloc do if they could control the second-period vote of the acquittal bloc?
Answer. The acquittal bloc would commit to execution, inducing the banishment bloc to vote for acquittal in the first round, and acquittal would win. The execution bloc would order the acquittal bloc

to choose banishment in the second round to avoid making the banishment bloc switch to acquittal.[3]

(4.3c) The normal Roman procedure began with a vote on execution versus no execution, and then voted on the alternatives in a second round if execution failed to gain a majority. Draw a tree to represent this. What would happen in this case?

Answer. Execution would fail by a vote of 20 to 80, and banishment would then win by 55 to 45. See figure A.4.

Figure A.4 Roman Legal Procedure

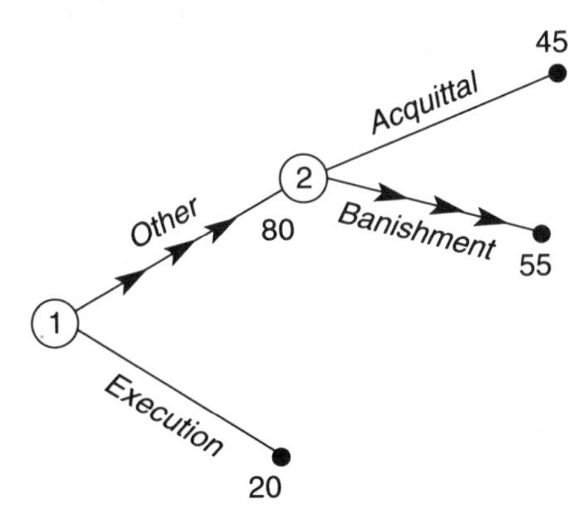

(4.3d) Pliny proposed that the Senators divide into three groups, depending on whether they supported acquittal, banishment, or execution, and that the outcome with the most votes should win. This proposal caused a roar of protest. Why did he propose it?

Answer. Pliny favored acquittal and hoped that all senators would vote their own preference. Acquittal would then win 45 to 35 to 20.

(4.3e) Pliny did not get the result he wanted with his voting procedure. Why not?

Answer. Pliny said that his arguments were so convincing that the senator who made the motion for the death penalty changed his mind, along with his supporters, and voted for banishment, which won (by 55 to 45 in our hypothesized numbers). He forgot that people do not always vote for their first preference. The execution bloc saw that acquittal would win unless they switched to banishment.

(4.3f) Suppose that personal considerations made it most important to a senator that he show his stand by his vote, even if he had to sacrifice his prefer-

[3] Note that preferences do not always work out this way. In Athens, six centuries before the Pliny episode, Socrates was found guilty in a first round of voting and then sentenced to death (instead of to a lesser punishment, banishment) by a bigger margin in the second round. This would imply that the ranking of the acquittal bloc there was AEB, with the complicating factor that Socrates was rather insulting in his sentencing speech.

ence for a particular outcome. If there were a vote over whether to use the traditional Roman procedure or Pliny's procedure, who would vote with Pliny, and what would happen to the freedmen?

Answer. Traditional procedure would win by capturing the votes of the execution bloc and the banishment bloc, and the freedmen would be banished. In this case, the voting procedure would matter to the result, because each senator would vote for his preference.

Chapter 5

5.1: Overlapping Generations.[4] There is a long sequence of players. One player is born in each period t, and he lives throughout periods t and $t + 1$. Thus, two players are alive in any one period, a youngster and an oldster. Each player is born with one unit of chocolate, which cannot be stored. Utility is increasing in chocolate consumption, and a player is very unhappy if he consumes less than 0.3 units of chocolate in a period: the per-period utility functions are $U(C) = -1$ for $C < 0.3$ and $U(C) = C$ for $C \geq 0.3$, where C is consumption. Players can give away their chocolate, but, since chocolate is the only good, they cannot sell it. A player's action is to consume X units of chocolate as a youngster and give away $1 - X$ to some oldster.

(5.1a) If there is finite number of generations, what is the unique Nash equilibrium?

Answer. $X = 1$. The Chainstore Paradox applies. Youngster T, the last one, has no incentive to give anything to Oldster $T - 1$. Therefore, Youngster $T - 1$ has no incentive either, and so on for every t.

(5.1b) If there are an infinite number of generations, what are two Pareto-ranked perfect equilibria?

Answer. (i) ($X = 1$, *regardless of what others do*), and (ii) ($X = 0.5$, *unless some player has deviated, in which case* $X = 1$). Equilibrium (ii) is Pareto superior.

(5.1c) If there is a probability θ at the end of each period (after consumption takes place) that barbarians will invade and steal all the chocolate (leaving the civilized people with payoffs of -1 for any C), what is the highest value of θ that still allows for an equilibrium with $X = 0.5$?

Answer. The payoff from the equilibrium strategy is $0.5 + (1 - \theta)0.5 + \theta(-1) = 1 - 1.5\theta$. The payoff from deviating to $X = 1$ is $1 - 1 = 0$. These are equal if $1 - 1.5\theta = 0$; that is, if $\theta = 2/3$. Hence, θ can take values up to $2/3$ and the $X = 0.5$ equilibrium can still be maintained.

[4] See Samuelson (1958).

5.3: Repeated Games.[5] Players Benoit and Krishna repeat the game in Table 5.4 three times, with discounting:

(5.3a) Why is there no equilibrium in which the players play *Deny* in all three periods?

Answer. A player could profitably deviate to *Confess* in the last period if the equilibrium specified *Deny* for the last period, receiving 15 instead of 10 for that period.

(5.3b) Describe a perfect equilibrium in which both players pick *Deny* in the first two periods.

Answer. (Play *Deny* for the first two periods and *Waffle* for the last period. If in the first or second periods the other player deviates, pick *Confess* thereafter.)

This is an equilibrium, because a player who deviated to *Confess* would gain 5 (= 15 − 10) from that deviation, but lose 8 (= 0 − 8) in the last period as a result. The punishment is self-enforcing because (*Confess*, *Confess*) is a Nash equilibrium.

(5.3c) Adapt your equilibrium to the twice-repeated game.

Answer. (Play *Deny* for the first period and *Waffle* for the second period. If in the first period the other player deviates, pick *Confess* in the second period.)

(5.3d) Adapt your equilibrium to the T-repeated game.

Answer. (Play *Deny* for the first $T - 1$ periods and *Waffle* for the last period. If anytime in the first $T - 1$ periods the other player deviates, pick *Confess* thereafter.)

(5.3e) What is the greatest discount rate for which your equilibrium still works in the 3-period game?

Answer. The equilibrium payoff is $\pi(eq) = 10 + 10/(1 + r) + 8/(1 + r)^2$. The deviation payoff from the most profitable deviation (*Confess* in the second period) is $\pi(dev) = 10 + 15/(1 + r) + 0$. Equating these and solving for r, we get $10/(1 + r) + 8/(1 + r)^2 = 15/(1 + r)$, and $r = 3/5$.

5.5: Repeated Prisoner's Dilemma. Set $P = 0$ in the general Prisoner's Dilemma in table 1.9, and assume that $2R > S + T$.

(5.5a) Show that the Grim Strategy, when played by both players, is a perfect equilibrium for the infinitely repeated game. What is the maximum discount rate for which the Grim Strategy remains an equilibrium?

Answer. The Grim Strategy is a perfect equilibrium because the payoff from continued cooperation is $R + R/r$, which for low discount rates is greater than the payoff from (*Confess*, *Deny*) once and (*Confess*, *Confess*) forever after, which is $T + 0/r$. To find the maximum discount rate, equate these two payoffs: $R + R/r = T$. This means that $r = (T - R)/R$ is the maximum.

[5] See Benoit & Krishna (1985).

(5.5b) Show that Tit-for-Tat is not a perfect equilibrium in the infinitely repeated Prisoner's Dilemma with no discounting.

Answer. Suppose Row has played *Confess*. Will Column retaliate? If both follow Tit-for-Tat after the deviation, retaliation results in a cycle of (*Confess*, *Deny*), (*Deny*, *Confess*) forever. Row's payoff is $T + S + T + S + \ldots$. If Column forgives, and they go back to cooperating, on the other hand, Row's payoff is $R + R + R + R + \ldots$. Comparing the first four periods, forgiveness has the higher payoff because $4R > 2S + 2T$. The payoffs of the first four periods simply repeat an infinite number of times to give the total payoff, so forgiveness dominates retaliation, and Tit-for-Tat is not perfect.[6]

[6] See Kalai, Samet and Stanford (1988).

Chapter 6

6.1: Cournot Duopoly Under Incomplete Information about Costs. This problem introduces incomplete information into the Cournot model of chapter 3 and allows for a continuum of player types.

(6.1a) Modify the Cournot Game of chapter 3 by specifying that Apex' average cost of production is c per unit, while Brydox' remains zero. What are the outputs of each firm if the costs are common knowledge? What are the numerical values if $c = 10$?

Answer. The payoff functions are

$$\begin{aligned} \pi_{Apex} &= (120 - q_a - q_b - c)q_a \\ \pi_{Brydox} &= (120 - q_a - q_b)\, q_b. \end{aligned} \tag{12}$$

The first-order conditions are then

$$\begin{aligned} \frac{\partial \pi_{Apex}}{\partial q_a} &= 120 - 2q_a - q_b - c = 0 \\ \frac{\partial \pi_{Brydox}}{\partial q_b} &= 120 - q_a - 2q_b = 0. \end{aligned} \tag{13}$$

Solving the first-order conditions together gives

$$\begin{aligned} q_a &= 40 - \frac{2c}{3} \\ q_b &= 40 + \frac{c}{3}. \end{aligned} \tag{14}$$

If $c = 10$, Apex produces $33\frac{1}{3}$ and Brydox produces $43\frac{1}{3}$. Apex' higher costs force it to cut back on its output, which encourages Brydox to produce more.

(6.1b) Let Apex' cost c be c_{max} with probability θ and 0 with probability $1 - \theta$, so Apex is one of two types. Brydox does not know Apex' type. What are the outputs of each firm?

Answer. Apex' payoff function is the same as in part (a), because

$$\pi_{Apex} = (120 - q_a - q_b - c)q_a, \tag{15}$$

which yields the reaction function

$$q_a = 60 - \frac{q_b + c}{2}. \tag{16}$$

Brydox' expected payoff is

$$\begin{aligned} \pi_{Brydox} = & (1 - \theta)(120 - q_a(c = 0) - q_b)q_b \\ & + \theta(120 - q_a(c = c_{max}) - q_b)q_b. \end{aligned} \tag{17}$$

The first-order condition is

$$\begin{aligned} \frac{\partial \pi_{Brydox}}{\partial q_b} = & (1 - \theta)(120 - q_a(c = 0) - 2q_b) \\ & + \theta(120 - q_a(c = c_{max}) - 2q_b) = 0. \end{aligned} \tag{18}$$

Now substitute the reaction function of Apex, equation (16), into (18) and condense a few terms to obtain

$$120 - 2q_b - [1 - \theta]\left[60 - \frac{q_b + 0}{2}\right] - \theta\left[60 - \frac{q_b + c_{max}}{2}\right] = 0. \tag{19}$$

Solving for q_b yields

$$q_b = 40 + \frac{\theta c_{max}}{3}. \tag{20}$$

One can then use equations (16) and (20) to find

$$q_a = 40 - \frac{\theta c_{max}}{6} - \frac{c}{2}. \tag{21}$$

Note that the outputs do not depend on θ or c_{max} separately, only on the expected value of Apex' cost, θc_{max}.

(6.1c) Let Apex' cost c be drawn from the interval $[0, c_{max}]$ using the uniform distribution, so there is a continuum of types. Brydox does not know Apex' type. What are the outputs of each firm?

Answer. Apex' payoff function is the same as in parts (a) and (b),

$$\pi_{Apex} = (120 - q_a - q_b - c)q_a, \tag{22}$$

which yields the reaction function

$$q_a = 60 - \frac{q_b + c}{2}. \tag{23}$$

Brydox' expected payoff is (letting the density of possible values of c be $f(c)$)

$$\pi_{Brydox} = \int_0^{c_{max}} (120 - q_a(c) - q_b) q_b f(c) dc. \tag{24}$$

The probability density is uniform, so $f(c) = 1/c_{max}$. Substituting this into (24), the first-order condition is

$$\frac{\partial \pi_{Brydox}}{\partial q_b} = \int_0^{c_{max}} (120 - q_a(c) - 2q_b) \left(\frac{1}{c_{max}}\right) dc = 0. \tag{25}$$

Now substitute in the reaction function of Apex, equation (23), which gives

$$\int_0^{c_{max}} (120 - [60 - \frac{q_b + c}{2}] - 2q_b) \left(\frac{1}{c_{max}}\right) dc = 0. \tag{26}$$

Simplifying by integrating out the terms in (26) which depend on c only through the probability density yields

$$60 - \frac{3q_b}{2} + \int_0^{c_{max}} \left(\frac{c}{2c_{max}}\right) dc = 0. \tag{27}$$

Integrating and rearranging yields

$$q_b = 40 + \frac{c_{max}}{6}. \tag{28}$$

One can then use equations (23) and (28) to find

$$q_a = 40 - \frac{c_{max}}{12} - \frac{c}{2}. \tag{29}$$

(6.1d) Outputs were 40 for each firm in the zero-cost game in chapter 3. Check your answers in parts (b) and (c) by seeing what happens if $c_{max} = 0$.
Answer. If $c_{max} = 0$, then in part (b), $q_a = 40 - 0/6 - 0/2 = 40$ and $q_b = 40 + 0/3 = 40$, which is as it should be.
If $c_{max} = 0$, then in part (c), $q_a = 40 - 0/12 - 0/2 = 40$ and $q_b = 40 + 0/6 = 40$, which is as it should be.

(6.1e) Let $c_{max} = 20$ and $\theta = 0.5$, so the expectation of Apex' average cost is 10 in parts (a), (b), and (c). What are the average outputs for Apex in each case?

Answer. In part (a), under full information, the outputs were $q_a = 33\frac{1}{3}$ and $q_b = 43\frac{1}{3}$. In part (b), with two types, $q_b = 43\frac{1}{3}$ from equation (20), and the average value of q_a is

$$Eq_a = (1-\theta)\left(40 - \frac{0.5(20)}{6} - \frac{0}{2}\right) + \theta\left(40 - \frac{0.5(20)}{6} - \frac{20}{2}\right) = 33\frac{1}{3}. \quad (30)$$

In part (c), with a continuum of types, $q_b = 43\frac{1}{3}$ and q_a is found from

$$\begin{aligned} Eq_a &= \int_0^{c_{max}} \left(40 - \frac{c_{max}}{8} - \frac{c}{2}\right)\left(\frac{1}{c_{max}}\right) dc \\ &= 40 - \frac{20}{8} - \frac{c_{max}^2}{4c_{max}} = 33\frac{1}{3}. \end{aligned} \quad (31)$$

(6.1f) Modify the model of part (b) so that $c_{max} = 20$ and $\theta = 0.5$, but somehow $c = 30$. What outputs do your formulas from part (b) generate? Is there anything this could sensibly model?

Answer. The purpose of Nature's move is to represent Brydox' beliefs about Apex, not necessarily to represent reality. Here, Brydox believes that Apex' costs are either 0 or 20 but he is wrong and they are actually 30. In this game that does not cause problems for the analysis. Using equations (20) and (21), the outputs are $q_b = 43\frac{1}{3}\left(= 40 + \frac{0.5(20)}{3}\right)$ and $q_a = 26\frac{2}{3}\left(= 40 - \frac{0.5(20)}{6} - \frac{30}{2}\right)$.

If the game were dynamic, however, a problem would arise. When Brydox observes the first-period output of $q_a = 26\frac{2}{3}$, what is he to believe about Apex' costs? Should he deduce that $c = 30$, or increase his belief that $c = 20$, or believe something else entirely? This departs from standard modelling.

6.3: Symmetric Information and Prior Beliefs. In the Expensive-Talk Game of figure A.5, the Battle of the Sexes is preceded by a communication move in which the man chooses *Silence* or *Talk*. *Talk* costs 1 payoff unit and consists of a declaration by the man that he is going to the prize fight. This declaration is just talk; it is not binding on him.

(6.3a) Draw the extensive form for this game, putting the man's move first in the simultaneous-move subgame.

Answer. See figure A.5.

Figure A.5 Extensive Form for the Expensive Talk Game

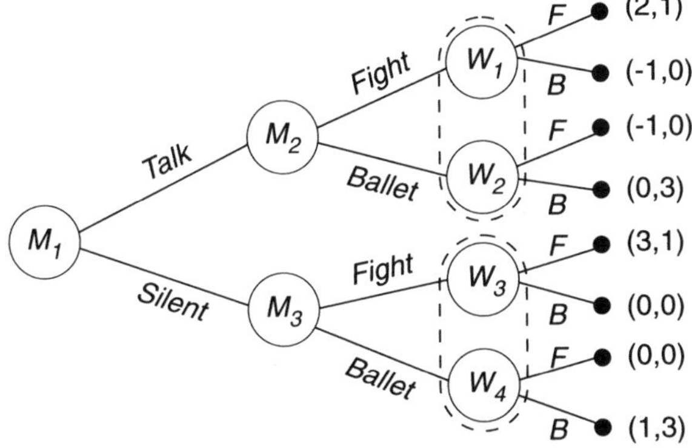

(6.3b) What are the strategy sets for the game? (start with the woman's)
Answer. The woman has two information sets in which to choose moves, and the man has three. Table A.10 shows the woman's four strategies.

Table A.10 The Woman's Strategies in the Expensive Talk Game

Strategy	W_1, W_2	W_3, W_4
1	F	F
2	F	B
3	B	F
4	B	B

Table A.11 shows the man's eight strategies, of which only the boldfaced four are important, since the others differ only in portions of the game tree that the man knows he will never reach unless he trembles at M_1.

(6.3c) What are the three perfect pure-strategy equilibrium outcomes in terms of observed actions? (remember: strategies are not the same thing as outcomes)
Answer. *SFF*, *SBB*, *TFF*.[7]

(6.3d) Describe the equilibrium strategies for a perfect equilibrium in which the man chooses to talk.
Answer. Woman: ($F|T$, $B|S$) and Man: (T, $F|T$, $B|S$).

7 The equilibrium that supports *SBB* is [(S, B), ($B|S$, $B|T$)].

Table A.11 The Man's Strategies in the Expensive Talk Game

Strategy	M_1	M_2	M_3
1	***T***	***F***	*F*
2	*T*	*F*	*B*
3	***T***	***B***	*B*
4	*T*	*B*	*F*
5	***S***	*F*	***F***
6	*S*	*B*	*F*
7	***S***	*B*	***B***
8	*S*	*F*	*B*

(6.3e) The idea of "forward induction" says that an equilibrium should remain an equilibrium even if strategies that are dominated in that equilibrium are removed from the game and the procedure is iterated. Show that this procedure rules out *SBB* as an equilibrium outcome.[8]

Answer. First delete the man's strategy of (*T*, *B*), which is dominated by (*S*, *B*) whatever the woman's strategy may be. Without this strategy in the game, if the woman sees the man deviate and choose *Talk*, she knows that the man must choose *Fight*. Her strategies of $(B|T, F|S)$ and $(B|T, B|S)$ are now dominated, so let us drop those. But then the man's strategy of (*S*, *B*) is dominated by $(T, F|T, B|S)$. The man will therefore choose to *Talk*, and the *SBB* equilibrium is broken.

This is a strange result. More intuitively: if the equilibrium is *SBB*, but the man chooses *Talk*, the argument is that the woman should think that the man would not do anything purposeless, so it must be that he intends to choose *Fight*. She therefore will choose *Fight* herself, and the man is quite happy to choose *Talk* in anticipation of her response. Taking forward induction one step further: *TFF* is not an equilibrium, because now that *SBB* has been ruled out, if the man chooses *Silence*, the woman should conclude it is because he thinks he can thereby get the *SFF* payoff. She decides that he will choose *Fight*, and so she will choose it herself. This makes it profitable for the man to deviate to *SFF* from *TFF*.

Chapter 7

7.1: First-Best Solutions in a Principal-Agent Model. Suppose an agent has the utility function $U = \sqrt{w} - e$, where e can assume the levels 0 or 1. Let the

[8] See Van Damme (1989). In fact, this procedure rules out *TFF* also.

reservation utility level be $\bar{U} = 3$. The principal is risk-neutral. Denote the agent's wage, conditioned on output, as $\underline{w}$ if output is 0 and $\bar{w}$ if output is 100. Table 7.5 shows the outputs.

(7.1a) What would the agent's effort choice and utility be if he owned the firm?

Answer. The agent gets everything in this case. His utility is either

$$U(High) = 0.1(0) + 0.9\sqrt{100} - 1 = 8 \tag{32}$$

or

$$U(Low) = 0.3(0) + 0.7\sqrt{100} - 0 = 7. \tag{33}$$

So the agent chooses high effort and a utility of 8.

(7.1b) If agents are scarce and principals compete for them, what will the agent's contract be under full information? His utility?

Answer. The efficient effort level is *High*, which produces an expected output of 90. The principal's profit is zero, because of competition. Since the agent is risk averse, he should be fully insured in equilibrium: $\bar{w} = \underline{\mathrm{w}} = 90$. But he should get this only if his effort is high. Thus, the contract is $w = 90$ if effort is high, $w = 0$ if effort is low. The agent's utility is 8.5 ($= \sqrt{90} - 1$, rounded).

(7.1c) If principals are scarce and agents compete to work for them, what would the contract be under full information? What will the agent's utility and the principal's profit be in this situation?

Answer. The efficient effort level is high. Since the agent is risk averse, he should be fully insured in equilibrium: $\bar{w} = \underline{\mathrm{w}} = w$. The contract must satisfy a participation constraint for the agent, so $\sqrt{w} - 1 = 3$. This yields $w = 16$, and a utility of 3 for the agent. The actual contract specified a wage of 16 for high effort and 0 for low effort. This is incentive compatible, because the agent would get only 0 in utility if he took low effort. The principal's profit is 74 ($= 90 - 16$).

(7.1d) Suppose that $U = w - e$. If principals are the scarce factor and agents compete to work for principals, what would the contract be when the principal cannot observe effort? (Negative wages are allowed.) What will the agent's utility and the principal's profit be in this situation?

Answer. The contract must satisfy a participation constraint for the agent, so $U = 3$. Since effort is 1, the expected wage must equal 4. One way to produce this result is to allow the agent to keep all the output, plus 4 extra for his labor, but to make him pay the expected output of 90 for this privilege ("selling the store"). Let $\bar{w} = 14$ and $\underline{w} = -86$ (other contracts also work). Then expected utility is 3 ($= 0.1(-86) + 0.9(14) - 1 = -8.6 + 12.6 - 1$). Expected profit is 86 ($= 0.1(0-(-86)) + 0.9(100 - 14) = 8.6 + 77.4$).

7.3: Why Entrepreneurs Sell Out. Suppose an agent has a utility function of $U = \sqrt{w} - e$, where e can assume the levels 0 or 2.4, and his reservation utility is $\bar{U} = 7$. The principal is risk-neutral. Denote the agent's wage, conditioned on output, as $w(0)$, $w(49)$, $w(100)$, or $w(225)$. Table 7.7 shows the output.

(7.3a) What would the agent's effort choice and utility be if he owned the firm?

Answer. $U(safe) = 0 + 0.1\sqrt{49} + 0.8\sqrt{100} + 0 - 0 = 0.7 + 8 = 8.7$. $U(risky) = 0 + 0.5\sqrt{49} + 0.5\sqrt{225} - 2.4 = 3.5 + 7.5 - 2.4 = 8.6$. Therefore he will choose the safe method, $e = 0$, and utility is 8.7.

(7.3b) If agents are scarce and principals compete for them, what will the agent's contract be under full information? His utility?

Answer. Agents are scarce, so $\pi = 0$. Since agents are risk averse, it is efficient to shield them from risk. If the risky method is chosen, then $w = 0.5(49) + 0.5(225) = 24.5 + 112.5 = 137$. Utility is 9.3 ($= \sqrt{137} - 2.4$, rounded). If the safe method is chosen, then $w = 0.1(49) + 0.8(100) = 84.9$. Utility is $U = \sqrt{84.9} \approx 9.21$. Therefore, the optimal contract specifies a wage of 137 if the risky method is used and zero (or any wage less than 49) if the safe method is used. This is better for the agent than if he ran the firm by himself and used the safe method.

(7.3c) If principals are scarce and agents compete to work for principals, what will the contract be under full information? What will the agent's utility and the principal's profit be in this situation?

Answer. Principals are scarce, so $U = \bar{U} = 7$, but the efficient effort level does not depend on who is scarce, so it is still high. The agent is risk averse, so he is paid a flat wage. The wage satisfies the participation constraint $\sqrt{w} - 2.4 = 7$ if the method is risky. The contract specifies a wage of 88.4 (rounded) for the risky method and zero for the safe. Profit is 48.6 ($= 0.5(49) + 0.5(225) - 88.4$).

(7.3d) If agents are the scarce factor and principals compete for them, what will the contract be when the principal cannot observe effort? What will the agent's utility and the principal's profit be in this situation?

Answer. A boiling-in-oil contract can be used. Set either $w(0) = -1000$ or $w(100) = -1000$, which induces the agent to pick the risky method. In order to protect the agent from risk, the wage should be flat except for those outputs, so $w(49) = w(225) = 137$. $\pi = 0$, since agents are scarce. $U = 9.3$, from part (b).

7.5: Worker Effort. A worker can be *Careful* or *Careless*, efforts which generate mistakes with probabilities 0.25 and 0.75. His utility function is $U = 100 - 10/w - x$, where w is his wage and x takes the value 2 if he is careful, and zero otherwise. Whether a mistake is made is contractible, but effort is not. Risk-neutral employers compete for the worker, and his output is worth zero if a mistake is made and 20 otherwise. No computation is needed for any part of this problem.

(7.5a) Will the worker be paid anything if he makes a mistake?

Answer. Yes. He is risk averse, unlike the principal, so his wage should be even across states.

(7.5b) Will the worker be paid more if he does not make a mistake?
Answer. Yes. Careful effort is efficient, and lack of mistakes is a good statistic for careful effort, which makes it useful for incentive compatibility.

(7.5c) How would the contract be affected if employers were also risk averse?
Answer. The wage would vary more across states, because the workers should be less insured—and perhaps should even be insuring the employer.

(7.5d) What would the contract look like if a third category, "slight mistake," with an output of 19, occurs with probability 0.1 after *Careless* effort, and probability zero after *Careful* effort?
Answer. The contract would pay equal amounts whether or not a mistake was made, but zero if a slight mistake was made, a "boiling-in-oil" contract.

Chapter 8

8.1: Monitoring with Error. An agent has a utility function $U = \sqrt{w} - \alpha e$, where $\alpha = 1$ and e is either 0 or 5. His reservation utility level is $\bar{U} = 9$, and his output is 100 with low effort and 250 with high effort. Principals are risk neutral and scarce, and agents compete to work for them. The principal cannot condition the wage on effort or output, but he can, if he wishes, spend five minutes of his time, worth 10 dollars, to drop in and watch the agent. If he does that, he observes the agent *Daydreaming* or *Working*, with probabilities that differ depending on the agent's effort. He can condition the wage on those two things, so the contract will be $\{\underline{w}, \bar{w}\}$. The probabilities are given by Table 8.1.

(8.1a) What are the profits in the absence of monitoring, if the agent is paid enough to make him willing to work for the principal?
Answer. Without monitoring, effort is low. The participation constraint is $9 \leq \sqrt{w} - 0$, so $w \geq 81$. Output is 100, so profit is 19.

(8.1b) Show that high effort is efficient under full information.
Answer. High effort yields output of 250. $\bar{U} \leq \sqrt{w} - \alpha e$ or $9 \leq \sqrt{w} - 5$ is the participation constraint, so $14 = \sqrt{w}$ and $w = 196$. Profit is then 54. This is superior to the profit of 19 from low effort (and the agent is no worse off), so high effort is more efficient.

(8.1c) If $\alpha = 1.2$, is high effort still efficient under full information?
Answer. If $\alpha = 1.2$, the wage must rise to 225, for profits of 25, so high effort is still efficient. The wage must rise to 225 because the participation constraint becomes $9 \leq \sqrt{w} - 1.2(5)$.

(8.1d) Under asymmetric information, with $\alpha = 1$, what are the participation and incentive compatibility constraints?
Answer. The incentive compatibility constraint is

$$0.6\sqrt{\underline{w}} + 0.4\sqrt{\bar{w}} \leq 0.1\underline{w} + 0.9\sqrt{\bar{w}} - 5.$$

The participation constraint is $9 \leq 0.1\sqrt{\underline{w}} + 0.9\sqrt{\bar{w}} - 5$.

(8.1e) Under asymmetric information, with $\alpha = 1$, what is the optimal contract?

Answer. From the participation constraint, $14 = 0.1\sqrt{\underline{w}} + 0.9\sqrt{\bar{w}}$, and $\sqrt{\bar{w}} = \frac{14}{0.9} - \left(\frac{1}{9}\right)\sqrt{\underline{w}}$. The incentive compatibility constraint tells us that $0.5\sqrt{\bar{w}} = 5 + 0.5\sqrt{\underline{w}}$, so $\sqrt{\bar{w}} = 10 + \sqrt{\underline{w}}$. Thus,

$$10 + \sqrt{\underline{w}} = 15.6 - 0.11\sqrt{\underline{w}} \tag{34}$$

and $\sqrt{\underline{w}} = 5.6/1.11 = 5.05$. Thus, $\underline{w} = 25.5$. It follows that $\sqrt{\bar{w}} = 10 + 5.05$, so $\bar{w} = 226.5$.

8.3: Unravelling. A prospector owns a gold mine worth an amount θ, drawn from the uniform distribution $U[0, 100]$, which nobody knows, including himself. He will certainly sell the mine, since he is too old to work it and it has no value to him if he does not sell it. The several prospective buyers are all risk neutral. The prospector can, if he desires, dig deeper into the hill and collect a sample of gold ore that will reveal the value of θ. If he shows the ore to the buyers, however, he must show genuine ore, since an unwritten Law of the West says that fraud is punished by hanging offenders from joshua trees as food for buzzards.

(8.3a) For how much can he sell the mine if he is clearly too feeble to have dug into the hill and examined the ore? What is the price in this situation if, in fact, the true value is $\theta = 70$?

Answer. The price is 50—the expected value of the uniform distribution from 0 to 100. Even if the mine is actually worth $\theta = 70$, the price remains at 50.

(8.3b) For how much can he sell the mine if he can dig a test tunnel at zero cost? Will he show the ore? What is the price in this situation if, in fact, the true value is $\theta = 70$?

Answer. The expected price is 50. Unravelling occurs, so he will show the ore, and the buyer can discover the true value, which is 50 on average. If the true value is $\theta = 70$, the buyer receives 70.

(8.3c) For how much can he sell the mine if, after digging a tunnel at zero cost and discovering θ, it costs him an additional 10 to verify the results for the buyers? What is his expected payoff?

Answer. He shows the ore iff $\theta \in [20, 100]$. This is because if the minimum quality ore he shows is b, the price at which he can sell the mine without showing the ore is $b/2$. If $b = 20$ and the true value is 20, he can sell the mine for 10, and showing the ore to raise the price to 20 would not increase his net profit, given the display cost of 10.

With probability 0.2, his price is 10 because $\theta \in [0,20]$, and with probability 0.8, it is an average price of 60 but he pays 10 to display the ore. Thus, the prospector's expected payoff is 42 $(= 0.2(10) + 0.8\,(60 - 10) = 2 + 40 = 42)$.

(8.3d) What is the prospector's expected payoff if with probability 0.5 digging a tunnel is costless, but with probability 0.5 it costs 120?

Answer. In equilibrium there exists some number c such that if the prospector has dug the tunnel and found the value of the mine to be $\theta \geq c$ he will show the ore. If he does not show any ore, the buyers' expected value for the mine is 0.5(100 − 0/2) + 0.5 (c − 0/2) = $c/4 + 25$. Having dug the tunnel, he will therefore show the ore if $\theta \geq c/4 + 25$, because then he can get a price of θ instead. Since c is defined as the minimal level he will disclose, it follows that $c = c/4 + 25$, which implies that c = 33⅓ (and the price is 1(¼)(33⅓) + 25 = 33⅓ if he does not show the ore).

With probability 0.5, the prospector will not dig the tunnel, and will receive a price of 33⅓. With probability 0.5 he will dig the tunnel, and will refuse to disclose with probability ⅓, for a price of 33⅓, and disclose with probability ⅔, for an average price of 66⅔, for an expected payoff of about 44.4.

8.5: Efficiency Wages and Risk Aversion.[9] In each of two periods of work, a worker decides whether to steal amount v, and is detected with probability α and suffers legal penalty p if he, in fact, did steal. A worker who is caught stealing can also be fired, after which he earns the reservation wage w_0. If the worker does not steal, his utility in the period is $U(w)$; if he steals, it is $U(w + v) - \alpha p$, where $U(w_0 + v) - \alpha p > U(w_0)$. The worker's marginal utility of income is diminishing: $U' > 0$, $U'' < 0$, and $\lim_{x \to \infty} U'(x) = 0$. There is no discounting. The firm definitely wants to deter stealing in each period, if at all possible.

(8.5a) Show that the firm can indeed deter theft, even in the second period, and, in fact, does so with a second-period wage w_2^* that is higher than the reservation wage w_0.

Answer. It is easiest to deter theft in the first period, since a high second-period wage increases the penalty of being fired. If w_2 is increased enough, however, the marginal utility of income becomes so low that $U(w_2 + v)$ and $U(w_2)$ become almost identical, with a difference less than αP, so theft is deterred even in the second period.

(8.5b) Show that the equilibrium second-period wage w_2^* is higher than the first-period wage w_1^*.

Answer. We already determined that $w_2 > w_0$. Hence, the worker looks hopefully towards being employed in period two, and in period one he is reluctant to risk his job by stealing. This means that he can be paid less in period one, even though he may still have to be paid more than the reservation wage.

[9] See Rasmusen (1992c).

Chapter 9

9.1: Insurance with Equations and Diagrams. The text analyzes Insurance Game III using diagrams. Here, let us use equations too. Let $U(t) = \log(t)$.

(9.1a) Give the numeric values (x, y) for the full-information separating contracts C_3 and C_4 from figure 9.6. What are the coordinates of C_3 and C_4?

Answer. C_3: $0.25x + 0.75(y - x) = 0$, and $12 - x = y - x$. Put together, these give $y = 4x/3$ and $y = 12$, so $x^* = 9$ and $y^* = 12$.

$C_3 = (3, 3)$ because $12 - 9 = 3$.

C_4 is such that $0.5x + 0.5(y - x) = 0$, and $12 - x = y - x$. Put together, these give $y = 2x$ and $y = 12$, so $x^* = 6$ and $y^* = 12$.

$C_4 = (6, 6)$ because $12 - 6 = 6$.

(9.1b) Why is it not necessary to use the $U(t) = \log(t)$ function to find the values?

Answer. We know there is full insurance at the first-best with any risk-averse utility function, so the precise function does not matter.

(9.1c) At the separating contract under incomplete information, C_5, $x = 2.01$. What is y? Justify the value 2.01 for x. What are the coordinates of C_5?

Answer. At C_5, the incentive-compatibility constraints require that $0.5x + 0.5(x - y) = 0$, so $y = 2x$; and $\pi_u(C_5) = \pi_u(C_3)$, so $0.25\log(12 - x) + 0.75\log(y - x) = 0.25\log(3) + 0.75\log(3)$. Solving these equations yields $x^* = 2.01$ and $y = 4.02$.

$C_5 = (9.99, 2.01)$ because $9.99 = 12 - 2.01$ and $2.01 = 4.02 - 2.01$.

(9.1d) Find a contract C_6 that might be profitable and that would lure both types away from C_3 and C_5.

Answer. One possibility is $C_6 = (8, 3)$, or $(x = 4, y = 7)$. The utility of this to the *Safes* is 1.59 ($= 0.5\log(8) + 0.5\log(3)$), compared to 1.57 ($= 0.5\log(10.99) + 0.5\log(2.01)$), so the *Safes* prefer it to C_5, and that means the *Unsafes* will certainly prefer it. If there are not many *Unsafes*, the contract can make a profit, because if it attracts only *Safes*, the profit is 0.5 ($= 0.5(4) + 0.5(4 - 7)$).

9.3: Finding the Mixed-Strategy Equilibrium in a Testing Game. Half of all high school graduates are talented, producing output $a = x$, and half are untalented, producing output $a = 0$. Both types have a reservation wage of 1 and are risk neutral. At a cost of 2 to himself and 1 to the job applicant, an employer can test a graduate and discover his true ability. Employers compete with each other in offering wages but they cooperate in revealing test results, so that an employer knows if an applicant has already been tested and failed. There is just one period of work. The employer cannot commit to testing every applicant.

(9.3a) Why is there no equilibrium in which either untalented workers do not apply or the employer tests every applicant?

Answer. If no untalented workers apply, the employer would deviate and save 2 by skipping the test and just hiring everybody who applies. Then the untalented workers would start to apply. If the employer tests every applicant, however, and pays only w_H, no untalented worker will apply. Again, the employer would deviate and skip the test.

(9.3b) In equilibrium, the employer tests workers with probability γ and pays those who pass the test w, the talented workers all present themselves for testing, and the untalented workers present themselves with probability α. Find an expression for the equilibrium value of α in terms of w. Explain why α is independent of x.

Answer. Using the payoff equating method of calculating a mixed strategy, and remembering that one must equate player 1's payoffs to find player 2's mixing probability, we must focus on the employer's profits. In the mixed-strategy equilibrium, the employer's profits are the same whether or not he tests a particular worker. Fraction $0.5 + 0.5\alpha$ of the workers will take the test, and the employer's cost for each worker that applies is 2, whether he is hired or not, so

$$\begin{aligned}\pi(test) &= \left(\frac{0.5}{0.5 + 0.5\alpha}\right)(x - w) - 2 \\ &= \pi(no\ test) = \left(\frac{0.5}{0.5 + 0.5\alpha}\right)(x - w) \\ &\qquad + \left(\frac{0.5\alpha}{0.5 + 0.5\alpha}\right)(0 - w),\end{aligned} \tag{35}$$

which yields

$$\alpha = \frac{2}{w - 2}. \tag{36}$$

The naive answer to why expression (36) does not depend on x is that α is the worker's strategy, so there is no reason why it should depend on a parameter that enters only into the employer's payoffs. That is wrong, because in a mixed-strategy equilibrium the worker is choosing his probability in a way that makes the employer indifferent between the employer's payoffs. Rather, what is going on here is that a talented worker's productivity is irrelevant to the decision of whether or not to test. The employer already knows he will hire all the talented workers, and the questions for him in deciding whether to test are how costly it is to test and how costly it is to hire untalented workers.

(9.3c) If $x = 8$, what are the equilibrium values of α, γ, and w?
Answer. We already have an expression for α from part (b). The next step is to find the the wage. Profits are zero in equilibrium, which requires that

$$\pi(no\ test) = \left(\frac{0.5}{0.5 + 0.5\alpha}\right)(x - w) + \left(\frac{0.5\alpha}{0.5 + 0.5\alpha}\right)(0 - w) = 0. \quad (37)$$

This implies that

$$\alpha = \frac{x - w}{w}. \quad (38)$$

Substituting $x = 8$ and solving (36) and (37) together yields $w^* = 4$ and $\alpha^* = 0.5$.

In the mixed-strategy equilibrium, the untalented worker's profits are the same whether he applies or not, so

$$\pi(apply) = \gamma(-1 + 1) + (1 - \gamma)(-1 + w) = 1. \quad (39)$$

Substituting $w = 4$ and solving for γ yields $\gamma^* = \frac{2}{3}$.

9.5: Insurance and State-Space Diagrams. Two types of risk-averse people, clean-living and dissolute, would like to buy health insurance. Clean-living people become sick with probability 0.3, and the dissolute with probability 0.9. In state-space diagrams with the person's wealth if he is healthy on the vertical axis and if he is sick on the horizontal, every person's initial endowment is (5, 10), because his initial wealth is 10 and the cost of medical treatment is 5.

(9.5a) What is the expected wealth of each type of person?
Answer. $E(W_c) = 8.5$ $(= 0.7(10) + 0.3(5))$. $E(W_d) = 5.5$ $(= 0.1(10) + 0.9(5))$.

(9.5b) Draw a state-space diagram with the indifference curves for a risk-neutral insurance company that insures each type of person separately. Draw in the post-insurance allocations C_1 for the dissolute and C_2 for the clean-living under the assumption that a person's type is contractible.
Answer. See figure A.6.

(9.5c) Draw a new state-space diagram with the initial endowment and the indifference curves of the two types of people that go through that point.
Answer. See figure A.7.

(9.5d) Explain why, under asymmetric information, no pooling contract C_3 can be part of a Nash equilibrium.
Answer. Because indifference curves for the clean-living are flatter than for the dissolute, a contract C_4 can be found which yields positive profits because it attracts the clean-living but not the dissolute. See figure A.8.

Figure A.6 A State-Space Diagram Showing Indifference Curves for the Insurance Company

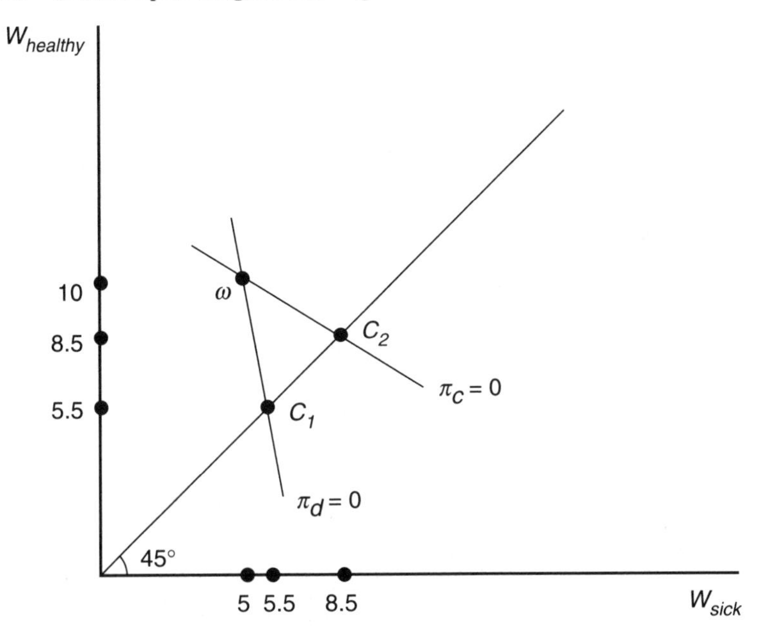

Figure A.7 A State-Space Diagram Showing Indifference Curves for the Customers

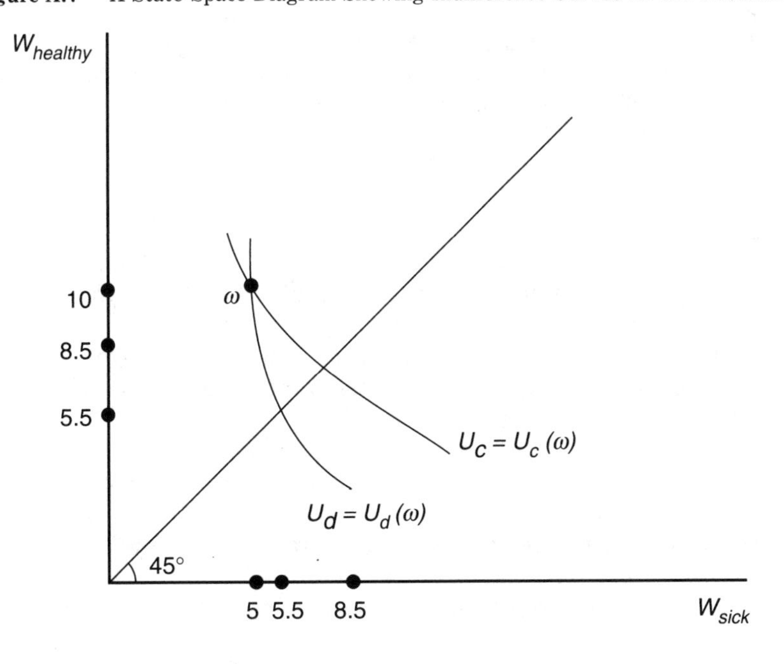

Figure A.8 Why A Pooling Contract Cannot be Part of an Equilibrium

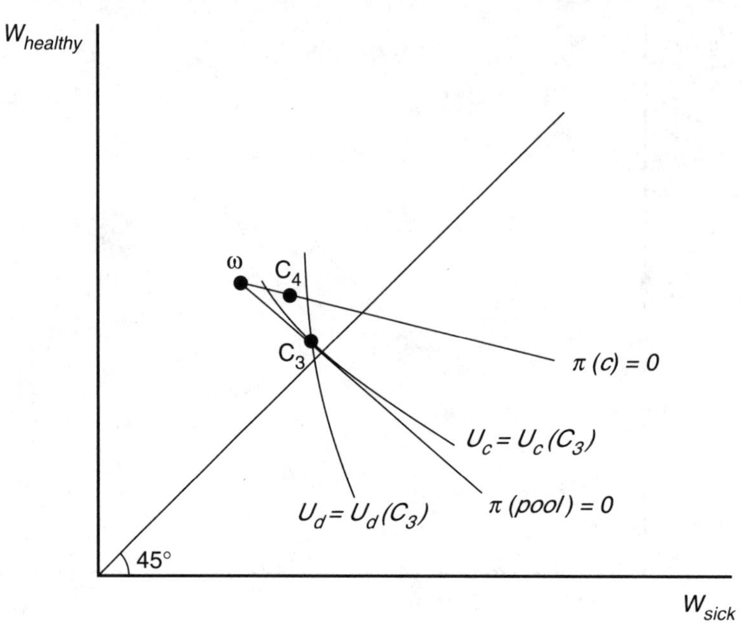

(9.5e) If the insurance company is a monopoly, can a pooling contract be part of a Nash equilibrium?
Answer. Yes. If the insurance company is a monopoly, a pooling contract can be part of a Nash equilibrium because there is no other player who might deviate by offering a cream-skimming contract.

Chapter 10

10.1: Is Lower Ability Better? Change Education I so that the two possible worker abilities are $a \in \{1, 4\}$.

(10.1a) What are the equilibria of this game? What are the payoffs of the workers (and the payoff averaged across workers) in each equilibrium?
Answer. The pooling equilibrium is

$$y_L = y_H = 0, \; w_0 = w_1 = 2.5, \; Pr(L|y = 1) = 0.5, \tag{40}$$

which uses passive conjectures. The payoffs are $U_L = U_H = 2.5$, for an average payoff of 2.5.

The separating equilibrium is

$$y_L = 0, \; y_H = 1, \; w_0 = 1, \; w_1 = 4. \tag{41}$$

The payoffs are $U_L = 1$ and $U_H = 2$, for an average payoff of 1.5. This equilibrium can be justified by the self-selection constraints

$$U_L(y = 0) = 1 > U_L(y = 1) = 4 - \frac{8}{1} = -4 \tag{42}$$

and

$$U_H(y = 0) = 1 < U_H(y = 1) = 4 - \frac{8}{4} = 2. \tag{43}$$

(10.1b) Apply the Intuitive Criterion (see N6.2). Are the equilibria the same?
Answer. Yes. The Intuitive Criterion does not rule out the pooling equilibrium in the game with $a_h = 4$. There is no incentive for *either* type to deviate from $y = 0$ even if the deviation makes the employers think that the deviator is high-ability. The payoff to a persuasive high-ability deviator is only 2, compared to the 2.5 that he can get in the pooling equilibrium.

(10.1c) What happens to the equilibrium worker payoffs if the high ability is 5 instead of 4?
Answer. The pooling equilibrium is

$$y_L = y_H = 0,\ w_0 = w_1 = 3,\ Pr(L|y = 1) = 0.5, \tag{44}$$

which uses passive conjectures. The payoffs are $U_L = U_H = 3$, with an average payoff of 3.

The separating equilibrium is

$$y_L = 0,\ y_H = 1,\ w_0 = 1,\ w_1 = 5. \tag{45}$$

The payoffs are $U_L = 1$ and $U_H = 3.4$, with an average payoff of 2.2. The self-selection constraints are

$$U_H(y = 0) = 1 < U_H(y = 1) = 5 - \frac{8}{5} = 3.4 \tag{46}$$

and

$$U_L(y = 0) = 1 > U_L(y = 1) = 5 - \frac{8}{1} = -3. \tag{47}$$

(10.1d) Apply the Intuitive Criterion to the new game. Are the equilibria the same?
Answer. No. The strategy of choosing $y = 1$ is dominated for the *Lows*, since its maximum payoff is -3, even if the employer is persuaded that he is *High*. So only the separating equilibrium survives.

(10.1e) Could it be that a rise in the maximum ability reduces the average worker's payoff? Can it hurt all the workers?

Answer. Yes. Rising ability would reduce the average worker payoff if the shift was from a pooling equilibrium when $a_h = 4$ to a separating equilibrium when $a_h = 5$. Since the Intuitive Criterion rules out the pooling equilibrium when $a_h = 5$, it is plausible that the equilibrium is separating when $a_h = 5$. Since the pooling equilibrium is Pareto dominant when $a_h = 4$, it is plausible that it is the equilibrium played out. So the average payoff may well fall from 2.5 to 2.2 when the high ability rises from 4 to 5. This cannot make every player worse off, however; the high-ability workers see their payoffs rise from 2.5 to 3.4.

10.3: Price and Quality. Consumers have prior beliefs that Apex produces low-quality goods with probability 0.4 and high-quality with probability 0.6. A unit of output costs 1 to produce in either case, and it is worth 10 to the consumer if it is high-quality and 0 if low-quality. The consumer, who is risk neutral, decides whether to buy in each of two periods, but he does not know the quality until he buys. There is no discounting.

(10.3a) What is Apex' price and profit if it must choose one price, p^*, for both periods?

Answer. A consumer's expected consumer surplus is

$$CS = 0.4(0 - p^*) + 0.6(10 - p^*) + 0.6(10 - p^*) = -1.6p^* + 12. \quad (48)$$

Apex maximizes its profits by setting $CS = 0$, in which case $p^* = 7.5$ and profit is $\pi_H = 13$ $(= 2(7.5 - 1))$ or $\pi_L = 6.5$ $(= 7.5 - 1)$.

(10.3b) What is Apex' price and profit if it can choose two prices, p_1 and p_2, for the two periods, but it cannot commit ahead to p_2?

Answer. If Apex is high-quality, it will choose $p_2 = 10$, since the consumer, having learned the quality in the first period, is willing to pay that much. Thus consumer surplus is

$$CS = 0.4(0 - p_1) + 0.6(10 - p_1) + 0.6(10 - 10) = -p_1 + 6, \quad (49)$$

and, setting this equal to zero, $p_1 = 6$, for a profit of $\pi_H = 14$ $(= (6 - 1) + (10 - 1))$ or $\pi_L = 5$ $(= 6 - 1)$.

(10.3c) What is the answer to part (b) if the discount rate is $r = 0.1$?

Answer. Apex cannot do better than the prices suggested in part (b).

(10.3d) Returning to $r = 0$, what if Apex can commit to p_2?

Answer. Commitment makes no difference in this problem, since Apex wants to charge a higher price in the second period anyway if it has high quality—a high price in the first period would benefit the low-quality Apex too, at the expense of the high-quality Apex.

(10.3e) How do the answers to (a) and (b) change if the probability of low-quality is 0.95 instead of 0.4? (There is a twist to this question.)

Answer. With a constant price, a consumer's expected consumer surplus is

$$\begin{aligned} CS &= 0.95(0 - p^*) + 0.05(10 - p^*) + 0.05(10 - p^*) \\ &= -1.05p^* + 0.1. \end{aligned} \quad (50)$$

Apex would set $CS = 0$, in which case $p^* \approx 0.95$, but since this is less than cost, Apex, in fact, would not sell anything at all, and would earn zero profit.

With changing prices, high-quality Apex will choose $p_2 = 10$, since the consumer, having learned the quality in the first period, is willing to pay that much. Thus, consumer surplus is

$$CS = 0.95(0 - p_1) + 0.05(10 - p_1) + 0.05(10 - 10) = -p_1 + 0.5, \quad (51)$$

and, setting this equal to zero, you might think that $p_1 = 0.5$, for a profit of $\pi_H = 8.5$ $(= (0.5 - 1) + (10 - 1))$. But notice that if the low-quality Apex tries to follow this strategy, his payoff is $\pi_L = 0.5 - 1 < 0$. Hence, only the high-quality Apex will try it. But then the consumers will know the product is high-quality, and they are willing to pay 10 even in the first period. What the high-quality Apex can do is charge up to $p_1 = 1$ in the first period, for profits of 9 $(= (1 - 1) + (10 - 1))$.

10.5: Advertising. Brydox introduces a new shampoo which is actually very good, but is believed by consumers to be good with only a probability of 0.5. A consumer would pay 10 for high quality and 0 for low quality, and the shampoo costs 6 per unit to produce. The firm may spend as much as it likes on stupid TV commercials showing happy people washing their hair, but the potential market consists of 100 cold-blooded economists who are not taken in by psychological tricks. The market can be divided into two periods.

(10.5a) If advertising is banned, will Brydox go out of business?

Answer. No. It can sell at a price of 5 in the first period and 10 in the second period. This would yield profits of 300 $(= (100)(5 - 6) + (100)(10 - 6))$.

(10.5b) If there are two periods of consumer purchase, and consumers discover the quality of the shampoo if they purchase in the first period, show that Brydox might spend substantial amounts on stupid commercials.

Answer. If the seller produces high quality, it can expect repeat purchases. This makes expenditure on advertising useful if it increases the number of initial purchases, even if the firm suffers losses in the first period. If the seller produces low-quality, there will be no repeat purchases. Hence, advertising expenditure can act as a signal of quality: consumers can view it as a signal that the seller intends to stay in business two periods.

(10.5c) What is the minimum and maximum that Brydox will spend on advertising, if it spends a positive amount?

Answer. If there is a separating signalling equilibrium, it will be as follows. Brydox will spend nothing on advertising if its shampoo is low-quality, and consumers will not buy from any company that advertises less than some amount X, because such a company is believed to produce low quality. Brydox will spend X on advertising if its quality is high, and charge a price of 10 in both periods.

Amount X is between 400 and 500. If a low-quality firm spends X on advertising, consumers do buy from it for one period, and it earns profits of $(100)(10 - 6) - X = 400 - X$. Thus, the high-quality firm must spend at least 400 to distinguish itself. If a high-quality firm spends X on advertising, consumers buy from it in both periods, and it earns profits of $(2)(100)(10 - 6) - X = 800 - X$. Since it can make profits of 300 even without advertising, a high-quality firm will spend up to 500 on advertising.

Chapter 11

11.1: A Fixed Cost of Bargaining and Grudges. Smith and Jones are trying to split 100 dollars. In bargaining round 1, Smith makes an offer at cost 0, proposing to keep S_1 for himself, and Jones either accepts (ending the game) or rejects. In round 2, Jones makes an offer at cost 10 of S_2 for Smith and Smith either accepts or rejects. In round 3, Smith makes an offer of S_3 at cost c, and Jones either accepts or rejects. If no offer is ever accepted, the 100 dollars goes to a third player, Dobbs.

(11.1a) If $c = 0$, what is the equilibrium outcome?

Answer. $S_1 = 100$ and Jones accepts it. If Jones refused, he would have to pay 10 to make a proposal that Smith would reject, and then Smith would propose $S_3 = 100$ again. $S_1 < 100$ would not be an equilibrium, because Smith could deviate to $S_1 = 100$ and Jones would still be willing to accept.

(11.1b) If $c = 80$, what is the equilibrium outcome?

Answer. If the game goes to Round 3, Smith will propose $S_3 = 100$ and Jones will accept, but this will cost Smith 80. Hence, if Jones proposes $S_2 = 20$, Smith will accept it, leaving 80 for Jones—who would, however, pay 10 to make his offer. Hence, in Round 1 Smith must offer $S_1 = 30$ to induce Jones to accept, and that will be the equilibrium outcome.

(11.1c) If $c = 10$, what is the equilibrium outcome?

Answer. If the game goes to Round 3, Smith will propose $S_3 = 100$ and Jones will accept, but this will cost Smith 10. Hence, if Jones proposes $S_2 = 90$, Smith will accept it, leaving 10 for Jones—who would, however, pay 10 to make his offer. Hence, in Round 1 Smith need only offer $S_1 = 100$ to induce Jones to accept, and that will be the equilibrium outcome.

(11.1d) What happens if $c = 0$, but Jones is very emotional and would spit in Smith's face and throw the 100 dollars to Dobbs if Smith proposes $S = 100$? Assume that Smith knows Jones' personality perfectly.

Answer. However emotional Jones may be, there is some minimum offer M that he would accept, which probably is less than 50 (but you never know—some people think they are entitled to everything, and one could imagine a utility function such that Jones would refuse $S = 5$ and prefer to bear the cost 10 in the second round in order to get the whole 100 dollars). The equilibrium will be for Smith to propose exactly $S - M$ in Round 1, if $S - M \geq 0$, and for Jones to accept.

11.3: The Nash Bargaining Solution. Smith and Jones, shipwrecked on a desert island, are trying to split 100 pounds of cornmeal and 100 pints of molasses, their only supplies. Smith's utility function is $U_s = C + 0.5M$ and Jones' is $U_j = 3.5C + 3.5M$. If they cannot agree, they fight to the death, with $U = 0$ for the loser. Jones wins with probability 0.8.

(11.3a) What is the threat point?

Answer. The threat point gives the expected utility for Smith and Jones if they fight. This is 560 for Jones ($= 0.8(350 + 350) + 0$), and 30 for Smith ($= 0.2(100 + 50) + 0$).

(11.3b) With a 50-50 split of the supplies, what are the utilities if the two players do not recontract? Is this efficient?

Answer. The split would give the utilities $U_s = 75$ ($= 50 + 25$) and $U_j = 350$. If Smith then traded 10 pints of molasses to Jones for 8 pounds of cornmeal, the utilities would become $U_s = 78$ ($= 58 + 20$) and $U_j = 357$ ($= 3.5(60) + 3.5(42)$), so both would have gained. The 50-50 split is not efficient.

(11.3c) Draw the threat point and the Pareto frontier in utility space (put U_s on the horizontal axis).

Answer. See figure A.9 on the next page.

To draw the diagram, first consider the extreme points. If Smith gets everything, his utility is 150 and Jones' is zero. If Jones gets everything, his utility is 700 and Smith's is zero. If we start at (150, 0) and wish to efficiently help Jones at the expense of Smith, this is done by giving Jones some molasses, since Jones puts a higher relative value on molasses. This can be done until Jones has all the molasses, at utility point (100, 350). Beyond there, one must take cornmeal away from Smith if one is to help Jones further, so the Pareto frontier acquires a flatter slope.

(11.3d) According to the Nash bargaining solution, what are the utilities? How are the goods split?

Answer. To find the Nash bargaining solution, maximize $(U_s - 30)(U_j - 560)$. Note from the diagram that it seems the solution will be on the upper part of the Pareto frontier, above (100, 350), where Jones is consuming all the molasses, and where, if Smith loses one utility unit, Jones gains 3.5. If we let X denote

Figure A.9 The Threat Point and Pareto Frontier

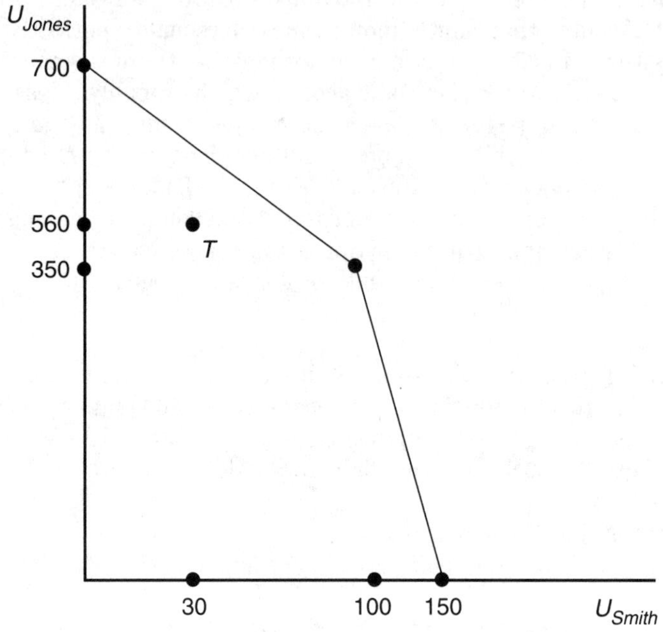

the amount of cornmeal that Jones gets, we can rewrite the problem as

$$\underset{X}{Maximize}\ (100 - X - 30)(350 + 3.5X - 560). \tag{52}$$

This maximand equals $(70 - X)(3.5X - 210) = -14{,}700 + 455X - 3.5X^2$. The first-order condition is $455 - 7X = 0$, so $X^* = 65$. Thus, Smith gets 35 pounds of cornmeal, Jones gets 65 pounds of cornmeal and 100 of molasses, and $U_s = 35$ and $U_j = 577.5$.

(11.3e) Suppose Smith discovers a cookbook full of recipes for a variety of molasses candies and corn muffins, and his utility function becomes $U_s = 10C + 5M$. Show that the split of goods in part (d) remains the same despite his improved utility function.

Answer. The utility point at which Jones has all the molasses and Smith has the molasses is now (1000, 350), since Smith's utility is (10)(100). Smith's new threat-point utility is 300 (= 0.2((10)(100) + (5)(100)) . Thus, the Nash problem of equation (52) becomes

$$\underset{X}{Maximize}\ (1000 - 10X - 300)(350 + 3.5X - 560). \tag{53}$$

But this maximand is the same as (10)(100 − X − 30)(350 + $3.5X$ − 560), so it must have the same solution as was found in part (d).

11.5: A Fixed Cost of Bargaining and Incomplete Information. Smith and Jones are trying to split 100 dollars. In bargaining round 1, Smith makes an offer at cost c, proposing to keep S_1 for himself. Jones either accepts (ending the game) or rejects. In round 2, Jones makes an offer of S_2 for Smith, at cost 10, and Smith either accepts or rejects. In round 3, Smith makes an offer of S_3 at cost c, and Jones either accepts or rejects. If no offer is ever accepted, the 100 dollars goes to a third player, Parker.

(11.5a) If $c = 0$, what is the equilibrium outcome?

Answer. $S_1 = 100$ and Jones accepts it. If Jones refused, he would have to pay 10 to make a proposal that Smith would reject, and then Smith would propose $S_3 = 100$ again. $S_1 < 100$ would not be an equilibrium, because Smith could deviate to $S_1 = 100$ and Jones would still be willing to accept.

(11.5b) If $c = 80$, what is the equilibrium outcome?

Answer. If the game goes to Round 3, Smith will propose $S_3 = 100$ and Jones will accept, but this will cost Smith 80. Hence, if Jones proposes $S_2 = 20$, Smith will accept it, leaving 80 for Jones—who would, however, pay 10 to make his offer. Hence, in Round 1 Smith must offer $S_1 = 30$ to induce Jones to accept, which will be the equilibrium outcome.

(11.5c) If Jones's priors are that $c = 0$ and $c = 80$ are equally likely, but only Smith knows the true value, what is the equilibrium outcome? (Hint: the equilibrium uses mixed strategies.)

Answer. Smith's equilibrium strategy is to offer $S_1 = 100$ with probability 1 if $c = 0$ and probability $1/7$ if $c = 80$; and to offer $S_1 = 30$ with probability $6/7$ if $c = 80$. He accepts $S_2 \geq 20$ if $c = 80$ and $S_2 = 100$ if $c = 0$, and proposes $S_3 = 100$ regardless of c. Jones accepts $S_1 = 100$ with probability $1/8$, rejects $S_1 \in (30, 100)$, and accepts $S_1 \leq 30$. He proposes $S_2 = 20$ and accepts $S_3 = 100$. Out of equilibrium, a supporting belief for Jones to believe is that if S_1 equals neither 30 nor 100, $Prob(c = 80) = 1$.

If $c = 0$, the equilibrium outcome is for Smith to propose $S_1 = 100$, for Jones to accept with probability $1/8$ and to propose $S_2 = 20$ otherwise and be rejected, and for Smith to then propose $S_3 = 100$ and be accepted. If $c = 80$, the equilibrium outcome is, with probability $6/7$, for Smith to propose $S_1 = 30$ and be accepted, with probability $(1/7)(1/8)$, for Smith to propose $S_1 = 100$ and be accepted, and, with probability $(1/7)(7/8)$, for Smith to propose $S_1 = 100$, be rejected, and then to be offered $S_2 = 20$ and to accept.

The rationale behind the equilibrium strategies is as follows. In Round 3, either type of Smith does best by proposing a share of 100, and Jones might as well accept. In Round 2, anything but $S_2 = 100$ would be rejected by Smith if $c = 0$, so Jones should give up on that and offer $S_2 = 20$, which would be accepted if $c = 80$ because if that type of Smith were to wait, he would have to pay 80 to propose $S_3 = 100$. In Round 1, if $c = 0$, Smith should propose $S_1 = 100$, since he can wait until Round 3 and get 100 anyway at zero extra cost. There is no pure-strategy equilibrium, because if $c = 80$, Smith would pretend that $c = 0$ and propose $S_1 = 100$ if Jones would accept that. But if Jones accepts only with probability θ, Smith runs the risk of only getting 20 in the second period, less than $S_1 = 30$, which would be accepted by Jones with probability 1. Similarly, if Smith proposes $S_1 = 100$ with probability γ when $c = 80$, Jones can either accept it, or wait, in which case Jones might either pay a cost of 10 and end up with $S_3 = 100$ anyway, or get Smith to accept $S_2 = 20$.

The probability γ must equate Jones's two pure-strategy payoffs,

$$\pi_j(accept\ S_1 = 100) = 0 \tag{54}$$

and, using Bayes' Rule,

$$\pi_j(reject\ S_1 = 100) = -10 + \left(\frac{0.5\gamma}{0.5\gamma + 0.5}\right)(80) + \left(\frac{0.5}{0.5\gamma + 0.5}\right)(0), \tag{55}$$

which yields $\gamma = \frac{1}{7}$.

The probability θ must equate Smith's two pure-strategy payoffs:

$$\pi_s(S_1 = 30) = 30 \tag{56}$$

and

$$\pi_s(S_1 = 100) = \theta \cdot 100 + (1 - \theta)(20), \tag{57}$$

which yields $\theta = \frac{1}{8}$.

Chapter 12

12.1: Rent-Seeking. Two risk-neutral neighbors in 16th century England, Smith and Jones, have gone to court and are considering bribing a judge. Each of them makes a gift, and the one whose gift is the largest is awarded property worth £2000. If both bribe the same amount, the chances are 50 percent for each of them to win the lawsuit. Gifts must be either £0, £900, or £2000.

(12.1a) What is the unique pure-strategy equilibrium for this game?

Answer. Each bids £900, for expected profits of 100 each (= −900 + 0.5(2000)). Table A.12 shows the payoffs (but also includes the payoffs for the case when the strategy of a bid of £1500 is allowed). A player who deviates to zero has a payoff of zero; a player who deviates to £2000 has a payoff of zero. But (0, 0) is not an equilibrium, because the expected payoff is £1000, whereas a player who deviated to 900 would have a payoff of £1100.

Table A.12 Bribes I

		Jones			
		£0	**£900**	**£1500**	**£2000**
Smith:	£0	1000,1000	0, 1100	0, 500	0, 0
	£900	1100,0	100, 100	−900, 500	−900, 0
	£1500	500,0	500, −900	−500, −500	−1500, 0
	£2000	0,0	0, −900	0, −1500	−1000, −1000

Payoffs to: (Smith, Jones)

(12.1b) Suppose that it is also possible to give a £1500 gift. Why does there no longer exist a pure-strategy equilibrium?

Answer. If Smith bids 0 or £900, Jones would bid £1500. If Smith bids £1500, Jones would bid £2000. If both bid £2000, then one can profit by deviating to 0. But if Smith bids £2000 and Jones bids 0, Smith will deviate to 900. This exhausts all the possibilities.

(12.1c) What is the symmetric mixed-strategy equilibrium for the expanded game? What is the judge's expected payoff?

Answer. Let $(\theta_0, \theta_{900}, \theta_{1500}, \theta_{2000})$ be the probabilities. It is pointless ever to bid £2000, because it can only yield zero or negative profits, so $\theta_{2000} = 0$. In a symmetric mixed-strategy equilibrium, the returns to the pure strategies are equal, and the probabilities add up to one, so

$$\pi_{Smith}(0) = \pi_{Smith}(900) = \pi_{Smith}(1500)$$

$$0.5\theta_0(2000) = -900 + \theta_0(2000) + 0.5\theta_{900}(2000) \quad (58)$$
$$= -1500 + \theta_0(2000) + \theta_{900}(2000) + 0.5\theta_{1500}(2000),$$

and

$$\theta_0 + \theta_{900} + \theta_{1500} = 1. \quad (59)$$

Solving out these three equations for three unknowns, the equilibrium is (0.4, 0.5, 0.1, 0.0).

The judge's expected payoff is £1200 (= 2(0.5(900) + 0.1(1500))).

Note: The results are sensitive to the bids allowed. Can you speculate as to what might happen if the strategy space were the whole continuum from 0 to 2000?

(12.1d) In the expanded game, if the losing litigant gets back his gift, what are the two equilibria? Would the judge prefer this rule?

Answer. Table A.13 shows the new outcome matrix. There are three equilibria: x_1 = (900, 900), x_2 = (1500, 1500), and x_3 = (2000, 2000).

Table A.13 Bribes II

		Jones			
		£0	£900	£1500	£2000
	£0	[1000], [1000]	0,1100	0,500	0,0
Smith:	£900	1100,0	[550], [550]	0,500	0,0
	£1500	500,0	500,0	[250], [250]	0,0
	£2000	0,0	0,0	0,0	[0], [0]

Payoffs to: (Smith, Jones)

The judge's payoff was 1200 under the unique mixed-strategy equilibrium in the original game. Now, his payoff is either 900, 1500, or 2000. Thus, whether he prefers the new rules depends on which equilibrium is played out under them.

12.3: Government and Monopoly. Incumbent Apex and potential entrant Brydox are bidding for government favors in the widget market. Apex wants to defeat a bill that would require it to share its widget patent rights with Brydox. Brydox wants the bill to pass. Whoever offers the chairman of the House Telecommunications Committee more campaign contributions wins, and the loser pays nothing. The market demand curve is $P = 25 - Q$, and marginal cost is constant at 1.

(12.3a) Who will bid higher if duopolists follow Bertrand behavior? How much will the winner bid?

Answer. Apex bids higher, because it gets monopoly profits from winning, and Bertrand profits equal zero. Apex can bid some small quantity ε and win.

(12.3b) Who will bid higher if duopolists follow Cournot behavior? How much will the winner bid?

Answer. Monopoly profits are found from the problem

$$\underset{Q_a}{Maximize}\; Q_a(25 - Q_a - 1), \tag{60}$$

which has the first-order condition $25 - 2Q_a - 1 = 0$, so that $Q_a = 12$ and $\pi_a = 144$ $(= 12(25 - 12 - 1))$.

Each firm's Cournot duopoly profits are found from the problem

$$\underset{Q_a}{Maximize}\; Q_a(25 - Q_a - Q_b - 1), \tag{61}$$

which has the first-order condition $25 - 2Q_a - Q_b - 1 = 0$. If the equilibrium is symmetric and $Q_b = Q_a$, then $Q_a = 8$ and $\pi_a = 64$ $(= 8(25 - 8 - 8 - 1))$.

Brydox will bid up to 64. Apex will bid up to 80 $(= 144 - 64)$, and so Apex will win the auction at a price of 64.

(12.3c) What happens under Cournot behavior if Apex can commit to giving away its patent freely to everyone in the world if the entry bill passes? How much will Apex bid?

Answer. Apex will bid some small ε and win. It will commit to giving away its patent if the bill succeeds, which means that if the bill succeeds, the industry will have zero profits and Brydox has no incentive to bid a positive amount to secure entry.

Chapter 13

13.1: Differentiated Bertrand with Advertising. Two firms that produce substitutes are competing with demand curves

$$q_1 = 10 - \alpha p_1 + \beta p_2 \tag{62}$$

and

$$q_2 = 10 - \alpha p_2 + \beta p_1. \tag{63}$$

Marginal cost is constant at $c = 3$. A player's strategy is his price. Assume that $\alpha > \beta/2$.

(13.1a) What is the reaction function for Firm 1? Draw the reaction curves for both firms.

Answer. Firm 1's profit function is

$$\pi_1 = (p_1 - c)q_1 = (p_1 - 3)(10 - \alpha p_1 + \beta p_2). \tag{64}$$

Differentiating with respect to p_1 and solving the first-order condition gives the reaction function

$$p_1 = \frac{10 + \beta p_2 + 3\alpha}{2\alpha}. \tag{65}$$

This is shown in Figure A.10.

Figure A.10 Reaction Curves in a Bertrand Game with Advertising

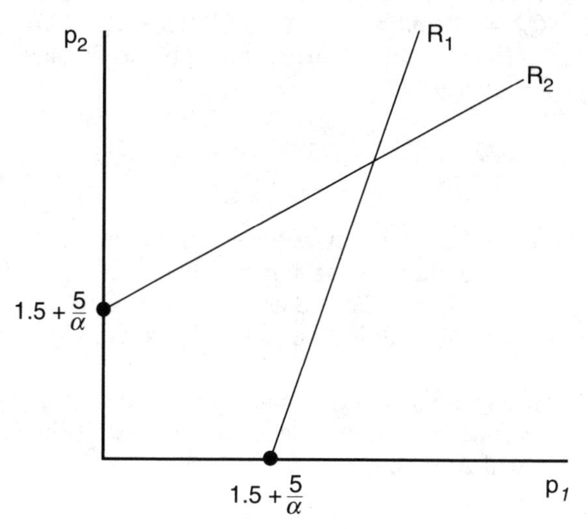

(13.1b) What is the equilibrium? What is the equilibrium quantity for Firm 1?
<u>Answer.</u> Using the symmetry of the problem, set $p_1 = p_2$ in the reaction function for Firm 1 and solve, to give $p_1^* = p_2^* = (10 + 3\alpha)/(2\alpha - \beta)$. Using the demand function for Firm 1, $q_1 = (10\alpha + 3\alpha(\beta - \alpha))/(2\alpha - \beta)$.

(13.1c) Show how Firm 2's reaction function changes when β increases. What happens to the reaction curves in the diagram?
<u>Answer.</u> The slope of Firm 2's reaction curve is $\partial p_2/\partial p_1 = \beta/2\alpha$. The change in this when β changes is $\partial^2 p_2/\partial p_1 \partial\beta = 1/2\alpha > 0$. Thus, Firm 2's reaction curve becomes steeper, as shown in Figure A.11.

(13.1d) Suppose that an advertising campaign could increase the value of β by one, and that this would increase the profits of each firm by more than the cost of the campaign. What does this mean? If either firm could pay for this campaign, what game would result between them?
<u>Answer.</u>. The meaning of an increase in β is that a firm's quantity demanded becomes more responsive to the other firm's price. The goods become closer substitutes and total demand for the two goods increases.

Figure A.11 How Reaction Curves Change When β Increases

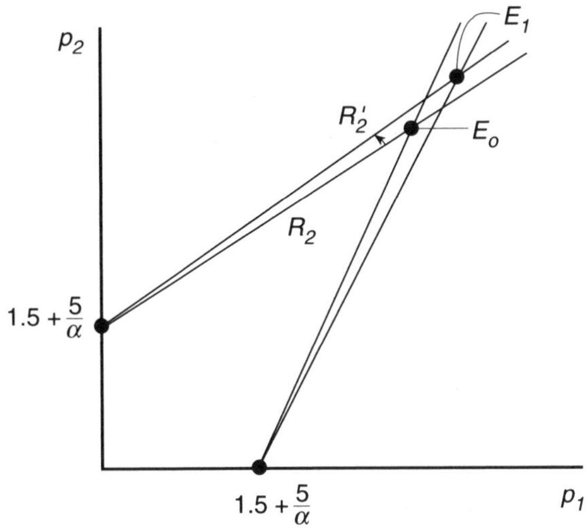

If either firm could pay, a game of Chicken results, with payoffs something like those shown in table A.14, where the advertising campaign costs 1 and yields extra profits of 4 to each firm.

Table A.14 An Advertising Chicken Game

		Firm 2		
		Advertise		*Do not advertise*
	Advertise	3,3	→	**3,4**
Firm 1:		↓		↑
	Do not advertise	**4,3**	←	0,0

Payoffs to: (Firm 1, Firm 2)

13.3: Differentiated Bertrand. Two firms that produce substitutes have the demand curves

$$q_1 = 1 - \alpha p_1 + \beta(p_2 - p_1) \tag{66}$$

and

$$q_2 = 1 - \alpha p_2 + \beta(p_1 - p_2), \tag{67}$$

where $\alpha > \beta$. Marginal cost is constant at c, where $c < 1/\alpha$. A player's strategy is his price.

(13.3a) What are the equations for the reaction curves $p_1(p_2)$ and $p_2(p_1)$? Draw them.

Answer. Firm 1 solves the problem of maximizing $\pi_1 = (p_1 - c)q_1 = (p_1 - c)(1 - \alpha p_1 + \beta[p_2 - p_1])$ by his choice of p_1. The first-order condition is $1 - 2(\alpha + \beta)p_1 + \beta p_2 + (\alpha + \beta)c = 0$, which gives the reaction function

$$p_1 = \frac{1 + \beta p_2 + (\alpha + \beta)c}{2(\alpha + \beta)}.$$

For p_2:

$$p_2 = \frac{1 + \beta p_1 + (\alpha + \beta)c}{2(\alpha + \beta)}.$$

Figure A.12 shows the reaction curves. Note that $\beta > 0$, because the goods are substitutes.

Figure A.12 Reaction Curves for the Differentiated Bertrand Game

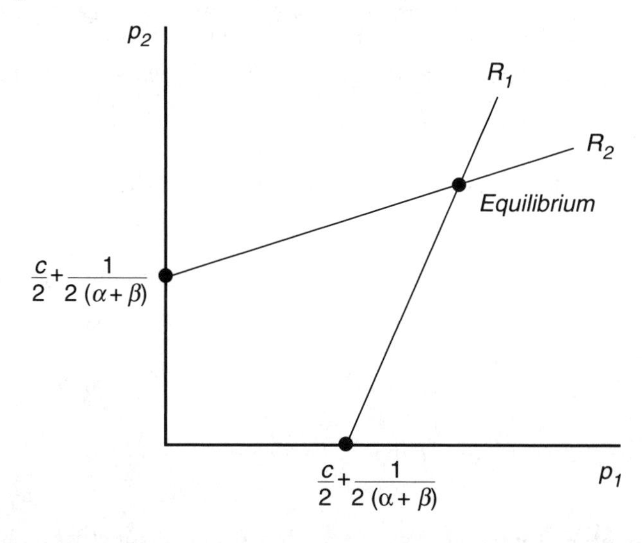

(13.3b) What is the pure-strategy equilibrium for this game?

Answer. This game is symmetric, so we can guess that $p_1^* = p_2^*$. In that case, using the reaction curves, $p_1^* = p_2^* = [1 + (\alpha + \beta)c]/[2\alpha + \beta]$.

(13.3c) What happens to prices if α, β, or c increase?

Answer. The response of p^* to an increase in α is

$$\begin{aligned} \frac{\partial p^*}{\partial \alpha} &= \frac{c}{2\alpha + \beta} - \frac{2[1 + (\alpha + \beta)c]}{(2\alpha + \beta)^2} \\ &= \left(\frac{1}{(2\alpha + \beta)^2}\right)(2\alpha c + \beta c - 2 - 2\alpha c - 2\beta c) < 0. \end{aligned} \quad (68)$$

The derivative has the same sign as $-\beta c - 2 < 0$, so, since $\beta > 0$, the price falls as α rises. This makes sense because α represents the responsiveness of the quantity demanded to the firm's own price.

The increase in p^* when β increases is

$$\begin{aligned}\frac{\partial p^*}{\partial \beta} &= \frac{c}{(2\alpha + \beta)} - \frac{1 + (\alpha + \beta)c}{(2\alpha + \beta)^2} \\ &= \left(\frac{1}{(2\alpha + \beta)^2}\right)(2\alpha c + \beta c - 1 - \alpha c - \beta c) < 0.\end{aligned} \tag{69}$$

The price falls with β, because $c < 1/\alpha$.

The increase in p^* when c increases is:

$$\frac{\partial p^*}{\partial c} = \frac{\alpha + \beta}{2\alpha + \beta} > 0. \tag{70}$$

When the marginal cost rises, so does the price.

(13.3d) What happens to each firm's price if α increases, but only Firm 2 realizes it (and Firm 2 knows that Firm 1 is uninformed)? Would Firm 2 reveal the change to Firm 1?

Answer. From the equation for the reaction curve of Firm 1, it can be seen that the reaction curve will shift and swivel as in figure A.13. This is because

$$\frac{\partial p_2}{\partial p_1} = \frac{\beta}{2(\alpha + \beta)},$$

Figure A.13 Changes in the Reaction Curves

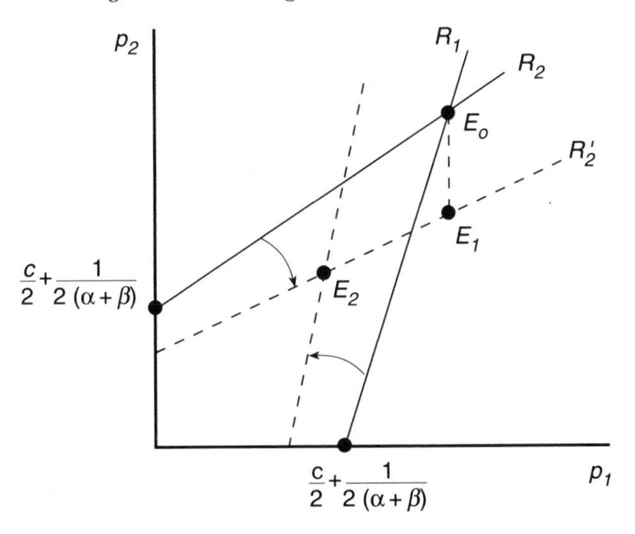

so

$$\frac{\partial^2 p_2}{\partial p_1 \partial \beta} = -\frac{\beta}{2(\alpha + \beta)^2} < 0.$$

Firm 2's reaction curve does not change, and it believes that Firm 1's reaction curve has not changed either, so Firm 2 has no reason to change its price. The equilibrium changes from E_0 to E_1: Firm 1 maintains its price, but Firm 2 reduces its price. Firm 2 would not want to reveal the change to Firm 1, because then Firm 1 would also reduce its price (and Firm 2 would reduce its price still further), and the new equilibrium would be E_2.

Appendix B Mathematics

This appendix has three purposes: to remind some readers of the definitions of terms they have seen before, to give other readers an idea of what the terms mean, and to list a few theorems for reference. In accordance with these limited purposes, some terms such as "boundary point" are left undefined. For fuller exposition, see Rudin (1964) on real analysis, Debreu's *Theory of Value* (1959), and Chiang (1984) and Takayama (1985) on mathematics for economists. Intriligator (1971) and Varian (1992) both have good mathematical appendices and are strong in discussing optimization, and Seierstad & Sydsaeter (1987) and Kamien & Schwartz (1991) cover maximizing by choice of functions. Border's 1985 book is entirely about fixed-point theorems. Stokey & Lucas (1989) is about dynamic programming. Fudenberg & Tirole (1991a) is the best source of mathematical theorems for use in game theory.

Notation

Σ Summation. $\sum_{i=1}^{3} x_i = x_1 + x_2 + x_3$.

Π Product. $\prod_{i=1}^{3} x_i = x_1 x_2 x_3$.

$|\;|$ Absolute value. If $x \geq 0$, then $|x| = x$ and if $x < 0$, then $|x| = -x$.

$|$ "Such that," "given that," or "conditional upon." $\{x|x < 3\}$ denotes the set of real numbers less than 3. *Prob*$(x|y < 5)$ denotes the probability of x given that y is less than 5.

$:$ "Such that." $\{x : x < 3\}$ denotes the set of real numbers less than 3. The colon is a synonym for $|$.

$\boldsymbol{R}^n$ The set of n-dimensional vectors of real numbers (integers, fractions, and the least upper bounds of any subsets thereof).

$\{\;\}$ A set of elements. The set $\{3, 5\}$ consists of two elements, 3 and 5.

$\in$ "Is an element of." $a \in \{2, 5\}$ means that a takes either the value 2 or 5.

$\subset$ Set inclusion. If $X = \{2, 3, 4\}$ and $Y = \{2, 4\}$, then $Y \subset X$ because Y is a subset of X.

$[a, b]$ A closed interval. The interval $[0, 1000]$ is the set $\{x|0 \leq x \leq 1000\}$. Square brackets are also used as delimiters.

(a, b) An open interval. The interval $(0, 1000)$ is the set $\{x|0 < x < 1000\}$. $(0, 1000]$ would be half-open interval, the set $\{x|0 < x \leq 1000\}$. Parentheses are also used as delimiters.

$\times$ The Cartesian product. $X \times Y$ is the set of points $\{x, y\}$, where $x \in X$ and $y \in Y$.

ε An arbitrarily small positive number. If my payoff from both *Left* and *Right* equals 10, I am indifferent between them; if my payoff from *Left* is changed to $10 + \epsilon$, I prefer *Left*.

$\sim$ We say that $X \sim F$ if the random variable X is distributed according to the distribution F.

$\exists$ "There exists. . ."

$\forall$ "For all. . ."

$\equiv$ "Equals by definition."

$\rightarrow$ If f maps space X into space Y, we write $f: X \rightarrow Y$.

$\frac{df}{dx}, \frac{d^2f}{dx^2}$ The first and second derivatives of a function. If $f(x) = x^2$, then $\frac{df}{dx} = 2x$ and $\frac{d^2f}{dx^2} = 2$.

f', f'' The first and second derivatives of a function. If $f(x) = x^2$, then $f' = 2x$ and $f'' = 2$. Primes are also used on variables (not functions) for other purposes: x' and x'' might denote two particular values of x.

$\frac{\partial f}{\partial x}, \frac{\partial^2 f}{\partial x \partial y}$ Partial derivatives of a function. If $f(x, y) = x^2y$ then $\frac{\partial f}{\partial x} = 2xy$ and $\frac{\partial^2 f}{\partial x \partial y} = 2x$.

y_{-i} The set y minus element i. If $y = \{y_1, y_2, y_3\}$, then $y_{-2} = \{y_1, y_3\}$.

Max(,) The maximum of two numbers. $Max(x, y)$ denotes the greater of x and y.

Min(,) The minimum of two numbers. $Min(5, 3) = 3$.

Sup X The supremum (least upper bound) of set X. If $X = \{x|0 \leq x < 1000\}$, then $sup\ X = 1000$. The supremum is useful because sometimes, as here, no maximum exists.

Inf X The infimum (greatest lower bound) of set X. If $X = \{x|0 \leq x < 1000\}$, then $inf\ X = 0$.

Argmax The argument that maximizes a function. If $e^* = argmax\ EU(e)$, then e^* is the value of e that maximizes the function $EU(e)$. The argmax of $f(x) = x - x^2$ is 1/2.

Maximum The greatest value that a function can take. $Maximum(x - x^2) = 1/4$.

Minimum The least value that a function can take. $Minimum(-5 + x^2) = -5$.

The Greek Alphabet

A	α	alpha	N	ν	nu
B	β	beta	Ξ	ξ	xi
Γ	γ	gamma	O	o	omicron
Δ	δ	delta	Π	π	pi
E	ϵ or ε	epsilon	P	ρ	rho
Z	ζ	zeta	Σ	σ	sigma
H	η	eta	T	τ	tau
Θ	θ	theta	Υ	υ	upsilon
I	ι	iota	Φ	ϕ	phi
K	κ	kappa	X	χ	chi
Λ	λ	lambda	Ψ	ψ	psi
M	μ	mu	Ω	ω	omega

Simple Formulas

The Quadratic Formula:

Let

$$ax^2 + bx + c = 0.$$

Then

$$x = \frac{-b \pm \sqrt{b^2 - 4ac}}{2a}.$$

$\log(xy) = \log(x) + \log(y)$.
$\log(x^2) = 2\log(x)$.
$a^x = (e^{\log(a)})^x$
$e^{rt} = (e^r)^t$.
$e^{a+b} = e^a e^b$.
$a > b \Rightarrow ka < kb$, if $k < 0$.

Derivatives

$f(x)$	$f'(x)$
x^a	ax^{a-1}
$\frac{1}{x}$	$-\frac{1}{x^2}$
$\frac{1}{x^2}$	$-\frac{2}{x^3}$
e^x	e^x
e^{rx}	re^{rx}
$\log(ax)$	$1/x$
$\log(x)$	$1/x$
a^x	$a^x \log(a)$
$f(g(x))$	$f'(g(x)) \cdot g'(x)$
$f(x)g(x)$	$f'(x)g(x) + f(x)g'(x)$

Determinants

$$\begin{vmatrix} a_{11} & a_{12} \\ a_{21} & a_{22} \end{vmatrix} = a_{11}a_{22} - a_{21}a_{12}.$$

$$\begin{vmatrix} a_{11} & a_{12} & a_{13} \\ a_{21} & a_{22} & a_{23} \\ a_{31} & a_{32} & a_{33} \end{vmatrix} = a_{11}a_{22}a_{33} - a_{11}a_{23}a_{32} + a_{12}a_{23}a_{31} - a_{12}a_{21}a_{33} + a_{13}a_{21}a_{32} - a_{13}a_{22}a_{31}.$$

Some Useful Functional Forms for x>o

$f(x)$	$f'(x)$	$f''(x)$	Slope	Curvature
$\log(x)$	$\frac{1}{x}$	$-\frac{1}{x^2}$	increasing	concave
$\sqrt{x}$	$\frac{1}{2\sqrt{x}}$	$-\frac{1}{4x^{(3/2)}}$	increasing	concave
x^2	$2x$	2	increasing	convex
$\frac{1}{x}$	$-\frac{1}{x^2}$	$\frac{2}{x^3}$	decreasing	convex
$7 - x^2$	$-2x$	-2	decreasing	concave
$7x - x^2$	$7 - 2x$	-2	increasing/ decreasing	concave

The signs of the derivatives can be confusing. The convex function $f(x) = x^2$ is increasing at an *increasing* rate, but the convex function $f(x) = \frac{1}{x}$ is decreasing at a *decreasing* rate, even though $f'' > 0$ in both cases—i.e., in both cases the slope is becoming more positive.

Glossary

affine An affine function $f(x)$ is a function of the form $f(x) = \alpha + \beta x$.

almost always See "generically."

annuity A riskless security paying a constant amount each year for a given period of years, with the amount conventionally paid at the end of each year.

closed A closed set in $\mathbf{R}^n$ includes its boundary points. The set $\{x : 0 \leq x \leq 1000\}$ is closed.

compact If a set X in $\mathbf{R}^n$ is closed and bounded, X is compact.

concave function The continuous function $f(x)$ defined on interval X is concave if for all elements w and z of X, $f(0.5w + 0.5z) \geq 0.5f(w) + 0.5f(z)$. If f maps $\mathbf{R}$ into $\mathbf{R}$ and f is differentiable and concave, $f'' \leq 0$. See figure B.1.

continuous function Let $d(x, y)$ represent the distance between points x and y. The function f is continuous if for every $\varepsilon > 0$ there exists a number $\delta(\epsilon) > 0$ such that $d(x, y) < \delta(\epsilon)$ implies $d(f(x), f(y)) < \epsilon$.

continuum A continuum is a closed interval of the real line, or a set that can be mapped one-to-one onto such an interval.

Figure B.1 Concavity and Convexity

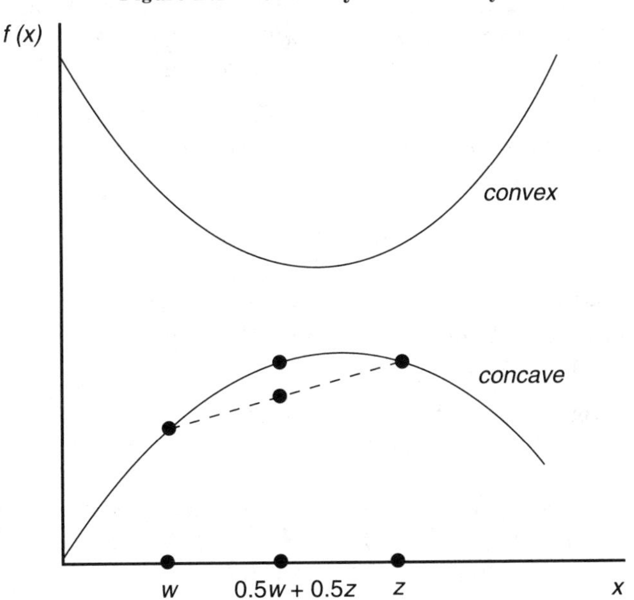

contraction The mapping $f(x)$ is said to be a contraction if there exists a number $c < 1$ such that for the metric d of the space X,

$$d(f(x), f(y)) \leq cd(x, y) \text{ for all } x, y \in X. \tag{1}$$

convex function The continuous function $f(x)$ is convex if for all elements w and z of X, $f(0.5w + 0.5z) \leq 0.5f(w) + 0.5f(z)$. See Figure B.1. Convex functions are only loosely related to convex sets.

convex set If set X is convex, if you take any two of its elements w and z and a real number t, $0 \leq t \leq 1$, then $tw + (1 - t)z$ is also in X.

correspondence A correspondence is a mapping that maps each point to one or more other points, as opposed to a function, which only maps to one.

function If f maps each point in X to exactly one point in Y, f is called a function. The two mappings in figure B.1 are functions, but the mapping in figure B.2 is not.

generically If an assertion is true on a set X generically, "except on a set of measure zero," or "almost always," it is false only on a subset of points Z that have the following property. If a point is randomly chosen using a density function with support X, a point in Z is chosen with probability zero. This implies that if the assertion is false on $z \in \mathbf{R}^n$ and z is perturbed by adding a random amount ϵ, the assertion is true on $z + \epsilon$ with probability one.

Lagrange multiplier The Lagrange multiplier λ is the marginal value of relaxing a constraint in an optimization problem. If the problem is

$$\{\underset{x}{Maximize}\ x^2 \text{ subject to } x \leq 5\}, \text{ then } \lambda = 2x^* = 10.$$

lattice A lattice is a partially ordered set (the $\geq$ ordering is defined) where for any two elements a and b, the values $inf(a, b)$ and $sup(a, b)$ are also in the set. A lattice is complete if the infimum and supremum of each of its subsets are in the lattice.

lower semicontinuous correspondence The correspondence ϕ is lower semicontinuous at the point x_0 if

$$x_n \rightarrow x_0,\ y_0 \in \phi(x_0), \text{ implies } \exists y_n \in \phi(x_n) \text{ such that } y_n \rightarrow y_0, \tag{2}$$

which means that associated with every x sequence leading to x_0 there is a y sequence leading to its image. See figure B.2. This idea is not as important as upper semicontinuity.

mean-preserving spread See discussion of **Risk** below.

measure zero See "generically."

metric The function $d(w, z)$ defined over elements of set X is a metric if (1) $d(w, z) > 0$ if $w \neq z$ and $d(w, z) = 0$ if $w = z$; (2) $d(w, z) = d(z, w)$; and (3) $d(w, z) \leq d(w, y) + d(y, z)$ for points $w, y, z \in X$.

Figure B.2 Upper Semicontinuity

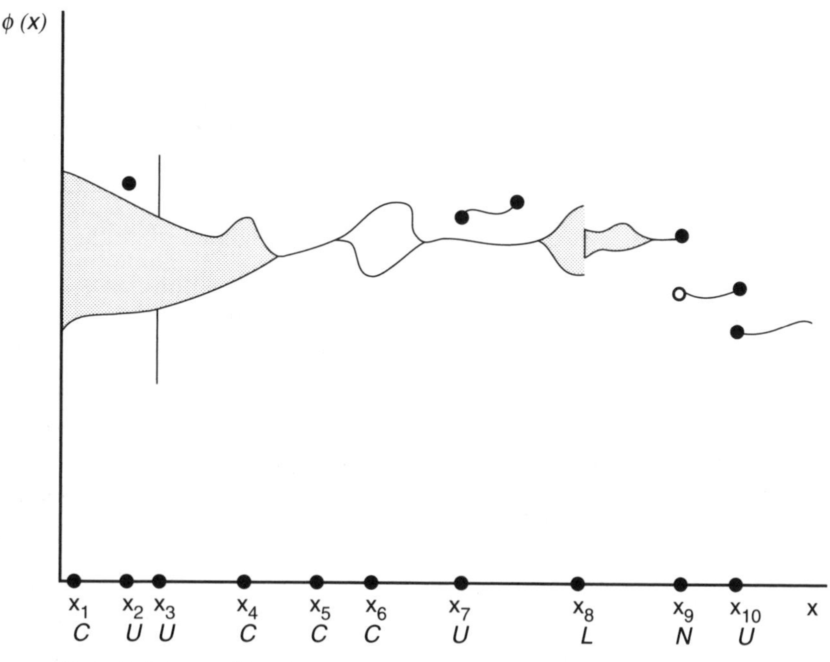

Note: Points at which the correspondence is just upper semicontinuous are labelled *U;* just lower semicontinuous, *L*; both, *C*; and neither, *N*.

metric space Set X is a metric space if it is associated with a metric that defines the distance between any two of its elements. Every compact metric space is complete.

monotonic A function $f(x)$ is monotonic and increasing if $x > y$ implies $f(x) > f(y)$. It is monotonic and decreasing if $x > y$ implies $f(x) < f(y)$.

one-to-one The mapping f: $X \rightarrow Y$ is one-to-one if every point in set X is mapped into one point in set Y.

onto The mapping f: $X \rightarrow Y$ is onto Y if every point in Y is mapped onto by some point in X.

open In the space $\mathbf{R}^n$, an open set is one that does not include all its boundary points. The set $\{x : 0 \leq x < 1000\}$ is open. In more general spaces, an open set is a member of a topology.

perpetuity A riskless security paying a constant amount each year in perpetuity, with the amount conventionally paid at the end of each year.

quasi-concave The continuous function f is quasi-concave if for $w \neq z$, $f(0.5w + 0.5z) > min[f(w), f(z)]$, or, equivalently, if the set $\{x \in X | f(x) > b\}$ is convex for any number b. Every concave function is quasi-concave, but not every quasi-concave function is concave.

risk See discussion of **Risk** below.

stochastic dominance See the discussion of **Risk** below.

strict The word "strict" is used in a variety of contexts to mean that a relationship does not hold with equality or is not arbitrarily close to being violated. If function f is concave and $f' > 0$, then $f'' \leq 0$, but if f is strictly concave, $f'' < 0$. The opposite of "strictly" is "weakly." The word "strong" is sometimes used as a synonym for "strict."

supermodular See discussion of **Supermodularity** below.

support The support of a probability distribution $F(x)$ is the closure of the set of values of x such that the density is positive. If each output between 0 and 20 has a positive probability density, and no other output does, the support of the output distribution is [0, 20].

topology Besides denoting a field of mathematics, a topology is a collection of subsets of a space called "open sets" that includes (1) the entire space and the empty set, (2) the intersection of any finite number of open sets, and (3) the union of any number of open sets. In a metric space, the metric "induces" a topology by defining an open set. Imposing a topology on a space is something like defining which elements are close to each other, which is easy to do for $\mathbf{R}^n$ but not for every space (e.g., spaces consisting of functions or of game trees).

upper semicontinuous correspondence The correspondence ϕ: $X \rightarrow Y$ is upper semicontinuous at point x_0 if

$$x_n \rightarrow x_0,\ y_n \in \phi(x_n),\ y_n \rightarrow y_0, \text{ implies } y_0 \in \phi(x_0), \tag{3}$$

which means that every sequence of points in $\phi(x)$ leads to a point also in $\phi(x)$. See figure B.2. An alternative definition, appropriate only if Y is compact, is that ϕ is upper semicontinuous if the set of points $\{x, \phi(x)\}$ is closed.

vector A vector is an ordered set of real numbers, a point in $\mathbf{R}^n$. The point $(2.5, 3, -4)$ is a vector in $\mathbf{R}^3$.

weak The word "weak" is used in a variety of contexts to mean that a relationship might hold with equality or be on a borderline. If f is concave and $f' > 0$, then $f'' \leq 0$, but to say that f is weakly concave, while technically adding nothing to the meaning, emphasizes that $f'' = 0$ under some or all parameters. The opposite of "weak" is "strict" or "strong."

Risk

We say that a player is **risk averse** if his utility function is strictly concave in money, which means that he has diminishing marginal utility of money. He is **risk neutral** if his utility function is linear in money. The qualifier, "in money," is used because utility may be a function of other variables too, such as effort.

We say that probability distribution F **dominates** distribution G in the sense of **first-order stochastic dominance** if the cumulative probability that the variable will take a value less than x is greater for G than for F, i.e. if for any x,

$$F(x) \leq G(x), \tag{4}$$

and if (4) is a strong inequality for at least one value of x. The distribution F dominates G in the sense of **second-order stochastic dominance** if the area under the cumulative distribution G up to $G(x)$ is greater than the area under F, i.e. if for any x,

$$\int_{-\infty}^{x} F(y)dy \leq \int_{-\infty}^{x} G(y)dy, \tag{5}$$

and if (5) is a strong inequality for some value of x. Equivalently, F dominates G if, for all functions U, limiting U to increasing functions for first-order dominance and increasing concave functions for second-order dominance,

$$\int_{-\infty}^{+\infty} U(x)dF(x) > \int_{-\infty}^{+\infty} U(x)dG(x). \tag{6}$$

If F is a first-order dominant gamble, it is preferred by all players; if F is a second-order dominant gamble, it is preferred by all risk-averse players. If F is first-order dominant, it is second-order dominant, but not vice versa. See Milgrom (1981b) and Copeland & Weston (1988) for further details.

Rothschild & Stiglitz (1970) shows how two gambles can be related in other ways equivalent to second-order dominance, the most important of which is the **mean-preserving spread**. Informally, a mean-preserving spread is a density function which transfers probability mass from the middle of a distribution to its tails. More formally, for discrete distributions placing sufficient probability on the four points a_1, a_2, a_3, and a_4,

> *A* **mean-preserving spread** *is a set of four locations* $a_1 < a_2 < a_3 < a_4$ *and four probabilities* $\gamma_1 \geq 0$, $\gamma_2 \leq 0$, $\gamma_3 \leq 0$, *and* $\gamma_4 \geq 0$ *such that* $-\gamma_1 = \gamma_2$, $\gamma_3 = -\gamma_4$, *and* $\sum_i \gamma_i a_i = 0$.

This definition can be extended to continuous distributions, and can be alternatively defined by taking probability mass from one point in the middle and moving it to the sides. See also Rasmusen & Petrakis (1992).

Milgrom (1981b) has used stochastic dominance to carefully define what we mean by **good news**. Let θ be a parameter about which the news is received in the form of message x or y, and let utility be increasing in θ. The message x is more favorable than y (is "good news") if for every possible nondegenerate prior for $F(\theta)$, the posterior $F(\theta|x)$ first-order dominates $F(\theta|y)$.

Probability Distributions

The definitive listing of probability distributions and their characteristics is the three volume series of Johnson & Kotz (1970). A few major distributions are listed here. A **probability distribution** is the same as a **cumulative density function** for a continuous distribution.

Exponential Distribution The exponential distribution, which has the set of nonnegative real numbers as its support, has the density function

$$f(x) = \frac{e^{-x/\lambda}}{\lambda}. \tag{7}$$

The cumulative density function is

$$F(x) = 1 - e^{-x/\lambda}. \tag{8}$$

Uniform Distribution A variable is uniformly distributed over support X if each point in X has equal probability. The density function for the support $X = [\alpha, \beta]$ is

$$f(x) = \begin{cases} 0 & x < \alpha \\ \dfrac{1}{\beta - \alpha} & \alpha \leq x \leq \beta \\ 0 & x > \beta \end{cases} \tag{9}$$

and the cumulative density function is

$$F(x) = \begin{cases} 0 & x < \alpha \\ \dfrac{x - \alpha}{\beta - \alpha} & \alpha \leq x \leq \beta \\ 1 & x > \beta \end{cases} \tag{10}$$

Normal Distribution The normal distribution is a two-parameter single-peaked distribution which has as its support the entire real line. The density function for mean μ and variance σ^2 is

$$f(x) = \frac{1}{\sqrt{2\pi\sigma^2}} e^{\frac{-(x-\mu)^2}{2\sigma^2}}. \tag{11}$$

The cumulative density function is the integral of this, often denoted $\Phi(x)$, which cannot be simplified analytically. Refer to a computer program such as *Mathematica* or to tables in statistics texts for its values.

Lognormal Distribution If $\log(x)$ has a normal distribution, it is said that x has a lognormal distribution. It is a skewed distribution which has the set of positive real numbers as its support, since the logarithm of a negative number is not defined.

Supermodularity[1]

Suppose that there are N players in a game, subscripted by m and n, and that player n has a strategy consisting of k_n elements, subscripted by i and j, so his strategy is the vector $x_n = (x_{n1}, \ldots, x_{nk_n})$. Let his strategy set be S_n and his payoff function be $\pi_n(x_n, x_{-n}; \tau)$, where τ represents a fixed parameter of the game. We say that the game is a **supermodular game** if the following four conditions are satisfied:

(A1) S_n is a complete lattice.

(A2) $\pi_n : S \rightarrow R \cup \{-\infty\}$ is order semi-continuous in x_n for fixed x_{-n}, and order continuous in x_{-n} for fixed x_n, and has a finite upper bound.

(A3) π_n is supermodular in x_n for fixed x_{-n}. For all strategy profiles x and y in S,

$$\pi_n(x) + \pi_n(y) \leq \pi_n(supremum\{x, y\}) + \pi_n(infimum\{x, y\}). \tag{12}$$

(A4) π_n has increasing differences in x_n and x_{-n}. For all $x_n \geq x_n'$, the difference $\pi_n(x_n, x_{-n}) - \pi_n(x_n', x_{-n})$ is nondecreasing in x_{-n}.

(A5) π_n has increasing differences in x_n and τ, for fixed x_{-n}. For all $x_n \geq x_n'$, the difference $\pi_n(x_n, x_{-n}, \tau) - \pi_n(x_n', x_{-n}, \tau)$ is nondecreasing in τ.

The conditions for **smooth supermodularity** are:

(A1′) The strategy set is an interval in $\boldsymbol{R}^{k_n}$:

$$S_n = [\,\underline{x_n}, \bar{x}_n]. \tag{13}$$

(A2′) π_n is twice continuously differentiable on S_n.

(A3′) (supermodularity) Increasing one component of player n's strategy does not decrease the net marginal benefit of any other component. For all n, and all i and j such that $1 \leq i < j \leq k_n$,

$$\frac{\partial^2 \pi_n}{\partial x_{ni} \partial x_{nj}} \geq 0. \tag{14}$$

(A4′) (increasing difference in own and other strategies) Increasing one component of n's strategy does not decrease the net marginal benefit of increasing any component of player m's strategy. For all $n \neq m$, and all i and j such that $1 \leq i \leq k_n$ and $1 \leq j \leq k_m$,

$$\frac{\partial^2 \pi_n}{\partial x_{ni} \partial x_{mj}} \geq 0. \tag{15}$$

[1] This discussion is drawn from Milgrom & Roberts (1990).

(A5′) (increasing differences in own strategies and parameters) Increasing parameter τ does not decrease the net marginal benefit to player n of any component of his strategy: for all n, and all i such that $1 \leq i \leq k_n$,

$$\frac{\partial^2 \pi_n}{\partial x_{ni} \partial \tau} \geq 0. \tag{16}$$

Theorem 13.1

If the game is supermodular, there exists a largest and smallest Nash equilibrium in pure strategies.

Theorem 13.1 is useful because it shows (a) existence of an equilibrium in pure strategies, and (b) if there are at least two equilibria (note that the largest and smallest equilibria might be the same strategy profile), two of them can be ranked in terms of the magnitudes of the components of each player's equilibrium strategy.

Theorem 13.2

If the game is supermodular and assumption (A5′) is satisfied, the largest and smallest equilibria are nondecreasing functions of the parameter τ.

Theorem B.1

If a game is supermodular, for each player n there is a largest and smallest serially undominated strategy, where both of these strategies are pure.

Theorem B.2

Let $\underline{x_n}$ denote the smallest element of player n's strategy set S_n in a supermodular game. Let y and z denote two equilibria, with $y \leq z$. Then:
(1) If $\pi_n(\underline{x_n}, x_{-n})$ is increasing in x_{-n}, then $\pi_n(y) \geq \pi_n(z)$.
(2) If $\pi_n(\underline{x_n}, x_{-n})$ is decreasing in x_{-n}, then $\pi_n(y) \leq \pi_n(z)$.
(3) If the condition in (1) holds for a subset N_1 of players and the condition in (2) holds for the remainder of the players, the big equilibrium y is the best equilibrium for players in N_1 and the worst for the remaining player, and the small equilibrium z is the worst equilibrium for players in N_1 and the best for the remaining players.

The theorems are taken from Milgrom & Roberts (1990). Theorem 13.1 is their corollary to Theorem 5. Theorem 13.2 is their Theorem 6 and corollary. Theorem B.1 is their Theorem 5, and B.2 is their Theorem 7. For more on supermodularity, see Milgrom & Roberts (1990) or pp. 489–97 of Fudenberg & Tirole (1991).

References and Name Index

Forthcoming and unpublished articles and books have been assigned the years (forthcoming) and (unpublished). The page numbers where a reference is mentioned in the text are listed after the reference. The date of first publication, which may differ from the date of the printing cited, follows the author's name. Not every publication is mentioned in the text; some are listed for reference only.

Abreu, Dilip, David Pearce & Ennio Stacchetti (1986) "Optimal Cartel Equilibria with Imperfect Monitoring" *Journal of Economic Theory*. June 1986. 39: 251–69. 106

Abreu, Dilip, David Pearce & Ennio Stacchetti (1990) "Toward a Theory of Discounted Repeated Games with Imperfect Monitoring" *Econometrica*. September 1990. 58: 1041–64. 138n

Akerlof, George (1970) "The Market for Lemons: Quality Uncertainty and the Market Mechanism" *Quarterly Journal of Economics*. August 1970. 84: 488–500. 224, 242n, 243n

Akerlof, George (1976) "The Economics of Caste and of the Rat Race and Other Woeful Tales" *Quarterly Journal of Economics*. November 1976. 90: 599–617. 242n

Akerlof, George (1980) "A Theory of Social Custom, of which Unemployment may be One Consequence" *Quarterly Journal of Economics*. June 1980. 94: 749–75. 241

Akerlof, George (1983) "Loyalty Filters" *American Economic Review*. March 1983. 73: 54–63. 210, 241

Akerlof, George & Janet Yellen, eds (1986) *Efficiency Wage Models of the Labor Market*. Cambridge: Cambridge University Press, 1986. 217n

Alchian, Armen. See Klein *et al.* (1978). 188n, 212

Alchian, Armen & Harold Demsetz (1972) "Production, Information Costs and Economic Organization" *American Economic Review*. December 1972. 62: 777–95. 188n

Antle, Rick & Abbie Smith (1986) "An Empirical Investigation of the Relative Performance Evaluation of Corporate Executives" *Journal of Accounting Research*. Spring 1986. 24: 1–39. 217n

Arrow, Kenneth (1985) "The Economics of Agency" In: *Principals and Agents:*

The Structure of Business, edited by John Pratt & Richard Zeckhauser. Boston: Harvard Business School Press, 1985, 37–51. 187n

Arrow, Kenneth & Debreu, Gerard (1954) "Existence of an Equilibrium for a Competitive Economy" *Econometrica*. July 1954. 22: 265–90.

Ashenfelter, Orley & David Bloom (1984) "Models of Arbitrator Behavior: Theory and Evidence" *American Economic Review*. March 1984. 74: 111–24.

Aumann, Robert (1964a) "Markets with a Continuum of Traders" *Econometrica*. January/April 1964. 32: 39–50. 87n

Aumann, Robert (1964b) "Mixed and Behavior Strategies in Infinite Extensive Games" *Annals of Mathematics Studies*, No. 52. Princeton: Princeton University Press, 1964. 627–50. 87n

Aumann, Robert (1974) "Subjectivity and Correlation in Randomized Strategies" *Journal of Mathematical Economics*. March 1974. 1: 67–96. 76

Aumann, Robert (1976) "Agreeing to Disagree" *Annals of Statistics*. November 1976. 4: 1236–9. 62n

Aumann, Robert (1981) "Survey of Repeated Games" In: *Essays in Game Theory and Mathematical Economics in Honor of Oscar Morgenstern*, edited by Robert Aumann. Mannheim: Bibliographisches Institut, 1981. 138n

Aumann, Robert (1987) "Correlated Equilibrium as an Expression of Bayesian Rationality" *Econometrica*. January 1987. 55: 1–18.

Aumann, Robert (1988) *Lectures on Game Theory* (Underground Classics in Economics). Boulder, Colorado: Westview Press, 1988. xii

Aumann, Robert & Sergiu Hart (1992) *Handbook of Game Theory with Economic Applications*. New York: North-Holland 1992. xii

Axelrod, Robert (1984) *The Evolution of Cooperation*. New York: Basic Books, 1984. 156–157, 159n

Axelrod, Robert & William Hamilton (1981) "The Evolution of Cooperation" *Science*. March 1981. 211: 1390–6. 116n

Ayres, Ian (1990) "Playing Games with the Law" *Stanford Law Review*. May 1990. 42: 1291–1317. 99

Ayres, Ian (1991) "Fair Driving: Gender and Race Discrimination in Retail Car Negotiations" *Harvard Law Review*. February 1991. 104: 817–872. 289

Bagchi, Arunabha (1984) *Stackelberg Differential Games in Economic Models*. Berlin: Springer-Verlag, 1984.

Bagehot, Walter (1971) "The Only Game in Town" *Financial Analysts Journal*. March/April 1971. 27: 12–22. 245n, 369

Baiman, Stanley (1982) "Agency Research in Managerial Accounting: A Survey" *Journal of Accounting Literature*. Spring 1982. 1: 154–213. 188n

Baird, Douglas, Gertner, Robert & Randal Picker (forthcoming) *Strategic Behavior and the Law: The Role of Game Theory and Information Economics in Legal Analysis*. xii, 99

Baker, George, Michael Jensen, & Kevin J. Murphy (1988) "Compensation and Incentives: Practice vs. Theory" *Journal of Finance*. July 1988. 43: 593–616. 209

Baldwin, B. & G. Meese (1979) "Social Behavior in Pigs Studied by Means of Operant Conditioning" *Animal Behavior*. August 1979. 27: 947–57. 31n

Bamberg, Gunter & Klaus Spremann, eds (1987) *Agency Theory, Information, and Incentives*. Berlin: Springer-Verlag, 1987. xi

Banks, Jeffrey (1990) *Signalling Games in Political Science*. Chur, Switzerland: Harwood Publishers, 1990. xii, 269n

Banks, Jeffrey & Joel Sobel (1987) "Equilibrium Selection in Signaling Games" *Econometrica*. May 1987. 55: 647–61. 158n

Baron, David (1989) "Design of Regulatory Mechanisms and Institutions" In: Schmalensee & Willig (1989). 217n

Baron, David & David Besanko (1984) "Regulation, Asymmetric Information, and Auditing" *Rand Journal of Economics*. Winter 1984. 15: 447–70. 88n

Baron, David & Robert Myerson (1982) "Regulating a Monopolist with Unknown Costs" *Econometrica*. July 1982. 50: 911–30. 374

Barzel, Yoram (1968) "Optimal Timing of Innovations" *Review of Economics and Statistics*. August 1968. 50: 348–55. 359n

Basil Blackwell (1985) *Guide for Authors*. Oxford: Basil Blackwell, 1985.

Basu, Kaushik (1993) *Lectures in Industrial Organization Theory*. Oxford: Blackwell Publishers, 1993. xii

Baumol, William & Stephen Goldfeld (1968) *Precursors in Mathematical Economics: An Anthology*. London: London School of Economics and Political Science, 1968. xii

Becker, Gary (1968) "Crime and Punishment: An Economic Approach" *Journal of Political Economy*. March/April 1968. 76: 169–217. 180

Becker, Gary & George Stigler (1974) "Law Enforcement, Malfeasance and Compensation of Enforcers" *Journal of Legal Studies*. January 1974. 3: 1–18. 206

Benoit, Jean-Pierre & Vijay Krishna (1985) "Finitely Repeated Games" *Econometrica*. July 1985. 17: 317–20. 137n, 395n

Bernanke, Benjamin (1983) "Nonmonetary Effects of the Financial Crisis in the Propagation of the Great Depression" *American Economic Review*. June 1983. 73: 257–76. 241

Bernheim, B. Douglas (1984a) "Rationalizable Strategic Behavior" *Econometrica*. July 1984. 52: 1007–28. 31n

Bernheim, B. Douglas (1984b) "Strategic Deterrence of Sequential Entry into an Industry" *Rand Journal of Economics*. Spring 1984. 15: 1–11. 216

Bernheim, B. Douglas, Bezalel Peleg, & Michael Whinston (1987) "Coalition-Proof Nash Equilibria I: Concepts" *Journal of Economic Theory*. June 1987. 42: 1–12. 106

Bernheim, B. Douglas & Michael Whinston (1987) "Coalition-Proof Nash Equilibria II: Applications" *Journal of Economic Theory*. June 1987. 42: 13–29. 106

Bertrand, Joseph (1883) "Recherches sur la théorie mathématique de la richesse" *Journal des savants*. September 1883. 48: 499–508. 314, 317

Besanko, David. See Baron & Besanko (1984). 88n

Bewley, Truman, ed (1987) *Advances in Economic Theory, Fifth World Congress*. Cambridge: Cambridge University Press, 1987.

Bierman, H. Scott & Fernandez, Luis (1993) *Game Theory with Economic Applications*. Reading, Massachusetts: Addison Wesley, 1993. xii, 381

Bikhchandani, Sushil (1988) "Reputations in Repeated Second Price Auctions" *Journal of Economic Theory*. October 1988. 46: 97–119. 302

Bikhchandani, Sushil, David Hirshleifer & Ivo Welch (1992) "A Theory of Fads, Fashion, Custom, and Cultural Change as Informational Cascades" *Journal of Political Economy*. October 1992. 100: 992–1026.

Binmore, Ken (1990) *Essays on the Foundations of Game Theory.* Oxford: Basil Blackwell Ltd., 1990. 158n
Binmore, Ken (1992) *Fun and Games: A Text on Game Theory.* Lexington, Mass.: D.C. Heath, 1992. 381, xii
Binmore, Ken & Partha Dasgupta, eds (1986) *Economic Organizations as Games.* Oxford: Basil Blackwell, 1986. xi
Binmore, Ken, Ariel Rubinstein, & Asher Wolinsky (1986) "The Nash Bargaining Solution in Economic Modelling" *Rand Journal of Economics.* Summer 1986. 17: 176–88. 289n
Blanchard, Olivier (1979) "Speculative Bubbles, Crashes, and Rational Expectations" *Economics Letters.* 1979. 3: 387–9. 139n
Bloom, David. See Ashenfelter & Bloom (1984).
Bond, Eric (1982) "A Direct Test of the 'Lemons' Model: The Market for Used Pickup Trucks" *American Economic Review.* September 1982. 72: 836–40. 243n
Border, Kim (1985) *Fixed Point Theorems with Applications to Economics and Game Theory.* Cambridge: Cambridge University Press, 1985. 429
Border, Kim & Joel Sobel (1987) "Samurai Accountant: A Theory of Auditing and Plunder" *Review of Economic Studies.* October 1987. 54: 525–40. 88n
Bowersock, G. (1985) "The Art of the Footnote" *American Scholar.* Winter 1983/84. 52: 54–62. 6n
Bowley, Arthur (1924) *The Mathematical Groundwork of Economics.* Oxford: Clarendon Press, 1924. 336n
Boyd, Robert & Jeffrey Lorberbaum (1987) "No Pure Strategy is Evolutionarily Stable in the Repeated Prisoner's Dilemma Game" *Nature.* May 1987. 327: 58–9. 138n
Boyd, Robert & Peter Richerson (1985) *Culture and the Evolutionary Process.* Chicago: University of Chicago Press, 1985. 116n
Brams, Steven (1980) *Biblical Games: A Strategic Analysis of Stories in the Old Testament.* Cambridge, Mass.: MIT Press, 1980.
Brams, Steven (1983) *Superior Beings: If They Exist, How Would We Know?* New York: Springer-Verlag, 1983. 30n
Brams, Steven & D. Marc Kilgour (1988) *Game Theory and National Security.* Oxford: Basil Blackwell, 1988. 32n
Brandenburger, Adam (1992) "Knowledge and Equilibrium in Games" *Journal of Economic Perspectives.* Fall 1992. 6: 83–102. 44, 62n
Bresnahan, Timothy & Peter Reiss (1991) "Entry and Competition in Concentrated Markets" *Journal of Political Economy.* October 1991. 99: 977–1009. 336n
Bulow, Jeremy (1982) "Durable-Goods Monopolists" *Journal of Political Economy.* April 1982. 90: 314–32. 331
Bulow, Jeremy, John Geanakoplos & Paul Klemperer (1985) "Multimarket Oligopoly: Strategic Substitutes and Complements" *Journal of Political Economy.* June 1985. 93: 488–511. 337n
Calfee, John. See Craswell & Calfee (1986).
Campbell, Richmond & Lanning Sowden (1985) *Paradoxes of Rationality and Cooperation: Prisoner's Dilemma and Newcomb's Problem.* Vancouver: University of British Columbia Press, 1985. 30n
Campbell, W. See Capen *et al.* (1971). 301, 302

Canzoneri, Matthew & Dale Henderson (1991) *Monetary Policy in Interdependent Economies*. Cambridge: MIT Press, 1991. 32n, 115n

Capen, E., R. Clapp, & W. Campbell (1971) "Competitive Bidding in High-Risk Situations" *Journal of Petroleum Technology*. June 1971. 23: 641–53. 301, 302

Cass, David & Karl Shell (1983) "Do Sunspots Matter?" *Journal of Political Economy*. April 1983. 91: 193–227. 76

Cassady, Ralph (1967) *Auctions and Auctioneering*. Berkeley: California University Press, 1967. 294, 298, 304n

Chammah, Albert. See Rapoport & Chammah (1965). 157

Chiang, Alpha (1984) *Fundamental Methods of Mathematical Economics*, 2nd ed. New York: McGraw Hill (1967). 429

Cho, In-Koo & David Kreps (1987) "Signaling Games and Stable Equilibria" *Quarterly Journal of Economics*. May 1987. 102: 179–221. 158n, 253

Clapp, R. See Capen *et al.* (1971). 301, 302

Clavell, James (1966) *Tai-pan*. New York: Dell Publishing Company, 1966. 306

Coase, Ronald (1960) "The Problem of Social Cost" *Journal of Law & Economics*. October 1960. 3: 1–44. 173

Coase, Ronald (1972) "Durability and Monopoly" *Journal of Law and Economics*. April 1972. 15: 143–9. 331, 337n

Cooter, Robert & Peter Rappoport (1984) "Were the Ordinalists Wrong about Welfare Economics?" *Journal of Economic Literature*. June 1984. 22: 507–30. 31n

Cooter, Robert & Daniel Rubinfeld (1989) "Economic Analysis of Legal Disputes and Their Resolution" *Journal of Economic Literature*. September 1989. 27: 1067–97. 101

Copeland, Thomas & Dan Galai (1983) "Information Effects of the Bid-Ask Spread" *Journal of Finance*. December 1983. 38: 1457–69. 245n

Copeland, Thomas & J. Fred Weston (1988) *Financial Theory and Corporate Policy*, 3rd ed. Reading, Mass.: Addison-Wesley, 1988 (1st ed. 1979). 109, 266, 437

Cornell, Bradford & Richard Roll (1981) "Strategies for Pairwise Competitions in Markets and Organizations" *Bell Journal of Economics*. Spring 1981. 12: 210–16. 116n

Cournot, Augustin (1838) *Recherches sur les principes mathematiques de la théorie des richesses*. Paris: M. Rivière & Cie., 1838. (Translated by N. T. Bacon as *Researches into the Mathematical Principles of the Theory of Wealth*. New York: A.M. Kelly, 1960.) 83, 309

Cox, D. & David Hinkley (1974) *Theoretical Statistics*. London: Chapman and Hall, 1974. 189n

Cramton, Peter (1984) "Bargaining with Incomplete Information: An Infinite Horizon Model with Two-Sided Uncertainty" *Review of Economic Studies*. October 1984. 51: 579–93. 290n

Craswell, Richard & John Calfee (1986) "Deterrence and Uncertain Legal Standards" *Journal of Law, Economics, and Organization*. Fall 1986. 2: 279–304.

Crawford, Robert. See Klein *et al.* (1978). 188n, 212

Crawford, Vincent (1982) "Compulsory Arbitration, Arbitral Risk and Negotiated Settlements: A Case Study in Bargaining under Imperfect Information" *Review of Economic Studies*. January 1982. 49: 69–82.

Crawford, Vincent & Hans Haller (1990) "Learning How to Cooperate: Optimal

Play in Repeated Coordination Games" *Econometrica.* May 1990. 58: 571–97. 32n

Crawford, Vincent & Joel Sobel (1982) "Strategic Information Transmission" *Econometrica.* November 1982. 50: 1431–52. 76

Crocker, Keith. See Masten & Crocker (1985). 188n

Dalkey, Norman (1953) "Equivalence of Information Patterns and Essentially Determinate Games" In: Kuhn & Tucker (1953) 217–43.

Dasgupta, Partha. See Binmore & Dasgupta (1986). xi

Dasgupta, Partha, Peter Hammond, & Eric Maskin (1979) "The Implementation of Social Choice Rules; Some General Rules on Incentive Compatibility" *Review of Economic Studies.* April 1979. 46: 185–216. 216n

Dasgupta, Partha & Eric Maskin (1986a) "The Existence of Equilibrium in Discontinuous Economic Games, I: Theory" *Review of Economic Studies.* January 1986. 53: 1–26. 235, 317, 324

Dasgupta, Partha & Eric Maskin (1986b) "The Existence of Equilibrium in Discontinuous Economic Games, II: Applications" *Review of Economic Studies.* January 1986. 53: 27–41. 235, 317, 324

Dasgupta, Partha & Joseph Stiglitz (1980) "Uncertainty, Industrial Structure, and the Speed of R&D" *Bell Journal of Economics.* Spring 1980. 11: 1–28. 359n

d'Aspremont, Claude, J. Gabszewicz, & Jacques Thisse (1979) "On Hotelling's 'Stability of Competition'" *Econometrica.* September 1979. 47: 1145–50. 323, 337n

David, Paul (1985) "CLIO and the Economics of QWERTY" *AEA Papers and Proceedings.* May 1985. 75: 332–7. 31n

Davis, Morton (1970) *Game Theory: A Nontechnical Introduction.* New York: Basic Books, 1970. xi

Davis, Philip & Reuben Hersh (1981) *The Mathematical Experience.* Boston: Birkhauser, 1981. 6n

Dawkins, Richard (1989) *The Selfish Gene*, 2nd ed. Oxford: Oxford University Press. 116n

Debreu, Gerard (1952) "A Social Equilibrium Existence Theorem" *Proceedings of the National Academy of Sciences.* 1952. 38: 886–93.

Debreu, Gerard (1959) *Theory of Value: An Axiomatic Analysis of Economic Equilibrium.* New Haven: Yale University Press, 1959. 429

Debreu, Gerard. See Arrow & Debreu (1954).

Debreu, Gerard & Herbert Scarf (1963) "A Limit Theorem on the Core of an Economy" *International Economic Review.* September 1963. 4: 235–46. 2

DeBrock, Lawrence & J. Smith (1983) "Joint Bidding, Information Pooling, and the Performance of Petroleum Lease Auctions" *Bell Journal of Economics.* Autumn 1983. 14: 395–404.

Demsetz, Harold. See Alchian & Demsetz (1972). 188n

Dewatripont, M. (1989) "Renegotiation and Information Revelation over Time in Optimal Labor Contracts" *Quarterly Journal of Economics.* August 1989. 104: 589–620. 203

Diamond, Douglas (1984) "Financial Intermediation and Delegated Monitoring" *Review of Economic Studies.* July 1984. 51: 393–414. 88n

Diamond, Douglas (1989) "Reputation Acquisition in Debt Markets" *Journal of Political Economy.* August 1989. 97: 828–62. 240, 362

Diamond, Peter & Michael Rothschild, eds (1978) *Uncertainty in Economics: Readings and Exercises.* New York: Academic Press, 1978. xi

DiMona, Joseph. See Haldeman & DiMona (1978). 116n

Dixit, Avinash (1979) "A Model of Duopoly Suggesting a Theory of Entry Barriers" *Bell Journal of Economics.* Spring 1979. 10: 20–32. 337n

Dixit, Avinash (1980) "The Role of Investment in Entry Deterrence" *Economic Journal.* March 1980. 90: 95–106. 355

Dixit, Avinash & Barry Nalebuff (1991) *Thinking Strategically: The Competitive Edge in Business, Politics, and Everyday Life.* New York: Norton, 1991. xii, 100

Dresher, Melvin, Albert Tucker, & Philip Wolfe, eds (1957) *Contributions to the Theory of Games,* Volume II. *Annals of Mathematics Studies,* No. 39. Princeton: Princeton University Press, 1957.

Dybvig, Philip & Chester Spatt (unpublished) "Does it Pay to Maintain a Reputation? Consumer Information and Product Quality" Yale Working Paper, May 1983. 132n

Eaton, C. & Richard Lipsey (1975) "The Principle of Minimum Differentiation Reconsidered: Some New Developments in the Theory of Spatial Competition" *Review of Economic Studies.* January 1975. 42: 27–49. 324

Eatwell, John, Murray Milgate & Peter Newman (1989) *The New Palgrave: Game Theory.* New York: W.W. Norton & Co., 1989. xii

Edgeworth, Francis (1881) *Mathematical Psychics.* London: Kegan Paul, 1881.

Edgeworth, Francis (1897) "La teoria pura del monopolio" *Giornale degli economisti.* 40: 13–31. Translated in Edgeworth, Francis, *Papers Relating to Political Economy,* Vol. I. London: Macmillan, 1925, 111–42. 317

Engers, Maxim (1987) "Signalling with Many Signals" *Econometrica.* May 1987. 55: 633–74. 265

Engers, Maxim & Luis Fernandez (1987) "Market Equilibrium with Hidden Knowledge and Self-Selection" *Econometrica.* March 1987. 55: 425–39. 236, 244n

Fama, Eugene (1980) "Banking in the Theory of Finance" *Journal of Monetary Economics.* January 1980. 6: 39–57. 188n

Farrell, Joseph (unpublished) "Monopoly Slack and Competitive Rigor: A Simple Model" MIT mimeo. February 1983. 207

Farrell, Joseph (1987) "Cheap Talk, Coordination, and Entry" *Rand Journal of Economics.* Spring 1987. 18: 34–9. 76

Farrell, Joseph & Garth Saloner (1985) "Standardization, Compatibility, and Innovation" *Rand Journal of Economics.* Spring 1985. 16: 70–83. 32n

Farrell, Joseph & Carl Shapiro (1988) "Dynamic Competition with Switching Costs" *Rand Journal of Economics.* Spring 1988. 19: 123–37. 135

Fernandez, Luis. See Engers & Fernandez (1987). 236

Fernandez, Luis & Eric Rasmusen (unpublished) "Perfectly Contestable Monopoly and Adverse Selection" UCLA AGSM Business Economics Working Paper #89–3, March 1989. 236

Fisher, Franklin (1989) "Games Economists Play: A Noncooperative View" *RAND Journal of Economics.* Spring 1989. 20: 113–24. 3, 86n

Fishman, Michael (1988) "A Theory of Pre-emptive Takeover Bidding" *Rand Journal of Economics.* Spring 1988. 19: 88–101.

Forgo, F. See Szep & Forgo (1985). xi

Fowler, H. (1965) *A Dictionary of Modern English Usage*, 2nd ed. New York: Oxford University Press, 1965. 6n

Fowler, H. & F. Fowler (1931) *The King's English*, 3rd ed. Oxford: Clarendon Press, 1949. 6n

Frank, Robert (1988) *Passions within Reason: The Strategic Role of the Emotions*. New York: Norton, 1988. 279

Franks, Julian, Robert Harris, & Colin Mayer (1988) "Means of Payment in Takeovers: Results for the U.K. and U.S." In: *Corporate Takeovers: Causes and Consequences*. Alan Auerbach, ed. Chicago: University of Chicago Press, 1988.

Freixas, Xavier, Roger Guesnerie, & Jean Tirole (1985) "Planning under Incomplete Information and the Ratchet Effect" *Review of Economic Studies*. April 1985. 52: 173–91. 217n

Friedman, Daniel (1991) "Evolutionary Games in Economics" *Econometrica*. May 1991. 59: 637–66. 114

Friedman, James (1990) *Game Theory with Applications to Economics*, 2nd ed. New York: Oxford University Press, 1990 (1st ed., 1986). xii

Friedman, Milton (1953) *Essays in Positive Economics*. Chicago: University of Chicago Press, 1953. 6n

Fudenberg, Drew & David Levine (1983) "Subgame-Perfect Equilibria of Finite- and Infinite-Horizon Games" *Journal of Economic Theory*. December 1983. 31: 251–68.

Fudenberg, Drew & David Levine (1986) "Limit Games and Limit Equilibria" *Journal of Economic Theory*. April 1986. 38: 261–79. 88n, 137n

Fudenberg, Drew & Eric Maskin (1986) "The Folk Theorem in Repeated Games with Discounting or with Incomplete Information" *Econometrica*. May 1986. 54: 533–54. 138n, 156

Fudenberg, Drew & Jean Tirole (1983) "Sequential Bargaining with Incomplete Information" *Review of Economic Studies*. April 1983. 50: 221–47. 284, 290n

Fudenberg, Drew & Jean Tirole (1986a) *Dynamic Models of Oligopoly*. Chur, Switzerland: Harwood Academic Publishers, 1986. xii, 360n

Fudenberg, Drew & Jean Tirole (1986b) "A Theory of Exit in Duopoly" *Econometrica*. July 1986. 54: 943–60. 87n

Fudenberg, Drew & Jean Tirole (1986c) "A Signal-Jamming Theory of Predation" *Rand Journal of Economics*. Autumn 1986. 17: 366–76. 349

Fudenberg, Drew & Jean Tirole (1990) "Moral Hazard and Renegotiation in Agency Contracts" *Econometrica*. November 1990. 58: 1279–1320.

Fudenberg, Drew & Jean Tirole (1991a) *Game Theory*. Cambridge, Massachusetts: MIT Press, 1991. xii, 62n, 75, 106, 128n, 138n, 216n, 381, 429, 440

Fudenberg, Drew & Jean Tirole (1991b) "Perfect Bayesian Equilibrium and Sequential Equilibrium" *Journal of Economic Theory*. 53: 236–60. 158n

Gabszewicz, J. See d'Aspremont *et al.* (1979). 323, 337n

Galai, Dan. See Copeland & Galai (1983). 245n

Galbraith, John Kenneth (1954) *The Great Crash, 1929*. Boston: Houghton Mifflin, 1954.

Gal-Or, Esther (1985) "First Mover and Second Mover Advantages" *International Economic Review*. October 1985. 26: 649–53. 337n

Gaskins, Darius (1974) "Alcoa Revisited: The Welfare Implications of a Second-Hand Market" *Journal of Economic Theory*. March 1974. 7: 254–71. 337n

Gaudet, Gerard & Stephen Salant (1991) "Uniqueness of Cournot Equilibrium: New Results from Old Methods" *Review of Economic Studies*, April 1991, 58: 399–404. 88n

Gaver, Kenneth & Jerold Zimmerman (1977) "An Analysis of Competitive Bidding on BART Contracts" *Journal of Business*. July 1977. 50: 279–95. 218n

Geanakoplos, John. See Bulow *et al.* (1985). 337n

Geanakoplos, John (1992) "Common Knowledge" *Journal of Economic Perspectives*. Fall 1992. 6: 53–82. 62n

Geanakoplos, John & Heraklis Polemarchakis (1982) "We Can't Disagree Forever" *Journal of Economic Theory*. October 1982. 28: 192–200. 63n

Gelman, Judith R. & Steven C. Salop (1983) "Judo Economics: Capacity Limitation and Coupon Competition" *Bell Journal of Economics*. Autumn 1983. 14: 315–25. 359

Gertner, Robert. See Baird *et al.* (forthcoming). xii, 99

Ghemawat, Pankaj & Barry Nalebuff (1985) "Exit" *Rand Journal of Economics*. Summer 1985. 16: 185–94. 87n

Gibbard, Allan (1973) "Manipulation of Voting Schemes: A General Result" *Econometrica*. July 1973. 41: 587–601. 216n

Gibbons, Robert (1992) *Game Theory for Applied Economists*. Princeton: Princeton University Press, 1992. xii, 115n

Gilbert, Richard & David Newberry (1982) "Preemptive Patenting and the Persistence of Monopoly" *American Economic Review*. June 1982. 72: 514–26. 345

Gillies, Donald (1953) "Locations of Solutions" In: *Report of an Informal Conference on the Theory of n-Person Games*. Princeton Mathematics mimeo. 1953. 11–12. 1

Gjesdal, Froystein (1982) "Information and Incentives: The Agency Information Problem" *Review of Economic Studies*. July 1982. 49: 373–90. 189n

Glicksberg, Irving (1952) "A Further Generalization of the Kakutani Fixed Point Theorem with Application to Nash Equilibrium Points" *Proceedings of the American Mathematical Society*. February 1952. 3: 170–74.

Glosten, Lawrence & Paul Milgrom (1985) "Bid, Ask, and Transaction Prices in a Specialist Model with Heterogeneously Informed Traders" *Journal of Financial Economics*, 14: 71–100. 369

Gonik, Jacob (1978) "Tie Salesmen's Bonuses to their Forecasts" *Harvard Business Review*. May/June 1978. 56: 116–23. 203

Gordon, David. See Rapaport, Guyer & Gordon (1976). 30n

Green, Edward (1984) "Continuum and Finite-Player Noncooperative Models of Competition" *Econometrica*. July 1984. 52: 975–93.

Green, Jerry & Jean-Jacques Laffont (1979) *Incentives in Public Decision-Making*. Amsterdam: North-Holland, 1979. xi

Greenhut, Melvin & Hiroshi Ohta (1975) *Theory of Spatial Pricing and Market Areas*. Durham, N.C.: Duke University Press, 1975. 337n

Grinblatt, Mark & Chuan-Yank Hwang (1989) "Signalling and the Pricing of New Issues" *The Journal of Finance*. June 1989. 44: 393–420. 40, 266

Grossman, Sanford & Oliver Hart (1980) "Takeover Bids, the Free-Rider Problem, and the Theory of the Corporation" *Bell Journal of Economics*. Spring 1980. 11: 42–64. 216, 364

Grossman, Sanford & Oliver Hart (1983) "An Analysis of the Principal Agent Problem" *Econometrica*. January 1983. 51: 7–45. 176, 189n, 190n

Grossman, Gene & Michael Katz (1983) "Plea Bargaining and Social Welfare" *American Economic Review*. September 1983. 73: 749–57. 269n

Grossman, Sanford & Motty Perry (1986) "Perfect Sequential Equilibria" *Journal of Economic Theory*. June 1986. 39: 97–119. 158n

Groves, Theodore (1973) "Incentives in Teams" *Econometrica*. July 1973. 41: 617–31. 245n, 297

Guasch, J. Luis & Andrew Weiss (1980) "Wages as Sorting Mechanisms in Competitive Markets with Asymmetric Information: A Theory of Testing" *Review of Economic Studies*. July 1980. 47: 653–64. 245n

Guesnerie, Roger. See Freixas *et al.* (1985). 217n

Gul, Faruk (1989) "Bargaining Foundations of Shaply Value" *Econometrica*. January 1989. 57: 81–96. 289n

Guth, Wener, Rold Schmittberger, & Bernd Schwarze (1982) "An Experimental Analysis of Ultimatum Bargaining" *Journal of Economic Behavior and Organization*. December 1982. 3: 367–88. 104

Guyer, Melvin. See Rapoport & Guyer (1966) 30n and Rapaport, Guyer & Gordon (1976). 30n

Guyer, Melvin & Henry Hamburger (1968) "A Note on 'A Taxonomy of 2×2 Games'" *General Systems*. 1968. 13: 205–8. 30n

Haldeman, H. R. & Joseph DiMona (1978) *The Ends of Power*. New York: Times Books, 1978. 116n

Haller, Hans (1986) "Noncooperative Bargaining of $N \geq 3$ Players" *Economic Letters*. 1986. 22: 11–13. 290n

Haller, Hans. See Crawford & Haller (1990). 32n

Halmos, Paul (1970) "How to Write Mathematics" *L'enseignement mathématique*. May/June 1970. 16: 123–52. 6n

Haltiwanger, John & Michael Waldman (unpublished) "Responders versus Nonresponders: A New Perspective of Heterogeneity" UCLA Economics Working Paper #436. February 1987. 336n

Hamburger, Henry. See Guyer & Hamburger (1968). 30n

Hamilton, William. See Axelrod & Hamilton (1981). 116n

Hammond, Peter. See Dasgupta *et al.* (1979). 216n

Han Fei Tzu, *Basic Writings* (Translated by Burton Watson). New York: Columbia University Press, 1964. 210

Harrington, Joseph (1987) "Collusion in Multiproduct Oligopoly Games under a Finite Horizon" *International Economic Review*. February 1987. 28: 1–14. 137n

Harris, Milton (1987) *Dynamic Economic Analysis*. Oxford, Oxford University Press, 1987. xii

Harris, Milton & Bengt Holmstrom (1982) "A Theory of Wage Dynamics" *Review of Economic Studies*. July 1982. 49: 315–34. 48

Harris, Milton & Arthur Raviv (1992) "Financial Contracting Theory" In: *Advances in Economic Theory: Sixth World Congress*. Jean-Jacques Laffont, ed. Cambridge: Cambridge University Press. 188n

Harris, Robert. See Franks *et al.* (1988).

Harsanyi, John (1967) "Games with Incomplete Information Played by 'Bayesian' Players, I: The Basic Model" *Management Science.* November 1967. 14: 159–82. 2, 145
Harsanyi, John (1968a) "Games with Incomplete Information Played by 'Bayesian' Players, II: Bayesian Equilibrium Points" *Management Science.* January 1968. 14: 320–34. 49, 63n
Harsanyi, John (1968b) "Games with Incomplete Information Played by 'Bayesian' Players, III: The Basic Probability Distribution of the Game" *Management Science.* March 1968. 14: 486–502. 49, 63n
Harsanyi, John (1973) "Games with Randomly Disturbed Payoffs: A New Rationale for Mixed Strategy Equilibrium Points" *International Journal of Game Theory.* 1973. 2: 1–23. 49, 63n
Harsanyi, John (1977) *Rational Behavior and Bargaining Equilibrium in Games and Social Situations.* New York: Cambridge University Press, 1977. xi, 289n
Harsanyi, John & Reinhard Selten (1988) *A General Theory of Equilibrium Selection in Games.* Cambridge, Mass.: MIT Press, 1988. 27
Hart, Oliver. See Grossman & Hart (1980) 216, 364 (1983), 176, 189n, 190n
Hart, Oliver & Bengt Holmstrom (1987) "The Theory of Contracts" In Bewley (1987). 187n, 188n, 189n, 190n
Hart, Sergiu. See Aumann & Hart (1992). xii
Hausman, J. & James Poterba (1987) "Household Behavior and the Tax Reform Act of 1986" *Journal of Economic Perspectives.* 1: 101–19.
Haywood, O. (1954). "Military Decisions and Game Theory" *Journal of the Operations Research Society of America.* November 1954. 2: 365–85. 31n
Henderson, Dale. See Canzoneri & Henderson (1991). 32n, 115n
Henry, O. (1945) *Best Stories of O. Henry.* Garden City, NY: The Sun Dial Press, 1945. 32n
Herodotus *The Persian Wars.* George Rawlinson, trans.. New York: Modern Library, 1947. 30n
Hersh, Reuben. See Davis & Hersh (1981). 6n
Hess, James (1983) *The Economics of Organization.* Amsterdam: North-Holland, 1983. xi, 188n
Hines, W. (1987) "Evolutionary Stable Strategies: A Review of Basic Theory" *Theoretical Population Biology.* April 1987. 31: 195–272. 116n
Hinkley, David. See Cox & Hinkley (1974). 189n
Hirshleifer, David. See Png & Hirshleifer (1987). 338n
Hirshleifer, David & Eric Rasmusen (1989) "Cooperation in a Repeated Prisoner's Dilemma with Ostracism" *Journal of Economic Behavior and Organization.* August 1989. 12: 87–106. 137n
Hirshleifer, David & Sheridan Titman (1990) "Share Tendering Strategies and the Success of Hostile Takeover Bids" *Journal of Political Economy.* April 1990. 98: 295–324. 158n
Hirshleifer, Jack (1982) "Evolutionary Models in Economics and Law: Cooperation versus Conflict Strategies" *Research in Law and Economics.* 1982. 4: 1–60. 29n, 31n, 116n
Hirshleifer, Jack (1987) "On the Emotions as Guarantors of Threats and Promises" In: *The Latest on the Best: Essays on Evolution and Optimality,* edited by John Dupre. Cambridge, Mass.: MIT Press, 1987. 139n
Hirshleifer, Jack & Juan Martinez-Coll (1988) "What Strategies can Support the

Evolutionary Emergence of Cooperation?" *Journal of Conflict Resolution*. June 1988. 32: 367–98. 138n

Hirshleifer, Jack & Eric Rasmusen (1992) "Are Equilibrium Strategies Unaffected By Incentives?" *Journal of Theoretical Politics*. July 1992. 4: 343–57. 88n

Hirshleifer, Jack & John Riley (unpublished) "Elements of the Theory of Auctions and Contests" UCLA Economics Working Paper #118b. August 1978.

Hirshleifer, Jack & John Riley (1979) "The Analytics of Uncertainty and Information: An Expository Survey" *Journal of Economic Literature*. December 1979. 17: 1375–1421.

Hirshleifer, Jack & John Riley (1992) *The Economics of Uncertainty and Information*. Cambridge: Cambridge University Press, 1992. 44, 381

Hoffman, Elizabeth & Matthew Spitzer (1985) "Entitlements, Rights and Fairness: An Experimental Examination of Subjects' Concepts of Distributive Justice" *Journal of Legal Studies*. 14: 269–97. 279

Hofstadter, Douglas (1983) "Computer Tournaments of the Prisoner's Dilemma Suggest how Cooperation Evolves" *Scientific American*. May 1983. 248: 16–26. 159n

Holmes, Oliver Wendell (1881) *The Common Law*. Boston: Little, Brown and Co., 1923. 211

Holmstrom, Bengt (1979) "Moral Hazard and Observability" *Bell Journal of Economics*. Spring 1979. 10: 74–91. 189n

Holmstrom, Bengt (1982) "Moral Hazard in Teams" *Bell Journal of Economics*. Autumn 1982. 13: 324–40. 213, 216, 218n

Holmstrom, Bengt. See Harris & Holmstrom (1982) 48 and Hart & Holmstrom (1987). 187n, 188n, 189n, 190n

Holmstrom, Bengt & Roger Myerson (1983) "Efficient and Durable Decision Rules with Incomplete Information" *Econometrica*. November 1983. 51: 1799–1819. 190n

Holt, Charles & David Scheffman (1987) "Facilitating Practices: The Effects of Advance Notice and Best-Price Policies" *Rand Journal of Economics*. Summer 1987. 18: 187–97. 338n

Hotelling, Harold (1929) "Stability in Competition" *Economic Journal*. March 1929. 39: 41–57. 320, 323

Hughes, Patricia (1986) "Signalling by Direct Disclosure Under Asymmetric Information" *Journal of Accounting and Economics*. June 1986. 8: 119–42. 265

Intriligator, Michael (1971) *Mathematical Optimization and Economic Theory*. Englewood Cliffs, N.J.: Prentice-Hall, 1971. 429

Isoda, Kazuo. See Nikaido & Isoda (1955).

Jacquemin, Alex (1985) *The New Industrial Organization*. Cambridge, Mass.: MIT Press, 1987. Translated by Fatemeh Mehta from *Sélection et pouvoir dans la nouvelle économie industrielle*. Louvain-la-Neuve: Cabay Libraire-Editeur, 1985. xii, 116n, 336n, 360n

Jarrell, Gregg & Sam Peltzman (1985) "The Impact of Product Recalls on the Wealth of Sellers" *Journal of Political Economy*. June 1985. 93: 512–36. 139n

Jensen, Michael & William Meckling (1976) "Theory of the Firm: Managerial Behavior, Agency Costs and Ownership Structure" *Journal of Financial Economics*. October 1976. 3: 305–60.

Jensen, Michael. See Baker *et al.* (1988). 209

Johnson, Norman & Samuel Kotz (1970) *Distributions in Statistics.* 3 vols, New York: John Wiley and Sons, 1970. 437

Joskow, Paul (1985) "Vertical Integration and Longterm Contracts: the Case of Coal-Burning Electric Generating Plants" *Journal of Law, Economics and Organization.* Spring 1985. 1: 33–80. 188n

Joskow, Paul (1987) "Contract Duration and Relationship-Specific Investments: Empirical Evidence from Coal Markets" *American Economic Review.* March 1987. 77: 168–85. 188n

Kahneman, Daniel, Paul Slovic, & Amos Tversky, eds (1982) *Judgement Under Uncertainty: Heuristics and Biases.* Cambridge: Cambridge University Press, 1982. 62n

Kakutani, Shizuo (1941) "A Generalization of Brouwer's Fixed Point Theorem" *Duke Mathematical Journal.* September 1941. 8: 457–9.

Kalai, Ehud, Dov Samet, & William Stanford (1988) "Note on Reactive Equilibria in the Discounted Prisoner's Dilemma and Associated Games" *International Journal of Game Theory.* 17: 177–86. 397n

Kamien, Morton & Nancy Schwartz (1982) *Market Structure and Innovation.* Cambridge: Cambridge University Press, 1982. 342

Kamien, Morton & Nancy Schwartz (1991) *Dynamic Optimization: The Calculus of Variations and Optimal Control in Economics and Management*, 2nd ed. New York: North Holland, 1991 (1st ed., 1981). 429

Karlin, Samuel (1959) *Mathematical Methods and Theory in Games, Programming and Economics.* Reading, Mass.: Addison-Wesley, 1959. 75

Katz, Lawrence (1986) "Efficiency Wage Theory: A Partial Evaluation" In: *NBER Macroeconomics Annual 1986*, edited by Stanley Fischer. Cambridge, Mass.: MIT Press, 1986. 217n

Katz, Michael & Carl Shapiro (1985) "Network Externalities, Competition, and Compatibility" *American Economic Review.* June 1985. 75: 424–40. 32n

Katz, Michael. See Grossman & Katz (1983). 269n

Katz, Michael. See Moskowitz *et al.* (1980).

Kennan, John & Robert Wilson (1993) "Bargaining with Private Information" *Journal of Economic Literature.* March 1993. 31: 45–104. 101, 290n, 338n

Kennedy, Peter (1979) *A Guide to Econometrics.* 3rd ed., Cambridge, Mass.: 1992. (1st ed., 1979).

Keynes, John Maynard (1933) *Essays in Biography.* New York: Harcourt, Brace and Company, 1933.

Keynes, John Maynard (1936) *The General Theory of Employment, Interest and Money.* London: Macmillan, 1947. 32n

Kierkegaard, Søren (1938) *The Journals of Søren Kierkegaard.* Oxford: Oxford University Press, 1938. 122

Kihlstrom, Richard & Michael Riordan (1984) "Advertising as a Signal" *Journal of Political Economy.* June 1984. 92: 427–50. 269n

Kilgour, D. Marc. See Brams & Kilgour (1988). 32n

Kindleberger, Charles (1983) "Standards as Public, Collective and Private Goods" *Kyklos.* 1983. 36: 377–96. 32n

Klein, Benjamin, Robert Crawford, & Armen Alchian (1978) "Vertical Integration, Appropriable Rents, and the Competitive Contracting Process" *Journal of Law and Economics.* October 1978. 21: 297–326. 188n, 212

Klein, Benjamin & Keith Leffler (1981) "The Role of Market Forces in Assuring

Contractual Performance" *Journal of Political Economy.* August 1981. 89: 615–41. 132, 140n, 217n

Klein, Benjamin & Lester Saft (1985) "The Law and Economics of Franchise Tying Contracts" *Journal of Law and Economics.* May 1985. 28: 345–61. 190n

Klemperer, Paul (1987) "The Competitiveness of Markets with Switching Costs" *Rand Journal of Economics.* Spring 1987. 18: 138–50. 135

Klemperer, Paul. See Bulow *et al.* (1985). 337n

Kohlberg, Elon & Jean-Francois Mertens (1986) "On the Strategic Stability of Equilibria" *Econometrica.* September 1986. 54: 1003–7. 88n, 158n

Kreps, David. See Cho & Kreps (1987). 158n

Kreps, David (1990a) *A Course in Microeconomic Theory.* Princeton: Princeton University Press, 1990. xii, 62n

Kreps, David (1990b) *Game Theory and Economic Modeling.* New York: Oxford University Press, 1990. xii, 158n

Kreps, David, Paul Milgrom, John Roberts, & Robert Wilson (1982) "Rational Cooperation in the Finitely Repeated Prisoners' Dilemma" *Journal of Economic Theory.* August 1982. 27: 245–52. 2, 155

Kreps, David & Jose Scheinkman (1983) "Quantity Precommitment and Bertrand Competition Yield Cournot Outcomes" *Bell Journal of Economics.* Autumn 1983. 14: 326–37. 336n

Kreps, David & A. Michael Spence (1984) "Modelling the Role of History in Industrial Organization and Competition" In: *Issues in Contemporary Microeconomics and Welfare*, edited by George Feiwel. London: Macmillan, 1984. 3, 6n

Kreps, David & Robert Wilson (1982a) "Reputation and Imperfect Information" *Journal of Economic Theory.* August 1982. 27: 253–79. 352, 360n

Kreps, David & Robert Wilson (1982b) "Sequential Equilibria" *Econometrica.* July 1982. 50: 863–94. 2, 145, 146

Krishna, Vijay. See Benoit & Krishna (1985). 137n

Krouse, Clement (1990) *Theory of Industrial Economics.* Oxford: Blackwell Publishers, 1990. xii, 360n

Kuhn, Harold (1953) "Extensive Games and the Problem of Information" In: Kuhn & Tucker (1953). 87n

Kuhn, Harold & Albert Tucker, eds (1950) *Contributions to the Theory of Games*, Volume I. *Annals of Mathematics Studies*, No. 24. Princeton: Princeton University Press, 1950.

Kuhn, Harold & Albert Tucker, eds (1953) *Contributions to the Theory of Games*, Volume II. *Annals of Mathematics Studies*, No. 28. Princeton: Princeton University Press, 1953.

Kydland, Finn & Edward Prescott (1977) "Rules Rather than Discretion: The Inconsistency of Optimal Plans" *Journal of Political Economy.* June 1977. 85: 473–91. 115n

Kyle, Albert (1985) "Continuous Auctions and Insider Trading" *Econometrica.* 53: 1315–36. 369, 371

Lachman, Judith (1984) "Knowing and Showing Economics and Law" *Yale Law Journal.* July 1984. 93: 1587–624.

Laffont, Jean-Jacques & Michel Moreaux, eds, translated by Francois Laisney

(1991) *Dynamics, Incomplete Information, and Industrial Economics.* Oxford: Blackwell Publishing, 1991. xiii

Laffont, Jean-Jacques & Jean Tirole (1986) "Using Cost Observation to Regulate Firms" *Journal of Political Economy.* June 1986. 94: 614–41.

Laffont, Jean-Jacques & Jean Tirole (1993) *A Theory of Incentives in Procurement and Regulation.* Cambridge: MIT Press, 1993. xiii, 106, 216n, 360n, 374, 379

Lakatos, Imre (1976) *Proofs and Refutations: The Logic of Mathematical Discovery.* Cambridge: Cambridge University Press, 1976. 3, 6n

Lane, W. (1980) "Product Differentiation in a Market with Endogenous Sequential Entry" *Bell Journal of Economics.* Spring 1980. 11: 237–60. 337n

Layard, Richard & George Psacharopoulos (1974) "The Screening Hypothesis and the Returns to Education" *Journal of Political Economy.* September/October 1974. 82: 985–98. 258, 269n

Lazear, Edward & Sherwin Rosen (1981) "Rank-Order Tournaments as Optimum Labor Contracts" *Journal of Political Economy.* October 1981. 89: 841–64. 217n

Leffler, Keith. See Klein & Leffler (1981). 132, 140n, 217n

Leibenstein, Harvey (1950) "Bandwagon, Snob and Veblen Effects in the Theory of Consumers' Demand" *Quarterly Journal of Economics.* May 1950. 64: 183–207. 243n

Leland, Hayne & David Pyle (1977) "Informational Asymmetries, Financial Structure, and Financial Intermediation" *Journal of Finance.* May 1977. 32: 371–87. 266

Levering, Robert. See Moskowitz *et al.* (1980).

Levine, David. See Fudenberg & Levine (1983, 1986) 88n, 137n.

Levitan, Richard & Martin Shubik (1972) "Price Duopoly and Capacity Constraints" *International Economic Review.* February 1972. 13: 111–22. 317

Levmore, Saul (1982) "Self-Assessed Valuation for Tort and Other Law" *Virginia Law Review.* April 1982. 68: 771–861. 216n

Lewis, D. (1969) *Convention: A Philosophical Study.* Cambridge, Mass.: Harvard University Press. 62n

Liebowitz, S. & Stephen Margolis (1990) "The Fable of the Keys" *Journal of Political Economy.* April 1990. 33: 1–25. 32n

Lippman, Steven & Richard Rumelt (1982) "Uncertain Imitability: An Analysis of Interfirm Differences in Efficiency under Competition" *Bell Journal of Economics.* Autumn 1982. 13: 418–38.

Lipsey, Richard. See Eaton & Lipsey (1975). 324

Locke, E. (1949) "The Finan-Seer" *Astounding Science Fiction.* October 1949. 44: 132–40.

Lorberbaum, Jeffrey. See Boyd & Lorberbaum (1987). 138n

Loury, Glenn (1979) "Market Structure and Innovation" *Quarterly Journal of Economics.* August 1979. 93: 395–410. 359n

Luce, R. Duncan & Howard Raiffa (1957) *Games and Decisions: Introduction and Critical Survey.* New York: Wiley, 1957. x, xi, 31n, 128, 137n, 289n

Luce, Duncan & Albert Tucker, eds (1959) *Contributions to the Theory of Games,* Volume IV. *Annals of Mathematics Studies,* No. 40. Princeton: Princeton University Press, 1959.

Macaulay, Stewart (1963) "Non-Contractual Relations in Business" *American Sociological Review*. February 1963. 28: 55–70. 139n
Macey, Jonathan & Fred McChesney (1985) "A Theoretical Analysis of Corporate Greenmail" *Yale Law Journal*. November 1985. 95: 13–61. 116n, 365
Machina, Mark. (1982) "'Expected Utility' Analysis without the Independence Axiom" *Econometrica*. March 1982. 50: 277–323. 62n
Macrae, Norman (1992) *John von Neumann*. New York: Random House, 1992. 15
Maddala, G. (1977) *Econometrics*. New York: McGraw-Hill Inc., 1977.
Margolis, Stephen. See Liebowitz & Margolis (1990). 32n
Marschak, Jacob & Roy Radner (1972) *Economic Theory of Teams*. New Haven: Yale University Press, 1972. 218n
Marshall, Alfred (1890) *Principles of Economics*. 9th (variorum) ed. London, Macmillan, 1961 (1st ed., 1890). xi
Martinez-Coll. See J. Hirshleifer & Martinez-Coll (1988). 138n
Maskin, Eric. See Dasgupta & Maskin (1986a, 1986b) 235, 317, 324 Dasgupta *et al.* (1979) 216n and Fudenberg & Maskin (1986). 138n, 156
Maskin, Eric & John Riley (1984) "Optimal Auctions with Risk Averse Buyers" *Econometrica*. November 1984. 52: 1473–518. 305n
Maskin, Eric & John Riley (1985) "Input vs. Output Incentive Schemes" *Journal of Public Economics*. October 1985. 28: 1–23.
Maskin, Eric & Jean Tirole (1987) "Correlated Equilibria and Sunspots" *Journal of Economic Theory*. December 1987. 43: 364–73. 76
Masten, Scott & Keith Crocker (1985) "Efficient Adaptation in Long-Term Contracts: Take-or-Pay Provisions for Natural Gas" *American Economic Review*. December 1985. 75: 1083–93. 188n
Mathewson, G. Frank. See Stiglitz & Mathewson (1986). 360n
Mathewson, G. Frank & Ralph Winter (1985) "The Economics of Franchise Contracts" *Journal of Law and Economics*. October 1985. 28: 503–26. 190n
Matthews, Steven & John Moore (1987) "Monopoly Provision of Quality and Warranties: An Exploration in the Theory of Multidimensional Screening" *Econometrica*. March 1987. 55: 441–67. 265
Mayer, Colin. See Franks *et al.* (1988).
Maynard Smith, John (1974) "The Theory of Games and the Evolution of Animal Conflicts" *Journal of Theoretical Biology*. September 1974. 47: 209–21. 87n
Maynard Smith, John (1982) *Evolution and the Theory of Games*. Cambridge: Cambridge University Press, 1982. 116n
McAfee, R. Preston & John McMillan (1986) "Bidding for Contracts: A Principal-Agent Analysis" *Rand Journal of Economics*. Autumn 1986. 17: 326–38.
McAfee, R. Preston & John McMillan (1987) "Auctions and Bidding" *Journal of Economic Literature*. June 1987. 25: 699–754. 304n
McAfee, R. Preston & John McMillan (1988) *Incentives in Government Contracts*. Toronto: University of Toronto Press, 1988.
McChesney, Fred. See Macey & McChesney (1985). 116n, 365
McCloskey, Donald (1985) "Economical Writing" *Economic Inquiry*. April 1985. 24: 187–222. 6n
McCloskey, Donald (1987) *The Writing of Economics*. New York: Macmillan, 1987. 6n

McGee, John (1958) "Predatory Price Cutting: The Standard Oil (N.J.) Case" *Journal of Law and Economics*. October 1958. 1: 137–69. 96

McLennan, Andrew (1985) "Justifiable Beliefs in Sequential Equilibrium" *Econometrica*. July 1985. 53: 889–904. 158n

McMillan, John (1986) *Game Theory in International Economics*. Chur, Switzerland: Harwood Academic Publishers, 1986. xii, xiii

McMillan, John, *Games, Strategies, and Managers: How Managers can Use Game Theory to Make Better Business Decisions*. Oxford: Oxford University Press, 1992. xiii, 64n

McMillan, John. See McAfee & McMillan (1986, 1987) 304n.

Mead, Walter, Asbjorn Moseidjord, & Philip Sorenson (1984) "Competitive Bidding under Asymmetrical Information: Behavior and Performance in Gulf of Mexico Drainage Lease Sales 1959–1969" *Review of Economics and Statistics*. August 1984. 66: 505–8. 301

Meckling, William. See Jensen & Meckling (1976).

Meese, G. See Baldwin & Meese (1979). 31n

Mertens, Jean-Francois. See Kohlberg & Mertens (1986). 88n, 158n

Mertens, Jean-Francois & S. Zamir (1985) "Formulation of Bayesian Analysis for Games with Incomplete Information" *International Journal of Game Theory*. 1985. 14: 1–29. 63n

Milgate, Murray. See Eatwell *et al.* (1989). xii

Milgrom, Paul (1981a) "An Axiomatic Characterization of Common Knowledge" *Econometrica*. January 1981. 49: 219–22. 62n

Milgrom, Paul (1981b) "Good News and Bad News: Representation Theorems and Applications" *Bell Journal of Economics*. Autumn 1981. 12: 380–91. 190n, 216n, 301, 437

Milgrom, Paul (1981c) "Rational Expectations, Information Acquisition, and Competitive Bidding" *Econometrica*. July 1981. 49: 921–43. 302

Milgrom, Paul (1987) "Auction Theory" In: Bewley (1987). 304n

Milgrom, Paul. See Kreps *et al.* (1982) 2, 155 and Glosten & Milgrom (1985). 369

Milgrom, Paul & John Roberts (1982a) "Limit Pricing and Entry under Incomplete Information: An Equilibrium Analysis" *Econometrica*. March 1982. 50: 443–59. 160n, 349

Milgrom, Paul & John Roberts (1982b) "Predation, Reputation, and Entry Deterrence" *Journal of Economic Theory*. August 1982. 27: 280–312.

Milgrom, Paul & John Roberts (1986) "Price and Advertising Signals of Product Quality" *Journal of Political Economy*. August 1986. 94: 796–821. 269n

Milgrom, Paul & John Roberts (1990) "Rationalizability, Learning, and Equilibrium in Games with Strategic Complementaries" *Econometrica*. November 1990. 58: 1255–79. 439, 440

Milgrom, Paul & John Roberts (1992) *Economics, Organizations, and Management*. Englewood Cliffs, New Jersey: Prentice-Hall, 1992. xiii, 188n

Milgrom, Paul & Robert Weber (1982) "A Theory of Auctions and Competitive Bidding" *Econometrica*. September 1982. 50: 1089–1122. 298, 299, 303, 304n, 305n

Miller, Geoffrey (1986) "An Economic Analysis of Rule 68" *Journal of Legal Studies*. January 1986. 15: 93–125. 99

Mirrlees, James (1971) "An Exploration in the Theory of Optimum Income Taxation" *Review of Economic Studies*. 38: 175–208. 237, 242n, 374

Mirrlees, James (1974) "Notes on Welfare Economics, Information and Uncertainty" In: *Essays on Economic Behavior under Uncertainty*, edited by Michael Balch, Daniel McFadden, and Shih-yen Wu. Amsterdam: North Holland, 1974. 190n, 245n

Monteverde, K. & David Teece (1982) "Supplier Switching Costs and Vertical Integration in the Automobile Industry" *Bell Journal of Economics*. Spring 1982. 13: 206–13. 188n

Mookherjee, Dilip (1984) "Optimal Incentive Schemes with Many Agents" *Review of Economic Studies*. July 1984. 51: 433–46.

Mookherjee, Dilip & Ivan Png (1989) "Optimal Auditing, Insurance, and Redistribution" *The Quarterly Journal of Economics*. May 1989. 104: 399–415. 88n

Mookherjee, Dilip & D. Ray (1992) "Learning-By-Doing and Industrial Structure: An Overview" *Theoretical Issues in Development Economics*, ed. by B. Dutta, S. G., Dilip Mookherjee & D. Ray. New Delhi: Oxford University Press, 1992.

Moore, John. See Matthews & Moore (1987). 265

Moreaux, Michel (1985) "Perfect Nash Equilibria in Finite Repeated Games and Uniqueness of Nash Equilibrium in the Constituent Game" *Economics Letters*. 1985. 17: 317–20. 137n

Moreaux, Michel. See Laffont & Moreaux (1991). xiii

Morgenstern, Oskar. See von Neumann & Morgenstern (1944). 1, 6n, 46, 88n

Moseidjord, Asbjorn. See Mead *et al.* (1984). 301

Moskowitz, Milton, Michael Katz, & Robert Levering, eds (1980) *Everybody's Business: An Almanac*. San Francisco: Harper and Row, 1980.

Moulin, Herve (1986a) *Game Theory for the Social Sciences*, 2nd (revised) ed. New York: NYU Press, 1986. xi

Moulin, Herve (1986b) *Eighty-Nine Exercises with Solutions from Game Theory for the Social Sciences*, 2nd (revised) ed. New York: NYU Press, 1986. xi, 381

Mulherin, J. Harold (1986) "Complexity in Long-term Contracts: An Analysis of Natural Gas Contractual Provisions" *Journal of Law, Economics, and Organization*. Spring 1986. 2: 105–18.

Murphy, Kevin J. (1986) "Incentives, Learning, and Compensation: A Theoretical and Empirical Investigation of Managerial Labor Contracts" *Rand Journal of Economics*. Spring 1986. 17: 59–76. 188n

Murphy, Kevin J. See Baker *et al.* (1988). 209

Muzzio, Douglas (1982) *Watergate Games*. New York: New York University Press, 1982. 32n

Myerson, Roger (1978) "Refinements of the Nash Equilibrium Concept" *International Journal of Game Theory*. 1978. 7: 73–80. 158n

Myerson, Roger (1979) "Incentive Compatibility and the Bargaining Problem" *Econometrica*. January 1979. 47: 61–73. 216n

Myerson, Roger (1991) *Game Theory: Analysis of Conflict*. Cambridge, Mass.: Harvard University Press, 1991. xiii, 168, 216n

Myerson, Roger. See Holmstrom & Myerson (1983). 190n

Nalebuff, Barry. See Ghemawat & Nalebuff (1985). 87n

Nalebuff, Barry & John Riley (1985) "Asymmetric Equilibria in the War of Attrition" *Journal of Theoretical Biology*. April 1985. 113: 517–27. 87n

Nalebuff, Barry & David Scharfstein (1987) "Testing in Models of Asymmetric Information" *Review of Economic Studies*. April 1987. 54: 265–78. 245n

Nalebuff, Barry & Joseph Stiglitz (1983) "Prizes and Incentives: Towards a General Theory of Compensation and Competition" *Bell Journal of Economics*. Spring 1983. 14: 21–43. 218n

Nash, John (1950a) "The Bargaining Problem" *Econometrica*. January 1950. 18: 155–62. 1

Nash, John (1950b) "Equilibrium Points in n-Person Games" *Proceedings of the National Academy of Sciences, USA*. January 1950. 36: 48–9. 1

Nash, John (1951) "Non-Cooperative Games" *Annals of Mathematics*. September 1951. 54: 286–95. 1

Nelson, Philip (1974) "Advertising as Information" *Journal of Political Economy*. July/August 1974. 84: 729–54. 269n

Newbery, David. See Gilbert & Newbery (1982).

Newman, John. See Eatwell *et al.* (1989). xii

Nikaido, Hukukane & Kazuo Isoda (1955) "Note on Noncooperative Convex Games" *Pacific Journal of Mathematics*. 1955. 5: 807–15.

Novshek, William (1985) "On the Existence of Cournot Equilibrium" *Review of Economic Studies*. January 1985. 52: 85–98. 335n

Ohta, Hiroshi. See Greenhut & Ohta (1975). 337n

Ordeshook, Peter (1986) *Game Theory and Political Theory: An Introduction.* Cambridge: Cambridge University Press, 1986. xi, 32n

Owen, Guillermo (1982) *Game Theory*, 2nd ed. New York: Academic Press, 1982. xi

Palfrey, Thomas & Sanjay Srivastava (1993) *Bayesian Implementation*. New York: Harwood Academic Publishers, 1993. xiii

Pearce, David (1984) "Rationalizable Strategic Behavior and the Problem of Perfection" *Econometrica*. July 1984. 52: 1029–50. 31n

Pearce, David. See Abreu *et al.* (1986, 106 1990). 138n

Peleg, Bezalel. See Bernheim *et al.* (1987). 106

Peltzman, Sam (1991) "The Handbook of Industrial Organization: A Review Article" *Journal of Political Economy*. February 1991. 99: 201–17. 86n

Peltzman, Sam. See Jarrell & Peltzman (1985). 139n

Perry, Motty (1986) "An Example of Price Formation in Bilateral Situations: A Bargaining Model with Incomplete Information" *Econometrica*. March 1986. 54: 313–21. 290n

Perry, Motty. See Grossman & Perry (1986). 158n

Phlips, Louis (1983) *The Economics of Price Discrimination*. Cambridge: Cambridge University Press, 1983. 245n

Phlips, Louis (1988) *The Economics of Imperfect Information*. Cambridge: Cambridge University Press, 1988. xiii

Picker, Randal. See Baird *et al.* (forthcoming). xii, 99

Pliny, *The Letters of the Younger Pliny*, translated by Betty Radice. London: Penguin, 1963. 117, 392

Png, Ivan (1983) "Strategic Behaviour in Suit, Settlement, and Trial" *Bell Journal of Economics*. Autumn 1983. 14: 539–50. 57, 99

Png, Ivan. See Mookherjee & Png (1989). 88n

Png, Ivan & David Hirshleifer (1987) "Price Discrimination through Offers to Match Price" *Journal of Business*. July 1987. 60: 365–83. 338n

Polemarchakis, Heraklis. See Geanakoplos & Polemarchakis (1982). 63n

Polinsky, A. Mitchell & Yeon Koo Che (1991) "Decoupling Liability: Optimal Incentives for Care and Litigation" *Rand Journal of Economics*. Winter 1991. 22: 562–70. 180

Ponssard, Jean-Pierre (1981) *Competitive Strategies*. Amsterdam: North-Holland, 1981. (Translated by A. Heesterman from *Logique de la négociation et théorie des jeux*. Paris: Les Editions d'Organization, 1977.)

Popkin, Samuel (1979) *The Rational Peasant: The Political Economy of Rural Society in Vietnam*. Berkeley: University of California Press, 1979. 212

Porter, Robert (1983a) "Optimal Cartel Trigger Price Strategies" *Journal of Economic Theory*. April 1983. 29: 313–38. 138n, 216

Porter, Robert (1983b) "A Study of Cartel Stability: The Joint Executive Committee, 1880–1886" *Bell Journal of Economics*. Autumn 1983. 14: 301–14. 138n

Posner, Richard (1975) "The Social Costs of Monopoly and Regulation" *Journal of Political Economy*. August 1975. 83: 807–27. 74, 345

Posner, Richard (1992) *The Economic Analysis of the Law*, 4th ed. Boston: Little, Brown, 1992 (1st ed., 1972). 211

Poterba, James. See Hausman & Poterba (1987).

Prescott, Edward. See Kydland & Prescott (1977). 115n

Psacharopoulos, George. See Layard & Psacharopoulos (1974). 258, 269n

Pyle, David. See Leland & Pyle (1977). 266

Radner, Roy (1980) "Collusive Behavior in Oligopolies with Long but Finite Lives" *Journal of Economic Theory*. April 1980. 22: 136–56. 137n

Radner, Roy (1985) "Repeated Principal-Agent Games with Discounting" *Econometrica*. September 1985. 53: 1173–98. 210

Radner, Roy. See Marschak & Radner (1972). 218n

Raff, Daniel & Lawrence Summers (1987) "Did Henry Ford Pay Efficiency Wages?" *Journal of Labor Economics*. October 1987. 5: 57–86. 240

Raiffa, Howard. See Luce & Raiffa (1957). x, xi, 31n, 128, 137n, 289n

Rapoport, Anatol (1960) *Fights, Games and Debates*. Ann Arbor: University of Michigan Press, 1960.

Rapoport, Anatol (1970) *N-Person Game Theory: Concepts and Applications*. Ann Arbor: University of Michigan Press, 1970. xi

Rapoport, Anatol & Albert Chammah (1965) *Prisoner's Dilemma: A Study in Conflict and Cooperation*. Ann Arbor: University of Michigan Press, 1965. 157

Rapoport, Anatol & Melvin Guyer (1966) "A Taxonomy of 2x2 Games" *General Systems*. 1966. 11: 203–14. 30n

Rapoport, Anatol, Melvin Guyer, & David Gordon (1976) *The 2x2 Game*. Ann Arbor: University of Michigan Press, 1976. 30n

Rappoport, Peter. See Cooter & Rappoport (1984). 31n

Rasmusen, Eric (1987) "Moral Hazard in Risk-Averse Teams" *Rand Journal of Economics*. Fall 1987. 18: 428–35. 216, 218n, 219n

Rasmusen, Eric (1988a) "Entry for Buyout" *Journal of Industrial Economics*. March 1988. 36: 281–300. 356

Rasmusen, Eric (1988b) "Stock Banks and Mutual Banks" *Journal of Law and Economics*. October 1988. 31: 395–422. 188n, 209, 217n

Rasmusen, Eric (1989a) *Games and Information*. Oxford: Basil Blackwell, 1989. Japanese translation by Moriki Hosoe, Shozo Murata, and Yoshinobu Arisada,

Kyushu University Press, Vol. I (1990), Vol. 2 (1991). Italian translation by Alberto Bernardo, Milan: Ulrico Hoepli Editore (1993). xiii

Rasmusen, Eric (1989b) "A Simple Model of Product Quality with Elastic Demand" *Economics Letters*. 1989. 29: 281–3. 132n

Rasmusen, Eric (1992a) "Folk Theorems for the Observable Implications of Repeated Games" *Theory and Decision*. March 1992. 32: 147–64. 138n, 139n

Rasmusen, Eric (1992b) "Managerial Conservatism and Rational Information Acquisition" *Journal of Economics and Management Strategy*. Spring 1992. 1: 175–202. 63n

Rasmusen, Eric (1992c) "An Income-Satiation Model of Efficiency Wages" *Economic Inquiry*. July 1992. 30: 467–78. 222n, 407n

Rasmusen, Eric (unpublished) "Signal Jamming and Limit Pricing: A Unified Approach," Indiana University Working Paper in Economics #92-020.

Rasmusen, Eric & Emmanuel Petrakis (1992) "Defining the Mean-Preserving Spread: 3-pt versus 4-pt" In: *Decision Making Under Risk and Uncertainty: New Models and Empirical Findings*, edited by John Geweke. Amsterdam: Kluwer, 1992. 437

Rasmusen, Eric & Todd Zenger (1990) "Diseconomies of Scale in Employment Contracts" *Journal of Law, Economics and Organization*. June 1990. 6: 65–92. 219n

Rasmusen, Eric. See Fernandez & Rasmusen (unpublished) 236, D. Hirshleifer & Rasmusen (1989), 137n and J. Hirshleifer & Rasmusen (1992). 88n

Reinganum, Jennifer (1985) "Innovation and Industry Evolution" *Quarterly Journal of Economics*. February 1985. 100: 81–99. 359n

Reinganum, Jennifer (1988) "Plea Bargaining and Prosecutorial Discretion" *The American Economic Review*. September 1988. 78: 713–28. 269n

Reinganum, Jennifer (1989) "The Timing of Innovation: Research, Development and Diffusion" In: Schmalensee & Willig (1989). 342

Reinganum, Jennifer & Nancy Stokey (1985) "Oligopoly Extraction of a Common Property Natural Resource: the Importance of the Period of Commitment in Dynamic Games" *International Economic Review*. February 1985. 26: 161–74. 116n

Reiss, Peter. See Bresnahan & Reiss (1991). 336n

Reynolds, Robert. See Salant *et al.* (1983). 336n

Richerson, Peter. See Boyd & Richerson (1985). 116n

Riker, William (1986) *The Art of Political Manipulation*. New Haven: Yale University Press. 32n, 117, 392

Riley, John (1979a) "Evolutionary Equilibrium Strategies" *Journal of Theoretical Biology*. January 1979. 76: 109–23. 116n

Riley, John (1979b) "Informational Equilibrium" *Econometrica*. March 1979. 47: 331–59. 236

Riley, John (1980) "Strong Evolutionary Equilibrium and the War of Attrition" *Journal of Theoretical Biology*. February 1980. 82: 383–400. 87n

Riley, John. See Hirshleifer & Riley (1979, 1992) 44, 381, Maskin & Riley (1984, 1985), and Nalebuff & Riley (1985). 87n

Riordan, Michael. See Kihlstrom & Riordan (1984). 269n

Roberts, John. See Kreps *et al.* (1982) 2, 155 and Milgrom & Roberts (1982a 160n, 349, 1982b, 1986 269n, 1990 439, 440, 1992 xiii, 188n).

Roberts, John & Hugo Sonnenschein (1976) "On the Existence of Cournot Equilibrium without Concave Profit Functions" *Journal of Economic Theory*. August 1976. 13: 112–17. 335n
Robinson, Marc (1985) "Collusion and the Choice of Auction" *Rand Journal of Economics*. Spring 1985. 16: 141–5. 299
Rogerson, William (1982) "The Social Costs of Monopoly and Regulation: a Game-Theoretic Analysis" *Bell Journal of Economics*. Autumn 1982. 13: 391–401. 345
Roll, Richard. See Cornell & Roll (1981). 116n
Romer, David (1984) "The Theory of Social Custom: A Modification and Some Extensions" *Quarterly Journal of Economics*. November 1984. 99: 717–27.
Rosen, J. (1965) "Existence and Uniqueness of Equilibrium Points for Concave n-Person Games" *Econometrica*. July 1965. 33: 520–34.
Rosen, Leo (1968) *The Joys of Yiddish*. New York: Washington Square Press, 1968.
Rosen, Sherwin (1986) "Prizes and Incentives in Elimination Tournaments" *American Economic Review*. September 1986. 76: 701–15. 217n
Rosen, Sherwin. See Lazear & Rosen (1981). 217n
Rosenberg, David & Steven Shavell (1985) "A Model in Which Suits are Brought for their Nuisance Value" *International Review of Law and Economics*. June 1985. 5: 3–13. 101n
Ross, Steven (1977) "The Determination of Financial Structure: The Incentive-Signalling Approach" *Bell Journal of Economics*. Spring 1977. 8: 23–40. 269n
Roth, Alvin (1984) "The Evolution of the Labor Market for Medical Interns and Residents: A Case Study in Game Theory" *Journal of Political Economy*. December 1984. 92: 991–1016. 245n
Roth, Alvin, ed. (1985) *Game Theoretic Models of Bargaining*. Cambridge: Cambridge University Press, 1985.
Rothkopf, Michael (1980) "TREES: A Decision-Maker's Lament" *Operations Research*. January/February 1980. 28: 3.
Rothschild, Michael. See Diamond & Rothschild (1978). xi
Rothschild, Michael & Joseph Stiglitz (1970) "Increasing Risk I" *Journal of Economic Theory*. 225–43. Reprinted in Diamond & Rothschild (1978). 437
Rothschild, Michael & Joseph Stiglitz (1976) "Equilibrium in Competitive Insurance Markets: An Essay on the Economics of Imperfect Information" *Quarterly Journal of Economics*. November 1976. 90: 629–49. 231, 264
Rubin, Paul (1978) "The Theory of the Firm and the Structure of the Franchise Contract" *Journal of Law and Economics*. April 1978. 21: 223–33. 190n
Rubinfeld, Daniel. See Cooter & Rubinfeld (1989). 101
Rubinstein, Ariel (1979) "An Optimal Conviction Policy for Offenses that May Have Been Committed by Accident" In: *Applied Game Theory*, edited by Steven Brams, A. Schotter, and Gerhard Schrodiauer. Physica-Verlag, 1979. 406–13. 220n, 247n
Rubinstein, Ariel (1982) "Perfect Equilibrium in a Bargaining Model" *Econometrica*. January 1982. 50: 97–109. 282, 283, 290n, 357
Rubinstein, Ariel (1985a) "A Bargaining Model with Incomplete Information about Time Preferences" *Econometrica*. September 1985. 53: 1151–72. 290n
Rubinstein, Ariel (1985b) "Choice of Conjectures in a Bargaining Game with Incomplete Information" In: Roth (1985). 158, 290n

Rubinstein, Ariel, ed (1990) *Game Theory in Economics.* Brookfield, Vermont: Edward Elgar Publishing Company, 1990. xi
Rubinstein, Ariel. See Binmore *et al.* (1986). 289n
Rudin, Walter (1964) *Principles of Mathematical Analysis.* New York: McGraw-Hill, 1964. 429
Rumelt, Richard. See Lippman & Rumelt (1982).
Saft, Lester. See Klein & Saft (1985). 190n
Salant, Stephen. See Gaudet & Salant (1991). 88n
Salant, Stephen, Sheldon Switzer, & Robert Reynolds (1983) "Losses from Horizontal Merger: The Effects of an Exogenous Change in Industry Structure on Cournot-Nash Equilibrium" *Quarterly Journal of Economics.* May 1983. 98: 185–99. 336n
Saloner, Garth. See Farrell & Saloner (1985). 32n
Salop, Steven & Joseph Stiglitz (1977) "Bargains and Ripoffs: A Model of Monopolistically Competitive Price Dispersion" *Review of Economic Studies.* October 1977. 44: 493–510. 239
Samuelson, Paul (1958) "An Exact Consumption-Loan Model of Interest with or without the Social Contrivance of Money" *Journal of Political Economy.* December 1958. 66: 467–82. 135
Samuelson, William (1984) "Bargaining under Asymmetric Information" *Econometrica.* July 1984. 52: 995–1005. 290n
Savage, Leonard (1954) *The Foundations of Statistics.* New York: Wiley, 1954.
Scarf, Herbert. See Debreu & Scarf (1963). 2, 63n
Scharfstein, David. See Nalebuff & Scharfstein (1987). 245n
Scheffman, David. See Holt & Scheffman (1987). 338n
Scheinkman, Jose. See Kreps & Scheinkman (1983). 336n
Schelling, Thomas (1960) *The Strategy of Conflict.* Cambridge, Mass.: Harvard University Press, 1960. 2, 28, 32n
Schelling, Thomas (1966) *Arms and Influence.* New Haven: Yale University Press, 1966. 32n
Schelling, Thomas (1978) *Micromotives and Macrobehavior.* New York: W.W. Norton, 1978. 32n
Scherer, Frederick (1980) *Industrial Market Structure and Economic Performance*, 2nd ed. Chicago: Rand McNally, 1980.
Schmalensee, Richard (1982) "Product Differentiation Advantages of Pioneering Brands" *American Economic Review.* June 1982. 72: 349–65. 134
Schmalensee, Richard & Robert Willig, eds (1989) *The Handbook of Industrial Organization.* New York: North-Holland, 1989. xiii, 216n, 360n
Schmittberger, Rold. See Guth *et al.* (1982).
Schwartz, Nancy. See Kamien & Schwartz (1982 342, 1991 429).
Schwarze, Bernd. See Guth *et al.* (1982).
Seierstad, A. & K. Sydsaeter (1987) *Optimal Control Theory with Economic Applications.* Amsterdam: North-Holland, 1987. 429
Seligman, Daniel (1992) *A Question of Intelligence: The IQ Debate in America.* New York: Carol Publishing, 1992. 242n
Selten, Reinhard (1965) "Spieltheoretische Behandlung eines Oligopolmodells mit Nachfragetragheit" *Zeitschrift für die gesamte Staatswissenschaft.* October 1965. 121: 301–24, 667–89. 2, 115n
Selten, Reinhard (1975) "Reexamination of the Perfectness Concept for Equilib-

rium Points in Extensive Games" *International Journal of Game Theory.* 1975. 4: 25–55. 2, 115n, 145

Selten, Reinhard (1978) "The Chain-Store Paradox" *Theory and Decision.* April 1978. 9: 127–59. 121, 137n

Selten, Reinhard. See Harsanyi & Selten (1988). 27

Sen, Amenya (1967) "Isolation, Assurance, and the Social Rate of Discount" *Quarterly Journal of Economics.* February 1967. 81: 112–24. 32n

Shaked, Avner (1982) "Existence and Computation of Mixed Strategy Nash Equilibrium for 3-Firms Location Problem" *Journal of Industrial Economics.* September/December, 1982. 31: 93–6. 324

Shaked, Avner & John Sutton (1984) "Involuntary Unemployment as a Perfect Equilibrium in a Bargaining Model" *Econometrica.* November 1984. 52: 1351–64. 290n

Shapiro, Carl (1982) "Consumer Information, Product Quality and Seller Reputation" *Bell Journal of Economics.* Spring 1982. 13: 20–35.

Shapiro, Carl (1983) "Premiums for High Quality Products as Returns to Reputation" *Quarterly Journal of Economics.* November 1983. 98: 659–79. 140n

Shapiro, Carl (1989) "The Theory of Business Strategy" *RAND Journal of Economics.* Spring 1989. 20: 125–37. 86n

Shapiro, Carl. See Farrell & Shapiro (1988) 135 and Katz & Shapiro (1985) 32n.

Shapiro, Carl & Joseph Stiglitz (1984) "Equilibrium Unemployment as a Worker Discipline Device" *American Economic Review.* June 1984. 74: 433–44. 139n, 206

Shapley, Lloyd (1953a) "Open Questions" In: *Report of an Informal Conference on the Theory of n-Person Games.* Princeton Mathematics mimeo. 1953. 15. 1

Shapley, Lloyd (1953b) "A Value for n-Person Games" In: Kuhn & Tucker (1953) 307–17. 1, 289n

Shavell, Steven (1979) "Risk Sharing and Incentives in the Principal and Agent Relationship" *Bell Journal of Economics.* Spring 1979. 10: 55–73. 189n

Shavell, Steven. See Rosenberg & Shavell (1985). 101n

Shell, Karl. See Cass & Shell (1983). 76

Shleifer, Andrei & Robert Vishny (1986) "Greenmail, White Knights, and Shareholders' Interest" *Rand Journal of Economics.* Autumn 1986. 17: 293–309. 158n, 366

Shubik, Martin (1971) "The Dollar Auction Game: A Paradox in Noncooperative Behavior and Escalation" *Journal of Conflict Resolution.* March 1971. 15: 109–11. 304n

Shubik, Martin (1982) *Game Theory in the Social Sciences: Concepts and Solutions.* Cambridge, Mass.: MIT Press, 1982. xi, 62n, 289n

Shubik, Martin. See Levitan & Shubik (1972). 317

Simon, Leo (1987) "Games with Discontinuous Payoffs" *Review of Economic Studies.* October 1987. 54: 569–98. 324

Slade, Margaret (1987) "Interfirm Rivalry in a Repeated Game: An Empirical Test of Tacit Collusion" *Journal of Industrial Economics.* June 1987. 35: 499–516. 138n

Slatkin, Montgomery (1980) "Altruism in Theory" Review of Scott Boorman & Paul Levitt, *The Genetics of Altruism. Science.* November 7, 1980. 210: 633–4. 3

Smith, Abbie. See Antle & Smith (1986). 217n

Smith, Adam (1776) *An Inquiry into The Nature and Causes of the Wealth of Nations*. Chicago: University of Chicago Press, 1976. 207

Smith, J. See DeBrock & Smith (1983).

Sobel, Joel. See Banks & Sobel (1987) 158n, Border & Sobel (1987) 88n, and Crawford & Sobel (1982) 76.

Sobel, Joel & Ichiro Takahashi (1983) "A Multi-Stage Model of Bargaining" *Review of Economic Studies*. July 1983. 50: 411–26. 290n

Sonnenschein, Hugo (1983) "Economics of Incentives: An Introductory Account" In: *Technology, Organization, and Economic Structure: Essays in Honor of Prof. Isamu Yamada*, edited by Ryuzo Sato & Martin Beckmann. Berlin: Springer-Verlag, 1983.

Sonnenschein, Hugo. See Roberts & Sonnenschein (1976). 335n

Sorenson, Philip. See Mead *et al.* (1984). 301

Sowden, Lanning. See Campbell & Sowden (1985). 30n

Spence, A. Michael (1973) "Job Market Signalling" *Quarterly Journal of Economics*. August 1973. 87: 355–74. 249, 269n

Spence, A. Michael (1977) "Entry, Capacity, Investment, and Oligopolistic Pricing" *Bell Journal of Economics*. Autumn 1977. 8: 534–44. 355

Spence, A. Michael. See Kreps & Spence (1984). 3, 6n

Spitzer, Matthew. See Hoffman & Spitzer (1985). 279

Spremann, Klaus. See Bamberg & Spremann (1987). xi

Spulber, Daniel (1989) *Regulation and Markets*. Cambridge, Mass.: MIT Press, 1989. 216n, 360n, 374, 379

Srivastava, Sanjay. See Palfrey & Srivastava (1993). xiii

Stacchetti, Ennio. See Abreu *et al.* (1986 106, 1990). 138n

Stackelberg, Heinrich von (1934) *Marktform und Gleichgewicht*. Berlin: J. Springer, 1934. Translated by Alan Peacock as *The Theory of the Market Economy*. London: William Hodge, 1952. 88n

Staten, Michael & John Umbeck (1982) "Information Costs and Incentives to Shirk: Disability Compensation of Air Traffic Controllers" *American Economic Review*. December 1982. 72: 1023–37. 188n

Staten, Michael & John Umbeck (1986) "A Study of Signaling Behavior in Occupational Disease Claims" *Journal of Law and Economics*. October 1986. 29: 263–86. 269n

Stevenson, Robert L. (1987) *Island Nights' Entertainments*. London: Hogarth, 1987. 103n

Stigler, George (1964) "A Theory of Oligopoly" *Journal of Political Economy*. February 1964. 72: 44–61. 138n

Stigler, George. See Becker & Stigler (1974). 206

Stiglitz, Joseph (1977) "Monopoly, Non-linear Pricing and Imperfect Information: The Insurance Market" *Review of Economic Studies*. October 1977. 44: 407–30. 245n

Stiglitz, Joseph (1982a) "Self-Selection and Pareto-Efficient Taxation" *Journal of Public Economics*. March 1982. 17: 213–40.

Stiglitz, Joseph (1982b) "Utilitarianism and Horizontal Equity: The Case of Random Taxation" *Journal of Public Economics*. June 1982. 18: 1–33.

Stiglitz, Joseph (1987) "The Causes and Consequences of the Dependence of Quality on Price" *Journal of Economic Literature*. March 1987. 25: 1–48. 134, 139n, 243n, 269n.

Stiglitz, Joseph. See Dasgupta & Stiglitz (1980) 359n, Nalebuff & Stiglitz (1983) 218n, Rothschild & Stiglitz (1970 437, 1976 231, 264), Salop & Stiglitz (1977) 239, and Shapiro & Stiglitz (1984) 139n, 206.

Stiglitz, Joseph & G. Frank Mathewson, eds (1986) *New Developments in the Analysis of Market Structure.* Cambridge, Mass.: MIT Press, 1986. 360n

Stiglitz, Joseph & Andrew Weiss (1981) "Credit Rationing in Markets with Imperfect Information" *American Economic Review.* June 1981. 71: 393–410. 240

Stiglitz, Joseph & Andrew Weiss (1989) "Sorting out the Differences Between Screening and Signalling Models" In: *Papers in Commemoration of the Economic Theory Seminar at Oxford University,* Michael Dempster, ed., Oxford: Oxford University Press. 269n

Stokey, Nancy & Robert Lucas (1989) *Recursive Methods in Economic Dynamics.* Cambridge, Mass.: Harvard University Press, 1989. 429

Stokey, Nancy. See Reinganum & Stokey (1985). 116n

Straffin, Philip (1980) "The Prisoner's Dilemma" *UMAP Journal.* 1: 101–3. 30n

Strunk, William & E. B. White (1959) *The Elements of Style.* New York: Macmillan, 1959. 6n

Sugden, Robert (1986) *The Economics of Rights, Co-operation and Welfare.* Oxford: Basil Blackwell, 1986. 116n, 122

Sultan, Ralph (1974) *Pricing in the Electrical Oligopoly, Vol. I: Competition or Collusion.* Cambridge, Mass.: Harvard University Press, 1974. 337n

Summers, Larry. See Raff & Summers (1987). 240

Sutton, John (1986) "Non-Cooperative Bargaining Theory: An Introduction" *Review of Economic Studies.* October 1986. 53: 709–24.

Sutton, John (1991) *Sunk Costs and Market Structure: Price Competition, Advertising, and the Evolution of Concentration.* Cambridge, Mass.: The MIT Press, 1991. 336n

Sutton, John. See Shaked & Sutton (1984). 290n

Switzer, Sheldon. See Salant *et al.* (1983). 336n

Szep, J. & F. Forgo (1985) *Introduction to the Theory of Games.* Dordrecht: D. Reidel, 1985. xi

Takahashi, Ichiro. See Sobel & Takahashi (1983). 290n

Takayama, Akira (1985) *Mathematical Economics,* 2nd ed. Cambridge: Cambridge University Press, 1985. 429

Teece, David. See Monteverde & Teece (1982). 188n

Telser, Lester (1966) "Cutthroat Competition and the Long Purse" *Journal of Law and Economics.* October 1966. 9: 259–77. 362

Tenorio, Rafael (1993) "Revenue-Equivalence and Bidding Behavior in a Multi-Unit Auction Market: An Empirical Analysis" *Review of Economics and Statistics.* May 1993. 2: 302–. 305n

Thaler, Richard (1992) *The Winner's Curse: Paradoxes and Anomalies of Economic Life.* New York: The Free Press, 1992. 62n, 279

Thisse, Jacques. See d'Aspremont *et al.* (1979). 323, 337n

Thomas, L. (1984) *Games, Theory and Applications.* Chichester, England: Ellis Horwood Ltd., 1984. xi

Tirole, Jean (1986) "Hierarchies and Bureaucracies: On the Role of Collusion in Organizations" *Journal of Law, Economics, and Organization.* Fall 1986. 2: 181–214. 212

Weiss, Andrew (1990) *Efficiency Wages.* Princeton: Princeton University Press, 1990. 217n
Weiss, Andrew. See Guasch & Weiss (1980) 245n and Stiglitz & Weiss (1981 240, 1989 269n).
Weitzman, Martin (1974) "Prices vs. Quantities" *Review of Economic Studies.* October 1974. 41: 477–91. 336n
Weston, J. Fred. See Copeland & Weston (1988). 109, 266, 437
Whinston, Michael. See Bernheim *et al.* (1987) and Bernheim & Whinston (1987). 106
White, E.B. See Strunk & White (1959). 6n
Wicksteed, Philip (1885) *The Common Sense of Political Economy.* New York: Kelley, 1950. 4
Wiley, John, Eric Rasmusen, & Mark Ramseyer (1990) "The Leasing Monopolist" *UCLA Law Review.* April 1990, 37: 693–731. 338n
Williams, J. (1966) *The Compleat Strategyst: Being a Primer on the Theory of Games of Strategy.* New York: McGraw-Hill, 1966. xi
Williamson, Oliver (1975) *Markets and Hierarchies: Analysis and Antitrust Implications: A Study in the Economics of Internal Organization.* New York: Free Press, 1975. 212
Willig, Robert. See Schmalensee & Willig (1989). xiii, 216n, 360n
Wilson, Charles (1980) "The Nature of Equilibrium in Markets with Adverse Selection" *Bell Journal of Economics.* Spring 1980. 11: 108–30. 230, 235, 269n
Wilson, Robert (1971) "Computing Equilibria of n-Person Games" SIAM Journal of Applied Mathematics. July 1971. 21: 80–7.
Wilson, Robert (1979) "Auctions of Shares" *Quarterly Journal of Economics.* November 1979. 93: 675–89. 305n
Wilson, Robert (unpublished) Stanford University 311b course notes.
Wilson, Robert. See Kennan & Wilson (1993) 101, 290n, 338n, Kreps & Wilson (1982a 352, 360n, 1982b 2, 145, 146), and Kreps *et al.* (1982). 2, 155
Winter, Ralph. See Mathewson & Winter (1985). 190n
Wolfe, Philip. See Dresher *et al.* (1957).
Wolfson, M. (1985) "Empirical Evidence of Incentive Problems and their Mitigation in Oil and Tax Shelter Programs" In: *Principals and Agents: The Structure of Business,* edited by John Pratt & Richard Zeckhauser. Boston: Harvard Business School Press, 1985. 101–25. 188n
Wolinsky, Asher. See Binmore *et al.* (1986). 289n
Wydick, Richard (1978) "Plain English for Lawyers" *California Law Review.* 1978. 66: 727–64. 6n
Yellen, Janet. See Akerlof & Yellen (1986). 217n
Zahavi, Amotz (1975) "Mate Selection: A Selection for a Handicap" *Journal of Theoretical Biology.* September 1975. 53: 205–14. 269n
Zamir, S. See Mertens & Zamir (1985). 63n
Zenger, Todd. See Rasmusen & Zenger (1990).
von Zermelo, E. (1913) "Uber eine Anwendung der Mengenlehre auf die Theorie des Schachspiels" *Proceedings, Fifth International Congress of Mathematicians.* 1913. 2: 501–4.
Zimmerman, Jerold. See Gaver & Zimmerman (1977). 218n

Subject Index